Student Solutions Manual

to accompany

Intermediate Algebra

Fourth Edition

Julie Miller
Daytona State College—Daytona Beach

Molly O'Neill
Daytona State College—Daytona Beach

Nancy Hyde

Prepared by
diacriTech, Inc., DBA LaurelTech

The McGraw-Hill Companies

Connect
Learn
Succeed™

Student Solutions Manual to accompany
INTERMEDIATE ALGEBRA, FOURTH EDITION
JULIE MILLER, MOLLY O'NEILL, AND NANCY HYDE

Published by McGraw-Hill Higher Education, an imprint of The McGraw-Hill Companies, Inc., 1221 Avenue of the Americas, New York, NY 10020. Copyright © 2014, 2011, 2008 and 2004 by The McGraw-Hill Companies, Inc. All rights reserved. Printed in the United States of America.

This book is printed on acid-free paper.

1 2 3 4 5 6 7 8 9 0 QDB/QDB 1 0 9 8 7 6 5 4 3

ISBN: 978-0-07-754876-6
MHID: 0-07-754876-0

Contents

Chapter R   Review of Basic Algebraic Concepts

R.1	Study Skills	1
R.2	Sets of Numbers and Interval Notation	1
R.3	Operations on Real Numbers	3
R.4	Simplifying Algebraic Expressions	6
	Chapter R Review Exercises	8
	Chapter R Test	10

Chapter 1   Linear Equations and Inequalities in One Variable

1.1	Linear Equations in One Variable	12
	Problem Recognition Exercises	20
1.2	Applications of Linear Equations in One Variable	21
1.3	Applications to Geometry and Literal Equations	27
1.4	Linear Inequalities in One Variable	32
1.5	Compound Inequalities	38
1.6	Absolute Value Equations	42
1.7	Absolute Value Inequalities	45
	Problem Recognition Exercises	48
	Chapter 1 Group Activity	50
	Chapter 1 Review Exercises	51
	Chapter 1 Test	57

Chapter 2   Linear Equations in Two Variables and Functions

2.1	Linear Equations in Two Variables	60
2.2	Slope of a Line and Rate of Change	65
2.3	Equations of a Line	68
	Problem Recognition Exercises	73
2.4	Applications of Linear Equations and Modeling	73

2.5	Introduction to Relations	77
2.6	Introduction to Functions	78
2.7	Graphs of Functions	81
	Problem Recognition Exercises	84
	Chapter 2 Group Activity	85
	Chapter 2 Review Exercises	85
	Chapter 2 Test	89
	Chapters 1 – 2 Cumulative Review Exercises	92

Chapter 3   Systems of Linear Equations and Inequalities

3.1	Solving Systems of Linear Equations by the Graphing Method	94
3.2	Solving Systems of Linear Equations by the Substitution Method	98
3.3	Solving Systems of Linear Equations by the Addition Method	102
	Problem Recognition Exercises	108
3.4	Applications of Systems of Linear Equations in Two Variables	110
3.5	Linear Inequalities and Systems of Linear Inequalities in Two Variables	115
3.6	Systems of Linear Equations in Three Variables and Applications	127
3.7	Solving Systems of Linear Equations by Using Matrices	135
	Chapter 3 Group Activity	138
	Chapter 3 Review Exercises	138
	Chapter 3 Test	147
	Chapters 1 – 3 Cumulative Review Exercises	150

Chapter 4   Polynomials

4.1	Properties of Integer Exponents and Scientific Notation	153
4.2	Addition and Subtraction of Polynomials and Polynomial Functions	157
4.3	Multiplication of Polynomials	161
4.4	Division of Polynomials	166
	Problem Recognition Exercises	172
4.5	Greatest Common Factor and Factoring by Grouping	174
4.6	Factoring Trinomials	176

4.7	Factoring Binomials	181
	Problem Recognition Exercises	185
4.8	Solving Equations by Using the Zero Product Rule	188
	Chapter 4 Group Activity	195
	Chapter 4 Review Exercises	195
	Chapter 4 Test	201
	Chapters 1 – 4 Cumulative Review Exercises	203

Chapter 5   Rational Expressions and Rational Equations

5.1	Rational Expressions and Rational Functions	205
5.2	Multiplication and Division of Rational Expressions	210
5.3	Addition and Subtraction of Rational Expressions	212
5.4	Complex Fractions	218
	Problem Recognition Exercises	222
5.5	Solving Rational Equations	224
	Problem Recognition Exercises	232
5.6	Applications of Rational Equations and Proportions	234
5.7	Variation	240
	Chapter 5 Group Activity	243
	Chapter 5 Review Exercises	243
	Chapter 5 Test	249
	Chapters 1 – 5 Cumulative Review Exercises	252

Chapter 6   Radicals and Complex Numbers

6.1	Definition of an nth Root	254
6.2	Rational Exponents	258
6.3	Simplifying Radical Expressions	261
6.4	Addition and Subtraction of Radicals	264
6.5	Multiplication of Radicals	267
	Problem Recognition Exercises	270
6.6	Division of Radicals and Rationalization	271

6.7	Solving Radical Equations	276
6.8	Complex Numbers	285
	Chapter 6 Group Activity	289
	Chapter 6 Review Exercises	289
	Chapter 6 Test	294
	Chapters 1 – 6 Cumulative Review Exercises	296

Chapter 7   Quadratic Equations, Functions, and Inequalities

7.1	Square Root Property and Completing the Square	299
7.2	Quadratic Formula	304
7.3	Equations in Quadratic Form	311
	Problem Recognition Exercises	318
7.4	Graphs of Quadratic Functions	321
7.5	Vertex of a Parabola: Applications and Modeling	326
7.6	Polynomial and Rational Inequalities	332
	Problem Recognition Exercises	346
	Chapter 7 Group Activity	349
	Chapter 7 Review Exercises	349
	Chapter 7 Test	358
	Chapters 1 – 7 Cumulative Review Exercises	363

Chapter 8   Exponential and Logarithmic Functions and Applications

8.1	Algebra of Functions and Composition	367
8.2	Inverse Functions	370
8.3	Exponential Functions	374
8.4	Logarithmic Functions	376
	Problem Recognition Exercises	380
8.5	Properties of Logarithms	380
8.6	The Irrational Number e and Change of Base	383
	Problem Recognition Exercises	387
8.7	Logarithmic and Exponential Equations and Applications	387

Chapter 8 Group Activity   393

Chapter 8 Review Exercises   393

Chapter 8 Test   398

Chapters 1 – 8 Cumulative Review Exercises   400

Chapter 9   Conic Sections

9.1   Distance Formula, Midpoint Formula, and Circles   404

9.2   More on the Parabola   409

9.3   The Ellipse and Hyperbola   413

    Problem Recognition Exercises   417

9.4   Nonlinear Systems of Equations in Two Variables   418

9.5   Nonlinear Inequalities and Systems of Inequalities   423

    Chapter 9 Group Activity   434

    Chapter 9 Review Exercises   434

    Chapter 9 Test   440

    Chapters 1 – 9 Cumulative Review Exercises   443

Chapter 10   Binomial Expansions, Sequences, and Series

10.1   Binomial Expansions   448

10.2   Sequences and Series   450

10.3   Arithmetic Sequences and Series   453

10.4   Geometric Sequences and Series   457

    Problem Recognition Exercises   462

    Chapter 10 Group Activity   462

    Chapter 10 Review Exercises   463

    Chapter 10 Test   465

    Chapters 1 – 10 Cumulative Review Exercises   466

Additional Topics Appendix

A.1   Determinants and Cramer's Rule   472

Chapter R   Review of Basic Algebraic Concepts

Are You Prepared?

1. $36{,}636 \div 43 = 852$ T

2. $0.25 \times 6340 = 1585$ B

3. $24.0842 + 365.7 = 389.7842$ I

4. $\dfrac{4}{7} \times \dfrac{21}{10} = \dfrac{\cancel{2} \cdot 2}{\cancel{7}} \times \dfrac{\cancel{7} \cdot 3}{\cancel{2} \cdot 5} = \dfrac{6}{5}$ D

5. $\dfrac{7}{3} \div 14 = \dfrac{7}{3} \cdot \dfrac{1}{14} = \dfrac{\cancel{7}}{3} \cdot \dfrac{1}{\cancel{7} \cdot 2} = \dfrac{1}{6}$ R

6. $\dfrac{5}{3} - \dfrac{1}{2} = \dfrac{10}{6} - \dfrac{3}{6} = \dfrac{7}{6}$ Y

7. $\dfrac{2}{3} - \dfrac{7}{12} = \dfrac{8}{12} - \dfrac{7}{12} = \dfrac{1}{12}$ P

8. $4\dfrac{5}{9} - 1\dfrac{1}{3} = 4\dfrac{5}{9} - 1\dfrac{3}{9} = 3\dfrac{2}{9}$ E

9. $3\dfrac{1}{5} \times 2\dfrac{1}{2} = \dfrac{16}{5} \times \dfrac{5}{2} = \dfrac{\cancel{2} \cdot 8}{\cancel{5}} \times \dfrac{\cancel{5}}{\cancel{2}} = 8$ O

10. $3.75 + 8\dfrac{1}{5} = 3.75 + 8.2 = 11.95$ S

11. $582 \div 0.01 = 58{,}200$ V

12. $582 \times 0.01 = 5.82$ U

In this chapter we will show that $a(b+c) = ab + ac$. This important property is called the:
DISTRIBUTIVE PROPERTY.

Section R.1   Study Skills

1. Answers will vary.

3. Answers will vary.

5. Problem Recognition Exercises: pages 171 and 217
Chapter Summary: pages 219 – 224
Chapter Review Exercises: pages 225 – 230
Chapter Test: pages 230 – 233
Cumulative Review Exercises: pages 233 – 234

7. Answers will vary.

9. Answers will vary.

Section R.2   Sets of Numbers and Interval Notation

1.
  a. Set
  b. inequalities
  c. a is less than b
  d. c is greater than or equal to d
  e. 5 is not equal to 6
  f. infinity; negative infinity

3. $\left\{ 1.7, \pi, -5, 4.\overline{2} \right\}$

    g. $\{x \mid x > 5\}$; interval

    h.   excludes; includes

    i.   parenthesis

5. $-10 = \dfrac{-10}{1}$

7. $-\dfrac{3}{5} = \dfrac{-3}{5}$

9. $0 = \dfrac{0}{1}$

11.

	Real Numbers	Irrational Numbers	Rational Numbers	Integers	Whole Numbers	Natural Numbers
5	✓		✓	✓	✓	✓
$-\sqrt{9}$	✓		✓	✓		
-1.7	✓		✓			
$\dfrac{1}{2}$	✓		✓			
$\sqrt{7}$	✓	✓				
$\dfrac{0}{4}$	✓		✓	✓	✓	
$0.\overline{2}$	✓		✓			

13. $-9 < -1$

15. $0.1\overline{5} > 0.15$

17. $\dfrac{5}{3} \cdot \dfrac{7}{7} = \dfrac{35}{21}; \dfrac{10}{7} \cdot \dfrac{3}{3} = \dfrac{30}{21}$

    $\dfrac{35}{21} > \dfrac{30}{21}$ so $\dfrac{5}{3} > \dfrac{10}{7}$

19. $-\dfrac{5}{8} < -\dfrac{1}{8}$

21. $(2, \infty)$

23. $(-\infty, 0]$

25. $(-5, 0]$

27. $[-4.7, \infty)$

29. $(-1, \infty)$

31. $(-\infty, -2]$

33. $\left(-\infty, \dfrac{9}{2}\right)$

35. $(-2.5, 4.5]$

37. $(-\infty, -3)$

39. $\left(\dfrac{5}{2}, \infty\right)$

41. $[2, \infty)$

43. $(-4, 4)$

45. $[-3,0]$

47. All real numbers less than -4

49. All real numbers greater than -2 and less than or equal to 7

51. All real numbers between -180 and 90, inclusive

53. All real numbers greater than 3.2

55. $a \geq 18$

57. $c \leq 25$

59. $s \geq 261$

61. $r \leq 4.5$

63. $18 \leq a \leq 25$

65. $p < 130$ mm Hg

67. 130 mm Hg $\leq p \leq 139$ mm Hg

69. $2.2 \leq pH \leq 2.4$ acidic

71. $3.0 \leq pH \leq 3.5$ acidic

Section R.3   Operations on Real Numbers

1.
  a. opposites
  b. $|a|$; 0
  c. base; n
  d. radical; square
  e. negative; positive
  f. positive; negative
  g. $-b$; positive
  h. $\dfrac{1}{a}$; 1
  i. 0
  j. 0; undefined

3. $(-\infty, 4)$

5. $[-3, -1)$

7. Distance can never be negative.

9. Negative

11.

Number	Opposite	Reciprocal	Absolute Value
6	-6	$\dfrac{1}{6}$	6
$\dfrac{1}{11}$	$-\dfrac{1}{11}$	11	$\dfrac{1}{11}$
-8	8	$-\dfrac{1}{8}$	8
$-\dfrac{13}{10}$	$\dfrac{13}{10}$	$-\dfrac{10}{13}$	$\dfrac{13}{10}$
0	0	Undefined	0
-3	3	$-0.\overline{3}$	3

3

13. $-|6|=-6;\quad |-6|=6$

$-|6|<|-6|$

15. $|-4|=4;\quad |4|=4$

$|-4|=|4|$

17. $-|-1|=-1;\quad 1=1$

$-|-1|<1$

19. $|2+(-5)|=|-3|=3;\quad |2|+|-5|=2+5=7$

$|2+(-5)|<|2|+|-5|$

21. $-8+4=-4$

23. $-12+(-7)=-19$

25. $-17-(-10)=-17+10=-7$

27. $5-(-9)=5+9=14$

29. $-6.3-15.8=-6.3+(-15.8)=-22.1$

31. $1.5-9.6=1.5+(-9.6)=-8.1$

33. $\dfrac{2}{3}+\left(-2\dfrac{1}{3}\right)=\dfrac{2}{3}+\left(-\dfrac{7}{3}\right)=-\dfrac{5}{3}=-1\dfrac{2}{3}$

35. $-\dfrac{5}{9}-\dfrac{14}{15}=-\dfrac{5}{9}\cdot\dfrac{5}{5}-\dfrac{14}{15}\cdot\dfrac{3}{3}=-\dfrac{25}{45}-\dfrac{42}{45}=-\dfrac{67}{45}$

37. $4(-8)=-32$

39. $\dfrac{2}{9}\cdot\dfrac{12}{7}=\dfrac{24}{63}=\dfrac{8}{21}$

41. $\dfrac{-6}{-10}=\dfrac{3}{5}$

43. $-2\dfrac{1}{4}\div\dfrac{5}{8}=-\dfrac{9}{4}\cdot\dfrac{8}{5}=-\dfrac{72}{20}=-\dfrac{18}{5}$

45. $7\div 0$ is Undefined

47. $0\div(-3)=0$

49. $(-1.2)(-3.1)=3.72$

51. $\dfrac{-5}{-11}=\dfrac{5}{11}$

53. $4^3=4\cdot 4\cdot 4=64$

55. $-7^2=-(7\cdot 7)=-(49)=-49$

57. $(-7)^2=(-7)\cdot(-7)=49$

59. $\left(\dfrac{5}{3}\right)^3=\dfrac{5}{3}\cdot\dfrac{5}{3}\cdot\dfrac{5}{3}=\dfrac{125}{27}$

61. $\sqrt{9}=3$

63. $\sqrt{-4}$ is not a real number

65. $\sqrt{\dfrac{1}{4}}=\dfrac{1}{2}$

67. $-\sqrt{49}=-7$

69. $5+3^3=5+3\cdot 3\cdot 3=5+27=32$

71. $5\cdot 2^3=5\cdot 8=40$

73. $(2+3)^2=5^2=25$

75. $2^2+3^2=4+9=13$

77. $6 + 10 \div 2 \cdot 3 - 4 = 6 + 5 \cdot 3 - 4 = 6 + 15 - 4$
$\qquad = 21 - 4 = 17$

79. $4^2 - (5-2)^2 \cdot 3 = 4^2 - 3^2 \cdot 3 = 16 - 9 \cdot 3$
$\qquad = 16 - 27 = -11$

81. $2 - 5\left(9 - 4\sqrt{25}\right)^2 = 2 - 5\left(9 - 4 \cdot 5\right)^2 = 2 - 5\left(9 - 20\right)^2 = 2 - 5\left(-11\right)^2 = 2 - 5 \cdot 121 = 2 - 605 = -603$

83. $\left(-\dfrac{3}{5}\right)^2 - \dfrac{3}{5} \cdot \dfrac{5}{9} + \dfrac{7}{10} = \dfrac{9}{25} - \dfrac{3}{\cancel{5}} \cdot \dfrac{\cancel{5}}{9} + \dfrac{7}{10} = \dfrac{9}{25} - \dfrac{\cancel{3} \cdot 1}{\cancel{3} \cdot 3} + \dfrac{7}{10} = \dfrac{9}{25} - \dfrac{1}{3} + \dfrac{7}{10} = \dfrac{9}{25} \cdot \dfrac{6}{6} - \dfrac{1}{3} \cdot \dfrac{50}{50} + \dfrac{7}{10} \cdot \dfrac{15}{15}$

$\qquad = \dfrac{54}{150} - \dfrac{50}{150} + \dfrac{105}{150} = \dfrac{109}{150}$

85. $1.75 \div 0.25 - (1.25)^2 = 1.75 \div 0.25 - 1.5625$
$\qquad = 7 - 1.5625 = 5.4375$

87. $\dfrac{\sqrt{10^2 - 8^2}}{3^2} = \dfrac{\sqrt{100 - 64}}{9} = \dfrac{\sqrt{36}}{9} = \dfrac{6}{9} = \dfrac{2}{3}$

89. $-\left|-11 + 5\right| + \left|7 - 2\right| = -\left|-6\right| + \left|5\right|$
$\qquad = -6 + 5 = -1$

91. $25 - 2\left[(7-3)^2 \div 4\right] + \sqrt{18-2} = 25 - 2\left[(4)^2 \div 4\right] + \sqrt{16} = 25 - 2\left[16 \div 4\right] + 4 = 25 - 2 \cdot 4 + 4$
$\qquad = 25 - 8 + 4 = 17 + 4 = 21$

93.
$\dfrac{\left|(10-7) - 2^3\right|}{6 - 16 \div 8 \cdot 3}$

$= \dfrac{\left|3 - 2^3\right|}{6 - 2 \cdot 3}$

$= \dfrac{\left|3 - 8\right|}{6 - 6}$

$= \dfrac{\left|-5\right|}{0}$ undefined

95. $\left(\dfrac{1}{2}\right)^2 + \left(\dfrac{6-4}{5}\right)^2 + \left(\dfrac{5+2}{10}\right)^2 = \left(\dfrac{1}{2}\right)^2 + \left(\dfrac{2}{5}\right)^2 + \left(\dfrac{7}{10}\right)^2 = \dfrac{1}{4} + \dfrac{4}{25} + \dfrac{49}{100} = \dfrac{25}{100} + \dfrac{16}{100} + \dfrac{49}{100} = \dfrac{90}{100} = \dfrac{9}{10}$

97. $\dfrac{(-18) + (-16) + (-20) + (-11) + (-4) + (-3) + 1}{7} = \dfrac{-71}{7} \approx -10.1^\circ \text{C}$

99. $C = \frac{5}{9}(F - 32)$

   a. $C = \frac{5}{9}(77 - 32) = \frac{5}{9}(45) = 25°\text{C}$

   b. $C = \frac{5}{9}(212 - 32) = \frac{5}{9}(180) = 100°\text{C}$

   c. $C = \frac{5}{9}(32 - 32) = \frac{5}{9}(0) = 0°\text{C}$

   d. $C = \frac{5}{9}(-40 - 32) = \frac{5}{9}(-72) = -40°\text{C}$

101.
$$G_H = \frac{1}{22}c + \frac{1}{30}h$$
$$= \frac{1}{22}(33) + \frac{1}{30}(80)$$
$$= \frac{3}{2} + \frac{8}{3}$$
$$= \frac{9}{6} + \frac{16}{6}$$
$$= \frac{25}{6} = 4\frac{1}{6}\ \text{gal}$$

103. $A = \frac{1}{2}(b_1 + b_2)h$

$$A = \frac{1}{2}(5 + 4)(2) = 5 + 4 = 9\ \text{in}^2$$

105. $A = \frac{1}{2}bh$

$$A = \frac{1}{2}(5.2)(3.1) = 2.6(3.1) = 8.06\ \text{cm}^2$$

107. $V = \frac{4}{3}\pi r^3$

$$V = \frac{4}{3}\pi(1.5)^3 = \frac{4}{3}\pi(3.375) \approx 14.1\ \text{ft}^3$$

109. $V = \frac{1}{3}\pi r^2 h$

$$V = \frac{1}{3}\pi(2.5)^2(4.1) = \frac{1}{3}\pi(6.25)(4.1)$$
$$\approx 26.8\ \text{ft}^3$$

111. $V = \pi r^2 h$

$$V = \pi(3)^2(5) = \pi(9)(5) \approx 141.4\ \text{in}^3$$

Section R.4   Simplifying Algebraic Expressions

1.   **a.** constant

     b. coefficient

     c. 1; 1

     d. like

3.
$$\frac{-3^2 - |5 + (-7)|}{26 - 2^2} = \frac{-3^2 - |-2|}{26 - 2^2} = \frac{-3^2 - 2}{26 - 2^2}$$
$$= \frac{-9 - 2}{26 - 4} = \frac{-11}{22} = -\frac{1}{2}$$

5. $\{x \mid x > |-3|\} = \{x \mid x > 3\}$ $(3, \infty)$

7. $\left\{w \mid -\frac{5}{2} < w \le \sqrt{9}\right\} = \left\{w \mid -\frac{5}{2} < w \le 3\right\}$

   $\left(-\frac{5}{2}, 3\right]$

9. $2x^3 - 5xy + 6$

     a. There are three terms.

     b. The constant term is 6.

     c. The coefficients are 2, –5, and 6.

11. $pq - 7 + q^2 - 4q + p$

     a. There are five terms.

     b. The constant term is –7.

     c. The coefficients are 1, –7, 1, –4, and 1.

13. a. Commutative property of addition

15. f. Identity property of addition

17. e. Associative property of addition

19. i. Inverse property of multiplication

21. b. Associative property of multiplication

23. g. Identity property of multiplication

25. d. Commutative property of multiplication

27. h. Inverse property of addition

29. a. Commutative property of addition

31. $2(x-3y+8)=2x-6y+16$

33. $-10(4s-9t-3)=-40s+90t+30$

35. $-(-7w+5z)=7w-5z$

37. $-\dfrac{1}{5}\left(-\dfrac{5}{2}a+10b-8\right)=\dfrac{1}{2}a-2b+\dfrac{8}{5}$

39. $3(2.6x-4.1)=7.8x-12.3$

41. $2(7c-8)-5(6d-f)=14c-16-30d+5f$

43. $8y-2x+y+5y=8y+y+5y-2x$
$\qquad =14y-2x$

45. $4p^2-2p+3p-6+2p^2$
$\qquad =4p^2+2p^2-2p+3p-6$
$\qquad =6p^2+p-6$

47. $2p-7p^2-5p+6p^2=-7p^2+6p^2+2p-5p$
$\qquad =-p^2-3p$

49. $m-4n^3+3+5n^3-9$
$\qquad =-4n^3+5n^3+m+3-9$
$\qquad =n^3+m-6$

51. $5ab+2ab+8a=7ab+8a$

53. $14xy^2-5y^2+2xy^2=14xy^2+2xy^2-5y^2$
$\qquad =16xy^2-5y^2$

55. $8(x-3)+1=8x-24+1=8x-23$

57. $-2(c+3)-2c=-2c-6-2c=-4c-6$

59. $-(10w-1)+9+w=-10w+1+9+w$
$\qquad =-9w+10$

61. $-9-4(2-z)+1=-9-8+4z+1=4z-16$

63. $4(2s-7)-(s-2)=8s-28-s+2=7s-26$

65. $-3(-5+2w)-8w+2(w-1)$
$\qquad =15-6w-8w+2w-2$
$\qquad =-12w+13$

67. $8x-4(x-2)-2(2x+1)-6$
$\qquad =8x-4x+8-4x-2-6$
$\qquad =0$

69. $\dfrac{1}{2}(4-2c)+5c=2-c+5c=4c+2$

71. $3.1(2x+2)-4(1.2x-1)$
$\qquad =6.2x+6.2-4.8x+4$
$\qquad =1.4x+10.2$

73.
$$2\left[5\left(\frac{1}{2}a+3\right)-\left(a^2+a\right)+4\right]$$
$$=2\left[\frac{5}{2}a+15-a^2-a+4\right]$$
$$=2\left[-a^2+\frac{3}{2}a+19\right]$$
$$=-2a^2+3a+38$$

75.
$$\left[(2y-5)-2\left(y-y^2\right)\right]-3y$$
$$=\left[2y-5-2y+2y^2\right]-3y$$
$$=2y-5-2y+2y^2-3y$$
$$=2y^2-3y-5$$

77.
$$2.2\left\{4-8\left[6x-1.5(x+4)-6\right]+7.5x\right\}$$
$$=2.2\left\{4-8\left[6x-1.5x-6-6\right]+7.5x\right\}$$
$$=2.2\left\{4-8\left[4.5x-12\right]+7.5x\right\}$$
$$=2.2\left\{4-36x+96+7.5x\right\}$$
$$=2.2\left\{-28.5x+100\right\}$$
$$=-62.7x+220$$

79.
$$\frac{1}{8}(24n-16m)-\frac{2}{3}(3m-18n-2)+\frac{2}{3}$$
$$=3n-2m-2m+12n+\frac{4}{3}+\frac{2}{3}$$
$$=-4m+15n+2$$

81. The identity element for addition is 0. For example: $3+0=3$.

83. Another name for a multiplicative inverse is a reciprocal.

85. The operation of subtraction is not commutative. For example:
$$6-5\neq5-6$$
$$1\neq-1$$

87. **a.** $x(y+z)$
 b. xy
 c. xz
 d. $xy+xz$
 e. $x(y+z)=xy+xz$ the distributive property of multiplication over addition

Chapter R Review Exercises

Section R.2

1. The number that is a whole number but not a natural number is 0.

3. For example: $-2,-1,0,1,2$

5. $(0,2.6]$ All real numbers greater than 0 but less than or equal to 2.6.

7. $(8,\infty)$ All real numbers greater than 8.

9. $(-\infty,\infty)$ All real numbers.

11. $[0,\infty)$

13. True

Section R.3

15.
Opposite: $-\dfrac{4}{9}$

Reciprocal: $\dfrac{9}{4}$

Absolute value: $\dfrac{4}{9}$

17. $25^2 = 25 \cdot 25 = 625$

$\sqrt{25} = 5$

19. $(-2)-(-5) = -2+5 = 3$

21. $(-1.1)(7.41) = -8.151$

23.
$$\left(-\dfrac{1}{4}\right) \div \left(-\dfrac{11}{16}\right) = \left(-\dfrac{1}{4}\right)\cdot\left(-\dfrac{16}{11}\right)$$
$$= \left(-\dfrac{1}{\cancel{4}}\right)\cdot\left(-\dfrac{\cancel{4}\cdot 4}{11}\right) = \dfrac{4}{11}$$

25.
$$4\dfrac{2}{3} - 3\left(1\dfrac{1}{6}\right) = \dfrac{14}{3} - 3\left(\dfrac{7}{6}\right)$$
$$= \dfrac{14}{3} - \dfrac{21}{6}$$
$$= \dfrac{28}{6} - \dfrac{21}{6}$$
$$= \dfrac{7}{6} \text{ or } 1\dfrac{1}{6}$$

27. $\dfrac{12(2)-8}{4(-3)+2(5)} = \dfrac{24-8}{-12+10} = \dfrac{16}{-2} = -8$

29. $40 \div 5 \cdot 6 = 8 \cdot 6 = 48$

31.
$$-91 + \sqrt{4}\left(\sqrt{25}-13\right)^2 = -91 + \sqrt{4}\left(5-13\right)^2$$
$$= -91 + \sqrt{4}\left(-8\right)^2 = -91 + 2\cdot\left(-8\right)^2$$
$$= -91 + 2\cdot 64 = -91 + 128 = 37$$

33.
$$\dfrac{4(5-2)^2}{|3-7-5|} = \dfrac{4(3)^2}{|-9|} = \dfrac{4(9)}{9} = \dfrac{36}{9} = 4$$

35. $A = bh$

$A = 42(18)$

$A = 756\,\text{in}^2$

Section R.4

37. $\dfrac{1}{2}(x+8y-5) = \dfrac{1}{2}x + 4y - \dfrac{5}{2}$

39. $-(13a-b-5c) = -13a + b + 5c$

41. $18p + 3 - 17p + 8p = 9p + 3$

43.
$$\dfrac{3}{4}(8x-4) + \dfrac{1}{2}(6x+4) = 6x - 3 + 3x + 2$$
$$= 9x - 1$$

45. For example: $5(2y) = (5\cdot 2)y$

Chapter R Test

1. **a.** $-5, -4, -3, -2, -1, 0, 1, 2$

    b. For example: $\dfrac{3}{2}, \dfrac{5}{4}, \dfrac{8}{5}$

3. The interval $[4, \infty)$ includes all real numbers 4 and greater, whereas, the interval $(4, \infty)$ does not include the endpoint, 4.

5.

    $\left(-\infty, -\tfrac{4}{3}\right)$

7. $x \le 5$

9. $\begin{aligned} |-8| - 4(2-3)^2 \div \sqrt{4} &= 8 - 4(-1)^2 \div 2 \\ &= 8 - 4(1) \div 2 \\ &= 8 - 4 \div 2 \\ &= 8 - 2 \\ &= 6 \end{aligned}$

11. $\begin{aligned} \left(-\dfrac{1}{6} + \sqrt{\dfrac{4}{9}}\right)^2 &= \left(-\dfrac{1}{6} + \dfrac{2}{3}\right)^2 = \left(-\dfrac{1}{6} + \dfrac{4}{6}\right)^2 \\ &= \left(\dfrac{3}{6}\right)^2 = \left(\dfrac{1}{2}\right)^2 = \dfrac{1}{4} \end{aligned}$

13. $z = \dfrac{x - \mu}{\sigma / \sqrt{n}}$

    $z = \dfrac{18 - 17.5}{1.8 / \sqrt{16}} = \dfrac{0.5}{1.8 / 4} = \dfrac{0.5}{0.45} \approx 1.1$

15. $\begin{aligned} -3(4-x) + 9(x-1) - 5(2x-4) &= -12 + 3x + 9x - 9 - 10x + 20 \\ &= 2x - 1 \end{aligned}$

17. False

19. True

Chapter 1   Linear Equations and Inequalities in One Variable

Are You Prepared?

Across

2. 1307

4. $d = 60(33) = 1980$ mi

6. 7805

7. $0.10(64,780) = 6478$

9. $90° - 41° = 49°$

10. $180° - 70° = 110°$

Down

1. $0.40(32,640) = 13,056$

2. $180°$

3. $|-7729 + 262| = |-7467| = 7467$

4. $2(8707) + 50 = 17,414 + 50 = 17,464$

5. $l = A/w = 6370/65 = 98$ ft

8. $I = Prt = 4000(0.05)(4) = 800$

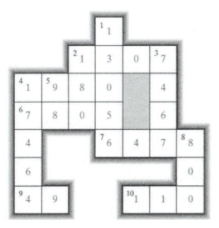

Section 1.1   Linear Equations in One Variable

1. **a.** equation
  b. solution
  c. linear
  d. first
  e. solution; set
  f. solution
  g. conditional
  h. contradiction
  i. empty set; { } or Ø
  j. identity

3. $8x - 3y + 2xy - 5x + 12xy = 3x - 3y + 14xy$

5. $2(3z - 4) - (z + 12) = 6z - 8 - z - 12$
$$= 5z - 20$$

7. $2x + 1 = 5$
$$2x - 4 = 0 \quad \text{Linear}$$

9. $x^2 + 7 = 9 \quad \text{Nonlinear}$

11. $-3 = x$
$$-x - 3 = 0 \quad \text{Linear}$$

13. $2x - 1 = 5$
  a. $2(2) - 1 = 5$
  $$4 - 1 = 5$$
  $$3 \neq 5$$
  2 is not a solution.
  b. $2(3) - 1 = 5$
  $$6 - 1 = 5$$
  $$5 = 5$$
  3 is a solution.
  c. $2(0) - 1 = 5$
  $$0 - 1 = 5$$
  $$-1 \neq 5$$
  0 is not a solution.

  $2(-1) - 1 = 5$
  $$-2 - 1 = 5$$
  $$-3 \neq 5$$
  d. −1 is not a solution.

15. $x + 7 = 19$
$$x + 7 - 7 = 19 - 7$$
$$x = 12 \qquad \{12\}$$
Check: $12 + 7 = 19$
$$19 = 19$$

17. $-x = 2$

$\quad\quad x = -2 \quad \{-2\}$

$\quad$ Check: $-(-2) = 2$

$\quad\quad\quad\quad\quad 2 = 2$

19.
$$-\frac{7}{8} = -\frac{5}{6}z$$

$$24\left(-\frac{7}{8}\right) = 24\left(-\frac{5}{6}z\right)$$

$$-21 = -20z$$

$$\frac{-21}{-20} = \frac{-20z}{-20}$$

$$z = \frac{21}{20} \quad \left\{\frac{21}{20}\right\}$$

Check: $-\dfrac{7}{8} = -\dfrac{5}{6}\left(\dfrac{21}{20}\right) = -\dfrac{105}{120} = -\dfrac{7}{8}$

21.
$$\frac{a}{5} = -8$$

$$5\left(\frac{a}{5}\right) = 5(-8)$$

$$a = -40 \quad \{-40\}$$

Check: $\dfrac{-40}{5} = -8$

$\quad\quad\quad -8 = -8$

23.
$$2.53 = -2.3t$$

$$\frac{2.53}{-2.3} = \frac{-2.3t}{-2.3}$$

$$-1.1 = t \quad \{-1.1\}$$

Check: $2.53 = -2.3(-1.1) = 2.53$

25.
$$p - 2.9 = 3.8$$

$$p - 2.9 + 2.9 = 3.8 + 2.9$$

$$p = 6.7 \quad \{6.7\}$$

Check: $6.7 - 2.9 = 3.8$

$\quad\quad\quad\quad 3.8 = 3.8$

27.
$$6q - 4 = 62$$

$$6q - 4 + 4 = 62 + 4 \quad\quad \text{Check:}$$

$$6q = 66 \quad\quad\quad\quad 6(11) - 4 = 62$$

$$\frac{6q}{6} = \frac{66}{6} \quad\quad\quad\quad 66 - 4 = 62$$

$$q = 11 \quad \{11\} \quad\quad\quad\quad 62 = 62$$

29.
$$4y - 17 = 35$$

$$4y - 17 + 17 = 35 + 17$$

$$4y = 52$$

$$\frac{4y}{4} = \frac{52}{4}$$

$$y = 13 \quad \{13\}$$

Check: $4(13) - 17 = 35$

$\quad\quad\quad\quad 52 - 17 = 35$

$\quad\quad\quad\quad\quad\quad 35 = 35$

31.
$$-b - 5 = 2$$

$$-b - 5 + 5 = 2 + 5$$

$$-b = 7$$

$$-1(-b) = -1(7)$$

$$b = -7 \quad \{-7\}$$

Check: $-(-7) - 5 = 2$

$\quad\quad\quad\quad 7 - 5 = 2$

$\quad\quad\quad\quad\quad 2 = 2$

33.
$$3(x-6)=2x-5$$
$$3x-18=2x-5$$
$$3x-18+18=2x-5+18$$
$$3x=2x+13$$
$$3x-2x=2x-2x+13$$
$$x=13 \quad \{13\}$$
$$\text{Check: } 3(13-6)=2(13)-5$$
$$3(7)=26-5$$
$$21=21$$

35.
$$6-(t+2)=5(3t-4)$$
$$6-t-2=15t-20$$
$$-t+4=15t-20$$
$$-t-15t+4=15t-15t-20$$
$$-16t+4=-20$$
$$-16t+4-4=-20-4$$
$$-16t=-24$$
$$\frac{-16t}{-16}=\frac{-24}{-16}$$
$$t=\frac{3}{2} \quad \left\{\frac{3}{2}\right\}$$
$$\text{Check: } 6-\left(\frac{3}{2}+2\right)=5\left(3\cdot\frac{3}{2}-4\right)$$
$$6-\frac{7}{2}=5\left(\frac{9}{2}-4\right)$$
$$\frac{5}{2}=5\left(\frac{1}{2}\right)$$
$$\frac{5}{2}=\frac{5}{2}$$

37.
$$6(a+3)-10=-2(a-4)$$
$$6a+18-10=-2a+8$$
$$6a+8=-2a+8$$
$$6a+2a+8=-2a+2a+8$$
$$8a+8=8$$
$$8a+8-8=8-8$$
$$8a=0$$
$$\frac{8a}{8}=\frac{0}{8}$$
$$a=0 \quad \{0\}$$
$$\text{Check: } 6(0+3)-10=-2(0-4)$$
$$6(3)-10=-2(-4)$$
$$18-10=8$$
$$8=8$$

39.
$$-2[5-(2z+1)]-4=2(3-z)$$
$$-2[5-2z-1]-4=6-2z$$
$$-10+4z+2-4=6-2z$$
$$4z-12=6-2z$$
$$4z+2z-12=6-2z+2z$$
$$6z-12=6$$
$$6z-12+12=6+12$$
$$6z=18$$
$$\frac{6z}{6}=\frac{18}{6}$$
$$z=3 \quad \{3\}$$
$$\text{Check: } -2[5-(2\cdot3+1)]-4=2(3-3)$$
$$-2[5-(6+1)]-4=2(0)$$
$$-2[5-7]-4=0$$
$$-2[-2]-4=0$$
$$4-4=0$$
$$0=0$$

41. $6(-y+4)-3(2y-3)=-y+5+5y$

$$-6y+24-6y+9=4y+5$$
$$-12y+33=4y+5$$
$$-12y+12y+33=4y+12y+5$$
$$33=16y+5$$
$$33-5=16y+5-5$$
$$28=16y$$
$$\frac{28}{16}=\frac{16y}{16}$$
$$\frac{7}{4}=y \qquad \left\{\frac{7}{4}\right\}$$

43. $14-2x+5x=-4(-2x-5)-6$

$$14+3x=8x+20-6$$
$$14+3x=8x+14$$
$$14+3x-8x=8x-8x+14$$
$$14-5x=14$$
$$14-14-5x=14-14$$
$$-5x=0$$
$$\frac{-5x}{-5}=\frac{0}{-5}$$
$$x=0 \qquad \{0\}$$

Check:

$$14-2\cdot0+5\cdot0=-4(-2\cdot0-5)-6$$
$$14-0+0=-4(0-5)-6$$
$$14=-4(-5)-6$$
$$14=20-6$$
$$14=14$$

45.

$$\frac{2}{3}x-\frac{1}{6}=-\frac{5}{12}x+\frac{3}{2}-\frac{1}{6}x$$
$$12\left(\frac{2}{3}x-\frac{1}{6}\right)=12\left(-\frac{5}{12}x+\frac{3}{2}-\frac{1}{6}x\right)$$
$$8x-2=-5x+18-2x$$
$$8x-2=-7x+18$$
$$8x+7x-2=-7x+7x+18$$
$$15x-2=18$$
$$15x-2+2=18+2$$
$$15x=20$$
$$\frac{15x}{15}=\frac{20}{15}$$
$$x=\frac{4}{3} \qquad \left\{\frac{4}{3}\right\}$$

47.

$$\frac{1}{5}(p-5)=\frac{3}{5}p+\frac{1}{10}p+1$$
$$\frac{1}{5}p-1=\frac{3}{5}p+\frac{1}{10}p+1$$
$$10\left(\frac{1}{5}p-1\right)=10\left(\frac{3}{5}p+\frac{1}{10}p+1\right)$$
$$2p-10=6p+p+10$$
$$2p-10=7p+10$$
$$2p-7p-10=7p-7p+10$$
$$-5p-10=10$$
$$-5p-10+10=10+10$$
$$-5p=20$$
$$\frac{-5p}{-5}=\frac{20}{-5}$$
$$p=-4 \qquad \{-4\}$$

49.

$$\frac{3x-7}{2}+\frac{3-5x}{3}=\frac{3-6x}{5}$$

$$30\left(\frac{3x-7}{2}+\frac{3-5x}{3}\right)=30\left(\frac{3-6x}{5}\right)$$

$$15(3x-7)+10(3-5x)=6(3-6x)$$

$$45x-105+30-50x=18-36x$$

$$-5x-75=18-36x$$

$$-5x+36x-75=18-36x+36x$$

$$31x-75=18$$

$$31x-75+75=18+75$$

$$31x=93$$

$$\frac{31x}{31}=\frac{93}{31}$$

$$x=3 \quad \{3\}$$

51.

$$\frac{4}{3}(2q+6)-\frac{5q-6}{6}-\frac{q}{3}=0$$

$$6\left[\frac{4}{3}(2q+6)-\frac{5q-6}{6}-\frac{q}{3}\right]=6(0)$$

$$8(2q+6)-(5q-6)-2q=0$$

$$16q+48-5q+6-2q=0$$

$$9q+54=0$$

$$9q+54-54=0-54$$

$$9q=-54$$

$$\frac{9q}{9}=\frac{-54}{9}$$

$$q=-6 \quad \{-6\}$$

53.

$$6.3w-1.5=4.8$$

$$10(6.3w-1.5)=10(4.8)$$

$$63w-15=48$$

$$63w-15+15=48+15$$

$$63w=63$$

$$\frac{63w}{63}=\frac{63}{63}$$

$$w=1 \quad \{1\}$$

55.

$$0.75(m-2)+0.25m=0.5$$

$$100\left[0.75(m-2)+0.25m\right]=100\left[0.5\right]$$

$$75(m-2)+25m=50$$

$$75m-150+25m=50$$

$$100m-150=50$$

$$100m-150+150=50+150$$

$$100m=200$$

$$\frac{100m}{100}=\frac{200}{100}$$

$$m=2 \quad \{2\}$$

57. A conditional equation is an equation that is true for some values of the variable but false for other values of the variable.

59.

$$4x+1=2(2x+1)-1$$

$$4x+1=4x+2-1$$

$$4x+1=4x+1$$

$$0=0$$

This is an identity. $\{x \mid x \text{ is a real number}\}$

61.
$$-11x + 4(x - 3) = -2x - 12$$
$$-11x + 4x - 12 = -2x - 12$$
$$-7x - 12 = -2x - 12$$
$$-7x + 2x - 12 = -2x + 2x - 12$$
$$-5x - 12 = -12$$
$$-5x - 12 + 12 = -12 + 12$$
$$-5x = 0$$
$$\frac{-5x}{-5} = \frac{0}{-5}$$
$$x = 0$$

This is a conditional equation. $\{0\}$

63.
$$2x - 4 + 8x = 7x - 8 + 3x$$
$$10x - 4 = 10x - 8$$
$$10x - 10x - 4 = 10x - 10x - 8$$
$$-4 = -8$$

This is a contradiction. $\{\ \}$

65.
$$-5b + 9 = -71$$
$$-5b + 9 - 9 = -71 - 9$$
$$-5b = -80$$
$$\frac{-5b}{-5} = \frac{-80}{-5}$$
$$b = 16 \qquad \{16\}$$

67.
$$16 = -10 + 13x$$
$$16 + 10 = -10 + 10 + 13x$$
$$26 = 13x$$
$$\frac{26}{13} = \frac{13x}{13}$$
$$2 = x \qquad \{2\}$$

69.
$$10c + 3 = -3 + 12c$$
$$10c - 12c + 3 = -3 + 12c - 12c$$
$$-2c + 3 = -3$$
$$-2c + 3 - 3 = -3 - 3$$
$$-2c = -6$$
$$\frac{-2c}{-2} = \frac{-6}{-2}$$
$$c = 3 \qquad \{3\}$$

71.
$$12b - 8b - 8 + 13 = 4b + 6 - 1$$
$$4b + 5 = 4b + 5$$
$$0 = 0$$

The equation is an identity. The solution set is $\{b \mid b \text{ is a real number}\}$.

73.
$$5(x - 2) - 2x = 3x + 7$$
$$5x - 10 - 2x = 3x + 7$$
$$3x - 10 = 3x + 7$$
$$3x - 3x - 10 = 3x - 3x + 7$$
$$-10 = 7$$
$$\{\ \}$$

75.
$$\frac{c}{2} - \frac{c}{4} + \frac{3c}{8} = 1$$
$$8\left(\frac{c}{2} - \frac{c}{4} + \frac{3c}{8}\right) = 8(1)$$
$$4c - 2c + 3c = 8$$
$$5c = 8$$
$$\frac{5c}{5} = \frac{8}{5}$$
$$c = \frac{8}{5} \qquad \left\{\frac{8}{5}\right\}$$

77.
$$0.75(8x-4)=\frac{2}{3}(6x-9)$$
$$6x-3=4x-6$$
$$6x-4x-3=4x-4x-6$$
$$2x-3=-6$$
$$2x-3+3=-6+3$$
$$2x=-3$$
$$\frac{2x}{2}=\frac{-3}{2}$$
$$x=-\frac{3}{2}\qquad\left\{-\frac{3}{2}\right\}$$

79.
$$7(p+2)-4p=3p+14$$
$$7p+14-4p=3p+14$$
$$3p+14=3p+14$$
$$3p-3p+14=3p-3p+14$$
$$14=14$$
$$\{p\,|\,p\ \text{is a real number}\}$$

81.
$$4[3+5(3-b)+2b]=6-2b$$
$$4[3+15-5b+2b]=6-2b$$
$$4[-3b+18]=6-2b$$
$$-12b+72=6-2b$$
$$-12b+2b+72=6-2b+2b$$
$$-10b+72=6$$
$$-10b+72-72=6-72$$
$$-10b=-66$$
$$\frac{-10b}{-10}=\frac{-66}{-10}$$
$$b=\frac{33}{5}=6.6\qquad\left\{\frac{33}{5}\right\}$$

83.
$$3-\frac{3}{4}x=3\left(3-\frac{1}{4}x\right)$$
$$3-\frac{3}{4}x=9-\frac{3}{4}x$$
$$3=9$$
The equation is a contradiction. The solution set is { }.

85.
$$\frac{5}{4}+\frac{y-3}{8}=\frac{2y+1}{2}$$
$$8\left(\frac{5}{4}+\frac{y-3}{8}\right)=8\left(\frac{2y+1}{2}\right)$$
$$10+y-3=4(2y+1)$$
$$y+7=8y+4$$
$$y-8y+7=8y-8y+4$$
$$-7y+7=4$$
$$-7y+7-7=4-7$$
$$-7y=-3$$
$$\frac{-7y}{-7}=\frac{-3}{-7}$$
$$y=\frac{3}{7}\qquad\left\{\frac{3}{7}\right\}$$

87.
$$\frac{2y-9}{10}+\frac{3}{2}=y$$
$$10\left(\frac{2y-9}{10}+\frac{3}{2}\right)=10y$$
$$2y-9+15=10y$$
$$2y+6=10y$$
$$2y-2y+6=10y-2y$$
$$6=8y$$
$$\frac{6}{8}=\frac{8y}{8}$$
$$\frac{3}{4}=y\qquad\left\{\frac{3}{4}\right\}$$

18

89.
$$0.48x - 0.08x = 0.12(260 - x)$$
$$100(0.48x - 0.08x) = 100\left[0.12(260 - x)\right]$$
$$48x - 8x = 12(260 - x)$$
$$40x = 3120 - 12x$$
$$40x + 12x = 3120 - 12x + 12x$$
$$52x = 3120$$
$$\frac{52x}{52} = \frac{3120}{52}$$
$$x = 60 \quad \{60\}$$

91.
$$0.5x + 0.25 = \frac{1}{3}x + \frac{5}{4}$$
$$\frac{1}{2}x + \frac{1}{4} = \frac{1}{3}x + \frac{5}{4}$$
$$12\left(\frac{1}{2}x + \frac{1}{4}\right) = 12\left(\frac{1}{3}x + \frac{5}{4}\right)$$
$$6x + 3 = 4x + 15$$
$$6x - 4x + 3 = 4x - 4x + 15$$
$$2x + 3 = 15$$
$$2x + 3 - 3 = 15 - 3$$
$$2x = 12$$
$$\frac{2x}{2} = \frac{12}{2}$$
$$x = 6 \quad \{6\}$$

93.
$$0.3b - 1.5 = 0.25(b + 2) + 0.05b$$
$$0.3b - 1.5 = 0.25b + 0.5 + 0.05b$$
$$0.3b - 1.5 = 0.3b + 0.5$$
$$1.5 = 0.5$$
The equation is a contradiction. The solution set is { }.

95.
$$-\frac{7}{8}y + \frac{1}{4} = \frac{1}{2}\left(5 - \frac{3}{4}y\right)$$
$$-\frac{7}{8}y + \frac{1}{4} = \frac{5}{2} - \frac{3}{8}y$$
$$8\left(-\frac{7}{8}y + \frac{1}{4}\right) = 8\left(\frac{5}{2} - \frac{3}{8}y\right)$$
$$-7y + 2 = 20 - 3y$$
$$-7y + 3y + 2 = 20 - 3y + 3y$$
$$-4y + 2 = 20$$
$$-4y + 2 - 2 = 20 - 2$$
$$-4y = 18$$
$$\frac{-4y}{-4} = \frac{18}{-4}$$
$$y = -\frac{9}{2} \quad \left\{-\frac{9}{2}\right\}$$

97.
$$0.12h + 14.89 = 137.77$$
$$0.12h = 137.77 - 14.89$$
$$0.12h = 122.88$$
$$h = \frac{122.88}{0.12}$$
$$h = 1024$$
The family used 1024 kWh.

99. a.
$$-2(y - 1) + 3(y + 2) = -2y + 2 + 3y + 6$$
$$= y + 8$$

b.
$$2(y - 1) + 3(y + 2) = 0$$
$$y + 8 = 0$$
$$y + 8 - 8 = 0 - 8$$
$$y = -8 \quad \{-8\}$$

c. To simplify an expression, clear parentheses and combine like terms. To solve an equation, isolate the variable to find a solution.

Problem Recognition Exercises

1. Expression
$$4x - 2 + 6 - 8x = 4x - 8x - 2 + 6 = -4x + 4$$

3. Equation
$$7b - 1 = 2b + 4$$
$$7b - 2b - 1 = 2b - 2b + 4$$
$$5b - 1 = 4$$
$$5b - 1 + 1 = 4 + 1$$
$$5b = 5$$
$$\frac{5b}{5} = \frac{5}{5}$$
$$b = 1 \quad \{1\}$$

5. Expression
$$4(a - 8) - 7(2a + 1) = 4a - 32 - 14a - 7$$
$$= -10a - 39$$

7. Equation
$$7(2 - w) = 5w + 8$$
$$14 - 7w = 5w + 8$$
$$14 - 7w - 5w = 5w - 5s + 8$$
$$14 - 12w = 8$$
$$14 - 14 - 12w = 8 - 14$$
$$-12w = -6$$
$$\frac{-12w}{-12} = \frac{-6}{-12}$$
$$w = \frac{1}{2} \quad \left\{\frac{1}{2}\right\}$$

9. Equation
$$2(3x - 4) - 4(5x + 1) = -8x + 7$$
$$6x - 8 - 20x - 4 = -8x + 7$$
$$-14x - 12 = -8x + 7$$
$$-14x + 8x - 12 = -8x + 8x + 7$$
$$-6x - 12 = 7$$
$$-6x - 12 + 12 = 7 + 12$$
$$-6x = 19$$
$$\frac{-6x}{-6} = \frac{19}{-6}$$
$$x = -\frac{19}{6} \quad \left\{-\frac{19}{6}\right\}$$

11. Expression
$$\frac{1}{2}v + \frac{3}{5} - \frac{2}{3}v - \frac{7}{10} = \frac{15}{30}v + \frac{18}{30} - \frac{20}{30}v - \frac{21}{30}$$
$$= -\frac{5}{30}v - \frac{3}{30} = -\frac{1}{6}v - \frac{1}{10}$$

13. Equation
$$20x - 8 + 7x + 28 = 27x - 9$$
$$27x + 20 = 27x - 9$$
$$27x - 27x + 20 = 27x - 27x - 9$$
$$20 = -9$$
$$\{\}$$

15. Equation

$$\frac{5}{6}y - \frac{7}{8} = \frac{1}{2}y + \frac{3}{4}$$
$$24\left(\frac{5}{6}y - \frac{7}{8}\right) = 24\left(\frac{1}{2}y + \frac{3}{4}\right)$$
$$20y - 21 = 12y + 18$$
$$20y - 12y - 21 = 12y - 12y + 18$$
$$8y - 21 = 18$$
$$8y - 21 + 21 = 18 + 21$$
$$8y = 39$$
$$\frac{8y}{8} = \frac{39}{8}$$
$$y = \frac{39}{8} \qquad \left\{\frac{39}{8}\right\}$$

17. Expression
$$0.29c + 4.495 - 0.12c = 0.17c + 4.495$$

19. Equation
$$0.125(2p - 8) = 0.25(p - 4)$$
$$0.25p - 1 = 0.25p - 1$$
$$0.25p - 0.25p - 1 = 0.25p - 0.25p - 1$$
$$-1 = -1$$
$$\{p \mid p \text{ is a real number}\}$$

Section 1.2   Applications of Linear Equations in One Variable

1. a. consecutive

b. even; odd

c. 1; 2; 2

d. $x + 1$

e. $x + 2$

f. $x + 2$; $x + 4$

g. Prt; interest

h. \$1300

i. 0.48 L; $0.12(x + 8)$

j. $\dfrac{d}{t}$; $\dfrac{d}{r}$

3.
$$7a - 2 = 11$$
$$7a - 2 + 2 = 11 + 2$$
$$7a = 13$$
$$\frac{7a}{7} = \frac{13}{7}$$
$$a = \frac{13}{7} \qquad \left\{\frac{13}{7}\right\}$$

5.
$$4(x-3)+7=19$$
$$4x-12+7=19$$
$$4x-5=19$$
$$4x-5+5=19+5$$
$$4x=24$$
$$\frac{4x}{4}=\frac{24}{4}$$
$$x=6 \quad \{6\}$$

7.
$$\frac{3}{8}p+\frac{3}{4}=p-\frac{3}{2}$$
$$8\left(\frac{3}{8}p+\frac{3}{4}\right)=8\left(p-\frac{3}{2}\right)$$
$$3p+6=8p-12$$
$$3p-8p+6=8p-8p-12$$
$$-5p+6=-12$$
$$-5p+6-6=-12-6$$
$$-5p=-18$$
$$\frac{-5p}{-5}=\frac{-18}{-5}$$
$$p=\frac{18}{5} \quad \left\{\frac{18}{5}\right\}$$

9. $x+5$

11. $x+5$

13. Let x = the smaller number
$2x+3$ = the larger number
(larger number) – (smaller number) = 8

The smaller number is 5 and the larger is 13.

15. Let x = the number
$3x+2$ = the sum
$x-4$ = the difference
(sum) = (difference)
$$3x+2=x-4$$
$$3x-x+2=x-x-4$$
$$2x+2=-4$$
$$2x+2-2=-4-2$$
$$2x=-6$$
$$\frac{2x}{2}=\frac{-6}{2}$$
$$x=-3$$
The number is –3.

17. Let x = the first page number
$x+1$ = the consecutive page number
(first) + (second) = 223
$$x+(x+1)=223$$
$$2x+1=223$$
$$2x+1-1=223-1$$
$$2x=222$$
$$\frac{2x}{2}=\frac{222}{2}$$
$$x=111$$
$$x+1=111+1=112$$
The consecutive page numbers are 111 and 112.

19. Let x = the first odd integer
$x+2$ = the consecutive odd integer
(first) + (second) = –148
$$x+(x+2)=-148$$
$$2x+2=-148$$
$$2x+2-2=-148-2$$
$$2x=-150$$
$$\frac{2x}{2}=\frac{-150}{2}$$
$$x=-75$$
$$x+2=-75+2=-73$$
The two consecutive odd integers are –75 and –73.

21. Let x = the smaller even integer
$x + 2$ = larger consecutive even integer
(3 times small) = (–146 minus 4 times larger)
$$3x = -146 - 4(x+2)$$
$$3x = -146 - 4x - 8$$
$$3x = -154 - 4x$$
$$3x + 4x = -154 - 4x + 4x$$
$$7x = -154$$
$$\frac{7x}{7} = \frac{-154}{7}$$
$$x = -22$$
$$x + 2 = -22 + 2 = -20$$
The two consecutive even integers are –22 and –20.

23. Let x = first odd integer
$x + 2$ = second consecutive odd integer
$x + 4$ = third consecutive odd integer
(2 times sum) = (23 more than 5 times third)
$$2(x + x + 2 + x + 4) = 5(x + 4) + 23$$
$$2(3x + 6) = 5x + 20 + 23$$
$$6x + 12 = 5x + 43$$
$$6x - 5x + 12 = 5x - 5x + 43$$
$$x + 12 = 43$$
$$x + 12 - 12 = 43 - 12$$
$$x = 31$$
$$x + 2 = 31 + 2 = 33$$
$$x + 4 = 31 + 4 = 35$$
The three consecutive odd integers are 31, 33, and 35.

25. Option 1:
Principal amount borrowed: $P = 15,000$
Interest rate: $r = 0.085$
Duration of loan: $t = 4$
$$x = Prt$$
$$x = 15,000(0.085)(4)$$
$$x = 5100$$

Option 2:
Principal amount borrowed: $P = 15,000$
Interest rate: $r = 0.0775$
Duration of loan: $t = 5$
$$x = Prt$$
$$x = 15,000(0.0775)(5)$$
$$x = 5812.50$$

She would pay $5100 for 4 yr at 8.5% and $5812.50 for 5 yr at 7.75%; the 8.5% option for 4 yr requires less interest.

27. Let x = the amount of sales
(earnings) = 600 + (sales amt)(comm. rate)
$$2400 = 600 + x(0.03)$$
$$2400 - 600 = 600 - 600 + 0.03x$$
$$1800 = 0.03x$$
$$\frac{1800}{0.03} = \frac{0.03x}{0.03}$$
$$60,000 = x$$
She needs to sell $60,000 to earn $2400.

29. Let c = the sales before tax
(total cash) = (sales) + (sales tax)
(sales tax) = (tax rate)(sales)
$$1293.38 = c + 0.0805c$$
$$1293.38 = 1.0805c$$
$$\frac{1293.38}{1.0805} = \frac{1.0805c}{1.0805}$$
$$1197.02 = c$$
$$t = 0.0805(1197.02) = 96.36$$

31. Let c = the cost before markup
(price) = (cost) + (markup)
(markup) = (markup rate)(cost)
$$43.08 = c + 0.20c$$
$$43.08 = 1.20c$$
$$\frac{43.08}{1.20} = \frac{1.20c}{1.20}$$
$$35.90 = c$$
The price before markup was $35.90.

The total merchandise was $1197.02 and the sales tax was $96.36.

33.

	2% Account	5% Account	Total
Amount Invested	x	12,500–x	12500
Interest Earned	0.02x	0.05(12,500–x)	370

(int at 2%) + (int at 5%) = (total int)

$$0.02x + 0.05(12,500 - x) = 370$$
$$0.02x + 625 - 0.05x = 370$$
$$-0.03x + 625 = 370$$
$$-0.03x + 625 - 625 = 370 - 625$$
$$-0.03x = -340$$
$$\frac{-0.03x}{-0.03} = \frac{-255}{-0.03}$$
$$x = 8500$$
$$12,500 - x = 12,500 - 8500$$
$$= 4000$$

$8500 was invested at 2% and $4000 was invested at 5%.

35.

	11% Loan	6% Loan	Total
Amount Borrowed	x	18,000–x	18,000
Interest Paid	0.11x	0.06(18,000–x)	1380

(int at 11%) + (int at 6%) = (total int)

$$0.11x + 0.06(18,000 - x) = 1380$$
$$0.11x + 1080 - 0.06x = 1380$$
$$0.05x + 1080 = 1380$$
$$0.05x + 1080 - 1080 = 1380 - 1080$$
$$0.05x = 300$$
$$\frac{0.05x}{0.05} = \frac{300}{0.05}$$
$$x = 6000$$
$$18,000 - x = 18,000 - 6000$$
$$= 12,000$$

$6000 was borrowed at 11% and $12,000 was borrowed at 6%.

37.

	4% Account	3% Account	Total
Amount Invested	x	x – 4000	
Interest Earned	0.04x	0.03(x – 4000)	720

(int at 4%) + (int at 3%) = (total int)

$$0.04x + 0.03(x - 4000) = 720$$
$$0.04x + 0.03x - 120 = 720$$
$$0.07x - 120 = 720$$
$$0.07x - 120 + 120 = 720 + 120$$
$$0.07x = 840$$
$$\frac{0.07x}{0.07} = \frac{840}{0.07}$$
$$x = 12,000$$
$$x - 4000 = 12,000 - 4000$$
$$= 8000$$

$12,000 was invested at 4% and $8000 was invested at 3%.

39.

	15% nitrogen	10% nitrogen	14% nitrogen
Amount of fertilizer	x	2	x + 2
Amount of nitrogen	0.15(x)	0.10(2)	0.14(x + 2)

(amt of 15%) + (amt of 10%) = (amt of 14%)

$$0.15x + 0.10(2) = 0.14(x + 2)$$
$$0.15x + 0.20 = 0.14x + 0.28$$
$$0.15x - 0.14x + 0.20 = 0.14x - 0.14x + 0.28$$
$$0.01x + 0.20 = 0.28$$
$$0.01x + 0.20 - 0.20 = 0.28 - 0.20$$
$$0.01x = 0.08$$
$$\frac{0.01x}{0.01} = \frac{0.08}{0.01}$$
$$x = 8$$

8 oz of 15% nitrogen fertilizer should be used.

41.

	50% antifreeze	75% antifreeze	60% antifreeze
Amount of fertilizer	3	x	$x + 3$
Amount of nitrogen	0.50(3)	0.75(x)	0.60($x + 3$)

(amt of 50%) + (amt of 75%) = (amt of 60%)

$$0.50(3) + 0.75x = 0.60(x+3)$$
$$1.5 + 0.75x = 0.60x + 1.8$$
$$0.75x - 0.60x + 1.5 = 0.60x - 0.60x + 1.8$$
$$0.15x + 1.5 = 1.8$$
$$0.15x + 1.5 - 1.5 = 1.8 - 1.5$$
$$0.15x = 0.3$$
$$\frac{0.15x}{0.15} = \frac{0.3}{0.15}$$
$$x = 2$$

2 L of the 75% antifreeze solution should be used.

43.

	18% Solution	10% Solution	15% Solution
Amount of Solution	x	20–x	20
Amount of Alcohol	0.18x	0.10(20–x)	0.15(20)

(amt of 18%) + (amt of 10%) = (amt of 15%)

$$0.18x + 0.10(20 - x) = 0.15(20)$$
$$0.18x + 2 - 0.10x = 3$$
$$0.08x + 2 = 3$$
$$0.08x + 2 - 2 = 3 - 2$$
$$0.08x = 1$$
$$\frac{0.08x}{0.08} = \frac{1}{0.08}$$
$$x = 12.5$$
$$20 - x = 20 - 12.5 = 7.5$$

12.5 L of 18% solution and 7.5 L of 10% solution must be mixed.

45.

	12% Super Grow	Pure Super Grow	17.5% Super Grow
Amount of Solution	32–x	x	32
Amount of Super Grow	0.12(32–x)	1.00x	0.175(32)

(amt of 12%) + (amt of pure) = (amt of 17.5%)

$$0.12(32 - x) + 1.00x = 0.175(32)$$
$$3.84 - 0.12x + x = 5.6$$
$$0.88x + 3.84 = 5.6$$
$$0.88x + 3.84 - 3.84 = 5.6 - 3.84$$
$$0.88x = 1.76$$
$$\frac{0.88x}{0.88} = \frac{1.76}{0.88}$$
$$x = 2$$

2 oz of pure Super Grow must be added.

47.

	Distance	Rate	Time
To FL	2(x + 60)	x + 60	2
Return	2.5x	x	2.5

(dist to FL) = (dist back to Atlanta)

$$2(x + 60) = 2.5x$$
$$2x + 120 = 2.5x$$
$$2x + 120 - 2x = 2.5x - 2x$$
$$120 = 0.5x$$
$$\frac{120}{0.5} = \frac{0.5x}{0.5}$$
$$240 = x$$
$$x + 60 = 240 + 60 = 300$$

The plane flies 300 mph from Atlanta to Fort Lauderdale and 240 mph on the return trip.

49.

	Distance	Rate	Time
Car A	2x	x	2
Car B	2(x + 4)	x + 4	2

(dist car A) + (dist car B) = (total dist)

51. Let x = the first integer

$30 - x$ = the second integer

(ten times first) = (five times second)

25

$$2x + 2(x+4) = 192$$
$$2x + 2x + 8 = 192$$
$$4x + 8 = 192$$
$$4x + 8 - 8 = 192 - 8$$
$$4x = 184$$
$$\frac{4x}{4} = \frac{184}{4}$$
$$x = 46$$
$$x + 4 = 46 + 4 = 50$$

The cars are traveling at 46 mph and 50 mph.

$$10x = 5(30 - x)$$
$$10x = 150 - 5x$$
$$10x + 5x = 150 - 5x + 5x$$
$$15x = 150$$
$$\frac{15x}{15} = \frac{150}{15}$$
$$x = 10$$
$$30 - x = 30 - 10 = 20$$

The integers are 10 and 20.

53. New Price = Original Price − Markdown
$$89.55 = x - 0.55x$$
$$89.55 = 0.45x$$
$$\frac{89.55}{0.45} = x$$
$$x = 199$$

The original price was $199.

55.

	Distance	Rate	Time
Boat A	$3x$	x	3
Boat B	$3(2x)$	$2x$	3

(dist boat B) − (dist boat A) = (dist between)
$$3(2x) - 3x = 60$$
$$6x - 3x = 60$$
$$3x = 60$$
$$\frac{3x}{3} = \frac{60}{3}$$
$$x = 20$$
$$2x = 2(20) = 40$$

The boat's rates are 20 mph and 40 mph.

57.

	5% Account	6% Account	Total
Amount Invested	x	$2x$	
Interest Earned	$0.05x$	$0.06(2x)$	765

(int at 5%) + (int at 6%) = (total int)
$$0.05x + 0.06(2x) = 765$$
$$0.05x + 0.12x = 765$$
$$0.17x = 765$$
$$\frac{0.17x}{0.17} = \frac{765}{0.17}$$
$$x = 4500$$
$$2x = 2(4500) = 9000$$

$4500 was invested at 5% and $9000 was invested at 6%.

59.

	Black Tea	Orange Pekoe Tea	Total
Pounds of Tea	x	$4-x$	4
Cost of Tea	$2.20x$	$3.00(4-x)$	$2.50(4)$

(cost black) + (cost orange) = (cost blend)
$$2.20x + 3.00(4 - x) = 2.50(4)$$
$$2.20x + 12 - 3x = 10$$
$$-0.80x + 12 = 10$$
$$-0.80x + 12 - 12 = 10 - 12$$
$$-0.80x = -2$$
$$\frac{-0.80x}{-0.80} = \frac{-2}{-0.80}$$
$$x = 2.5$$
$$4 - x = 4 - 2.5 = 1.5$$

2.5 lb of black tea and 1.5 lb of orange pekoe tea are used in the blend.

61.   New Price = Original Price − Price Drop

$$202,100 = x - 0.06x$$
$$202,100 = 0.94x$$
$$\frac{202,100}{0.94} = x$$
$$x = 215,000$$

The median price the previous year was $215,000.

Section 1.3   Applications to Geometry and Literal Equations

1.   **a.** $2l + 2w$

　　b. 90°

　　c.   supplementary

　　d. 180°

3.
$$\frac{3}{5}y - 3 + 2y = 5$$
$$5\left(\frac{3}{5}y - 3 + 2y\right) = 5(5)$$
$$3y - 15 + 10y = 25$$
$$13y - 15 = 25$$
$$13y - 15 + 15 = 25 + 15$$
$$13y = 40$$
$$y = \frac{40}{13} \qquad \left\{\frac{40}{13}\right\}$$

5.
$$2a - 4 + 8a = 7a - 8 + 3a$$
$$10a - 4 = 10a - 8$$
$$10a - 10a - 4 = 10a - 10a - 8$$
$$-4 = -8$$
$$\{\ \}$$

7.   Let w = the width of the rectangle
$l = 2w$ = the length of the rectangle
$$P = 2l + 2w$$
$$177 = 2(2w) + 2w$$
$$177 = 4w + 2w$$
$$177 = 6w$$
$$w = 29.5$$
$$l = 2w = 2(29.5) = 59$$
The court's dimensions are 29.5 ft by 59 ft.

7.   Let w = the width of the rectangle
$l = 2w$ = the length of the rectangle
$$P = 2l + 2w$$
$$177 = 2(2w) + 2w$$
$$177 = 4w + 2w$$
$$177 = 6w$$
$$w = 29.5$$
$$l = 2w = 2(29.5) = 59$$
The court's dimensions are 29.5 ft by 59 ft.

9.   Let x = the length of the one side
$x + 2$ = the length of the second side
$x + 4$ = the length of the third side

$$P = a + b + c$$
$$24 = x + x + 2 + x + 4$$
$$24 = 3x + 6$$
$$24 - 6 = 3x + 6 - 6$$
$$18 = 3x$$
$$6 = x$$
$$x + 2 = 6 + 2 = 8$$
$$x + 4 = 6 + 4 = 10$$

The lengths of the sides of the triangle are 6 m, 8 m, and 10 m.

11. **a.** Let $l =$ the length of the run
$$A = lw$$
$$100 = l\left(12\tfrac{1}{2}\right)$$
$$100 = \frac{25}{2}l$$
$$2(100) = 2\left(\frac{25}{2}l\right)$$
$$200 = 25l$$
$$8 = l$$

The dimensions are 8yd by 12.5 yd.

b. $P = 2l + 2w$
$$P = 2(8) + 2(12.5)$$
$$P = 16 + 25$$
$$P = 41$$

The perimeter is 41 yd.

13. Let $w =$ the width of the pen
$2w - 7 =$ the length of the pen
$$P = 2l + 2w$$
$$40 = 2(2w - 7) + 2w$$
$$40 = 4w - 14 + 2w$$
$$40 = 6w - 14$$
$$40 + 14 = 6w - 14 + 14$$
$$54 = 6w$$
$$9 = w$$
$$2w - 7 = 2(9) - 7 = 18 - 7 = 11$$

The width is 9 ft and the length is 11 ft.

15. Let $x =$ the measure of the two equal angles
$2(x + x) =$ the measure of the third angle
$$x + x + 2(x + x) = 180$$
$$x + x + 2x + 2x = 180$$
$$6x = 180$$
$$x = 30$$
$$2(x + x) = 2(30 + 30) = 120$$

The measures of the angles are 30º, 30º, and 120º.

17. Let $x =$ the measure of one angle
$5x =$ the measure of the other angle
$$x + 5x = 90$$
$$6x = 90$$
$$x = 15$$
$$5x = 5(15) = 75$$

The measures of the complementary angles are 15º and 75º.

19. $(7x-1)+(2x+1)=180$

$9x=180$

$x=20$

$7x-1=7(20)-1=139$

$2x+1=2(20)+1=41$

The measures of the angles are 139°, and 41°.

21. $(2x+5)+(x+2.5)=90$

$3x+7.5=90$

$3x+7.5-7.5=90-7.5$

$3x=82.5$

$x=27.5$

$2x+5=2(27.5)+5=60$

$x+2.5=27.5+2.5=30$

The measures of the angles are 60°, and 30°.

23. $(2x)+(5x+1)+(x+35)=180$

$8x+36=180$

$8x+36-36=180-36$

$8x=144$

$x=18$

$2x=2(18)=36$

$5x+1=5(18)+1=91$

$x+35=18+35=53$

The measures of the angles are 36°, 91°, and 53°.

25. $(2x-4)+3(x-7)=90$

$2x-4+3x-21=90$

$5x-25=90$

$5x-25+25=90+25$

$5x=115$

$x=23$

$2x-4=2(23)-4=42$

$3(x-7)=3(23-7)=3(16)=48$

The measures of the angles are 42° and 48°.

27. $d=rt$

a. $r=\dfrac{d}{t}$

b. $r=\dfrac{500}{3.094}\approx 161.6$ mph

29. a. $I=Prt$

$t=\dfrac{I}{Pr}$

b. $t=\dfrac{1400}{5000(0.04)}=7$ years

31. $A=lw$ for l

$\dfrac{A}{w}=\dfrac{lw}{w}$

$l=\dfrac{A}{w}$

33. $I=Prt$ for P

$\dfrac{I}{rt}=\dfrac{Prt}{rt}$

$P=\dfrac{I}{rt}$

33. $I=Prt$ for P

$\dfrac{I}{rt}=\dfrac{Prt}{rt}$

$P=\dfrac{I}{rt}$

35. $W=K_2-K_1$ for K_1

$W+K_1=K_2-K_1+K_1$

$W+K_1=K_2$

$W-W+K_1=K_2-W$

$K_1=K_2-W$

37.

$$F = \frac{9}{5}C + 32 \quad \text{for } C$$

$$5F = 5\left(\frac{9}{5}C + 32\right)$$

$$5F = 9C + 160$$

$$5F - 160 = 9C + 160 - 160$$

$$5F - 160 = 9C$$

$$\frac{5F - 160}{9} = \frac{9C}{9}$$

$$C = \frac{5F - 160}{9}$$

$$C = \frac{5(F - 32)}{9} = \frac{5}{9}(F - 32)$$

39.

$$K = \frac{1}{2}mv^2 \quad \text{for } v^2$$

$$2K = 2 \cdot \frac{1}{2}mv^2$$

$$2K = mv^2$$

$$\frac{2K}{m} = \frac{mv^2}{m}$$

$$v^2 = \frac{2K}{m}$$

41.

$$v = v_0 + at \quad \text{for } a$$

$$v - v_0 = v_0 - v_0 + at$$

$$v - v_0 = at$$

$$\frac{v - v_0}{t} = \frac{at}{t}$$

$$a = \frac{v - v_0}{t}$$

43.

$$w = p(v_2 - v_1) \quad \text{for } v_2$$

$$\frac{w}{p} = \frac{p(v_2 - v_1)}{p}$$

$$\frac{w}{p} = v_2 - v_1$$

$$\frac{w}{p} + v_1 = v_2 - v_1 + v_1$$

$$v_2 = \frac{w}{p} + v_1$$

$$v_2 = \frac{w + pv_1}{p}$$

45.

$$ax + by = c \quad \text{for } y$$

$$ax + by - ax = c - ax$$

$$by = c - ax$$

$$\frac{by}{b} = \frac{c - ax}{b}$$

$$y = \frac{c - ax}{b}$$

47.

$$V = \frac{1}{3}Bh \quad \text{for } B$$

$$3V = 3 \cdot \frac{1}{3}Bh$$

$$3V = Bh$$

$$\frac{3V}{h} = \frac{Bh}{h}$$

$$B = \frac{3V}{h}$$

49.

$$3x + y = 6$$

$$3x - 3x + y = 6 - 3x$$

$$y = -3x + 6$$

51.

$$5x - 4y = 20$$

$$5x - 5x - 4y = 20 - 5x$$

$$-4y = 20 - 5x$$

$$\frac{-4y}{-4} = \frac{20 - 5x}{-4}$$

$$y = \frac{5}{4}x - 5$$

53.
$$-6x - 2y = 13$$
$$-6x + 6x - 2y = 13 + 6x$$
$$-2y = 6x + 13$$
$$\frac{-2y}{-2} = \frac{6x + 13}{-2}$$
$$y = -3x - \frac{13}{2}$$

55.
$$3x - 3y = 6$$
$$3x - 3x - 3y = 6 - 3x$$
$$-3y = -3x + 6$$
$$\frac{-3y}{-3} = \frac{-3x + 6}{-3}$$
$$y = x - 2$$

57.
$$9x + \frac{4}{3}y = 5$$
$$3\left(9x + \frac{4}{3}y\right) = 3(5)$$
$$27x + 4y = 15$$
$$27x - 27x + 4y = 15 - 27x$$
$$4y = -27x + 15$$
$$\frac{4y}{4} = \frac{-27x + 15}{4}$$
$$y = -\frac{27}{4}x + \frac{15}{4}$$

59.
$$-x + \frac{2}{3}y = 0$$
$$-x + x + \frac{2}{3}y = 0 + x$$
$$\frac{2}{3}y = x$$
$$\frac{3}{2} \cdot \frac{2}{3}y = \frac{3}{2}x$$
$$y = \frac{3}{2}x$$

59.
$$-x + \frac{2}{3}y = 0$$
$$-x + x + \frac{2}{3}y = 0 + x$$
$$\frac{2}{3}y = x$$
$$\frac{3}{2} \cdot \frac{2}{3}y = \frac{3}{2}x$$
$$y = \frac{3}{2}x$$

61. a.
$$z = \frac{x - \mu}{\sigma}$$
$$z \cdot \sigma = \frac{x - \mu}{\sigma} \cdot \sigma$$
$$z\sigma = x - \mu$$
$$z\sigma + \mu = x - \mu + \mu$$
$$x = z\sigma + \mu$$

b. $x = 2.5(12) + 100 = 30 + 100 = 130$

63.
$$\frac{-5}{x - 3} = -\frac{5}{x - 3}$$
$$\frac{-5}{x - 3} = \frac{-1}{-1} \cdot \frac{-5}{x - 3} = \frac{5}{-x + 3} = \frac{5}{3 - x}$$
Expressions a, b, and c are equivalent.

65.
$$\frac{-x - 7}{y} = \frac{-1}{-1} \cdot \frac{-x - 7}{y} = \frac{x + 7}{-y} = -\frac{x + 7}{y}$$
Expressions a and b are equivalent.

67. $6t - rt = 12$ for t

$t(6-r) = 12$

$\dfrac{t(6-r)}{6-r} = \dfrac{12}{6-r}$

$t = \dfrac{12}{6-r}$

69. $ax + 5 = 6x + 3$ for x

$ax - 6x + 5 = 6x - 6x + 3$

$ax - 6x + 5 = 3$

$ax - 6x + 5 - 5 = 3 - 5$

$ax - 6x = -2$

$x(a-6) = -2$

$\dfrac{x(a-6)}{a-6} = \dfrac{-2}{a-6}$

$x = \dfrac{-2}{a-6}$ or $x = \dfrac{2}{6-a}$

71. $A = P + Prt$ for P

$A = P(1 + rt)$

$\dfrac{A}{1+rt} = \dfrac{P(1+rt)}{1+rt}$

$P = \dfrac{A}{1+rt}$

73. $T = mg - mf$ for m

$T = m(g - f)$

$\dfrac{T}{g-f} = \dfrac{m(g-f)}{g-f}$

$m = \dfrac{T}{g-f}$

75. $ax + by = cx + z$ for x

$ax - cx + by = cx - cx + z$

$x(a-c) + by = z$

$x(a-c) + by - by = z - by$

$x(a-c) = z - by$

$\dfrac{x(a-c)}{a-c} = \dfrac{z-by}{a-c}$

$x = \dfrac{z-by}{a-c}$ or $x = \dfrac{by-z}{c-a}$

Section 1.4   Linear Inequalities in One Variable

1. **a.** linear; inequality

   b. negative

   c. Both statements are correct.

	Set-Builder Notation	Interval Notation	Graph
3.	$\{x \mid x > 5\}$	$(5, \infty)$	
5.	$\{x \mid -3 < x \le 6\}$	$(-3, 6]$	
7.	$\{x \mid x \ge 4\}$	$[4, \infty)$	

9.a. $-2x + 4 = 10$
$-2x = 10 - 4$
$-2x = 6$
$x = -3 \qquad \{-3\} \qquad$ n/a

11. $2y + 6 \le 4$
$2y + 6 - 6 \le 4 - 6$
$2y \le -2$
$\dfrac{2y}{2} \le \dfrac{-2}{2}$
$y \le -1$
a. $\{y \mid y \le -1\}$
b. $(-\infty, -1]$

b. $-2x + 4 < 10$
$-2x < 10 - 4$
$-2x < 6$
$x < -3$
$\{x \mid x > -3\} \qquad (-3, \infty)$

c. $-2x + 4 > 10$
$-2x > 10 - 4$
$-2x > 6$
$x > -3$
$\{x \mid x < -3\} \qquad (-\infty, -3)$

13.
$$-2x - 5 \le -25$$
$$-2x - 5 + 5 \le -25 + 5$$
$$-2x \le -20$$
$$\frac{-2x}{-2} \ge \frac{-20}{-2}$$
$$x \ge 10$$

a. $\{x \mid x \ge 10\}$

b. $[10, \infty)$

15.
$$6z + 3 > 16$$
$$6z + 3 - 3 > 16 - 3$$
$$6z > 13$$
$$\frac{6z}{6} > \frac{13}{6}$$
$$z > \frac{13}{6}$$

a. $\left\{ z \mid z > \frac{13}{6} \right\}$

b. $\left(\frac{13}{6}, \infty \right)$

17.
$$-8 > \frac{2}{3}t$$
$$\frac{3}{2}(-8) > \frac{3}{2} \cdot \frac{2}{3}t$$
$$-12 > t$$
$$t < -12$$

a. $\{t \mid t < -12\}$

b. $(-\infty, -12)$

19.
$$\frac{3}{4}(8y - 9) < 3$$
$$\frac{4}{3}\left[\frac{3}{4}(8y - 9) \right] < \frac{4}{3}[3]$$
$$8y - 9 < 4$$
$$8y - 9 + 9 < 4 + 9$$
$$8y < 13$$
$$\frac{8y}{8} < \frac{13}{8}$$
$$y < \frac{13}{8}$$

a. $\left\{ y \mid y < \frac{13}{8} \right\}$

b. $\left(-\infty, \frac{13}{8} \right)$

21.
$$0.8a - 0.5 \le 0.3a - 11$$
$$10(0.8a - 0.5) \le 10(0.3a - 11)$$
$$8a - 5 \le 3a - 110$$
$$8a - 3a - 5 \le 3a - 3a - 110$$
$$5a - 5 \le -110$$
$$5a - 5 + 5 \le -110 + 5$$
$$5a \le -105$$
$$\frac{5a}{5} \le \frac{-105}{5}$$
$$a \le -21$$

23.
$$-5x + 7 < 22$$
$$-5x + 7 - 7 < 22 - 7$$
$$-5x < 15$$
$$\frac{-5x}{-5} > \frac{15}{-5}$$
$$x > -3$$

a. $\{x \mid x > -3\}$

b. $(-3, \infty)$

a. $\{a \mid a \le -21\}$

b. $(-\infty, -21]$

-21

25.
$$-\frac{5}{6}x \le -\frac{3}{4}$$
$$-\frac{6}{5}\left(-\frac{5}{6}x\right) \ge -\frac{6}{5}\left(-\frac{3}{4}\right)$$
$$x \ge \frac{18}{20}$$
$$x \ge \frac{9}{10}$$

a. $\left\{x \mid x \ge \dfrac{9}{10}\right\}$

b. $\left[\dfrac{9}{10}, \infty\right)$

$\frac{9}{10}$

27.
$$\frac{3p-1}{-2} > 5$$
$$-2\left(\frac{3p-1}{-2}\right) < -2(5)$$
$$3p-1 < -10$$
$$3p-1+1 < -10+1$$
$$3p < -9$$
$$\frac{3p}{3} < \frac{-9}{3}$$
$$p < -3$$

a. $\{p \mid p < -3\}$

b. $(-\infty, -3)$

-3

29.
$$0.2t+1 > 2.4t-10$$
$$10(0.2t+1) > 10(2.4t-10)$$
$$2t+10 > 24t-100$$
$$2t-24t+10 > 24t-24t-100$$
$$-22t+10 > -100$$
$$-22t+10-10 > -100-10$$
$$-22t > -110$$
$$\frac{-22t}{-22} < \frac{-110}{-22}$$
$$t < 5$$

a. $\{t \mid t < 5\}$

b. $(-\infty, 5)$

5

31.
$$3-4(y+2) \le 6+4(2y+1)$$
$$3-4y-8 \le 6+8y+4$$
$$-4y-5 \le 8y+10$$
$$-4y-8y-5 \le 8y-8y+10$$
$$-12y-5 \le 10$$
$$-12y-5+5 \le 10+5$$
$$-12y \le 15$$
$$\frac{-12y}{-12} \ge \frac{15}{-12}$$
$$y \ge -\frac{5}{4}$$

a. $\left\{y \mid y \ge -\dfrac{5}{4}\right\}$

b. $\left[-\dfrac{5}{4}, \infty\right)$

$-\frac{5}{4}$

33.
$$7.2k - 5.1 \geq 5.7$$
$$10(7.2k - 5.1) \geq 10(5.7)$$
$$72k - 51 \geq 57$$
$$72k - 51 + 51 \geq 57 + 51$$
$$72k \geq 108$$
$$\frac{72k}{72} \geq \frac{108}{72}$$
$$k \geq \frac{3}{2} \quad \text{or} \quad k \geq 1.5$$

a. $\{k \mid k \geq 1.5\}$

b. $[1.5, \infty)$

35.
$$\frac{3}{4}x - 8 \leq 1$$
$$\frac{3}{4}x - 8 + 8 \leq 1 + 8$$
$$\frac{3}{4}x \leq 9$$
$$\frac{4}{3}\left(\frac{3}{4}x\right) \leq \frac{4}{3}(9)$$
$$x \leq 12$$

a. $\{x \mid x \leq 12\}$

b. $(-\infty, 12]$

37.
$$-1.2b - 0.4 \geq -0.4b$$
$$-1.2b + 1.2b - 0.4 \geq -0.4b + 1.2b$$
$$-0.4 \geq 0.8b$$
$$\frac{-0.4}{0.8} \geq \frac{0.8b}{0.8}$$
$$-0.5 \geq b$$
$$b \leq -0.5$$

a. $\{b \mid b \leq -0.5\}$

b. $(-\infty, -0.5]$

39.
$$-\frac{3}{4}c - \frac{5}{4} \geq 2c$$
$$4\left(-\frac{3}{4}c - \frac{5}{4}\right) \geq 4(2c)$$
$$-3c - 5 \geq 8c$$
$$-3c + 3c - 5 \geq 8c + 3c$$
$$-5 \geq 11c$$
$$\frac{-5}{11} \geq \frac{11c}{11}$$
$$-\frac{5}{11} \geq c \quad \text{or} \quad c \leq -\frac{5}{11}$$

a. $\left\{c \mid c \leq -\frac{5}{11}\right\}$

b. $\left(-\infty, -\frac{5}{11}\right]$

41.
$$4 - 4(y - 2) < -5y + 6$$
$$4 - 4y + 8 < -5y + 6$$
$$-4y + 12 < -5y + 6$$
$$-4y + 5y + 12 < -5y + 5y + 6$$
$$y + 12 < 6$$
$$y + 12 - 12 < 6 - 12$$
$$y < -6$$

a. $\{y \mid y < -6\}$

b. $(-\infty, -6)$

43.
$$-6(2x + 1) < 5 - (x - 4) - 6x$$
$$-12x - 6 < 5 - x + 4 - 6x$$
$$-12x - 6 < -7x + 9$$
$$-12x + 7x - 6 < -7x + 7x + 9$$
$$-5x - 6 < 9$$
$$-5x - 6 + 6 < 9 + 6$$
$$-5x < 15$$
$$\frac{-5x}{-5} > \frac{15}{-5}$$
$$x > -3$$

-6

a. $\{x \mid x > -3\}$

b. $(-3, \infty)$

-3

45.
$$6a - (9a + 1) - 3(a - 1) \geq 2$$
$$6a - 9a - 1 - 3a + 3 \geq 2$$
$$-6a + 2 \geq 2$$
$$-6a + 2 - 2 \geq 2 - 2$$
$$-6a \geq 0$$
$$\frac{-6a}{-6} \leq \frac{0}{-6}$$
$$a \leq 0$$

a. $\{a \mid a \leq 0\}$

b. $(-\infty, 0]$

0

47.

a.
$$80 \leq \frac{80 + 86 + 73 + 91 + x}{5} < 90$$
$$5 \cdot 80 \leq 5 \cdot \frac{80 + 86 + 73 + 91 + x}{5} < 5 \cdot 90$$
$$400 \leq 330 + x < 450 \qquad \text{Nadia}$$
$$400 - 330 \leq 330 - 330 + x < 450 - 330$$
$$70 \leq x < 120$$
needs to score at least a 70% but less than 120% to get a B average.

$$\frac{80 + 86 + 73 + 91 + x}{5} \geq 90$$

b.
$$5 \cdot \frac{80 + 86 + 73 + 91 + x}{5} \geq 5 \cdot 90$$
$$330 + x \geq 450$$
$$330 - 330 + x \geq 450 - 330$$
$$x \geq 120$$
It would be impossible for Nadia to get an A because she would have to earn 120% on her last quiz and it is impossible to earn more than 100%.

49.
$$2.5a + 31 \geq 51$$
$$2.5a + 31 - 31 \geq 51 - 31$$
$$2.5a \geq 20$$
$$\frac{2.5a}{2.5} \geq \frac{20}{2.5}$$
$$a \geq 8$$
Boys 8 years old or older will be on average at least 51 in. tall.

51.
$$2.5a + 31 \leq 46$$
$$2.5a + 31 - 31 \leq 46 - 31$$
$$2.5a \leq 15$$
$$\frac{2.5a}{2.5} \leq \frac{15}{2.5}$$
$$a \leq 6$$
Boys 6 years old or younger will be on average no more than 46 in. tall.

51.
$$2.5a + 31 \leq 46$$
$$2.5a + 31 - 31 \leq 46 - 31$$
$$2.5a \leq 15$$
$$\frac{2.5a}{2.5} \leq \frac{15}{2.5}$$
$$a \leq 6$$
Boys 6 years old or younger will be on average no more than 46 in. tall.

53.
$$25,000 + 0.04x > 40,000$$
$$25,000 - 25,000 + 0.04x > 40,000 - 25,000$$
$$0.04x > 15,000 \qquad \text{Her}$$
$$\frac{0.04x}{0.04} > \frac{15,000}{0.04}$$
$$x > 375,000$$
sales must exceed $375,000.

$$25,000 + 0.04x > 80,000$$
$$25,000 - 25,000 + 0.04x > 80,000 - 25,000$$
$$0.04x > 55,000 \qquad \text{Her}$$
$$\frac{0.04x}{0.04} > \frac{55,000}{0.04}$$
$$x > 1,375,000$$

sales must exceed $1,375,000.
The base salary is still the same; the increase comes solely from commission.

55.
$$R > C$$
$$49.95x > 2300 + 18.50x$$
$$49.95x - 18.50x > 2300 + 18.50x - 18.50x$$
$$31.45x > 2300$$
$$\frac{31.45x}{31.45} > \frac{2300}{31.45}$$
$$x > 73.13$$

There will be a profit if more than 73 jackets are sold.

57.
$$a > b$$
$$a + c > b + c$$

59. $a > b$

$ac < bc$ for $c < 0$

Section 1.5   Compound Inequalities

1. **a.** union; $A \cup B$
  b. intersection; $A \cap B$
  c. intersection
  d. $a < x < b$
  e. union

3.
$$-6u + 8 > 2$$
$$-6u > -6$$
$$\frac{-6u}{-6} < \frac{-6}{-6}$$
$$u < 1 \qquad (-\infty, 1)$$

5.
$$-12 \le \frac{3}{4}p$$
$$\frac{4}{3}(-12) \le \frac{4}{3}\left(\frac{3}{4}p\right)$$
$$-16 \le p$$
$$p \ge -16 \qquad [-16, \infty)$$

7. **a.** $M \mid N = \{-3, -1\}$
  b. $M \cup N = \{-4, -3, -2, -1, 0, 1, 3, 5\}$

9. $A \mid C = [-7, -4)$

11. $A \cup B = (-\infty, -4) \cup (2, \infty)$

13. $A \mid B = \{\ \}$

15. $B \cup C = [-7, \infty)$

17. $C \mid D = [0, 5)$

19. $C \cup D = [-7, \infty)$

21. a. $(-2, 5) \cap [-1, \infty) = [-1, 5)$

   b. $(-2, 5) \cup [-1, \infty) = (-2, \infty)$

23. a. $\left(-\dfrac{5}{2}, 3\right) \cap \left(-1, \dfrac{9}{2}\right) = (-1, 3)$

   b. $\left(-\dfrac{5}{2}, 3\right) \cup \left(-1, \dfrac{9}{2}\right) = \left(-\dfrac{5}{2}, \dfrac{9}{2}\right)$

25. a. $(-4, 5] \cap (0, 2] = (0, 2]$

   b. $(-4, 5] \cup (0, 2] = (-4, 5]$

27. $y - 7 \geq -9$ and $y + 2 \leq 5$

$\qquad y \geq -2 \quad \cap \quad y \leq 3$

$\qquad\qquad [-2, \infty) \cap (-\infty, 3] = [-2, 3]$

29. $2t + 7 < 19$ and $5t + 13 > 28$

$\qquad 2t < 12 \quad \cap \quad 5t > 15$

$\qquad\quad t < 6 \quad \cap \quad\quad t > 3$

$\qquad\quad (-\infty, 6) \cap (3, \infty) = (3, 6)$

31. $\qquad 2.1k - 1.1 \leq 0.6k + 1.9 \qquad$ and $\qquad 0.3k - 1.1 < -0.1k + 0.9$

$\quad 10(2.1k - 1.1) \leq 10(0.6k + 1.9)$ and $10(0.3k - 1.1) < 10(-0.1k + 0.9)$

$\qquad\quad 21k - 11 \leq 6k + 19 \qquad\qquad$ and $\qquad 3k - 11 < -k + 9$

$\qquad\qquad 15k - 11 \leq 19 \qquad\qquad\quad | \qquad\qquad 4k - 11 < 9$

$\qquad\qquad\quad 15k \leq 30 \qquad\qquad\qquad | \qquad\qquad\quad 4k < 20$

$\qquad\qquad\qquad k \leq 2 \qquad\qquad\qquad\quad | \qquad\qquad\qquad k < 5$

$\qquad\qquad\qquad\quad (-\infty, 2] \cap (-\infty, 5) = (-\infty, 2]$

33. $\qquad \dfrac{2}{3}(2p - 1) \geq 10$ and $\dfrac{4}{5}(3p + 4) \geq 20$

$\quad \dfrac{3}{2} \cdot \dfrac{2}{3}(2p - 1) \geq \dfrac{3}{2} \cdot 10 \quad | \quad \dfrac{5}{4} \cdot \dfrac{4}{5}(3p + 4) \geq \dfrac{5}{4} \cdot 20$

$\qquad\quad 2p - 1 \geq 15 \quad | \quad\quad 3p + 4 \geq 25$

$\qquad\qquad 2p \geq 16 \quad | \quad\qquad 3p \geq 21$

$\qquad\qquad\quad p \geq 8 \quad | \quad\qquad\quad p \geq 7$

$\qquad\quad [8, \infty) \cap [7, \infty) = [8, \infty)$

35. $\quad -2 < -x - 12$ and $-14 < 5(x - 3) + 6x$

$\qquad\quad 10 < -x \quad | \quad -14 < 5x - 15 + 6x$

$\qquad\quad -10 > x \quad | \quad\quad -14 < 11x - 15$

$\qquad\quad\quad x < -10 \quad | \quad\qquad 1 < 11x$

$\qquad\quad\quad x < -10 \quad | \quad\qquad x > \dfrac{1}{11}$

$\qquad (-\infty, -10) \cap \left(\dfrac{1}{11}, \infty\right) = \{\ \}$

37. $-4 \leq t$ and $t < \dfrac{3}{4}$

39. The statement $6 < x < 2$ is equivalent to $6 < x$ and $x < 2$. However, no real number is greater than 6 and also less than 2.

41. The statement $-5 > y > -2$ is equivalent to $-5 > y$ and $y > -2$. However, no real number is less than -5 and also greater than -2.

43.
$$0 \le 2b - 5 < 9$$
$$5 \le 2b < 14$$
$$\frac{5}{2} \le b < 7 \qquad \left[\frac{5}{2}, 7\right)$$

45.
$$-1 < \frac{a}{6} \le 1$$
$$-6 < a \le 6 \qquad (-6, 6]$$

47.
$$-\frac{2}{3} < \frac{y - 4}{-6} < \frac{1}{3}$$
$$4 > y - 4 > -2$$
$$8 > y > 2 \qquad (2, 8)$$

49.
$$5 \le -3x - 2 \le 8$$
$$7 \le -3x \le 10$$
$$-\frac{7}{3} \ge x \ge -\frac{10}{3} \qquad \left[-\frac{10}{3}, -\frac{7}{3}\right]$$

51.
$$12 > 6x + 3 \ge 0$$
$$9 > 6x \ge -3$$
$$\frac{3}{2} > x \ge -\frac{1}{2} \qquad \left[-\frac{1}{2}, \frac{3}{2}\right)$$

53.
$$-0.2 < 2.6 + 7t < 4$$
$$-2.8 < 7t < 1.4$$
$$-0.4 < t < 0.2 \qquad (-0.4, 0.2)$$

55.
$$2y - 1 \ge 3 \text{ or } y < -2$$
$$2y \ge 4 \ \cup \ y < -2$$
$$y \ge 2 \ \cup \ y < -2$$
$$(-\infty, -2) \cup [2, \infty)$$

57.
$$1 > 6z - 8 \text{ or } 8z - 6 \le 10$$
$$9 > 6z \qquad \cup \qquad 8z \le 16$$
$$\frac{3}{2} > z \qquad \cup \qquad z \le 2$$
$$\left(-\infty, \frac{3}{2}\right) \cup (-\infty, 2] = (-\infty, 2]$$

59.
$$5(x - 1) \ge -5 \text{ or } 5 - x \le 11$$
$$5x - 5 \ge -5 \quad \cup \quad -x \le 6$$
$$5x \ge 0 \qquad \cup \qquad x \ge -6$$
$$x \ge 0 \qquad \cup \qquad x \ge -6$$
$$[0, \infty) \cup [-6, \infty) = [-6, \infty)$$

61.
$$\frac{5}{3}v \le 5 \text{ or } -v - 6 < 1$$
$$\frac{3}{5} \cdot \frac{5}{3}v \le \frac{3}{5} \cdot 5 \ \cup \ -v < 7$$
$$v \le 3 \qquad \cup \qquad v > -7$$
$$(-\infty, 3] \cup (-7, \infty) = (-\infty, \infty)$$

63.
$$0.5w + 5 < 2.5w - 4 \text{ or } 0.3w \le -0.1w - 1.6$$
$$-2w + 5 < -4 \qquad \cup \qquad 0.4w \le -1.6$$
$$-2w < -9 \qquad \cup \qquad w \le -4$$
$$w > \frac{9}{2} \qquad \cup \qquad w \le -4$$
$$\left(\frac{9}{2}, \infty\right) \cup (-\infty, -4]$$

65. a. $3x-5<19$ and $-2x+3<23$

$3x<24$ | $-2x<20$

$x<8$ | $x>-10$

$(-\infty,\,8)\,|\,(-10,\,\infty)=(-10,\,8)$

$3x-5<19$ or $-2x+3<23$

$3x<24$ ∪ $-2x<20$

b. $x<8$ ∪ $x>-10$

$(-\infty,\,8)\cup(-10,\,\infty)=(-\infty,\,\infty)$

67. a. $8x-4\geq6.4$ or $0.3(x+6)\leq-0.6$

$8x\geq10.4$ ∪ $x+6\leq-2$

$x\geq1.3$ ∪ $x\leq-8$

$[1.3,\,\infty)\cup(-\infty,\,-8]$

b. $8x-4\geq6.4$ and $0.3(x+6)\leq-0.6$

$8x\geq10.4$ | $x+6\leq-2$

$x\geq1.3$ | $x\leq-8$

$[1.3,\,\infty)\,|\,(-\infty,\,-8]=$ No Solution

69. $-4\leq\dfrac{2-4x}{3}<8$

$-12\leq2-4x<24$

$-14\leq-4x<22$

$\dfrac{7}{2}\geq x>-\dfrac{11}{2}$ $\left(-\dfrac{11}{2},\,\dfrac{7}{2}\right]$

71. $5\geq-4(t-3)+3t$ or $6<12t+8(4-t)$

$5\geq-4t+12+3t$ ∪ $6<12t+32-8t$

$5\geq-t+12$ ∪ $6<4t+32$

$-7\geq-t$ ∪ $-26<4t$

$7\leq t$ ∪ $-\dfrac{13}{2}<t$

$[7,\,\infty)\cup\left(-\dfrac{13}{2},\,\infty\right)=\left(-\dfrac{13}{2},\,\infty\right)$

73. $\dfrac{-x+3}{2}>\dfrac{4+x}{5}$ or $\dfrac{1-x}{4}>\dfrac{2-x}{3}$

$5(-x+3)>2(4+x)$ ∪ $3(1-x)>4(2-x)$

$-5x+15>8+2x$ ∪ $3-3x>8-4x$

$-7x+15>8$ ∪ $3+x>8$

$-7x>-7$ ∪ $x>5$

$x<1$ ∪ $x>5$

$(-\infty,\,1)\cup(5,\,\infty)$

75. a. $4800\leq x\leq10{,}800$

b. $x<4800$ or $x>10{,}800$

77. a. $44\%<x<48\%$

b. $x\leq44\%$ or $x\geq48\%$

79. $-3<2x<12$

$-\dfrac{3}{2}<x<6$

All real numbers between $-\dfrac{3}{2}$ and 6

81. $2x+1>5$ or $2x+1<-1$

$2x>4$ or $2x<-2$

$x>2$ or $x<-1$

All real numbers greater than 2 or
less than -1

83. a. $0.8(92)+0.2x\geq90$

$73.6+0.2x\geq90$

$0.2x\geq16.4$

$x\geq82$

Amy would need 82% or better on her
final exam.

b.

$$80 \le 0.8(92) + 0.2x < 90$$
$$80 \le 73.6 + 0.2x < 90$$
$$6.4 \le 0.2x < 16.4$$
$$32 \le x < 82$$

If Amy scores at least 32% and less than 82% on her final exam, she will receive a "B" in the class.

85.
$$0.0 \le \frac{5}{9}(F - 32) \le 5.6$$
$$9(0.0) \le 9 \cdot \frac{5}{9}(F - 32) \le 9(5.6)$$
$$0 \le 5(F - 32) \le 50.4$$
$$0 \le 5F - 160 \le 50.4$$
$$0 + 160 \le 5F - 160 + 160 \le 50.4 + 160$$
$$160 \le 5F \le 210.4$$
$$\frac{160}{5} \le \frac{5F}{5} \le \frac{210.4}{5}$$
$$32° \le F \le 42.08°$$

Section 1.6   Absolute Value Equations

1.
  a. absolute; $\{a, -a\}$
  b. Subtract 5 from both sides.
  c. $y; -y$
  d. $\{\ \}; \{-4\}$

3.
$$3x - 5 \ge 7x + 3 \quad \text{or} \quad 2x - 1 \le 4x - 5$$
$$-4x - 5 \ge 3 \quad \text{U} \quad -2x - 1 \le -5$$
$$-4x \ge 8 \quad \text{U} \quad -2x \le -4$$
$$x \le -2 \quad \text{U} \quad x \ge 2$$
$$(-\infty, -2] \text{U} [2, \infty)$$

5.
$$5 \ge \frac{x - 4}{-2} > -3$$
$$-2(5) \le -2\left(\frac{x - 4}{-2}\right) < -2(-3)$$
$$-10 \le x - 4 < 6$$
$$-6 \le x < 10 \quad [-6, 10)$$

7.
$$|p| = 7$$
$$p = 7 \text{ or } p = -7 \quad \{7, -7\}$$

9.
$$|x| + 5 = 11$$
$$|x| = 6$$
$$x = 6 \text{ or } x = -6 \quad \{6, -6\}$$

11.
$$|y| + 8 = 5$$
$$|y| = -3 \quad \{\ \}$$

13.
$$|w| - 3 = -1$$
$$|w| = 2$$
$$w = -2 \text{ or } w = 2 \quad \{-2, 2\}$$

15.
$$|y| = \sqrt{2}$$
$$y = \sqrt{2} \text{ or } y = -\sqrt{2} \quad \{\sqrt{2}, -\sqrt{2}\}$$

17. $|w| - 3 = -5$

$|w| = -2$ $\{\ \}$

19. $|3q| = 0$

$3q = 0$ or $3q = -0$

$q = 0$ $\{0\}$

21. $|3x - 4| = 8$

$3x - 4 = 8$ or $3x - 4 = -8$

$3x = 12$ or $3x = -4$

$x = 4$ or $x = -\dfrac{4}{3}$ $\left\{4, -\dfrac{4}{3}\right\}$

23. $5 = |2x - 4|$

$2x - 4 = 5$ or $2x - 4 = -5$

$2x = 9$ or $2x = -1$

$x = \dfrac{9}{2}$ or $x = -\dfrac{1}{2}$ $\left\{\dfrac{9}{2}, -\dfrac{1}{2}\right\}$

25. $\left|\dfrac{7z}{3} - \dfrac{1}{3}\right| + 3 = 6$

$\left|\dfrac{7z}{3} - \dfrac{1}{3}\right| = 3$

$\dfrac{7z}{3} - \dfrac{1}{3} = 3$ or $\dfrac{7z}{3} - \dfrac{1}{3} = -3$

$7z - 1 = 9$ or $7z - 1 = -9$

$7z = 10$ or $7z = -8$

$z = \dfrac{10}{7}$ or $z = -\dfrac{8}{7}$ $\left\{\dfrac{10}{7}, -\dfrac{8}{7}\right\}$

27. $|0.2x - 3.5| = -5.6$

$\{\ \}$

29. $1 = -4 + \left|2 - \dfrac{1}{4}w\right|$

$\left|2 - \dfrac{1}{4}w\right| = 5$

$2 - \dfrac{1}{4}w = 5$ or $2 - \dfrac{1}{4}w = -5$

$8 - w = 20$ or $8 - w = -20$

$-w = 12$ or $-w = -28$

$w = -12$ or $w = 28$ $\{-12, 28\}$

31. $10 = 4 + |2y + 1|$

$|2y + 1| = 6$

$2y + 1 = 6$ or $2y + 1 = -6$

$2y = 5$ or $2y = -7$

$y = \dfrac{5}{2}$ or $y = -\dfrac{7}{2}$ $\left\{\dfrac{5}{2}, -\dfrac{7}{2}\right\}$

33. $-2|3b - 7| - 9 = -9$

$-2|3b - 7| = 0$

$|3b - 7| = 0$

$3b - 7 = 0$ or $3b - 7 = -0$

$3b = 7$

$b = \dfrac{7}{3}$ $\left\{\dfrac{7}{3}\right\}$

35. $-2|x + 3| = 5$

$|x + 3| = -\dfrac{5}{2}$ $\{\ \}$

37.
$$0 = |6x - 9|$$
$$6x - 9 = 0 \quad \text{or} \quad 6x - 9 = -0$$
$$6x = 9$$
$$x = \frac{3}{2} \qquad \left\{\frac{3}{2}\right\}$$

39.
$$\left|-\frac{1}{5} - \frac{1}{2}k\right| = \frac{9}{5}$$
$$-\frac{1}{5} - \frac{1}{2}k = \frac{9}{5} \quad \text{or} \quad -\frac{1}{5} - \frac{1}{2}k = -\frac{9}{5}$$
$$-2 - 5k = 18 \quad \text{or} \quad -2 - 5k = -18$$
$$-5k = 20 \quad \text{or} \qquad -5k = -16$$
$$k = -4 \text{ or} \qquad k = \frac{16}{5} \quad \left\{-4, \frac{16}{5}\right\}$$

41.
$$-3|2 - 6x| + 5 = -10$$
$$-3|2 - 6x| = -15$$
$$|2 - 6x| = 5$$
$$2 - 6x = 5 \quad \text{or} \quad 2 - 6x = -5$$
$$-6x = 3 \quad \text{or} \quad -6x = -7$$
$$x = -\frac{1}{2} \text{ or} \qquad x = \frac{7}{6} \quad \left\{-\frac{1}{2}, \frac{7}{6}\right\}$$

43.
$$|4x - 2| = |-8|$$
$$|4x - 2| = 8$$
$$4x - 2 = 8 \quad \text{or} \quad 4x - 2 = -8$$
$$4x = 10 \quad \text{or} \qquad 4x = -6$$
$$x = \frac{5}{2} \quad \text{or} \qquad x = -\frac{3}{2} \quad \left\{\frac{5}{2}, -\frac{3}{2}\right\}$$

45.
$$|4w + 3| = |2w - 5|$$
$$4w + 3 = 2w - 5 \quad \text{or} \quad 4w + 3 = -(2w - 5)$$
$$4w + 3 = 2w - 5 \quad \text{or} \quad 4w + 3 = -2w + 5$$
$$2w + 3 = -5 \qquad \text{or} \quad 6w + 3 = 5$$
$$2w = -8 \qquad \text{or} \qquad 6w = 2$$
$$w = -4 \qquad \text{or} \qquad w = \frac{1}{3} \quad \left\{-4, \frac{1}{3}\right\}$$

47.
$$|2y + 5| = |7 - 2y|$$
$$2y + 5 = 7 - 2y \quad \text{or} \quad 2y + 5 = -(7 - 2y)$$
$$2y + 5 = 7 - 2y \quad \text{or} \quad 2y + 5 = -7 + 2y$$
$$4y + 5 = 7 \qquad \text{or} \qquad 5 = -7$$
$$4y = 2 \qquad \text{or} \qquad \text{contradiction}$$
$$y = \frac{1}{2} \qquad \left\{\frac{1}{2}\right\}$$

49.
$$\left|\frac{4w - 1}{6}\right| = \left|\frac{2w}{3} + \frac{1}{4}\right|$$
$$\frac{4w - 1}{6} = \frac{2w}{3} + \frac{1}{4} \quad \text{or} \quad \frac{4w - 1}{6} = -\left(\frac{2w}{3} + \frac{1}{4}\right)$$
$$\frac{4w - 1}{6} = \frac{2w}{3} + \frac{1}{4} \quad \text{or} \quad \frac{4w - 1}{6} = -\frac{2w}{3} - \frac{1}{4}$$
$$2(4w - 1) = 8w + 3 \quad \text{or} \quad 2(4w - 1) = -8w - 3$$
$$8w - 2 = 8w + 3 \quad \text{or} \quad 8w - 2 = -8w - 3$$
$$-2 = 3 \qquad \text{or} \quad 16w - 2 = -3$$
$$\text{contradiction} \qquad \text{or} \qquad 16w = -1$$
$$w = -\frac{1}{16} \quad \left\{-\frac{1}{16}\right\}$$

51.
$$|x + 2| = |-x - 2|$$
$$x + 2 = -x - 2 \quad \text{or} \quad x + 2 = -(-x - 2)$$
$$2x = -4 \quad \text{or} \qquad x + 2 = x + 2$$
$$x = -2 \qquad \text{or} \qquad x = x$$
$$\{x \,|\, x \text{ is a real number}\}$$

53. $|3.5m - 1.2| = |8.5m + 6|$

$3.5m - 1.2 = 8.5m + 6$ or

$\qquad 3.5m - 1.2 = -(8.5m + 6)$

$3.5m - 1.2 = 8.5m + 6$ or

$\qquad 3.5m - 1.2 = -8.5m - 6$

$-5m - 1.2 = 6$ or $12m - 1.2 = -6$

$\quad -5m = 7.2$ or $\quad 12m = -4.8$

$\quad m = -1.44$ or $\quad m = -0.4$

$\qquad\qquad \{-1.44, -0.4\}$

55. $|4x - 3| = -|2x - 1|$

$\{\ \}$ - A positive number cannot equal a negative number.

57. $|8 - 7w| = |7w - 8|$

$8 - 7w = 7w - 8$ or $8 - 7w = -(7w - 8)$

$-14w = -16$ or $8 - 7w = -7w + 8$

$\quad w = \dfrac{8}{7}$ or $\quad w = w$

$\qquad \{w \mid w \text{ is a real number}\}$

59. $|x + 2| + |x - 4| = 0$

$|x + 2| = -|x - 4|$

$\{\ \}$ - A positive number cannot equal a negative number.

61. $|x| = 6$

63. $|x| = \dfrac{4}{3}$

Section 1.7   Absolute Value Inequalities

1. **a.** $-a\,;\ a$

 b. $-a\,;\ >$

 c. $\{\ \};\ (-\infty, \infty)$

 d. includes; excludes

3. $2 = |5 - 7x| + 1$

$1 = |5 - 7x|$

$5 - 7x = 1$ or $5 - 7x = -1$

$-7x = -4$ or $-7x = -6$

$x = \dfrac{4}{7}$ or $x = \dfrac{6}{7}$ $\left\{\dfrac{4}{7}, \dfrac{6}{7}\right\}$

5. $-15 < 3w - 6 \le -9$

$-9 < 3w \le -3$

$-3 < w \le -1$ $(-3, -1]$

7. $m - 7 \le -5$ or $m - 7 \ge -10$

$m \le 2$ $\cup$ $m \ge -3$

$(-\infty, 2] \cup [-3, \infty) = (-\infty, \infty)$

9. **a.** $|x| = 5$

 $x = -5$ or $x = 5$ $\{-5, 5\}$

 b. $|x| > 5$

 $x < -5$ or $x > 5$ $(-\infty, -5) \cup (5, \infty)$

 c.

11. **a.** $|x - 3| = 7$

 $x - 3 = -7$ or $x - 3 = 7$

 $x = -4$ or $x = 10$ $\{-4, 10\}$

$|x| < 5$

$-5 < x < 5$ $(-5, 5)$

b. $|x - 3| > 7$

$x - 3 < -7$ or $x - 3 > 7$

$x < -4$ or $x > 10$

$(-\infty, -4) \cup (10, \infty)$

c. $|x - 3| < 7$

$-7 < x - 3 < 7$

$-4 < x < 10$ $(-4, 10)$

13. a. $|p| = -2$ $\{\ \}$

$|p| > -2$

b. All real numbers $(-\infty, \infty)$

c. $|p| < -2$ $\{\ \}$

15. a. $|y + 1| = -6$ $\{\ \}$

$|y + 1| > -6$

b. All real numbers $(-\infty, \infty)$

c. $|y + 1| < -6$ $\{\ \}$

17. a. $|x| = 0$

$x = 0$ $\{0\}$

$|x| > 0$

$x < 0$ or $x > 0$ $(-\infty, 0) \cup (0, \infty)$

b.
c.

$|x| < 0$ $\{\ \}$

19. a. $|k - 7| = 0$

$k - 7 = 0$

$k = 7$ $\{7\}$

b. $|k - 7| > 0$

$k - 7 < 0$ or $k - 7 > 0$

$k < 7$ or $k > 7$ $(-\infty, 7) \cup (7, \infty)$

c.

$|k - 7| < 0$ $\{\ \}$

21. $|x| > 6$

$x < -6$ or $x > 6$ $(-\infty, -6) \cup (6, \infty)$

23. $|t| \le 3$

$-3 \le t \le 3$ $[-3, 3]$

25. $|y + 2| \ge 0$

All real numbers $(-\infty, \infty)$

27. $5 \le |2x - 1|$

$|2x - 1| \ge 5$

$2x - 1 \le -5$ or $2x - 1 \ge 5$

$2x \le -4$ or $2x \ge 6$

$x \le -2$ or $x \ge 3$

$(-\infty, -2] \cup [3, \infty)$

29. $|k-7| < -3$ $\{\ \}$

31.
$$\left|\frac{w-2}{3}\right| - 3 \le 1$$
$$\left|\frac{w-2}{3}\right| \le 4$$
$$-4 \le \frac{w-2}{3} \le 4$$
$$-12 \le w-2 \le 12$$
$$-10 \le w \le 14 \qquad \left[-10, 14\right]$$

33.
$$12 \le |9-4y| - 2$$
$$|9-4y| \ge 14$$
$$9-4y \le -14 \ \text{ or } \ 9-4y \ge 14$$
$$-4y \le -23 \ \text{ or } \quad -4y \ge 5$$
$$y \ge \frac{23}{4} \quad \text{ or } \qquad y \le -\frac{5}{4}$$
$$\left(-\infty, -\frac{5}{4}\right] \cup \left[\frac{23}{4}, \infty\right)$$

35.
$$4 > -1 + \left|\frac{2x+1}{4}\right|$$
$$5 > \left|\frac{2x+1}{4}\right|$$
$$-5 < \frac{2x+1}{4} < 5$$
$$-20 < 2x+1 < 20$$
$$-21 < 2x < 19$$
$$-\frac{21}{2} < x < \frac{19}{2} \qquad \left(-\frac{21}{2}, \frac{19}{2}\right)$$

37.
$$8 < |4-3x| + 12$$
$$-4 < |4-3x|$$
$$|4-3x| > -4$$
All real numbers $\left(-\infty, \infty\right)$

39.
$$5 - |2m+1| > 5$$
$$-|2m+1| > 0$$
$$|2m+1| < 0 \quad \{\ \}$$

41.
$$|p+5| \le 0$$
$$-0 \le p+5 \le 0$$
$$-5 \le p \le -5$$
$$\{-5\}$$

43.
$$|z-6| + 5 > 5$$
$$|z-6| > 0$$
$$z-6 < -0 \ \text{ or } \ z-6 > 0$$
$$z < 6 \quad \text{ or } \quad z > 6$$
$$(-\infty, 6) \cup (6, \infty)$$

45. $5|2y-6|+3\geq 13$

$\qquad 5|2y-6|\geq 10$

$\qquad |2y-6|\geq 2$

$\qquad\quad 2y-6\leq -2 \ \text{ or } \ 2y-6\geq 2$

$\qquad\qquad 2y\leq 4 \quad \text{ or } \qquad 2y\geq 8$

$\qquad\qquad\quad y\leq 2 \quad \text{ or } \qquad y\geq 4$

$\qquad\qquad (-\infty, 2]\cup[4, \infty)$

47. $-3|6-t|+1>-5$

$\qquad -3|6-t|>-6$

$\qquad |6-t|<2$

$\qquad\quad -2<6-t<2$

$\qquad\quad -8<-t<-4$

$\qquad\qquad 8>t>4 \qquad (4, 8)$

49. $|0.02x+0.06|-0.1<0.05$

$\qquad |0.02x+0.06|<0.15$

$\qquad\quad -0.15<0.02x+0.06<0.15$

$\qquad\quad -0.21<0.02x<0.09$

$\qquad\quad -10.5<x<4.5$

$\qquad\qquad (-10.5, 4.5)$

51. $|x|>7$

53. $|x-2|\leq 13$

55. $|x-32|\leq 0.05$

57. $\left|x-6\dfrac{3}{4}\right|\leq\dfrac{1}{8}$

59. $|w-2|\leq 0.01$

$\qquad -0.01\leq w-2\leq 0.01$

$\qquad\quad 1.99\leq w\leq 2.01 \qquad [1.99, 2.01]$

The solution set is $\{w\,|\,1.99\leq w\leq 2.01\}$ or equivalently in interval notation, $[1.99, 2.01]$. This means that the actual width of the bolt could be between 1.99 cm and 2.01 cm, inclusive.

61. b

63. a

Problem Recognition Exercises

1. **a.** $3x-9=18$

$\qquad 3x=27$

$\qquad\quad x=9 \quad \{9\}$

b. $|3x-9|=18$

$\qquad 3x-9=18 \text{ or } 3x-9=-18$

$\qquad\quad 3x=27 \text{ or } \qquad 3x=-9$

$\qquad\qquad x=9 \ \text{ or } \qquad x=-3 \quad \{9, -3\}$

c.

3. **a.** $-2t-14=0$

$\qquad -2t=14$

$\qquad\quad t=-7 \quad \{-7\}$

b. $-2t-14>0$

$\qquad -2t>14$

$\qquad\quad t<-7 \quad (-\infty, -7)$

$\qquad -2t-14\leq 0$

$\qquad\quad -2t\leq 14$

$\qquad\qquad t\geq -7 \quad [-7, \infty)$

c.

d.
$$|3x-9|<18$$
$$-18<3x-9<18$$
$$-9<3x<27$$
$$-3<x<9 \quad (-3, 9)$$
$$|3x-9|\geq 18$$
$$3x-9\geq 18 \text{ or } 3x-9\leq -18$$
$$3x\geq 27 \text{ or } \quad 3x\leq -9$$
$$x\geq 9 \text{ or } \quad x\leq -3$$
$$(-\infty, -3]\cup[9, \infty)$$

5. a.
$$|8t-2|=|-2t+3|$$
$$8t-2=-2t+3 \text{ or } 8t-2=-(-2t+3)$$
$$10t-2=3 \quad \text{ or } 8t-2=2t-3$$
$$10t=5 \quad \text{ or } 6t-2=-3$$
$$t=\frac{5}{10} \quad \text{ or } \quad 6t=-1$$
$$t=\frac{1}{2} \quad \text{ or } \quad t=-\frac{1}{6} \left\{\frac{1}{2}, -\frac{1}{6}\right\}$$
$$8t-2=-2t+3$$
$$10t=5$$
$$t=\frac{1}{2} \quad \left\{\frac{1}{2}\right\}$$

7. a.
$$-4x-9<11 \text{ or } 2\leq x+1$$
$$-4x<20 \text{ or } 1\leq x$$
$$x>-5 \text{ or } x\geq 1 \quad (-5, \infty)$$

b.
$$-4x-9<11 \text{ and } 2\leq x+1$$
$$-4x<20 \text{ and } 1\leq x$$
$$x>-5 \text{ and } x\geq 1 \quad [1, \infty)$$

9. a. linear equation
b.
$$-0.5y+0.7=3.7$$
$$-0.5y=3$$
$$y=-6 \quad \{-6\}$$

11. a. absolute value inequality
b.
$$|2t+8|\leq 4$$
$$-4\leq 2t+8\leq 4$$
$$-12\leq 2t\leq -4$$
$$-6\leq t\leq -2 \quad [-6, -2]$$

13. a. compound inequality
b.
$$-11<2t+1<19$$
$$-12<2t<18$$
$$-6<t<9 \quad (-6, 9)$$

15. a. absolute value equation
b.
$$\left|\frac{1}{2}y+3\right|=5$$
$$\frac{1}{2}y+3=5 \text{ or } \frac{1}{2}y+3=-5$$
$$\frac{1}{2}y=2 \text{ or } \quad \frac{1}{2}y=-8$$
$$y=4 \text{ or } \quad y=-16 \quad \{4, -16\}$$

17. a. linear inequality
b.

19. a. absolute value inequality
b.

$$-\frac{3}{4}p \geq -9$$

$$-\frac{4}{3}\left(-\frac{3}{4}p\right) \leq -\frac{4}{3}(-9)$$

$$p \leq 12 \quad (-\infty, 12]$$

$$\left|\frac{2x-9}{3}\right| \geq 5$$

$$\frac{2x-9}{3} \geq 5 \text{ or } \frac{2x-9}{3} \leq -5$$

$$2x - 9 \geq 15 \text{ or } 2x - 9 \leq -15$$

$$2x \geq 24 \text{ or } 2x \leq -6$$

$$x \geq 12 \text{ or } x \leq -3 \quad (-\infty, -3] \cup [12,)$$

21. a. absolute value equation

b. $|2 - c| + 5 = 3$

$$|2 - c| = -2 \quad \{ \ \}$$

23. a. linear equation

$$\frac{w-4}{5} - \frac{w+1}{3} = 1$$

$$15\left(\frac{w-4}{5} - \frac{w+1}{3}\right) = 15(1)$$

$$3(w-4) - 5(w+1) = 15$$

$$3w - 12 - 5w - 5 = 15$$

$$-2w - 17 = 15$$

b.
$$-2w = 32$$

$$w = -16 \quad \{-16\}$$

25. a. compound inequality

b. $2x - 7 > 9$ and $3x \leq 36$

$$2x > 16 \text{ and } x \leq 12$$

$$x > 8 \text{ and } x \leq 12 \quad (8, 12]$$

27. a. linear equation

b. $5(x-2) + 7 = 2x + 3(x-1)$

$$5x - 10 + 7 = 2x + 3x - 3$$

$$5x - 3 = 5x - 3$$

$$-3 = -3 \quad (-\infty, \infty)$$

Chapter 1 Group Activity

1. False

3. True

5. True

7. True

9. True

11. False

13. False

15. False

17. False

19. False

21. True

23. False

Chapter 1 Review Exercises

Section 1.1

1. The empty set; no solution

3.
$$x - 27 = -32$$
$$x - 27 + 27 = -32 + 27$$
$$x = -5 \quad \{-5\}$$
A conditional equation

5.
$$7.23 + 0.6x = 0.2x$$
$$7.23 + 0.6x - 0.6x = 0.2x - 0.6x$$
$$7.23 = -0.4x$$
$$\frac{7.23}{-0.4} = \frac{-0.4x}{-0.4}$$
$$-18.075 = x \quad \{-18.075\}$$
A conditional equation

7.
$$-(4 + 3m) = 9(3 - m)$$
$$-4 - 3m = 27 - 9m$$
$$-4 - 3m + 9m = 27 - 9m + 9m$$
$$-4 + 6m = 27$$
$$-4 + 4 + 6m = 27 + 4$$
$$6m = 31$$
$$\frac{6m}{6} = \frac{31}{6}$$
$$m = \frac{31}{6} \quad \left\{\frac{31}{6}\right\}$$
A conditional equation

9.
$$\frac{x-3}{5} - \frac{2x+1}{2} = 1$$
$$10\left(\frac{x-3}{5} - \frac{2x+1}{2}\right) = 10(1)$$
$$2(x-3) - 5(2x+1) = 10$$
$$2x - 6 - 10x - 5 = 10$$
$$-8x - 11 = 10$$
$$-8x = 21$$
$$x = -\frac{21}{8} \quad \left\{-\frac{21}{8}\right\}$$
A conditional equation

11.
$$\frac{10}{8}m + 18 - \frac{7}{8}m = \frac{3}{8}m + 25$$
$$\frac{3}{8}m + 18 = \frac{3}{8}m + 25$$
$$\frac{3}{8}m - \frac{3}{8}m + 18 = \frac{3}{8}m - \frac{3}{8}m + 25$$
$$18 = 25 \quad \{\ \}$$
This is a contradiction.

Section 1.2

13. $x, x+1, x+2$

15. $D = rt$ Distance equals rate times time

17. **a.** Let x = the amount of tax
$$\text{tax} = (\text{tax rate})(\text{income})$$
$$x = 0.28(85,200)$$
$$x = 23,856$$
The tax is $23,856.
b. Let y = net income

19. Let x = the number of alcohol deaths in 1999
(recent year) = (1999 deaths) + (increase)
(increase) = (increase rate)(1999 deaths)

$$\text{net income} = (\text{income}) - (\text{tax})$$
$$y = 85,200 - 23,856$$
$$y = 61,344$$

The net income is $61,344.

$$17,430 = x + 0.05x$$
$$17,430 = 1.05x$$
$$\frac{17,430}{1.05} = \frac{1.05x}{1.05}$$
$$16,600 = x$$

The number of alcohol-related deaths in 1999 was 16,600.

21. Let x = the length of the first piece
$\frac{1}{3}x$ = the length of the second piece
(length of first) + (length of second) = (total)

$$x + \frac{1}{3}x = 2\frac{2}{3}$$
$$\frac{4}{3}x = \frac{8}{3}$$
$$\frac{3}{4} \cdot \frac{4}{3}x = \frac{3}{4} \cdot \frac{8}{3}$$
$$x = 2$$
$$\frac{1}{3}x = \frac{1}{3}(2) = \frac{2}{3}$$

The lengths are 2 ft and $\frac{2}{3}$ ft.

23.

	10% Acid Solution	25% Acid Solution	15% Acid Solution
Amount of Solution	x	1	$x + 1$
Amount of Alcohol	$0.10x$	$0.25(1)$	$0.15(x + 1)$

(amt of 10%) + (amt of 25%)=(amt of 15%)

$$0.10x + 0.25(1) = 0.15(x + 1)$$
$$0.10x + 0.25 = 0.15x + 0.15$$
$$0.10x - 0.15x + 0.25 = 0.15x - 0.15x + 0.15$$
$$-0.05x + 0.25 = 0.15$$
$$-0.05x + 0.25 - 0.25 = 0.15 - 0.25$$
$$-0.05x = -0.10$$
$$\frac{-0.05x}{-0.05} = \frac{-0.10}{-0.05}$$
$$x = 2$$

2 L of 10% solution should be used.

Section 1.3

25. Let w = the width of the rectangle
$w + 2$ = the length of the rectangle

$$P = 2l + 2w$$
$$40 = 2(w + 2) + 2w$$
$$40 = 2w + 4 + 2w$$
$$40 = 4w + 4$$
$$40 - 4 = 4w + 4 - 4$$
$$36 = 4w$$
$$9 = w$$
$$w + 2 = 9 + 2 = 11$$

The width is 9 ft and the length is 11 ft.

27. Let w = the width of the rectangle
$w + 2$ = the length of the rectangle

$$P = 2l + 2w$$
$$40 = 2(w + 2) + 2w$$
$$40 = 2w + 4 + 2w$$
$$40 = 4w + 4$$
$$40 - 4 = 4w + 4 - 4$$
$$36 = 4w$$
$$9 = w$$
$$w + 2 = 9 + 2 = 11$$

The width is 9 ft and the length is 11 ft.

29.
$$-6x + y = 12 \quad \text{for } y$$
$$-6x + 6x + y = 6x + 12$$
$$y = 6x + 12$$

31.
$$A = \frac{1}{2}bh \quad \text{for } b$$
$$2A = 2\left(\frac{1}{2}bh\right)$$
$$2A = bh$$
$$\frac{2A}{h} = \frac{bh}{h}$$
$$\frac{2A}{h} = b$$

Section 1.4

33.
$$-6x - 2 > 6$$
$$-6x - 2 + 2 > 6 + 2$$
$$-6x > 8$$
$$\frac{-6x}{-6} < \frac{8}{-6}$$
$$x < -\frac{4}{3}$$

a. $\left\{x \mid x < -\dfrac{4}{3}\right\}$

b. $\left(-\infty, -\dfrac{4}{3}\right)$

35.
$$5 - 7(x + 3) > 19x$$
$$5 - 7x - 21 > 19x$$
$$-7x - 16 > 19x$$
$$-7x + 7x - 16 > 19x + 7x$$
$$-16 > 26x$$
$$\frac{-16}{26} > \frac{26x}{26}$$
$$-\frac{8}{13} > x \quad \text{or} \quad x < -\frac{8}{13}$$

a. $\left\{x \mid x < -\dfrac{8}{13}\right\}$

b. $\left(-\infty, -\dfrac{8}{13}\right)$

37.
$$\frac{5 - 4x}{8} \geq 9$$
$$8\left(\frac{5 - 4x}{8}\right) \geq 8 \cdot 9$$
$$5 - 4x \geq 72$$
$$5 - 5 - 4x \geq 72 - 5$$
$$-4x \geq 67$$
$$\frac{-4x}{-4} \leq \frac{67}{-4}$$
$$x \leq -\frac{67}{4}$$

39.
$$\frac{82 + 88 + 92 + 93 + x}{5} \geq 90$$
$$5 \cdot \frac{82 + 88 + 92 + 93 + x}{5} \geq 5 \cdot 90$$
$$355 + x \geq 450$$
$$355 - 355 + x \geq 450 - 355$$
$$x \geq 95$$
Dave must earn at least 95% on his 5th test.

a. $\left\{ x \mid x \le -\dfrac{67}{4} \right\}$

b. $\left(-\infty, -\dfrac{67}{4} \right]$

Section 1.5

41. $X \mid Y = [-10, 1)$

43. $Y \cup Z = (-\infty, \infty)$

45. $Z \cup W = (-\infty, -3] \cup (-1, \infty)$

47. $4m > -11$ and $4m - 3 \le 13$

$m > -\dfrac{11}{4}$ $\mid$ $4m \le 16$

$m > -\dfrac{11}{4}$ $\mid$ $m \le 4$

$\left(-\dfrac{11}{4}, \infty \right) \mid (-\infty, 4] = \left(-\dfrac{11}{4}, 4 \right]$

49. $-3y + 1 \ge 10$ and $-2y - 5 \le -15$

$-3y \ge 9$ $\mid$ $-2y \le -10$

$y \le -3$ $\mid$ $y \ge 5$

$(-\infty, -3] \mid [5, \infty) = \{ \ \}$

51. $\dfrac{2}{3}t - 3 \le 1$ or $\dfrac{3}{4}t - 2 > 7$

$\dfrac{2}{3}t \le 4$ U $\dfrac{3}{4}t > 9$

$\dfrac{3}{2} \cdot \dfrac{2}{3}t \le \dfrac{3}{2} \cdot 4$ U $\dfrac{4}{3} \cdot \dfrac{3}{4}t > \dfrac{4}{3} \cdot 9$

$t \le 6$ U $t > 12$

$(-\infty, 6] \cup (12, \infty)$

53. $-7 < -7(2w + 3)$ or $-2 < -4(3w - 1)$

$-7 < -14w - 21$ U $-2 < -12w + 4$

$14 < -14w$ U $-6 < -12w$

$-1 > w$ U $\dfrac{1}{2} > w$

$(-\infty, -1) \cup \left(-\infty, \dfrac{1}{2} \right) = \left(-\infty, \dfrac{1}{2} \right)$

55. $2 \ge -(b - 2) - 5b \ge -6$

$2 \ge -b + 2 - 5b \ge -6$

$2 \ge -6b + 2 \ge -6$

$0 \ge -6b \ge -8$

$0 \le b \le \dfrac{4}{3}$ $\left[0, \dfrac{4}{3} \right]$

57. $-1 < \dfrac{1}{3}(x + 3) < 5$

$-3 < x + 3 < 15$

$-6 < x < 12$ $(-6, 12)$

All real numbers between –6 and 12

59. **a.** $125 \le x \le 200$

b. $x < 125$ or $x > 200$

Section 1.6

61. $|x| = 10$

$x = 10$ or $x = -10$ $\{10, -10\}$

63. $|8.7 - 2x| = 6.1$

$8.7 - 2x = 6.1$ or $8.7 - 2x = -6.1$

$-2x = -2.6$ or $-2x = -14.8$

$x = 1.3$ or $x = 7.4$

$\{1.3, 7.4\}$

65. $16 = |x + 2| + 9$

$7 = |x + 2|$

$x + 2 = 7$ or $x + 2 = -7$

$x = 5$ or $x = -9$ $\{5, -9\}$

67. $|4x - 1| + 6 = 4$

$|4x - 1| = -2$ $\{\ \}$

69. $\left|\dfrac{7x - 3}{5}\right| + 4 = 4$

$\left|\dfrac{7x - 3}{5}\right| = 0$

$\dfrac{7x - 3}{5} = 0$ or $\dfrac{7x - 3}{5} = -0$

$7x - 3 = 0$

$7x = 3$

$x = \dfrac{3}{7}$ $\left\{\dfrac{3}{7}\right\}$

71. $|3x - 5| = |2x + 1|$

$3x - 5 = 2x + 1$ or $3x - 5 = -(2x + 1)$

$3x - 5 = 2x + 1$ or $3x - 5 = -2x - 1$

$x - 5 = 1$ or $5x - 5 = -1$

$x = 6$ or $5x = 4$

$x = 6$ or $x = \dfrac{4}{5}$ $\left\{6, \dfrac{4}{5}\right\}$

73. $|2 + 7d| = |-7d - 2|$

$2 + 7d = -7d - 2$ or $2 + 7d = -(-7d - 2)$

$2 + 7d = -7d - 2$ or $2 + 7d = 7d + 2$

$14d = -4$ or $7d = 7d$

$d = -\dfrac{2}{7}$ or $d = d$

$\{d \mid d$ is a real number$\}$

75. Both expressions give the distance between 3 and –2.

Section 1.7

77. $|x| < 4$

79. $|x| > \dfrac{2}{3}$

81. $|x+8| \le 3$

$-3 \le x+8 \le 3$

$-11 \le x \le -5$

$[-11, -5]$

83. $4|5x+1| - 3 > -3$

$4|5x+1| > 0$

$|5x+1| > 0$

$5x+1 < -0$ or $5x+1 > 0$

$5x < -1$ or $5x > -1$

$x < -\dfrac{1}{5}$ or $x > -\dfrac{1}{5}$

$\left(-\infty, -\dfrac{1}{5}\right) \cup \left(-\dfrac{1}{5}, \infty\right)$

85. $|5x-3| + 3 \le 6$

$|5x-3| \le 3$

$-3 \le 5x-3 \le 3$

$0 \le 5x \le 6$

$0 \le x \le \dfrac{6}{5}$ $\left[0, \dfrac{6}{5}\right]$

87. $\left|\dfrac{x}{3} + 2\right| < 2$

$-2 < \dfrac{x}{3} + 2 < 2$

$-4 < \dfrac{x}{3} < 0$

$-12 < x < 0$ $(-12, 0)$

89. $|9+3x| + 1 \ge 1$

$|9+3x| \ge 0$

$9+3x \le -0$ or $9+3x \ge 0$

$3x \le -9$ or $3x \ge -9$

$x \le -3$ or $x \ge -3$

$(-\infty, -3] \cup [-3, \infty) = (-\infty, \infty)$

91. $-|2.5x+15| < 7$

$|2.5x+15| > -7$

All real numbers $(-\infty, \infty)$

93. $|x+5| < -4$ $\{\ \}$

95. If an absolute value is greater than a negative number, then all real numbers are solutions.

97. $\left|L - 3\dfrac{3}{8}\right| \le \dfrac{1}{4}$

$-\dfrac{1}{4} \le L - 3\dfrac{3}{8} \le \dfrac{1}{4}$

$3\dfrac{1}{8} \le L \le 3\dfrac{5}{8}$ $\left[3\dfrac{1}{8}, 3\dfrac{5}{8}\right]$

This means that the actual length of the screw may be between $3\frac{1}{8}$ in and $3\frac{5}{8}$ in, inclusive.

Chapter 1 Test

1.
$$\frac{x}{7} + 1 = 20$$
$$7\left(\frac{x}{7} + 1\right) = 7 \cdot 20$$
$$x + 7 = 140$$
$$x + 7 - 7 = 140 - 7$$
$$x = 133 \quad \{133\}$$

3.
$$0.12(x) + 0.08(60{,}000 - x) = 10{,}500$$
$$0.12x + 4800 - 0.08x = 10{,}500$$
$$0.04x + 4800 = 10{,}500$$
$$0.04x + 4800 - 4800 = 10{,}500 - 4800$$
$$0.04x = 5700$$
$$\frac{0.04x}{0.04} = \frac{5700}{0.04}$$
$$x = 142{,}500$$
$$\{142{,}500\}$$

5.
$$\left|\frac{1}{2}x + 3\right| - 4 = 4$$
$$\left|\frac{1}{2}x + 3\right| = 8$$
$$\frac{1}{2}x + 3 = 8 \quad \text{or} \quad \frac{1}{2}x + 3 = -8$$
$$\frac{1}{2}x = 5 \quad \text{or} \quad \frac{1}{2}x = -11$$
$$x = 10 \quad \text{or} \quad x = -22 \quad \{10, -22\}$$

7.
$$-5 = -8 + |2y - 3|$$
$$3 = |2y - 3|$$
$$2y - 3 = 3 \quad \text{or} \quad 2y - 3 = -3$$
$$2y = 6 \quad \text{or} \quad 2y = 0$$
$$y = 3 \quad \text{or} \quad y = 0 \quad \{3, 0\}$$

7.
$$-5 = -8 + |2y - 3|$$
$$3 = |2y - 3|$$
$$2y - 3 = 3 \quad \text{or} \quad 2y - 3 = -3$$
$$2y = 6 \quad \text{or} \quad 2y = 0$$
$$y = 3 \quad \text{or} \quad y = 0 \quad \{3, 0\}$$

9.
$$|8x + 11| = |8x + 5|$$
$$8x + 11 = 8x + 5 \quad \text{or} \quad 8x + 11 = -(8x + 5)$$
$$8x + 11 = 8x + 5 \quad \text{or} \quad 8x + 11 = -8x - 5$$
$$11 = 5 \quad \text{or} \quad 16x = -16$$
$$\text{contradiction} \quad \text{or} \quad x = -1 \quad \{-1\}$$

11. Let x = the smaller number
   $5x$ = the larger number
$$5x - x = 72$$
$$4x = 72$$
$$\frac{4x}{4} = \frac{72}{4}$$
$$x = 18$$
$$5x = 5(18) = 90$$
The numbers are 18 and 90.

13.

	5% Account	3.5% Account
Total Amount Invested	x	$x{-}100$
Interest Earned	$0.05x$	$0.035(x{-}100)$

81.50
(int at 5%) + (int at 3.5%) = (total int)

$$0.05x + 0.035(x - 100) = 81.50$$
$$0.05x + 0.035x - 3.50 = 81.50$$
$$0.085x - 3.50 = 81.50$$
$$0.085x - 3.50 + 3.50 = 81.50 + 3.50$$
$$0.085x = 85.00$$
$$\frac{0.085x}{0.085} = \frac{85.00}{0.085}$$
$$x = 1000$$

Shawnna invested $1000 in the CD.

15. Let x = first odd integer
$x + 2$ = second consecutive odd integer
$x + 4$ = third consecutive odd integer
(sum) = (4 times third less 41)
$$x + (x + 2) + (x + 4) = 4(x + 4) - 41$$
$$3x + 6 = 4x + 16 - 41$$
$$3x + 6 = 4x - 25$$
$$3x - 3x + 6 = 4x - 3x - 25$$
$$6 = x - 25$$
$$6 + 25 = x - 25 + 25$$
$$31 = x$$
$$x + 2 = 31 + 2 = 33$$
$$x + 4 = 31 + 4 = 35$$

The three consecutive odd integers are 31, 33, and 35.

17.
$$4x + 2y = 6 \quad \text{for } y$$
$$4x - 4x + 2y = 6 - 4x$$
$$2y = 6 - 4x$$
$$\frac{2y}{2} = \frac{6 - 4x}{2}$$
$$y = -2x + 3$$

19.
$$x + 8 > 42$$
$$x + 8 - 8 > 42 - 8$$
$$x > 34$$

$(34, \infty)$

21.
$$-2 < 3x - 1 \leq 5$$
$$-2 + 1 < 3x - 1 + 1 \leq 5 + 1$$
$$-1 < 3x \leq 6$$
$$\frac{-1}{3} < \frac{3x}{3} \leq \frac{6}{3}$$
$$-\frac{1}{3} < x \leq 2$$

$\left(-\dfrac{1}{3}, 2\right]$

23.
$$-4 \leq \frac{6 - 2x}{5} < 2$$
$$-20 \leq 6 - 2x < 10$$
$$-26 \leq -2x < 4$$
$$13 \geq x > -2$$
$$-2 < x \leq 13 \qquad (-2, 13]$$

25. $-2x - 3 > -3$ and $x + 3 \geq 0$
$$-2x > 0 \quad | \quad x \geq -3$$
$$x < 0 \quad | \quad x \geq -3$$
$$(-\infty, 0) | \; [-3, \infty) = [-3, 0)$$

27. $2x - 3 > 1$ and $x + 4 < -1$

$2x > 4$ | $x < -5$

$x > 2$ | $x < -5$

$(2, \infty)$∣ $(-\infty, -5) = \{\ \}$

29. $|3x - 8| \geq 9$

$3x - 8 \leq -9$ or $3x - 8 \geq 9$

$3x \leq -1$ or $3x \geq 17$

$x \leq -\dfrac{1}{3}$ or $x \geq \dfrac{17}{3}$

$\left(-\infty, -\dfrac{1}{3}\right] \cup \left[\dfrac{17}{3}, \infty\right)$

31. $|7 - 3x| + 1 > -3$

$|7 - 3x| > -4$

All real numbers $(-\infty, \infty)$

33. Let x = the number of additional passengers

(weight of 4 pass)+(weight of add pass) $\leq$ 2000

$4(180) + x(180) \leq 2000$

$720 + 180x \leq 2000$

$720 - 720 + 180x \leq 2000 - 720$

$180x \leq 1280$

$\dfrac{180x}{180} \leq \dfrac{1280}{180}$

$x \leq 7.1$

The elevator can carry at most 7 additional passengers.

35. $|x - 15.41| \leq 0.01$

Chapter 2   Linear Equations in Two Variables and Functions

Are You Prepared?

1. **X**   44 in.

2. **I**   3 yr

3. **E** $h = 2.5a + 31$
$h = 2.5(6) + 31 = 15 + 31 = 46$ in.

4. **T** $50 = 2.5a + 31$
$19 = 2.5a$
$7.6 = a$

5. **C** $h = 2.5a + 31$
$h = 2.5(11) + 31 = 27.5 + 31 = 58.5$ in.

A graph intersects the x-axis at an $x - i n t e r c e p t$.

Section 2.1   Linear Equations in Two Variables

1. **a.** x; y-axis
    b.   ordered
    c.   origin; $(0, 0)$
    d.   quadrants
    e.   negative
    f.   III
    g. $Ax + By = C$
    h. x-intercept
    i. y-intercept
    j.   vertical
    k.   horizontal

3. For (x, y), if $x > 0$, $y > 0$, the point is in quadrant I. If $x < 0$, $y > 0$, the point is in quadrant II. If $x < 0$, $y < 0$, the point is in quadrant III. If $x > 0$, $y < 0$, the point is in quadrant IV.

5.

7. 0

9. A $(-4, 5)$, II
    B $(-2, 0)$, x-axis
    C $(1, 1)$, I
    D $(4, -2)$, IV
    E $(-5, -3)$, III

11. **a.** $2(0) - 3(-3) = 9$
$0 + 9 = 9$
$9 = 9$
$(0, -3)$ is a solution.

$2(-6) - 3(1) = 9$
$-12 - 3 = 9$
$-15 = 9$

b. $(-6, 1)$ is not a solution.

$$2(1)-3\left(-\frac{7}{3}\right)=9$$

c. $2+7=9$

$$9=9$$

$\left(1, -\frac{7}{3}\right)$ is a solution.

13.   **a.** $-1=\dfrac{1}{3}(0)+1$

$-1=0+1$

$-1=1$

$(-1, 0)$ is not a solution.

b. $2=\dfrac{1}{3}(3)+1$

$2=1+1$

$2=2$

$(2, 3)$ is a solution.

c. $-6=\dfrac{1}{3}(1)+1$

$-6=\dfrac{1}{3}+1$

$-6=\dfrac{4}{3}$

$(-6, 1)$ is not a solution.

15. $3x-2y=4$

x	y
0	−2
4	4
−1	$-\frac{7}{2}$

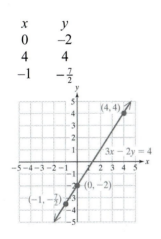

17. $y=-\dfrac{1}{5}x$

x	y
0	0
5	−1
−5	1

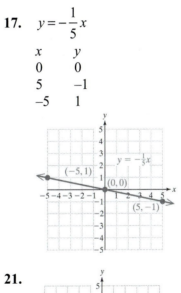

19.

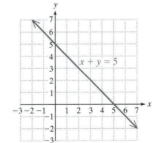

21.

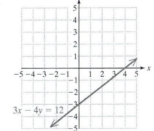

23.

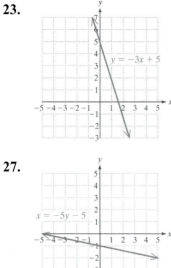

25.

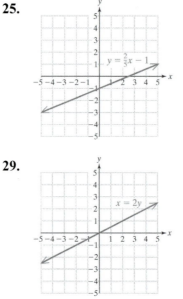

27.

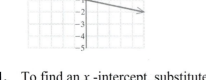

29.

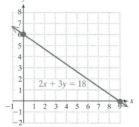

31. To find an x-intercept, substitute $y = 0$ and solve for x. To find a y-intercept, substitute $x = 0$ and solve for y.

33. $2x + 3y = 18$

a. $2x + 3(0) = 18$

$2x = 18$

$x = 9$

The x-intercept is (9, 0).

b. $2(0) + 3y = 18$

$3y = 18$

$y = 6$

The y-intercept is (0, 6).

c.

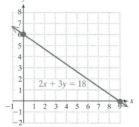

35. $x - 2y = 4$

a. $x - 2(0) = 4$

$x = 4$

The x-intercept is (4, 0).

b. $0 - 2y = 4$

$-2y = 4$

$y = -2$

The y-intercept is (0, –2).

37. $5x = 3y$

a. $5x = 3(0)$

$5x = 0$

$x = 0$

The x-intercept is (0, 0).

b. $5(0) = 3y$

$0 = 3y$

$0 = y$

The y-intercept is (0, 0)

c.

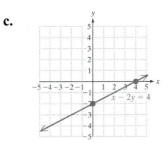

c.

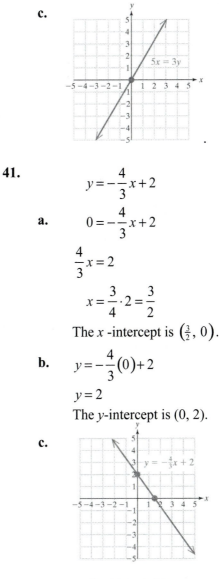

39.

$$y = 2x + 4$$

a.
$$0 = 2x + 4$$
$$-2x = 4$$
$$x = -2$$
The x-intercept is $(-2, 0)$.

b.
$$y = 2(0) + 4$$
$$y = 4$$
The y-intercept is $(0, 4)$.

c.

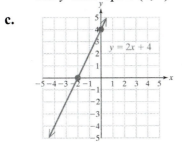

41.

$$y = -\frac{4}{3}x + 2$$

a.
$$0 = -\frac{4}{3}x + 2$$
$$\frac{4}{3}x = 2$$
$$x = \frac{3}{4} \cdot 2 = \frac{3}{2}$$
The x-intercept is $\left(\frac{3}{2}, 0\right)$.

b.
$$y = -\frac{4}{3}(0) + 2$$
$$y = 2$$
The y-intercept is $(0, 2)$.

c.

43.

$$x = \frac{1}{4}y$$

a.
$$x = \frac{1}{4}(0)$$
$$x = 0$$
The x-intercept is $(0, 0)$.

b.
$$0 = \frac{1}{4}y$$
$$0 = y$$
The y-intercept is $(0, 0)$.

45.

$$y = 15,000 + 0.08x$$

a.
$$y = 15,000 + 0.08(500,000)$$
$$= 15,000 + 40,000 = 55,000$$
The salary is $55,000.

$$y = 15,000 + 0.08(300,000)$$
$$= 15,000 + 24,000 = 39,000$$

b.
The salary is $39,000.

$$y = 15,000 + 0.08(0)$$
$$= 15,000 + 0 = 15,000$$

c.
The y-intercept is $(0, 15,000)$. For $0 in sales, the salary is $15,000. Total sales cannot be negative.

d.

c.

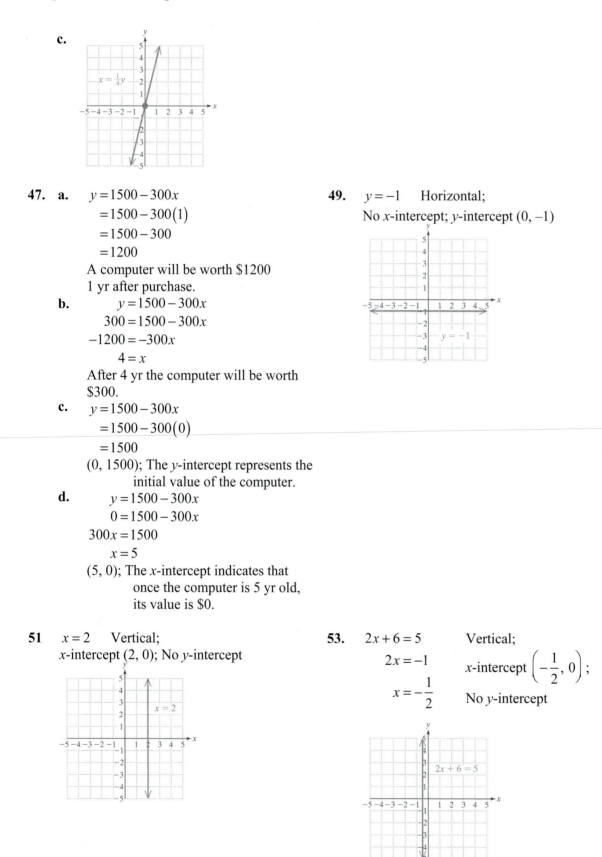

$x = \frac{1}{4}y$

47. a.
$$y = 1500 - 300x$$
$$= 1500 - 300(1)$$
$$= 1500 - 300$$
$$= 1200$$
A computer will be worth $1200
1 yr after purchase.

b.
$$y = 1500 - 300x$$
$$300 = 1500 - 300x$$
$$-1200 = -300x$$
$$4 = x$$
After 4 yr the computer will be worth
$300.

c.
$$y = 1500 - 300x$$
$$= 1500 - 300(0)$$
$$= 1500$$
(0, 1500); The y-intercept represents the
initial value of the computer.

d.
$$y = 1500 - 300x$$
$$0 = 1500 - 300x$$
$$300x = 1500$$
$$x = 5$$
(5, 0); The x-intercept indicates that
once the computer is 5 yr old,
its value is $0.

49. $y = -1$ Horizontal;
No x-intercept; y-intercept (0, −1)

51 $x = 2$ Vertical;
x-intercept (2, 0); No y-intercept

53. $2x + 6 = 5$ Vertical;
$$2x = -1$$ x-intercept $\left(-\frac{1}{2}, 0\right)$;
$$x = -\frac{1}{2}$$ No y-intercept

55. $-2y + 1 = 9$ Horizontal;
$\qquad -2y = 8$ No x-intercept;
$\qquad\quad y = -4$ y-intercept $(0, -4)$

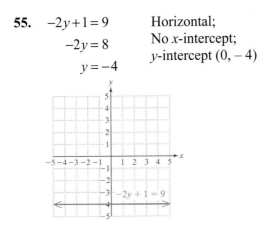

57. A horizontal line parallel to the x-axis will not have an x-intercept. A vertical line parallel to the y-axis will not have a y-intercept.

59. b, c, d

61.
$$\frac{x}{2} + \frac{y}{3} = 1 \qquad\qquad \frac{0}{2} + \frac{y}{3} = 1$$
$$\frac{x}{2} + \frac{0}{3} = 1 \qquad\qquad \frac{y}{3} = 1$$
$$\frac{x}{2} = 1 \qquad\qquad y = 3$$
$$x = 2$$

The x-intercept is $(2, 0)$ and the y-intercept is $(0, 3)$.

63
$$\frac{x}{a} + \frac{y}{b} = 1 \qquad\qquad \frac{0}{a} + \frac{y}{b} = 1$$
$$\frac{x}{a} + \frac{0}{b} = 1 \qquad\qquad \frac{y}{b} = 1$$
$$\frac{x}{a} = 1 \qquad\qquad y = b$$
$$x = a$$

The x-intercept is $(a, 0)$ and the y-intercept is $(0, b)$.

Section 2.2   Slope of a Line and Rate of Change

1. **a.** slope; $\dfrac{y_2 - y_1}{x_2 - x_1}$

  b. parallel; same
  c. right
  d. -1

3.
$$\frac{1}{2}x + y = 4$$

  a. $\dfrac{1}{2}(0) + y = 4$
  $\qquad\qquad y = 4$

  The ordered pair is $(0, 4)$.

b.
$$\frac{1}{2}x + 0 = 4$$
$$\frac{1}{2}x = 4$$
$$x = 8$$
The ordered pair is (8, 0).

c.
$$\frac{1}{2}(-4) + y = 4$$
$$-2 + y = 4$$
$$y = 6$$
The ordered pair is (−4, 6).

5.
$$4 - 2y = 0$$
$$-2y = -4$$
$$y = 2$$
There is no x-intercept.
The y-intercept is (0, 2).

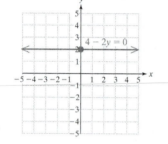

7. $m = \dfrac{24}{7} = \dfrac{24}{7}$

9. $m = \dfrac{8}{72} = \dfrac{1}{9}$

11. $m = \dfrac{4}{100} = \dfrac{1}{25}$

13. $m = \dfrac{y_2 - y_1}{x_2 - x_1} = \dfrac{-3 - 0}{0 - 6} = \dfrac{-3}{-6} = \dfrac{1}{2}$

15. $m = \dfrac{y_2 - y_1}{x_2 - x_1} = \dfrac{-7 - 3}{4 - (-2)} = \dfrac{-10}{6} = -\dfrac{5}{3}$

17. $m = \dfrac{y_2 - y_1}{x_2 - x_1} = \dfrac{-3 - 5}{2 - (-2)} = \dfrac{-8}{4} = -2$

19. $m = \dfrac{y_2 - y_1}{x_2 - x_1} = \dfrac{-0.8 - (-1.1)}{-0.1 - 0.3} = \dfrac{0.3}{-0.4} = -\dfrac{3}{4}$

21. $m = \dfrac{y_2 - y_1}{x_2 - x_1} = \dfrac{7 - 3}{2 - 2} = \dfrac{4}{0}$ Undefined

23. $m = \dfrac{y_2 - y_1}{x_2 - x_1} = \dfrac{-1 - (-1)}{-3 - 5} = \dfrac{0}{-8} = 0$

25. $m = \dfrac{y_2 - y_1}{x_2 - x_1} = \dfrac{6.4 - 4.1}{0 - (-4.6)} = \dfrac{2.3}{4.6} = \dfrac{1}{2}$

27. $m = \dfrac{y_2 - y_1}{x_2 - x_1} = \dfrac{1 - \dfrac{4}{3}}{\dfrac{7}{2} - \dfrac{3}{2}} = \dfrac{-\dfrac{1}{3}}{\dfrac{4}{2}} = -\dfrac{1}{3} \cdot \dfrac{1}{2} = -\dfrac{1}{6}$

29.

$$m = \frac{y_2 - y_1}{x_2 - x_1} = \frac{2\frac{1}{3} - \frac{7}{3}}{\frac{1}{2} - \frac{3}{4}} = \frac{\frac{7}{3} - \frac{7}{3}}{\frac{2}{4} - \frac{3}{4}} = \frac{0}{-\frac{1}{4}} = 0$$

31. The slope of a line is positive if the graph increases from left to right. The slope of a line is negative if the graph decreases from left to right. The slope of a line is zero if the graph is horizontal. The slope of a line is undefined if the graph is vertical.

33. $m = 0$

35. $m = \frac{1}{10}$

37. $m = -1$

39. **a.** $m = 5$

   b. $m = -\frac{1}{5}$

41. **a.** $m = -\frac{4}{7}$

   b. $m = \frac{7}{4}$

43. **a.** $m = 0$

   b. m is undefined.

45. No, because the product of the slopes of perpendicular lines must be −1. The product of two positive numbers is not negative.

47. $y = -5$ is the equation of a horizontal line; thus a perpendicular line will be a vertical line whose slope is undefined.

49. $m = 0$

51. undefined

53. $m_{L_1} = \frac{9 - 5}{4 - 2} = \frac{4}{2} = 2$

   $m_{L_2} = \frac{2 - 4}{3 - (-1)} = \frac{-2}{4} = -\frac{1}{2}$

   The lines are perpendicular.

55. $m_{L_1} = \frac{-1 - (-2)}{3 - 4} = \frac{1}{-1} = -1$

   $m_{L_2} = \frac{-16 - (-1)}{-10 - (-5)} = \frac{-15}{-5} = 3$

   The lines are neither parallel nor perpendicular.

57. $m_{L_1} = \frac{9 - 3}{5 - 5} = \frac{6}{0}$ Undefined

   $m_{L_2} = \frac{2 - 2}{0 - 4} = \frac{0}{-4} = 0$

   The lines are perpendicular. One line is horizontal and the other is vertical.

59. $m_{L_1} = \frac{3 - (-2)}{2 - (-3)} = \frac{5}{5} = 1$

   $m_{L_2} = \frac{5 - 1}{0 - (-4)} = \frac{4}{4} = 1$

   The lines are parallel.

61. **a.** $m = \frac{313 - 70}{2010 - 1998} = \frac{243}{12} = 20.25$

   b. The number of cell phone subscriptions increased at a rate of 20.25 million per year during this period.

63. **a.** $m = \frac{74.5 - 44.5}{10 - 5} = \frac{30}{5} = 6$

   b. The weight of boys tends to increase by 6 lb/yr during this period of growth.

65. **a.** $(-1, -4)$ and $(0, -2)$

67. **a.** $(-2, 0)$ and $(0, -4)$

$$m = \frac{-2-(-4)}{0-(-1)} = \frac{2}{1} = 2$$

$$m = \frac{-4-0}{0-(-2)} = \frac{-4}{2} = -2$$

b. $(0, -2)$ and $(3, 4)$

$$m = \frac{4-(-2)}{3-0} = \frac{6}{3} = 2$$

b. $(0, -4)$ and $(2, 0)$

$$m = \frac{0-(-4)}{2-0} = \frac{4}{2} = 2$$

c. $(-1, -4)$ and $(3, 4)$

$$m = \frac{4-(-4)}{3-(-1)} = \frac{8}{4} = 2$$

c. $(0, -4)$ and $(3, 5)$

$$m = \frac{5-(-4)}{3-0} = \frac{9}{3} = 3$$

69. For example: $(1, 2)$

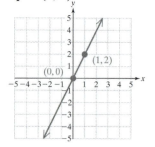

71. For example: $(2, 0)$

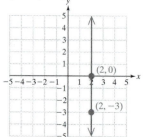

73. For example: $(2, 0)$

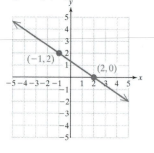

75.

$$\frac{6-y}{4-(-2)} = -\frac{3}{2}$$

$$\frac{6-y}{6} = -\frac{3}{2}$$

$$6-y = 6\left(-\frac{3}{2}\right)$$

$$6-y = -9$$

$$-y = -15$$

$$y = 15$$

77. a. $\text{Pitch} = \dfrac{4}{24} = \dfrac{1}{6}$

b. $m = \dfrac{4}{12} = \dfrac{1}{3}$

Section 2.3   Equations of a Line

1. a. $y = mx + b$

 b. standard

 c. horizontal

 d. vertical

 e. slope; y-intercept

 f. $y - y_1 = m(x - x_1)$

3. a. 0

 b. undefined

5. If the slope of one line is the opposite of the reciprocal of the slope of the other line, then the lines are perpendicular

7.

$$y = -\frac{2}{3}x - 4$$

Slope: $-\frac{2}{3}$

y-intercept: $(0, -4)$

9. $y = 2 + 3x$

$y = 3x + 2$

Slope: 3

y-intercept: $(0, 2)$

11. $17x + y = 0$

$y = -17x$

Slope: -17

y-intercept: $(0, 0)$

13. $18 = 2y$

$9 = y$

$y = 0x + 9$

Slope: 0

y-intercept: $(0, 9)$

15. $8x + 12y = 9$

$12y = -8x + 9$

$y = -\dfrac{8}{12}x + \dfrac{9}{12} = -\dfrac{2}{3}x + \dfrac{3}{4}$

Slope: $-\dfrac{2}{3}$

y-intercept: $\left(0, \dfrac{3}{4}\right)$

17. $y = 0.625x - 1.2$

Slope: 0.625

y-intercept: $(0, -1.2)$

19. d

21. f

23. b

25. $y - 2 = 4x$

$y = 4x + 2$

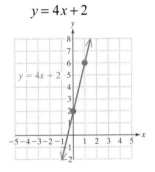

27. $3x + 2y = 6$

$2y = -3x + 6$

$y = -\dfrac{3}{2}x + 3$

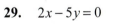

29. $2x - 5y = 0$

$-5y = -2x$

$y = \dfrac{2}{5}x$

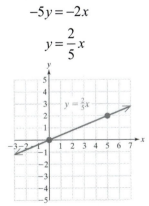

31. $Ax + By = C$

$By = -Ax + C$

$y = -\dfrac{A}{B}x + \dfrac{C}{B}$

The slope is given by $m = -\dfrac{A}{B}$.

The y-intercept is $\left(0, \dfrac{C}{B}\right)$.

33. $-3y = 5x - 1$

$$y = -\frac{5}{3}x + \frac{1}{3}$$

$$m_1 = -\frac{5}{3}$$

$6x = 10y - 12$

$10y = 6x + 12$

$$y = \frac{3}{5}x + \frac{6}{5}$$

$$m_2 = \frac{3}{5}$$

The lines are perpendicular.

35. $3x - 4y = 12$

$-4y = -3x + 12$

$$y = \frac{3}{4}x - 3$$

$$m_1 = \frac{3}{4}$$

$\frac{1}{2}x - \frac{2}{3}y = 1$

$6\left(\frac{1}{2}x - \frac{2}{3}y\right) = 6 \cdot 1$

$3x - 4y = 6$

$-4y = -3x + 6$

$$y = \frac{3}{4}x - \frac{3}{2}$$

$$m_2 = \frac{3}{4}$$

The lines are parallel.

37. $3y = 5x + 6$

$$y = \frac{5}{3}x + 2$$

$$m_1 = \frac{5}{3}$$

$5x + 3y = 9$

$3y = -5x + 9$

$$y = -\frac{5}{3}x + 3$$

$$m_2 = -\frac{5}{3}$$

The lines are neither parallel nor perpendicular.

39. $m = 3,$ point: $(0, 5)$

$y = mx + b$

$y = 3x + 5$

41. $m = 2,$ point: $(4, -3)$

$y = mx + b$

$-3 = 2(4) + b$

$-3 = 8 + b$

$-11 = b$

$y = 2x - 11$

43. $m = -\frac{4}{5},$ point: $(10, 0)$

$y = mx + b$

$0 = -\frac{4}{5}(10) + b$

$0 = -8 + b$

$8 = b$

$$y = -\frac{4}{5}x + 8$$

45. $m = 3,$ y-intercept: $(0, -2)$

$y = 3x - 2$ or $3x - y = 2$

47. $m = 2,$ point: $(2, 7)$

$y - y_1 = m(x - x_1)$

$y - 7 = 2(x - 2)$

$y - 7 = 2x - 4$

$y = 2x + 3$ or $2x - y = -3$

49. $m = -3,$ point: $(-2, -5)$

$y - y_1 = m(x - x_1)$

$y - (-5) = -3(x - (-2))$

$y + 5 = -3x - 6$

$y = -3x - 11$ or $3x + y = -11$

51. $m = -\frac{4}{5},$ point: $(6, -3)$

$$y - y_1 = m(x - x_1)$$
$$y - (-3) = -\frac{4}{5}(x - 6)$$
$$y + 3 = -\frac{4}{5}x + \frac{24}{5}$$
$$y = -\frac{4}{5}x + \frac{9}{5} \quad \text{or} \quad \frac{4}{5}x + y = \frac{9}{5}$$
$$4x + 5y = 9$$

53.
$$m = \frac{0 - 4}{3 - 0} = \frac{-4}{3} = -\frac{4}{3}$$
$$y - y_1 = m(x - x_1)$$
$$y - 4 = -\frac{4}{3}(x - 0)$$
$$y - 4 = -\frac{4}{3}x + 0$$
$$y = -\frac{4}{3}x + 4 \quad \text{or} \quad \frac{4}{3}x + y = 4$$
$$4x + 3y = 12$$

55.
$$m = \frac{10 - 12}{4 - 6} = \frac{-2}{-2} = 1$$
$$y - y_1 = m(x - x_1)$$
$$y - 12 = 1(x - 6)$$
$$y - 12 = x - 6$$
$$y = x + 6 \quad \text{or} \quad x - y = -6$$

57.
$$m = \frac{2 - 2}{-1 - (-5)} = \frac{0}{4} = 0$$
$$y - y_1 = m(x - x_1)$$
$$y - 2 = 0(x - (-5))$$
$$y - 2 = 0$$
$$y = 2$$

59.
$$m = -\frac{3}{4}, \quad \text{point: } (3, 2)$$
$$y - y_1 = m(x - x_1)$$
$$y - 2 = -\frac{3}{4}(x - 3)$$
$$y - 2 = -\frac{3}{4}x + \frac{9}{4}$$
$$y = -\frac{3}{4}x + \frac{17}{4} \quad \text{or} \quad \frac{3}{4}x + y = \frac{17}{4}$$
$$3x + 4y = 17$$

61.
$$m = \frac{4}{3}, \quad \text{point: } (3, 2)$$
$$y - y_1 = m(x - x_1)$$
$$y - 2 = \frac{4}{3}(x - 3)$$
$$y - 2 = \frac{4}{3}x - 4$$
$$y = \frac{4}{3}x - 2 \quad \text{or} \quad \frac{4}{3}x - y = 2$$
$$4x - 3y = 6$$

63.
$$3x - 4y = -7$$
$$-4y = -3x - 7$$
$$y = \frac{3}{4}x + \frac{7}{4}$$
$$m = \frac{3}{4}, \quad \text{point: } (2, -5)$$

$$y - y_1 = m(x - x_1)$$

$$y - (-5) = \frac{3}{4}(x - 2)$$

$$y + 5 = \frac{3}{4}x - \frac{3}{2}$$

$$y = \frac{3}{4}x - \frac{13}{2} \quad \text{or} \quad \frac{3}{4}x - y = \frac{13}{2}$$

$$3x - 4y = 26$$

65. $-15x + 3y = 9$

$$3y = 15x + 9$$

$$y = 5x + 3$$

$$m = 5, \quad m_\perp = -\frac{1}{5}, \quad \text{point: } (-8, -1)$$

$$y - y_1 = m(x - x_1)$$

$$y - (-1) = -\frac{1}{5}(x - (-8))$$

$$y + 1 = -\frac{1}{5}x - \frac{8}{5}$$

$$y = -\frac{1}{5}x - \frac{13}{5} \quad \text{or} \quad \frac{1}{5}x + y = -\frac{13}{5}$$

$$x + 5y = -13$$

67. $3x = 2y$

$$y = \frac{3}{2}x$$

$$m = \frac{3}{2}, \quad \text{point: } (4, 0)$$

$$y - y_1 = m(x - x_1)$$

$$y - 0 = \frac{3}{2}(x - 4)$$

$$y = \frac{3}{2}x - 6 \quad \text{or} \quad \frac{3}{2}x - y = 6$$

$$3x - 2y = 12$$

69. $3y + 2x = 21$

$$3y = -2x + 21$$

$$y = -\frac{2}{3}x + 7$$

$$m_\perp = \frac{3}{2}, \quad \text{point: } (2, 4)$$

$$y - y_1 = m(x - x_1)$$

$$y - 4 = \frac{3}{2}(x - 2)$$

$$y - 4 = \frac{3}{2}x - 3$$

$$y = \frac{3}{2}x + 1 \quad \text{or} \quad \frac{3}{2}x - y = -1$$

$$3x - 2y = -2$$

71. $\frac{1}{2}y = x$

$$y = 2x$$

$$m_\perp = -\frac{1}{2}, \quad \text{point: } (-3, 5)$$

$$y - y_1 = m(x - x_1)$$

$$y - 5 = -\frac{1}{2}(x - (-3))$$

$$y - 5 = -\frac{1}{2}x - \frac{3}{2}$$

$$y = -\frac{1}{2}x + \frac{7}{2} \quad \text{or} \quad \frac{1}{2}x + y = \frac{7}{2}$$

$$x + 2y = 7$$

73. $3x + y = 7$

$$y = -3x + 7$$

$$m_P = -3, \quad \text{point: } (0, 0)$$

75. $m = 0, \quad \text{point: } (2, -3)$

$$y - y_1 = m(x - x_1)$$
$$y - 0 = -3(x - 0)$$
$$y = -3x \ \text{ or } \ 3x + y = 0$$

$$y - y_1 = m(x - x_1)$$
$$y - (-3) = 0(x - 2)$$
$$y + 3 = 0$$
$$y = -3$$

77. A line with an undefined slope is a vertical line, which is in the form $x = c$. Therefore, a line containing $(2, -3)$ would have the equation $x = 2$.

79. A line parallel to the x-axis has the form $y = c$. Therefore, a line containing the point $(4, 5)$ would have the equation $y = 5$.

81. A line parallel to the line $x = 4$ is a vertical line and has the form $x = c$. Therefore, a line containing the point $(5, 1)$ would have the equation $x = 5$.

83. $x = -2$ is not in the slope-intercept form. It has no y-intercept and its slope is undefined.

85. $y = 3$ is in the slope-intercept form, $y = 0x + 3$. Its slope is 0 and the y-intercept is $(0, 3)$.

87. Two points on the line are $(0, 3)$ and $(1, 1)$.
$$m = \frac{1 - 3}{1 - 0} = \frac{-2}{1} = -2$$
$$y - y_1 = m(x - x_1) \qquad (x_1, y_1) = (0, 3)$$
$$y - 3 = -2(x - 0)$$
$$y = -2x + 3$$

89. This is a horizontal line of the form $y = k$. The y-intercept is 2, so the line is $y = 2$.

Problem Recognition Exercises

1. b, f

3. a

5. c, e

7. c, h

9. e

11. c, h

13. g

15. h

17. e

19. d, h

Section 2.4   Applications of Linear Equations and Modeling

1. a. model

3. a. $m = \dfrac{-2 - 0}{3 - (-3)} = \dfrac{-2}{6} = -\dfrac{1}{3}$

b.

$$y - 0 = -\frac{1}{3}\left(x - (-3)\right)$$

c.

$$y = -\frac{1}{3}x - 1 \quad \text{or} \quad \frac{1}{3}x + y = -1$$

$$x + 3y = -3$$

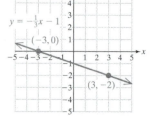

5. **a.**

$$m = \frac{3-3}{-2-(-4)} = \frac{0}{2} = 0$$

b.

$$y - 3 = 0\left(x - (-4)\right)$$

$$y - 3 = 0$$

$$y = 3$$

c.

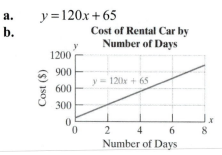

7. **a.** $y = 120x + 65$

b.

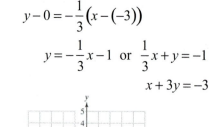

c. The y-intercept is $(0, 65)$. The base cost to rent the car is $65.

d.
$$y = 120(2) + 65$$
$$= 240 + 65$$
$$= 305$$
A 2-day rental costs $305.
$$y = 120(5) + 65$$
$$= 600 + 65$$
$$= 665$$
A 5-day rental costs $665.
$$y = 120(7) + 65$$
$$= 840 + 65$$
$$= 905$$
A 7-day rental costs $905.

e. Yes; $799 is less expensive than $905.

f.
$$y = 1.06(120x + 65)$$
$$= 1.06(120(4) + 65)$$
$$= 1.06(480 + 65)$$
$$= 1.06(545)$$
$$= 577.7$$
A 4-day rental with 6% sales tax costs $577.70.

g. No, the car cannot be driven for a negative number of days.

9. **a.** $y = 52x + 2742$

b.

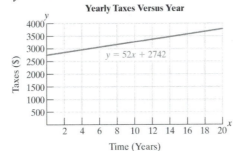

c. $m = 52$. The taxes increase $52 per year.

d. The y-intercept is (0, 2742). In the initial year ($x = 0$) the taxes were $2742.

e. $y = 52(10) + 2742 = 520 + 2742 = 3262$

After 10 years the taxes are $3262.

$y = 52(15) + 2742 = 780 + 2742 = 3522$

After 15 years the taxes are $3522.

13. **a.** The year 2010 is 4 years after the year 2006.

$y = 9.4x + 35.7$
$\quad = 9.4(4) + 35.7$
$\quad = 37.6 + 35.7$
$\quad = 73.3$

The average amount spent in the year 2010 was $73.30.

b. The year 2008 is 2 years after the year 2006.

$y = 9.4x + 35.7$
$\quad = 9.4(2) + 35.7$
$\quad = 18.8 + 35.7$
$\quad = 54.5$
$55.80 - 54.50 = 1.30$

The approximate value differs from the actual value by $1.30.

c. $m = 9.4$; The amount spent per person on video games increased by an average rate of $9.40 per year.

d. (0, 35.7); The y-intercept means that the average amount spent on video games per person was $35.70 in the year 2006.

11. **a.** $y = 0.2(4) = 0.8$ The storm is 0.8 mi away when the time difference is 4 sec.

$y = 0.2(12) = 2.4$

The storm is 2.4 mi away when the time difference is 12 sec.

$y = 0.2(16) = 3.2$

The storm is 3.2 mi away when the time difference is 16 sec.

b. $4.2 = 0.2x$
$21 = x$

When the storm is 4.2 miles away, the time difference is 21 sec.

15. **a.**

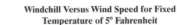
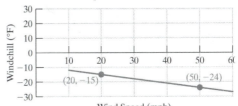

b. $m = \dfrac{-24 - (-15)}{50 - 20} = \dfrac{-9}{30} = -\dfrac{3}{10} = -0.3$

$y - (-15) = -0.3(x - 20)$
$\quad y + 15 = -0.3x + 6$
$\qquad y = -0.3x - 9$

c. $y = -0.3(40) - 9 = -12 - 9 = -21°\,F$

d. $y = -0.3(50) - 9 = -15 - 9 = -24°\,F$

e. $m = -0.3$. This means that temperature decreases at a rate of 0.3°F for every 1 mph increase in wind speed.

17. a.
$$m = \frac{665 - 455}{34 - 20} = \frac{210}{14} = 15$$
$$y - 455 = 15(x - 20)$$
$$y - 455 = 15x - 300$$
$$y = 15x + 155$$

b. The slope is 15 and means that the number of associate degrees awarded in the United States increased by 15 thousand per year.

c. $y = 15(45) + 155 = 675 + 155 = 830$

The number of associate degrees awarded in the United States in 2015 will be about 830 thousand.

19. a.

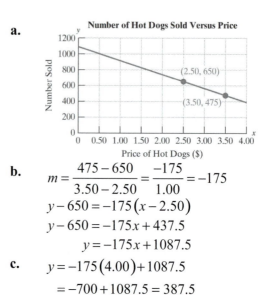

b.
$$m = \frac{475 - 650}{3.50 - 2.50} = \frac{-175}{1.00} = -175$$
$$y - 650 = -175(x - 2.50)$$
$$y - 650 = -175x + 437.5$$
$$y = -175x + 1087.5$$

c. $y = -175(4.00) + 1087.5$
$$= -700 + 1087.5 = 387.5$$

Approximately 388 hotdogs would be sold when the price of a hotdog is $4.00.

21. a.

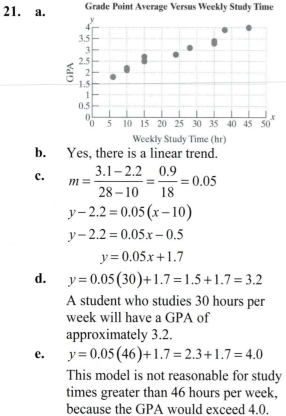

b. Yes, there is a linear trend.

c.
$$m = \frac{3.1 - 2.2}{28 - 10} = \frac{0.9}{18} = 0.05$$
$$y - 2.2 = 0.05(x - 10)$$
$$y - 2.2 = 0.05x - 0.5$$
$$y = 0.05x + 1.7$$

d. $y = 0.05(30) + 1.7 = 1.5 + 1.7 = 3.2$

A student who studies 30 hours per week will have a GPA of approximately 3.2.

e. $y = 0.05(46) + 1.7 = 2.3 + 1.7 = 4.0$

This model is not reasonable for study times greater than 46 hours per week, because the GPA would exceed 4.0.

23.
$$m = \frac{-5 - (-4)}{0 - 3} = \frac{-1}{-3} = \frac{1}{3}$$
$$m = \frac{-2 - (-5)}{9 - 0} = \frac{3}{9} = \frac{1}{3}$$
$$m = \frac{-2 - (-4)}{9 - 3} = \frac{2}{6} = \frac{1}{3}$$

Since the slopes are equal, the points are collinear.

25.
$$m = \frac{12 - 2}{-2 - 0} = \frac{10}{-2} = -5$$
$$m = \frac{6 - 12}{-1 - (-2)} = \frac{-6}{1} = -6$$

Since the slopes are not equal, the points are not collinear.

Section 2.5   Introduction to Relations

1. **a.** relation
    b. domain
    c. range

3. **a.**
$$2x - 3 = 4$$
$$2x = 7$$
$$x = \frac{7}{2} \quad \text{vertical line}$$

    b. The slope is undefined because this is a vertical line.

    c. The x-intercept is $\left(\frac{7}{2}, 0 \right)$.

    d. There is no y-intercept.

5. **a.**
$$3x - 2y = 4$$
$$-2y = -3x + 4$$
$$y = \frac{3}{2}x - 2 \quad \text{slanted line}$$

    b. $m = \dfrac{3}{2}$

    c.
$$3x - 2(0) = 4$$
$$3x = 4$$
$$x = \frac{4}{3}$$

The x-intercept is $\left(\frac{4}{3}, 0 \right)$.

    d. The y-intercept is $(0, -2)$.

7. **a.** $\{(\text{Northeast}, 54.1), (\text{Midwest}, 65.6), (\text{South}, 110.7), (\text{West}, 70.7)\}$

    b. Domain: $\{\text{Northeast, Midwest, South, West}\}$; Range: $\{54.1, 65.6, 110.7, 70.7\}$

9. **a.** $\{(\text{USSR}, 1961), (\text{USA}, 1962), (\text{Poland}, 1978), (\text{Vietnam}, 1980), (\text{Cuba}, 1980)\}$

    b. Domain: $\{\text{USSR, USA, Poland, Vietnam, Cuba}\}$; Range: $\{1961, 1962, 1978, 1980\}$

11. **a.** $\{(A, 1), (A, 2), (B, 2), (C, 3), (D, 5), (E, 4)\}$

    b. Domain: $\{A, B, C, D, E\}$; Range: $\{1, 2, 3, 4, 5\}$

13. **a.** $\{(-4, 4), (1, 1), (2, 1), (3, 1), (4, -2)\}$

    b. Domain: $\{-4, 1, 2, 3, 4\}$; Range: $\{-2, 1, 4\}$

15. Domain: $[0, 4]$; Range: $[-2, 2]$

17. Domain: $[-5, 3]$; Range: $[-2.1, 2.8]$

19. Domain: $(-\infty, 2]$

Range: $(-\infty, \infty)$

21. Domain: $(-\infty, \infty)$

Range: $(-\infty, \infty)$

23. Domain: $\{-3\}$

Range: $(-\infty, \infty)$

25. Domain: $(-\infty, 2)$

Range: $[-1.3, \infty)$

27. Domain: $\{-3, -1, 1, 3\}$

Range: $\{0, 1, 2, 3\}$

29. Domain: $[-4, 5)$

Range: $\{-2, 1, 3\}$

31. **a.** 2.85
b. 9.33
c. December
d. (November, 2.66)
e. (Sept., 7.63)
f. {Jan., Feb., Mar., Apr., May, June, July, Aug., Sept., Oct., Nov., Dec.}

33. **a.** $y = 0.146x + 31$

$y = 0.146(6) + 31 = 0.876 + 31$

$y = 31.876$ million or 31,876,000

b. $32.752 = 0.146x + 31$

$1.752 = 0.146x$

$x = 12$

The year 2012.

35. **a.** For example:
{(Julie, New York), (Peggy, Florida), (Stephen, Kansas), (Pat, New York)}
b. Domain: {Julie, Peggy, Stephen, Pat}
Range: {New York, Florida, Kansas}

37. $y = 2x - 1$

39. $y = x^2$

Section 2.6   Introduction to Functions

1. **a.** function
b. vertical
c. $2x + 1$
d. domain; range
e. denominator; negative
f. -2
g. 3
h. (1, 6)

3. **a.** {(Kevin, Kayla), (Kevin, Kira), (Kathleen, Katie), (Kathleen, Kira)}
b. Domain: {Kevin, Kathleen}
c. Range: {Kayla, Katie, Kira}

5. Function

7. Not a function

9. Function

11. Not a function

13. Function

15. Not a function

17. When x is 2, the function value y is 5.

19. $(0, -2)$

21. $g(2) = -(2)^2 - 4(2) + 1 = -4 - 8 + 1 = -11$

23. $g(0) = -(0)^2 - 4(0) + 1 = 0 - 0 + 1 = 1$

25. $k(0)=|0-2|=|-2|=2$

27. $f(t)=6(t)-2=6t-2$

29. $h(u)=7$

31. $g(-3)=-(-3)^2-4(-3)+1=-9+12+1=4$

33. $k(-2)=|-2-2|=|-4|=4$

35. $f(x+1)=6(x+1)-2=6x+6-2=6x+4$

37. $g(2x)=-(2x)^2-4(2x)+1$
$$=-\left(4x^2\right)-8x+1$$
$$=-4x^2-8x+1$$

39. $g(-\pi)=-(-\pi)^2-4(-\pi)+1=-\pi^2+4\pi+1$

41. $h(a+b)=7$

43. $f(-a)=6(-a)-2=-6a-2$

45. $k(-c)=|-c-2|$

47. $f\left(\dfrac{1}{2}\right)=6\left(\dfrac{1}{2}\right)-2=3-2=1$

49. $h\left(\dfrac{1}{7}\right)=7$

51. $f(-2.8)=6(-2.8)-2=-16.8-2=-18.8$

53. $p(2)=-7$

55. $p(3)=2\pi$

57. $q(2)=-5$

59. $q(6)=4$

61.
  a. $f(0)=3$
  b. $f(3)=1$
  c. $f(-2)=1$
  d. $x=-3$
  e. $x=0,\ x=2$
  f. Domain: $(-\infty,\ 3]$
  g. Range: $(-\infty,\ 5]$

63.
  a. $H(-3)=3$
  b. $H(4)=$ not defined (4 not in domain)
  c. $H(3)=4$
  d. $x=-3$ and $x=2$
  e. all x in the interval $[-2,\ 1]$
  f. Domain: $[-4,\ 4)$
  g. Range: $[2,\ 5)$

65.
  a. $p(2)=-4$
  b. $p(-1)=0$
  c. $p(1)=-3$
  d. $x=-1$
  e. There are no such values of x.
  f. $(-\infty,\ \infty)$
  g. $(-\infty,\ -3]\cup(-2,\ \infty)$

67. Domain: $\left\{-3,\ -7,\ -\tfrac{3}{2},\ 1.2\right\}$

69. Range: $\{6,\ 0\}$

71. -3 and 1.2

73. 6 and 1

75. $f(-7) = -3$

77. The domain is the set of all real numbers for which the denominator is not zero. Set the denominator equal to zero, and solve the resulting equation. The solution(s) to the equation must be excluded from the domain. In this case setting $x - 2 = 0$ indicates that $x = 2$ must be excluded from the domain, The domain is $(-\infty, 2) \cup (2, \infty)$.

79.
$$k(x) = \frac{x - 3}{x + 6}$$
$$x + 6 = 0$$
$$x = -6$$
Domain: $(-\infty, -6) \cup (-6, \infty)$

81.
$$f(t) = \frac{5}{t}$$
$$t = 0$$
Domain: $(-\infty, 0) \cup (0, \infty)$

83.
$$h(p) = \frac{p - 4}{p^2 + 1}$$
$p^2 + 1$ will never equal zero.
Domain: $(-\infty, \infty)$

85.
$$h(t) = \sqrt{t + 7}$$
$$t + 7 \geq 0$$
$$t \geq -7$$
Domain: $[-7, \infty)$

87.
$$f(a) = \sqrt{a - 3}$$
$$a - 3 \geq 0$$
$$a \geq 3$$
Domain: $[3, \infty)$

89.
$$m(x) = \sqrt{1 - 2x}$$
$$1 - 2x \geq 0$$
$$-2x \geq -1$$
$$x \leq \frac{1}{2}$$
Domain: $\left(-\infty, \frac{1}{2}\right]$

91. $p(t) = 2t^2 + t - 1$

There are no restrictions on the domain.

Domain: $(-\infty, \infty)$

93. $f(x) = x + 6$

There are no restrictions on the domain.

Domain: $(-\infty, \infty)$

95.
$$h(t) = -16t^2 + 80$$
a. $h(1) = -16(1)^2 + 80 = -16 + 80 = 64$

$h(1.5) = -16(1.5)^2 + 80$
$$= -16(2.25) + 80 = -36 + 80 = 44$$
b. After 1 sec, the height of the ball is 64 ft. After 1.5 sec, the height of the ball is 44 ft.

97.
$$d(t) = 11.5t$$
a. $d(1) = 11.5(1) = 11.5$

$d(1.5) = 11.5(1.5) = 17.25$

b. After 1 hr, the distance is 11.5 mi. After 1.5 hr, the distance is 17.25 mi.

99. $f(x) = 2x + 3$

101. $f(x) = |x| - 10$

103.

$$q(x) = \frac{2}{\sqrt{x+2}}$$

$$x + 2 > 0$$

$$x > -2$$

Domain: $(-2, \infty)$

Section 2.7   Graphs of Functions

1.　**a.**　linear

　　b.　constant

　　c.　quadratic

　　d.　parabola

　　e.　0

　　f.　y

3.　**a.**　Yes, the relation is a function.

　　b.　Domain: $\{7, 2, -5\}$

　　c.　Range: $\{3\}$

5.

$$f(x) = \sqrt{x+4}$$

a.　$f(0) = \sqrt{0+4} = \sqrt{4} = 2$

　　$f(-3) = \sqrt{-3+4} = \sqrt{1} = 1$

　　$f(-4) = \sqrt{-4+4} = \sqrt{0} = 0$

　　$f(-5) = \sqrt{-5+4} = \sqrt{-1}$ cannot be
　　　　　evaluated because $x = -5$ is
　　　　　not in the domain of f.

b.　Domain: $[-4, \infty)$

7.　Horizontal

9.　$f(x) = 2$

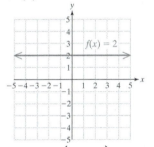

Domain: $(-\infty, \infty)$;　Range: $\{2\}$

11.

$$f(x) = \frac{1}{x}$$

x	$f(x)$
-2	$-\frac{1}{2}$
-1	-1
$-\frac{1}{2}$	-2
$-\frac{1}{4}$	-4
$\frac{1}{4}$	4
$\frac{1}{2}$	2
1	1
2	$\frac{1}{2}$

13. $h(x) = x^3$

x	$h(x)$
-2	-8
-1	-1
0	0
1	1
2	8

15. $q(x) = x^2$

x	$q(x)$
-2	4
-1	1
0	0
1	1
2	4

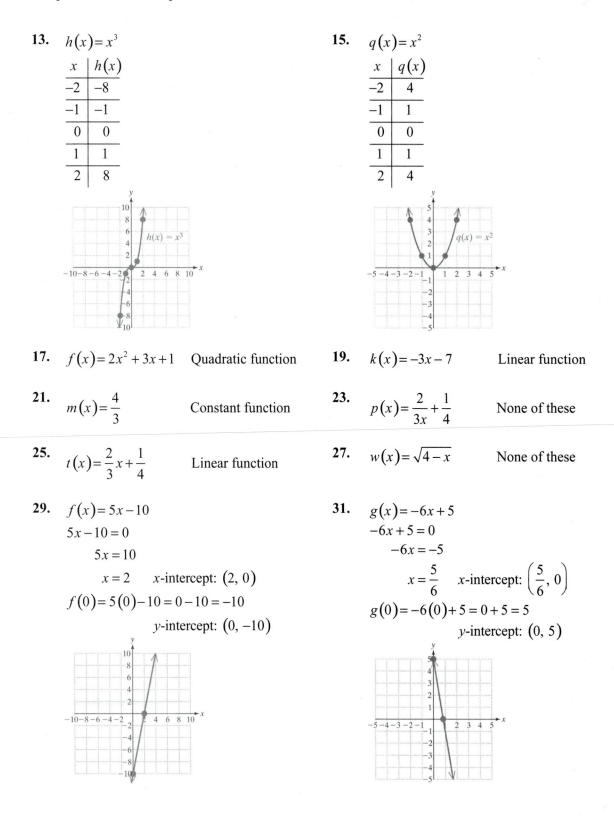

17. $f(x) = 2x^2 + 3x + 1$ Quadratic function

19. $k(x) = -3x - 7$ Linear function

21. $m(x) = \dfrac{4}{3}$ Constant function

23. $p(x) = \dfrac{2}{3x} + \dfrac{1}{4}$ None of these

25. $t(x) = \dfrac{2}{3}x + \dfrac{1}{4}$ Linear function

27. $w(x) = \sqrt{4 - x}$ None of these

29. $f(x) = 5x - 10$

$5x - 10 = 0$

$5x = 10$

$x = 2$ x-intercept: $(2, 0)$

$f(0) = 5(0) - 10 = 0 - 10 = -10$

y-intercept: $(0, -10)$

31. $g(x) = -6x + 5$

$-6x + 5 = 0$

$-6x = -5$

$x = \dfrac{5}{6}$ x-intercept: $\left(\dfrac{5}{6}, 0\right)$

$g(0) = -6(0) + 5 = 0 + 5 = 5$

y-intercept: $(0, 5)$

33. $f(x) = 18$

$18 \neq 0$ x-intercept: none

$f(0) = 18$ y-intercept: $(0, 18)$

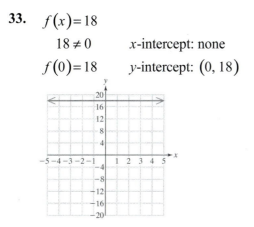

35.

$g(x) = \dfrac{2}{3}x + 2$

$\dfrac{2}{3}x + 2 = 0$

$\dfrac{2}{3}x = -2$

$2x = -6$

$x = -3$ x-intercept: $(-3, 0)$

$g(0) = \dfrac{2}{3}(0) + 2 = 0 + 2 = 2$

y-intercept: $(0, 2)$

37.
   a. $f(x) = 0$ when $x = -1$
   b. $f(0) = 1$

39.
   a. $f(x) = 0$ when $x = -2$ and $x = 2$
   b. $f(0) = -2$

41.
   a. $f(x) = 0$ There are none.
   b. $f(0) = 2$

43. $q(x) = 2x^2$
   a. $(-\infty, \infty)$
   b. $q(0) = 2(0)^2 = 2(0) = 0$
       y-intercept is $(0, 0)$.
   c.   Graph: vi

45. $h(x) = x^3 + 1$
   a. $(-\infty, \infty)$
   b. $h(0) = (0)^3 + 1 = 0 + 1 = 1$
       y-intercept is $(0, 1)$.
   c.   Graph: viii

47. $r(x) = \sqrt{x + 1}$
   a. $[-1, \infty)$
   b. $r(0) = \sqrt{0 + 1} = \sqrt{1} = 1$
       y-intercept is $(0, 1)$.
   c.   Graph: vii

49.

$$f(x) = \frac{1}{x-3}$$

a. $(-\infty, 3) \cup (3, \infty)$

b. $f(0) = \frac{1}{0-3} = -\frac{1}{3}$

y-intercept is $\left(0, -\frac{1}{3}\right)$.

c. Graph: ii

51.

$$k(x) = |x+2|$$

a. $(-\infty, \infty)$

b. $k(0) = |0+2| = |2| = 2$

y-intercept is $(0, 2)$.

c. Graph: iv

53. a. Linear

b. $G(90) = \frac{3}{4}(80) + \frac{1}{4}(90) = 60 + 22.5$

$= 82.5$

This means that if the student gets a 90% on her final exam, then her overall course average is 82.5%.

c. $G(50) = \frac{3}{4}(80) + \frac{1}{4}(50) = 60 + 12.5$

$= 72.5$

This means that if the student gets a 50% on her final exam, then her overall course average is 72.5%.

55. $f(x) = x^2 + 2$

57. $f(x) = 3$

59.

$$f(x) = \frac{1}{2}x + (-2)$$

$$= \frac{1}{2}x - 2$$

Problem Recognition Exercises

1. a, c, d, f, g

3. $c(-1) = 3(-1)^2 - 2(-1) - 1$
$= 3(1) - 2(-1) - 1 = 3 + 2 - 1 = 4$

5. $\left\{0, 1, \frac{1}{2}, -3, 2\right\}$

7. $[-2, 4]$

9. $(0, 3)$

11. $(0, 1)$ and $(0, -1)$

13. c

15. $5x - 9 = 6$
$5x = 15$
$x = 3$

Chapter 2 Group Activity: Deciphering a Coded Message

1. a. Message:
M A T H _ I S _ T H E _ K E Y _ T O _ T H E _ S C I E N C E S
Original:
13 1 20 8 27 9 19 27 20 8 5 27 11 5 25 27 20 15 27 20 8 5 27 19 3 9 5 14 3 5 19
Coded:
54,6,82,34,110,38,78,110,82,34,22,110,46,22,102,110,82,62,110,82,34,22,110,78,14,38,22,58,14,22,78

b. Message: **M A T H _ I S _ N O T _ A _ S P E C T A T O R _ S P O R T**
Original: 13 1 20 8 27 9 19 27 14 15 20 27 1 27 19 16 5 3 20 1 20 15 18 27 19 16 15 18 20
Coded: 38,2,59,23,80,26,56,80,41,44,59,80, 2,80,56,47,14, 8,59,2,59,44,53,80,56,47,44,53,59

Chapter 2 Review Exercises

Section 2.1

1.

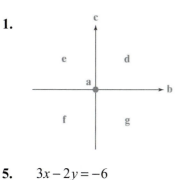

3. $5x = -15$

$5(-3) = -15 = -15$

$(-3, 4)$ is a solution of the equation.

5. $3x - 2y = -6$

x	y
0	3
−2	0
1	$\frac{9}{2}$

7. $6 - x = 2$

x	y
4	0
4	1
4	−2

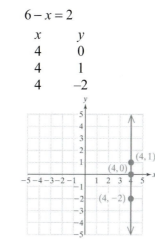

9. $5x - 2y = 0$

$5x - 2(0) = 0 \qquad 5(0) - 2y = 0$

$5x = 0 \qquad\qquad -2y = 0$

$x = 0 \qquad\qquad y = 0$

The x-intercept is $(0, 0)$.
The y-intercept is $(0, 0)$.
A second point is $(2, 5)$.
A third point is $(-2, -5)$.

11. $-3x = 6$

$x = -2$

The x-intercept is $(-2, 0)$.
There is no y-intercept.
A second point is $(-2, 5)$.
A third point is $(-2, -1)$.

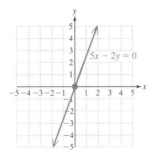

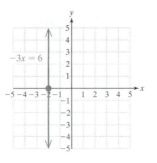

Section 2.2

13. For example: $y = 2x$

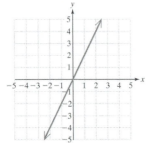

15.
$$m = \frac{0-6}{-1-2} = \frac{-6}{-3} = 2$$

17.
$$m = \frac{2-2}{3-8} = \frac{0}{-5} = 0$$

19.
$$m_1 = \frac{-2-(-6)}{3-4} = \frac{4}{-1} = -4$$
$$m_2 = \frac{0-(-1)}{7-3} = \frac{1}{4}$$
The lines are perpendicular.

21. The lines are neither parallel nor perpendicular.

23. **a.**
$$m = \frac{3080-2020}{2010-1990} = \frac{1060}{20} = 53$$

b. The enrollment increases at a rate of 53 students per year.

Section 2.3

25. a. $y = k$

b. $y - y_1 = m(x - x_1)$

c. $Ax + By = C$

d. $x = k$

e. $y = mx + b$

27.
$$m = -\frac{2}{3}, \quad x\text{-intercept: } (3, 0)$$
$$y - 0 = -\frac{2}{3}(x - 3)$$
$$y = -\frac{2}{3}x + 2 \quad \text{or} \quad \frac{2}{3}x + y = 2$$
$$2x + 3y = 6$$

29.
$$y = -\frac{1}{3}x + 2$$
$$m_\perp = 3, \quad \text{point: } (6, -2)$$
$$y - (-2) = 3(x - 6)$$
$$y + 2 = 3x - 18$$
$$y = 3x - 20 \quad \text{or} \quad 3x - y = 20$$

31. a. $y = -2$

b. $x = -3$

c. $x = -3$

d. $y = -2$

Section 2.4

33. a. $y = 0.75x + 50$

b.
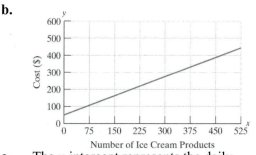

c. The y-intercept represents the daily fixed cost of $50 when no ice cream is sold

d.
$$y = 0.75(450) + 50$$
$$= 337.50 + 50 = 387.50$$

e. The cost is $387.50 when 450 ice cream products are sold.

f. The slope of the line is 0.75. The cost increases at a rate of $0.75 per ice cream product.

Section 2.5

35.
Domain: $\left\{ \dfrac{1}{3}, 6, \dfrac{1}{4}, 7 \right\}$

Range: $\left\{ 10, -\dfrac{1}{2}, 4, \dfrac{2}{5} \right\}$

37. Domain: $[-3, 9]$
Range: $[0, 60]$

Section 2.6

39. Answers will vary. For example:

41. a. Not a function

b. $[1, 3]$

c. $[-4, 4]$

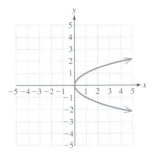

43. **a.** A function

    b. $\{1, 2, 3, 4\}$

    c. $\{3\}$

45 **a.** Not a function

    b. $\{4, 9\}$

    c. $\{2, -2, 3, -3\}$

47. $f(0) = 6(0)^2 - 4 = 6(0) - 4 = 0 - 4 = -4$

49. $f(-1) = 6(-1)^2 - 4 = 6(1) - 4 = 6 - 4 = 2$

51. $f(b) = 6(b)^2 - 4 = 6b^2 - 4$

53. $f(W) = 6(W)^2 - 4 = 6W^2 - 4$

55. $g(x) = 7x^3 + 1$

    Domain: $(-\infty, \infty)$

57. $k(x) = \sqrt{x - 8}$

    $x - 8 \geq 0$

    $x \geq 8$

    Domain: $[8, \infty)$

59. $p(x) = 48 + 5x$

    a. $p(10) = 48 + 5(10) = 48 + 50 = \$98$

    b. $p(15) = 48 + 5(15) = 48 + 75 = \$123$

    c. $p(20) = 48 + 5(20) = 48 + 100 = \$148$

Section 2.7

61. $f(x) = x^2$

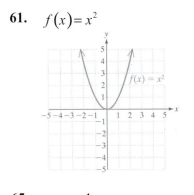

63. $w(x) = |x|$

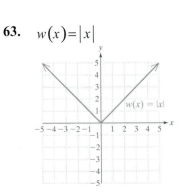

65. $r(x) = \dfrac{1}{x}$

67. $k(x) = 2x + 1$

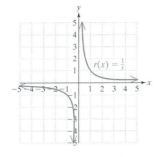

69. $r(x) = 2\sqrt{x-4}$

    a. $r(2) = 2\sqrt{2-4} = 2\sqrt{-2}$ not a real number

        $r(4) = 2\sqrt{4-4} = 2\sqrt{0} = 2 \cdot 0 = 0$

        $r(5) = 2\sqrt{5-4} = 2\sqrt{1} = 2 \cdot 1 = 2$

        $r(8) = 2\sqrt{8-4} = 2\sqrt{4} = 2 \cdot 2 = 4$

    b. $x - 4 \geq 0$

          $x \geq 4$

      Domain: $[4, \infty)$

71.   **a.** $k(x) = -|x+3|$

        $k(-5) = -|-5+3| = -|-2| = -2$

        $k(-4) = -|-4+3| = -|-1| = -1$

        $k(-3) = -|-3+3| = -|0| = 0$

        $k(2) = -|2+3| = -|5| = -5$

    b.   Domain: $(-\infty, \infty)$

73. $q(x) = -2x + 9$

    $-2x + 9 = 0$

       $-2x = -9$

         $x = \dfrac{9}{2}$ x-intercept: $\left(\dfrac{9}{2}, 0 \right)$

    $f(0) = -2(0) + 9 = 0 + 9 = 9$

           y-intercept: $(0, 9)$

75. $g(-2) = -1$

77. $g(x) = 0$ when $x = 0$ and $x = 4$.

79.   Domain: $(-4, \infty)$

Chapter 2 Test

1. $x - \dfrac{2}{3}y = 6$

    $0 - \dfrac{2}{3}y = 6$

       $y = -\dfrac{3}{2}(6) = -9$

        $(0, -9)$

    $x - \dfrac{2}{3}(0) = 6$

       $x = 6$

        $(6, 0)$

    $x - \dfrac{2}{3}(-3) = 6$

       $x + 2 = 6$

        $x = 4$

         $(4, -3)$

3.
$$y = -\frac{1}{2}x - 3$$

$$-1 = -\frac{1}{2}(-4) - 3 = 2 - 3 - 1$$

$(-4, -1)$ is a solution of the equation.

5.
$$6x - 8y = 24$$

$$6x - 8(0) = 24 \qquad 6(0) - 8y = 24$$

$$6x = 24 \qquad\qquad -8y = 24$$

$$x = 4 \qquad\qquad y = -3$$

The x-intercept is $(4, 0)$.
The y-intercept is $(0, -3)$.

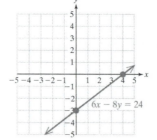

7.
$$3x = 5y$$

$$3x = 5(0) \qquad 3(0) = 5y$$

$$3x = 0 \qquad\quad 0 = 5y$$

$$x = 0 \qquad\quad 0 = y$$

The x-intercept is $(0, 0)$.
The y-intercept is $(0, 0)$.

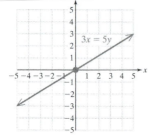

9. a.
$$m = \frac{-8 - (-3)}{-1 - 7} = \frac{-5}{-8} = \frac{5}{8}$$

b.
$$6x - 5y = 1$$

$$-5y = -6x + 1$$

$$y = \frac{-6}{-5}x + \frac{1}{-5}$$

$$y = \frac{6}{5}x - \frac{1}{5}$$

$$m = \frac{6}{5}$$

11. a. $m = -7$

b. $m = \frac{1}{7}$

13.
$$m_1 = \frac{-6 - (-4)}{1 - 4} = \frac{-2}{-3} = \frac{2}{3}$$

$$m_2 = \frac{3 - 0}{0 - (-2)} = \frac{3}{2}$$

The lines are neither parallel nor perpendicular

90

15. a. For example: $y = 3x + 2$
 b. For example: $x = 2$
 c. For example: $y = 3$ $m = 0$
 d. For example: $y = -2x$

17. $6x - 3y = 1$
$$-3y = -6x + 1$$
$$y = 2x - \frac{1}{3}$$
$$m_p = 2 \quad \text{point: } (4, -3)$$
$$y - (-3) = 2(x - 4)$$
$$y + 3 = 2x - 8$$
$$y = 2x - 11 \quad \text{or} \quad 2x - y = 11$$

19. $3x + y = 7$
$$y = -3x + 7$$
$$m_\perp = \frac{1}{3} \quad \text{point: } (-10, -3)$$
$$y - (-3) = \frac{1}{3}(x - (-10))$$
$$y + 3 = \frac{1}{3}x + \frac{10}{3}$$
$$y = \frac{1}{3}x + \frac{1}{3}$$

21. a. $(0, 66)$ For a woman born in 1940, the life expectancy was about 66 years.
 b. $m = \dfrac{75 - 66}{30 - 0} = \dfrac{9}{30} = \dfrac{3}{10}$
 Life expectancy increases by 3 years for every 10 years that elapse.
 c. $y = \dfrac{3}{10}x + 66$
 d. 1994 corresponds to $x = 54$.
 $$y = \frac{3}{10}(54) + 66 = 16.2 + 66 = 82.2$$
 Life expectancy is 82.2 years for a woman born in 1994. This is 3.2 years longer than 79 years reported.

23.

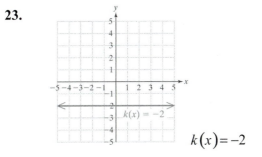

$$k(x) = -2$$

25.

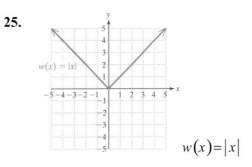

$$w(x) = |x|$$

27. a. A function.
 b. Domain: $(-\infty, \infty)$
 c. Range: $(-\infty, 0]$

29. $f(x) = \sqrt{x + 7}$
$$x + 7 \geq 0$$
$$x \geq -7$$
Domain: $[-7, \infty)$

31. $r(x) = x^2 - 2x + 1$
 a. $r(-2) = (-2)^2 - 2(-2) + 1 = 4 + 4 + 1 = 9$
 $r(0) = (0)^2 - 2(0) + 1 = 0 - 0 + 1 = 1$
 $r(3) = (3)^2 - 2(3) + 1 = 9 - 6 + 1 = 4$
 b. Domain: $(-\infty, \infty)$

33. $f(x) = -3x^2$ Quadratic function

35. $h(x) = -3$ Constant function

37. To find the x-intercept(s), solve for the real solutions of the equation $f(x) = 0$. To find the y-intercept, find $f(0)$.

39. $f(1) = 1$

41. Domain: $(-1, 7]$

43. False

45. $f(x) = 0$ when $x = 6$

Chapters 1 – 2   Cumulative Review Exercises

1.
$$\frac{5 - 2^3 \div 4 + 7}{-1 - 3(4 - 1)} = \frac{5 - 8 \div 4 + 7}{-1 - 3(3)}$$
$$= \frac{5 - 2 + 7}{-1 - 9} = \frac{10}{-10} = -1$$

3.
$$4\left[-3x - 5(y - 2x) + 3\right] - 7(6y + x)$$
$$= 4\left[-3x - 5y + 10x + 3\right] - 42y - 7x$$
$$= 4\left[7x - 5y + 3\right] - 42y - 7x$$
$$= 28x - 20y + 12 - 42y - 7x$$
$$= 21x - 62y + 12$$

5.
$$z - (3 + 2z) + 5 = -z - 5$$
$$z - 3 - 2z + 5 = -z - 5$$
$$-z + 2 = -z - 5$$
$$-z + z + 2 = -z + z - 5$$
$$2 = -5 \quad \{\,\}$$

7.
$$-4 \le \frac{x - 1}{2} < 3$$
$$-8 \le x - 1 < 6$$
$$-7 \le x < 7 \qquad [-7, 7)$$

9.
$$3x + 2 < 11 \text{ or } -4 < 2x$$
$$3x < 9 \text{ or } -2 < x$$
$$x < 3 \text{ or } \quad x > -2 \qquad (-\infty, \infty)$$

11.
$$\left|\frac{x - 3}{5}\right| \ge 2$$
$$\frac{x - 3}{5} \ge 2 \quad \text{or} \quad \frac{x - 3}{5} \le -2$$
$$x - 3 \ge 10 \text{ or } x - 3 \le -10$$
$$x \ge 13 \text{ or } x \le -7 \qquad (-\infty, -7] \cup [13, \infty)$$

13.
$$m = \frac{-3 - (-5)}{-6 - 4} = \frac{2}{-10} = -\frac{1}{5}$$

15. a.
$$3x - 5y = 10$$
$$3x - 5(0) = 10$$
$$3x = 10$$
$$x = \frac{10}{3}$$

$$3(0) - 5y = 10$$
$$-5y = 10$$
$$y = -2$$

The x-intercept is $\left(\frac{10}{3}, 0\right)$.

The y-intercept is $(0, -2)$.

b. $m = \dfrac{-2 - 0}{0 - \frac{10}{3}} = \dfrac{-2}{-\frac{10}{3}} = 2\left(\dfrac{3}{10}\right) = \dfrac{6}{10} = \dfrac{3}{5}$

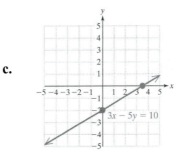

c.

17. $x = 7$

19. Domain: $\{3, 4\}$ Range: $\{-1, -5, -8\}$
This is not a function.

21.
$$y = \frac{1}{4}x - 2$$

$$m_\perp = -4 \qquad \text{point: } (1, -4)$$

$$y - (-4) = -4(x - 1)$$

$$y + 4 = -4x + 4$$

$$y = -4x$$

23. $x - 15 \neq 0$
$\qquad x \neq 15$
Domain: $(-\infty, 15) \cup (15, \infty)$

25. Let x = the yearly rainfall in Los Angeles
$\qquad 2x - 0.7$ = the yearly rainfall in Seattle

$$x + (2x - 0.7) = 50$$

$$3x - 0.7 = 50$$

$$3x = 50.7$$

$$x = 16.9$$

$$2x - 0.7 = 2(16.9) - 0.7 = 33.1$$

Los Angeles gets 16.9 in of rain per year and Seattle gets 33.1 in of rain per year.

Chapter 3   Systems of Linear Equations and Inequalities

Are You Prepared?

1. **O** $(1, 3)$

2. **T** $(-1, -2)$

3. **I** $(-3, -1)$

4. **S** Line 1

5. **L** Line 5

6. **U** Line 2

7. **N** no point of intersection

A $\underline{S\,O\,L\,U\,T\,I\,O\,N}$ to a system of linear equations is a point of intersection between two lines.

Section 3.1   Solving Systems of Linear Equations by the Graphing Method

1. a. system
    b. solution
    c. intersect
    d. consistent
    e. the empty set, $\{\ \}$
    f. dependent
    g. independent

3. $y = 8x - 5$
    $y = 4x + 3$
    Substitute $(-1,13)$:
    $13 = 8(-1) - 5 = -8 - 5 = -13$
      Not a solution.
    Substitute $(-1,1)$:
    $1 = 8(-1) - 5 = -8 - 5 = -13$
      Not a solution.
    Substitute $(2,11)$:
    $11 = 8(2) - 5 = 16 - 5 = 11$
    $11 = 4(2) + 3 = 8 + 3 = 11$
      $(2,11)$ is a solution.

5. $2x - 7y = -30$
    $y = 3x + 7$
    Substitute $(0,-30)$:
    $2(0) - 7(-30) = 0 + 210 = 210 \neq -30$
      Not a solution.
    Substitute $\left(\dfrac{3}{2}, 5 \right)$:
    $2\left(\dfrac{3}{2} \right) - 7(5) = 3 - 35 = -32 \neq -30$
      Not a solution.
    Substitute $(-1,4)$:
    $2(-1) - 7(4) = -2 - 28 = -30 = -30$
    $4 = 3(-1) + 7 = -3 + 7 = 4$
      $(-1,4)$ is a solution.

7. $x - y = 6$

$4x + 3y = -4$

Substitute $(4, -2)$:

$4 - (-2) = 4 + 2 = 6 = 6$

$4(4) + 3(-2) = 16 - 6 = 10 \neq -4$

Not a solution.

Substitute $(6, 0)$:

$6 - 0 = 6 = 6$

$4(6) + 3(0) = 24 + 0 = 24 \neq -4$

Not a solution.

Substitute $(2, 4)$:

$2 - 4 = -2 \neq 6$

Not a solution.

9.   **a.**   Consistent
   b.   Independent
   c.   One solution

11.   **a.**   Inconsistent
   b.   Independent
   c.   Zero solutions

13.   **a.**   Consistent
   b.   Dependent
   c.   Infinitely many solutions

15. $2x + y = -3$ $-x + y = 3$

$y = -2x - 3$ $y = x + 3$

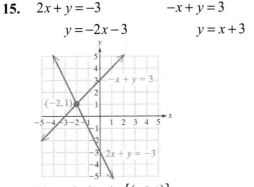

The solution is $\{(-2, 1)\}$.

17. $f(x) = -2x + 3$ $g(x) = 5x - 4$

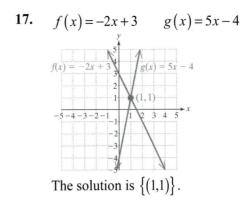

The solution is $\{(1, 1)\}$.

19. $k(x) = \dfrac{1}{3}x - 5$ $f(x) = -\dfrac{2}{3}x - 2$

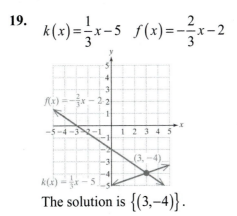

The solution is $\{(3, -4)\}$.

21. $x = 4$ $y = 2x - 3$

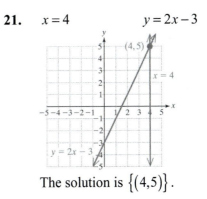

The solution is $\{(4, 5)\}$.

95

23. $y = -2x + 3$ $-2x = y + 1$

                                $y = -2x - 1$

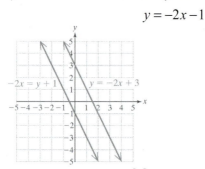

There is no solution; $\{\ \}$. Inconsistent system.

25. $y = \dfrac{2}{3}x - 1$ $2x = 3y + 3$

                                $3y = 2x - 3$

                                $y = \dfrac{2}{3}x - 1$

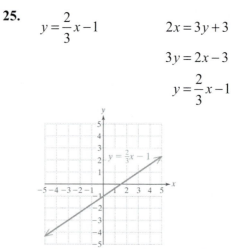

Infinitely many solutions of the form $\left\{ (x, y) \middle| y = \dfrac{2}{3}x - 1 \right\}$. Dependent equations.

27. $2x = 4$ $\dfrac{1}{2}y = -1$

        $x = 2$ $y = -2$

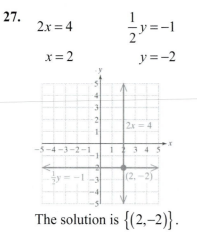

The solution is $\{(2, -2)\}$.

29. $-x + 3y = 6$ $6y = 2x + 12$

        $3y = x + 6$ $y = \dfrac{1}{3}x + 2$

        $y = \dfrac{1}{3}x + 2$

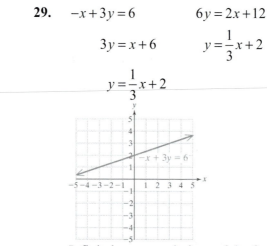

Infinitely many solutions of the form $\{(x, y) | -x + 3y = 6\}$. Dependent equations.

31. $2x - y = 4$ $4x + 2 = 2y$

        $-y = -2x + 4$ $2y = 4x + 2$

        $y = 2x - 4$ $y = 2x + 1$

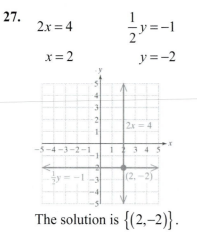

There is no solution; $\{\ \}$. Inconsistent system.

33.   False

35.   True

37.   For example: The system $\begin{array}{c} x+y=9 \\ 2x+y=13 \end{array}$

has solution $\{(4,5)\}$

39.
$$Cx+2y=11$$
$$C(1)+2(3)=11$$
$$C+6=11$$
$$C=5$$

$$-3x+Dy=9$$
$$3(1)+D(3)=9$$
$$-3+3D=9$$
$$3D=12$$
$$D=4$$

41. $y=5.62x+15.46$
$y=-1.96x-11.07$

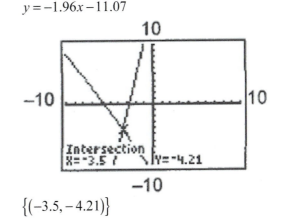

$\{(-3.5,-4.21)\}$

43. $2.4x-4.8y=-9.36$
$$-4.8y=-2.4x-9.36$$
$$y=0.5x+1.95$$
$$-1.8x+5.4y=12.456$$
$$5.4y=1.8x+12.456$$
$$y=\frac{1}{3}x+\frac{173}{75}$$

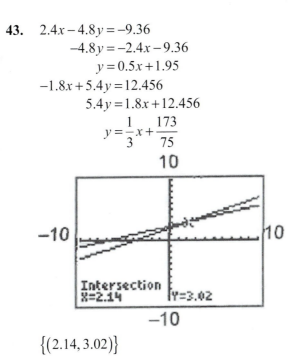

$\{(2.14,3.02)\}$

Section 3.2   Solving Systems of Linear Equations by the Substitution Method

1. $y = 8x - 1$ $2x - 16y = 3$

$-16y = -2x + 3$

$y = \dfrac{1}{8}x - \dfrac{3}{16}$

One solution

3. $2x - 4y = 0$ $x - 2y = 9$

$-4y = -2x$ $-2y = -x + 9$

$y = \dfrac{1}{2}x$ $y = \dfrac{1}{2}x - \dfrac{9}{2}$

No solution

5. $-x + 2y = 10$ $2x - y = 11$

$-(-4) + 2(3) = 10$ $2(-4) - 3 = 11$

$4 + 6 = 10$ $-8 - 3 = 11$

$10 = 10$ $-11 = 11$

$(-4, 3)$ is not a solution.

7. $y = 2x + 3$ $6x + 3y = 9$

$3y = -6x + 9$

$y = -2x + 3$

The solution is $\{(0,3)\}$.

9. $y = -3x - 1$

$2x - 3y = -8$

$2x - 3(-3x - 1) = -8$

$2x + 9x + 3 = -8$

$11x = -11$

$x = -1$

$y = -3x - 1 = -3(-1) - 1 = 3 - 1 = 2$

The solution is $\{(-1, 2)\}$.

11. $-3x + 8y = -1$

$4x - 11 = y$

$-3x + 8(4x - 11) = -1$

$-3x + 32x - 88 = -1$

$29x = 87$

$x = 3$

$y = 4x - 11 = 4(3) - 11 = 12 - 11 = 1$

The solution is $\{(3, 1)\}$.

13. $3x + 12y = 36$

$x - 5y = 12 \rightarrow x = 5y + 12$

$3(5y + 12) + 12y = 36$

$15y + 36 + 12y = 36$

$27y = 0$

$y = 0$

$x = 5y + 12 = 5(0) + 12 = 0 + 12 = 12$

The solution is $\{(12, 0)\}$.

15. $x - y = 8 \rightarrow x = y + 8$

$3x + 2y = 9$

$3(y + 8) + 2y = 9$

$3y + 24 + 2y = 9$

$5y = -15$

$y = -3$

$x = y + 8 = -3 + 8 = 5$

The solution is $\{(5, -3)\}$.

17. $2x - y = -1$

$y = -2x$

$2x - (-2x) = -1$

$4x = -1$

$x = -\dfrac{1}{4}$

$y = -2x = -2\left(-\dfrac{1}{4}\right) = \dfrac{1}{2}$

The solution is $\left\{\left(-\dfrac{1}{4}, \dfrac{1}{2}\right)\right\}$.

19. $2x + 3 = 7 \rightarrow 2x = 4 \rightarrow x = 2$

$3x - 4y = 6$

$3(2) - 4y = 6$

$6 - 4y = 6$

$4y = 0$

$y = 0$

The solution is $\{(2,0)\}$.

21. $4x - 5y = 14$

$3y = x - 7 \rightarrow y = \dfrac{x-7}{3}$

$4x - 5\left(\dfrac{x-7}{3}\right) = 14$

$3(4x) - 5(x-7) = 3(14)$

$12x - 5x + 35 = 42$

$7x + 35 = 42$

$7x = 7$

$x = 1$

$y = \dfrac{1-7}{3} = \dfrac{-6}{3} = -2$

The solution is $\{(1,-2)\}$.

23. $2x - 6y = -2$

$x = 3y - 1$

$2(3y - 1) - 6y = -2$

$6y - 2 - 6y = -2$

$-2 = -2$

Infinitely many solutions of the form

$\{(x,y) \mid x = 3y - 1\}$; dependent equations.

25. $y = \dfrac{1}{7}x + 3$

$x - 7y = -4$

$x - 7\left(\dfrac{1}{7}x + 3\right) = -4$

$x - x - 21 = -4$

$-21 \neq -4$

There is no solution; $\{\ \}$. This is an inconsistent system.

27. $5x - y = 10 \rightarrow y = 5x - 10$

$2y = 10x - 5$

$2(5x - 10) = 10x - 5$

$10x - 20 = 10x - 5$

$-20 \neq -5$

There is no solution; $\{\ \}$. This is an inconsistent system.

29. $3x - y = 7 \rightarrow y = 3x - 7$

$-14 + 6x = 2y$

$-14 + 6x = 2(3x - 7)$

$-14 + 6x = 6x - 14$

$-14 = -14$

Infinitely many solutions of the form

$\{(x,y) \mid 3x - y = 7\}$; dependent equations.

31. If you get an identity, such as $0 = 0$ or $5 = 5$ when solving a system of equations, then the equations are dependent.

33. $x = 1.3y + 1.5$

$y = 1.2x - 4.6$

$x = 1.3(1.2x - 4.6) + 1.5$

$x = 1.56x - 5.98 + 1.5$

$-0.56x = -4.48$

$x = 8$

$y = 1.2x - 4.6 = 1.2(8) - 4.6 = 9.6 - 4.6 = 5$

The solution is $\{(8,5)\}$.

35. $y = \dfrac{2}{3}x - \dfrac{1}{3}$

$x = \dfrac{1}{4}y + \dfrac{17}{4}$

$x = \dfrac{1}{4}\left(\dfrac{2}{3}x - \dfrac{1}{3}\right) + \dfrac{17}{4}$

$x = \dfrac{1}{6}x - \dfrac{1}{12} + \dfrac{17}{4}$

$\dfrac{5}{6}x = -\dfrac{1}{12} + \dfrac{51}{12}$

$\dfrac{5}{6}x = \dfrac{50}{12}$

$x = \dfrac{50}{12} \cdot \dfrac{6}{5} = \dfrac{300}{60} = 5$

$y = \dfrac{2}{3}x - \dfrac{1}{3} = \dfrac{2}{3}(5) - \dfrac{1}{3} = \dfrac{10}{3} - \dfrac{1}{3} = \dfrac{9}{3} = 3$

The solution is $\{(5,3)\}$.

37. $-2x + y = 4 \rightarrow y = 2x + 4$

$-\dfrac{1}{4}x + \dfrac{1}{8}y = \dfrac{1}{4}$

$-\dfrac{1}{4}x + \dfrac{1}{8}(2x + 4) = \dfrac{1}{4}$

$-\dfrac{1}{4}x + \dfrac{1}{4}x + \dfrac{1}{2} = \dfrac{1}{4}$

$\dfrac{1}{2} \neq \dfrac{1}{4}$

There is no solution; $\{\ \}$. This is an inconsistent system.

39. $3x + 2y = 6$

$y = x + 3$

$3x + 2(x + 3) = 6$

$3x + 2x + 6 = 6$

$5x + 6 = 6$

$5x = 0$

$x = 0$

$y = 0 + 3 = 3$

The solution is $\{(0,3)\}$.

41. $-300x - 125y = 1350$

$y + 2 = 8 \rightarrow y = 6$

$-300x - 125(6) = 1350$

$-300x - 750 = 1350$

$-300x = 2100$

$x = -7$

The solution is $\{(-7,6)\}$.

43. $2x - y = 6 \rightarrow y = 2x - 6$

$\dfrac{1}{6}x - \dfrac{1}{12}y = \dfrac{1}{2}$

$\dfrac{1}{6}x - \dfrac{1}{12}(2x - 6) = \dfrac{1}{2}$

$\dfrac{1}{6}x - \dfrac{1}{6}x + \dfrac{1}{2} = \dfrac{1}{2}$

$\dfrac{1}{2} = \dfrac{1}{2}$

Infinitely many solutions of the form $\{(x,y)\,|\,2x - y = 6\}$; dependent equations.

45. $y = -2.7x - 5.1$

$y = 3.1x - 63.1$

$3.1x - 63.1 = -2.7x - 5.1$

$5.8x = 58$

$x = 10$

$y = 3.1x - 63.1 = 3.1(10) - 63.1$

$= 31 - 63.1 = -32.1$

The solution is $\{(10, -32.1)\}$.

47. $4x + 4y = 5$

$x - 4y = -\dfrac{5}{2} \rightarrow x = 4y - \dfrac{5}{2}$

$4\left(4y - \dfrac{5}{2}\right) + 4y = 5$

$16y - 10 + 4y = 5$

$20y = 15$

$y = \dfrac{15}{20} = \dfrac{3}{4}$

$x = 4y - \dfrac{5}{2} = 4\left(\dfrac{3}{4}\right) - \dfrac{5}{2} = 3 - \dfrac{5}{2} = \dfrac{1}{2}$

The solution is $\left\{\left(\dfrac{1}{2}, \dfrac{3}{4}\right)\right\}$.

49. $2(x + 2y) = 12 \rightarrow x + 2y = 6 \rightarrow x = -2y + 6$

$-6x = 5y - 8$

$-6(-2y + 6) = 5y - 8$

$12y - 36 = 5y - 8$

$7y = 28$

$y = \dfrac{28}{7} = 4$

$x = -2(4) + 6 = -8 + 6 = -2$

The solution is $\{(-2, 4)\}$.

51. $5(3y - 2) = x + 4 \rightarrow 15y - 10 = x + 4$

$\rightarrow 15y - 14 = x$

$4y = 7x - 3$

$4y = 7(15y - 14) - 3$

$4y = 105y - 98 - 3$

$4y = 105y - 101$

$-101y = -101$

$y = 1$

$x = 15(1) - 14 = 15 - 14 = 1$

The solution is $\{(1, 1)\}$.

53. $2x - 5 = 7 \rightarrow 2x = 12 \rightarrow x = 6$

$4 = 3y + 1 \rightarrow 3 = 3y \rightarrow y = 1$

The solution is $\{(6, 1)\}$.

55. $0.01y = 0.02x - 0.11 \rightarrow y = 2x - 11$

$0.3x - 0.5y = 2 \rightarrow 3x - 5y = 20$

$3x - 5(2x - 11) = 20$

$3x - 10x + 55 = 20$

$-7x + 55 = 20$

$-7x = -35$

$x = 5$

$y = 2(5) - 11 = 10 - 11 = -1$

The solution is $\{(5, -1)\}$.

57. **a.** points $(-4, 1)$ and $(5, 5)$

$m = \dfrac{5 - 1}{5 - (-4)} = \dfrac{4}{5 + 4} = \dfrac{4}{9}$

59. **a.** $y = \dfrac{4}{9}x + \dfrac{25}{9}$

$y = \dfrac{4}{9}\left(\dfrac{1}{2}\right) + \dfrac{25}{9} = \dfrac{2}{9} + \dfrac{25}{9} = \dfrac{27}{9} = 3$

$$y - y_1 = m(x - x_1) \qquad (x_1, y_1) = (-4, 1)$$
$$y - 1 = \frac{4}{9}\left[x - (-4)\right]$$

b.
$$y - 1 = \frac{4}{9}x + \frac{16}{9}$$
$$y = \frac{4}{9}x + \frac{25}{9}$$

points $(-3, 5)$ and $(4, 1)$

$$m = \frac{1 - 5}{4 - (-3)} = \frac{-4}{7} = -\frac{4}{7}$$

$$y - y_1 = m(x - x_1) \qquad (x_1, y_1) = (-3, 5)$$
$$y - 5 = -\frac{4}{7}\left[x - (-3)\right]$$
$$y - 5 = -\frac{4}{7}x - \frac{12}{7}$$
$$y = -\frac{4}{7}x + \frac{23}{7}$$

c.
$$\frac{4}{9}x + \frac{25}{9} = -\frac{4}{7}x + \frac{23}{7}$$
$$63\left[\frac{4}{9}x + \frac{25}{9}\right] = 63\left[-\frac{4}{7}x + \frac{23}{7}\right]$$
$$28x + 175 = -36x + 207$$
$$64x = 32$$
$$x = \frac{1}{2}$$

$$y = \frac{4}{9}x + \frac{25}{9}$$
$$y = \frac{4}{9}\left(\frac{1}{2}\right) + \frac{25}{9} = \frac{2}{9} + \frac{25}{9} = \frac{27}{9} = 3$$

The centroid is $\left(\frac{1}{2}, 3\right)$.

b.
$$800x + 250 = 750x + 500$$
$$50x = 250$$
$$x = 5$$
The amount spent is the same for 5 months.

Section 3.3   Solving Systems of Linear Equations by the Addition Method

1.   a. -3

    b. 5

3. One solution – different slopes

5. Add the two equations and solve for y:
$$3x - y = -1$$
$$\underline{-3x + 4y = -14}$$
$$3y = -15$$
$$y = -5$$
Substitute into the first equation and solve

for x:

$$3x - (-5) = -1$$
$$3x + 5 = -1$$
$$3x = -6$$
$$x = -2$$

The solution is $\{(-2, -5)\}$.

7.
$$2x + 3y = 3$$
$$-10x + 2y = -32$$

Multiply the first equation by 5, add to the second equation and solve for y:

$$2x + 3y = 3 \xrightarrow{\times 5} 10x + 15y = 15$$
$$-10x + 2y = -32 \longrightarrow -10x + 2y = -32$$
$$\overline{\qquad\qquad 17y = -17}$$
$$y = -1$$

Substitute into the first equation and solve for x:

$$2x + 3(-1) = 3$$
$$2x - 3 = 3$$
$$2x = 6$$
$$x = 3$$

The solution is $\{(3, -1)\}$.

9.
$$3x + 7y = -20$$
$$-5x + 3y = -84$$

Multiply the first equation by 5 and the second equation by 3, add the results and solve for y:

$$3x + 7y = -20 \xrightarrow{\times 5} 15x + 35y = -100$$
$$-5x + 3y = -84 \xrightarrow{\times 3} -15x + 9y = -252$$
$$\overline{\qquad\qquad 44y = -352}$$
$$y = -8$$

Substitute into the first equation and solve for x:

$$3x + 7(-8) = -20$$
$$3x - 56 = -20$$
$$3x = 36$$
$$x = 12$$

The solution is $\{(12, -8)\}$.

11. Write in standard form:

$$3x = 10y + 13 \rightarrow 3x - 10y = 13$$
$$7y = 4x - 11 \rightarrow -4x + 7y = -11$$

Multiply the first equation by 4 and the second equation by 3, add the results and solve for y:

$$3x - 10y = 13 \xrightarrow{\times 4} 12x - 40y = 52$$
$$-4x + 7y = -11 \xrightarrow{\times 3} -12x + 21y = -33$$
$$\overline{\qquad\qquad -19y = 19}$$
$$y = -1$$

Substitute into the first equation and solve for x:

$$3x = 10(-1) + 13$$
$$3x = -10 + 13$$
$$3x = 3$$
$$x = 1$$

The solution is $\{(1, -1)\}$.

13. Multiply each equation by 10:

$$1.2x - 0.6y = 3 \rightarrow 12x - 6y = 30$$
$$0.8x - 1.4y = 3 \rightarrow 8x - 14y = 30$$

Multiply the first equation by 2 and the second equation by -3, add the results and solve for y:

$$12x - 6y = 30 \xrightarrow{\times 2} 24x - 12y = 60$$
$$8x - 14y = 30 \xrightarrow{\times -3} -24x + 42y = -90$$
$$\overline{\qquad\qquad 30y = -30}$$
$$y = -1$$

Substitute into the first equation and solve for x:

$$12x - 6(-1) = 30$$
$$12x + 6 = 30$$
$$12x = 24$$
$$x = 2$$

The solution is $\{(2, -1)\}$.

15. Write in standard form:
$$3x + 2 = 4y + 2 \rightarrow 3x - 4y = 0$$
$$7x = 3y \rightarrow \quad 7x - 3y = 0$$
Multiply the first equation by 3 and the second equation by –4, add the results and solve for x:
$$3x - 4y = 0 \xrightarrow{\times 3} \quad 9x - 12y = 0$$
$$7x - 3y = 0 \xrightarrow{\times -4} \underline{-28x + 12y = 0}$$
$$-19x = 0$$
$$x = 0$$
Substitute into the first equation and solve for y:
$$3(0) - 4y = 0$$
$$0 - 4y = 0$$
$$-4y = 0$$
$$y = 0$$
The solution is $\{(0.0)\}$.

17. $3x - 2y = 1$
$$-6x + 4y = -2$$
Multiply the first equation by 2, add to the second equation and solve for y:
$$3x - 2y = 1 \xrightarrow{\times 2} \quad 6x - 4y = 2$$
$$-6x + 4y = -2 \xrightarrow{} \underline{-6x + 4y = -2}$$
$$0 = 0$$
Infinitely many solutions of the form
$\{(x, y) \mid 3x - 2y = 1\}$; dependent equations.

19. Write in standard form:
$$6y = 14 - 4x \rightarrow 4x + 6y = 14$$
$$2x = -3y - 7 \rightarrow 2x + 3y = -7$$
Multiply the second equation by –2, add to the first equation and solve for y:
$$4x + 6y = 14 \xrightarrow{} \quad 4x + 6y = 14$$
$$2x + 3y = -7 \xrightarrow{\times -2} \underline{-4x - 6y = 14}$$
$$0 \neq 28$$
There is no solution; $\{\ \}$. This is an inconsistent system.

21. Write in standard form:
$$12x - 4y = 2 \quad \rightarrow \quad 12x - 4y = 2$$
$$6x = 1 + 2y \rightarrow \quad 6x - 2y = 1$$
Multiply the second equation by –2, add to the first equation and solve for y:
$$12x - 4y = 2 \xrightarrow{} \quad 12x - 4y = 2$$
$$6x - 2y = 1 \xrightarrow{\times -2} \underline{-12x + 4y = -2}$$
$$0 = 0$$
Infinitely many solutions of the form
$\{(x, y) \mid 12x - 4y = 2\}$; dependent equations.

23. $\dfrac{1}{2}x + y = \dfrac{7}{6}$
$$x + 2y = 4.5$$
Multiply the first equation by –2, add to the second equation and solve for y:
$$\frac{1}{2}x + y = \frac{7}{6} \xrightarrow{\times -2} \quad -x - 2y = -\frac{7}{3}$$
$$x + 2y = 4.5 \xrightarrow{} \underline{\quad x + 2y = 4.5}$$
$$0 \neq \frac{13}{6}$$
There is no solution; $\{\ \}$. This is an inconsistent system.

25. Use the substitution method if one equation has x or y already isolated.

27. False

29. True

104

31.   True

33.
$$2x - 4y = 8$$
$$y = 2x + 1$$
$$2x - 4(2x+1) = 8$$
$$2x - 8x - 4 = 8$$
$$-6x = 12$$
$$x = -2$$
$$y = 2x + 1 = 2(-2) + 1 = -4 + 1 = -3$$
The solution is $\{(-2, -3)\}$.

35.
$$2x + 5y = 9$$
$$4x - 7y = -16$$
Multiply the first equation by –2, add to the second equation and solve for y:
$$2x + 5y = 9 \xrightarrow{\times -2} -4x - 10y = -18$$
$$4x - 7y = -16 \xrightarrow{} \underline{4x - 7y = -16}$$
$$-17y = -34$$
$$y = 2$$
Substitute into the first equation and solve for x:
$$2x + 5(2) = 9$$
$$2x + 10 = 9$$
$$2x = -1$$
$$x = -\frac{1}{2}$$
The solution is $\left\{\left(-\frac{1}{2}, 2\right)\right\}$.

37.
$$0.2x - 0.1y = 0.8$$
$$0.1x - 0.1y = 0.4 \rightarrow 0.1x = 0.1y + 0.4$$
$$\rightarrow x = y + 4$$
$$0.2(y+4) - 0.1y = 0.8$$
$$0.2y + 0.8 - 0.1y = 0.8$$
$$0.1y + 0.8 = 0.8$$
$$0.1y = 0$$
$$y = 0$$
$$x = 0 + 4 = 4$$
The solution is $\{(4, 0)\}$.

39.
$$4x - 6y = 5$$
$$2x - 3y = 7$$
Multiply the second equation by –2, add to the first equation and solve for y:
$$4x - 6y = 5 \xrightarrow{} 4x - 6y = 5$$
$$2x - 3y = 7 \xrightarrow{\times -2} \underline{-4x + 6y = -14}$$
$$0 \neq -9$$
There is no solution; $\{\ \}$. This is an inconsistent system.

41.   Multiply each equation by the LCD:
$$\frac{1}{4}x - \frac{1}{6}y = -2 \xrightarrow{\times 12} 3x - 2y = -24$$
$$-\frac{1}{6}x + \frac{1}{5}y = 4 \xrightarrow{\times 30} -5x + 6y = 120$$
Multiply the first equation by 3, add to the second equation and solve for x:
$$3x - 2y = -24 \xrightarrow{\times 3} 9x - 6y = -72$$
$$-5x + 6y = 120 \xrightarrow{} \underline{-5x + 6y = 120}$$
$$4x = 48$$
$$x = 12$$
Substitute into the first equation and solve for y:

$$3(12) - 2y = -24$$
$$36 - 2y = -24$$
$$-2y = -60$$
$$y = 30$$

The solution is $\{(12, 30)\}$.

43.
$$\frac{1}{3}x - \frac{1}{2}y = 0$$

$$x = \frac{3}{2}y$$

$$\frac{1}{3}\left(\frac{3}{2}y\right) - \frac{1}{2}y = 0$$

$$\frac{1}{2}y - \frac{1}{2}y = 0$$

$$0 = 0$$

Infinitely many solutions of the form $\{(x, y) \mid x = \frac{3}{2}y\}$. The equations are dependent.

45. Write in standard form:
$$2(x + 2y) = 20 - y \;\rightarrow\; 2x + 4y = 20 - y$$
$$\rightarrow\; 2x + 5y = 20$$
$$-7(x - y) = 16 + 3y \rightarrow -7x + 7y = 16 + 3y$$
$$\rightarrow\; -7x + 4y = 16$$

Multiply the first equation by 4 and the second equation by −5, add the results and solve for x:
$$2x + 5y = 20 \xrightarrow{\times 4} \quad 8x + 20y = 80$$
$$-7x + 4y = 16 \xrightarrow{\times -5} 35x - 20y = -80$$
$$\overline{43x = 0}$$
$$x = 0$$

Substitute into the first equation and solve for y:
$$2(0) + 5y = 20$$
$$0 + 5y = 20$$
$$5y = 20$$
$$y = 4$$

The solution is $\{(0, 4)\}$.

47. Solve each equation:
$$-4y = 10 \qquad\qquad 4x + 3 = 1$$
$$\qquad\qquad\qquad 4x = -2$$
$$y = -\frac{10}{4} = -\frac{5}{2} \qquad x = -\frac{2}{4} = -\frac{1}{2}$$

The solution is $\left\{\left(-\frac{1}{2}, -\frac{5}{2}\right)\right\}$.

49.
$$0.04x = -0.05y + 1.7 \rightarrow 4x = -5y + 170$$
$$\rightarrow 4x + 5y = 170$$
$$-0.03y = -2.4 + 0.07x \rightarrow -3y = -240 + 7x$$
$$\rightarrow 7x + 3y = 240$$

Multiply the first equation by 3 and the second equation by −5, add the results and solve for x:
$$4x + 5y = 170 \xrightarrow{\times 3} \quad 12x + 15y = 510$$
$$7x + 3y = 240 \xrightarrow{\times -5} -35x - 15y = -1200$$
$$\overline{-23x = -690}$$
$$x = 30$$

Substitute into the first equation and solve for y:
$$4(30) + 5y = 170$$
$$120 + 5y = 170$$
$$5y = 50$$
$$y = 10$$

The solution is $\{(30, 10)\}$.

51. Write in standard form:

$$3x - 2 = \frac{1}{3}(11 + 5y) \rightarrow \qquad 3x - 2 = \frac{11}{3} + \frac{5}{3}y$$

$$\rightarrow 3x - \frac{5}{3}y = \frac{17}{3} \quad \rightarrow \quad 9x - 5y = 17$$

$$x + \frac{2}{3}(2y - 3) = -2 \quad \rightarrow \quad x + \frac{4}{3}y - 2 = -2$$

$$\rightarrow \quad x + \frac{4}{3}y = 0 \quad \rightarrow \quad 3x + 4y = 0$$

Multiply the second equation by –3, add to the first equation and solve for y:

$$9x - 5y = 17 \longrightarrow \qquad 9x - 5y = 17$$

$$3x + 4y = 0 \xrightarrow{\times -3} -9x - 12y = 0$$

$$\overline{\qquad\qquad -17y = 17}$$

$$y = -1$$

Substitute into the first equation and solve:

$$9x - 5(-1) = 17$$

$$9x + 5 = 17$$

$$9x = 12$$

$$x = \frac{12}{9} = \frac{4}{3}$$

The solution is $\left\{\left(\frac{4}{3}, -1\right)\right\}$.

53.

$$\frac{1}{4}x + \frac{1}{2}y = \frac{11}{4}$$

$$\frac{2}{3}x + \frac{1}{3}y = \frac{7}{3}$$

Multiply the first equation by 4 and the second equation by –6, add the results and solve for x:

$$\frac{1}{4}x + \frac{1}{2}y = \frac{11}{4} \xrightarrow{\times 4} \qquad x + 2y = 11$$

$$\frac{2}{3}x + \frac{1}{3}y = \frac{7}{3} \xrightarrow{\times -6} -4x - 2y = -14$$

$$\overline{\qquad\qquad -3x = -3}$$

$$x = 1$$

Substitute into the first equation above and solve for y:

$$1 + 2y = 11$$

$$2y = 10$$

$$y = 5$$

The solution is $\{(1, 5)\}$.

55.

$$4x = 3y \rightarrow x = \frac{3}{4}y$$

$$y = \frac{4}{3}x + 2$$

Substitute for x and solve for y:

$$y = \frac{4}{3}\left(\frac{3}{4}y\right) + 2$$

$$y = y + 2$$

$$0 = 2$$

There is no solution; $\{\ \}$. This is an inconsistent system.

57. Multiply each equation by the LCD:

$$\frac{1}{16}c + \frac{1}{24}h = 12 \xrightarrow{\times 48} \quad 3c + 2h = 576$$

$$\frac{1}{14}c + \frac{1}{20}h = 14 \xrightarrow{\times 140} 10c + 7h = 1960$$

Multiply the first equation by $-\frac{10}{3}$, add to the second equation and solve for h:

$$3c + 2h = 576 \xrightarrow{\times -\frac{10}{3}} -10c - \frac{20}{3}h = -192$$

$$10c + 7h = 1960 \longrightarrow \quad \underline{10c + 7h = 1960}$$

$$\frac{1}{3}h = 40$$

$$h = 120$$

Substitute into the first equation and solve for c:

$$3c + 2h = 576$$
$$3c + 2(120) = 576$$
$$3c + 240 = 576$$
$$3c = 336$$
$$c = 112$$

112 mi in the city and 120 mi on the highway.

59. $9x + 11y = 47$

$-5x + 3y = 23$

Multiply the first equation by -3 and the second equation by 11, add the results and solve for x:

$9x + 11y = 47 \xrightarrow{\times -3} -27x - 33y = -141$
$-5x + 3y = 23 \xrightarrow{\times 11} \underline{-55x + 33y = 253}$
$\phantom{-5x + 3y = 23 \xrightarrow{\times 11}} -82x = 112$
$\phantom{-5x + 3y = 23 \xrightarrow{\times 11}} x = -\dfrac{112}{82} = -\dfrac{56}{41}$

Multiply the first equation by 5 and the second equation by 9, add the results and solve for y:

$9x + 11y = 47 \xrightarrow{\times 5} 45x + 55y = 235$
$-5x + 3y = 23 \xrightarrow{\times 9} \underline{-45x + 27y = 207}$
$\phantom{-5x + 3y = 23 \xrightarrow{\times 9}} 82y = 442$
$\phantom{-5x + 3y = 23 \xrightarrow{\times 9}} y = \dfrac{442}{82} = \dfrac{221}{41}$

The solution is $\left\{ \left(-\dfrac{56}{41}, \dfrac{221}{41} \right) \right\}$.

61. $4x - 10y = 19$

$5x + 12y = -41$

Multiply the first equation by 6 and the second equation by 5, add the results and solve for x:

$4x - 10y = 19 \xrightarrow{\times 6} 24x - 60y = 114$
$5x + 12y = -41 \xrightarrow{\times 5} \underline{25x + 60y = -205}$
$\phantom{5x + 12y = -41 \xrightarrow{\times 5}} 49x = -91$
$\phantom{5x + 12y = -41 \xrightarrow{\times 5}} x = -\dfrac{91}{49} = -\dfrac{13}{7}$

Multiply the first equation by -5 and the second equation by 4, add the results and solve for y:

$4x - 10y = 19 \xrightarrow{\times -5} -20x + 50y = -95$
$5x + 12y = -41 \xrightarrow{\times 4} \underline{20x + 48y = -164}$
$\phantom{5x + 12y = -41 \xrightarrow{\times 4}} 98y = -259$
$\phantom{5x + 12y = -41 \xrightarrow{\times 4}} y = -\dfrac{259}{98} = -\dfrac{37}{14}$

The solution is $\left\{ \left(-\dfrac{13}{7}, -\dfrac{37}{14} \right) \right\}$.

Problem Recognition Exercises

1. a. $-3x + y = -2$ $\qquad$ $4x - y = 4$

$ y = 3x - 2$ $\qquad$ $-y = -4x + 4$

$ y = 4x - 4$

3. a. $5x = 2y \rightarrow y = \dfrac{5}{2}x$

$ y = \dfrac{5}{2}x + 1$

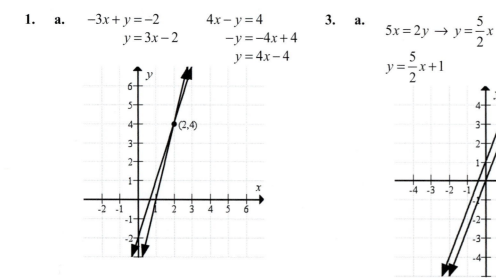

The solution is $\{(2,4)\}$.

b. $-3x + y = -2 \rightarrow y = 3x - 2$

$4x - y = 4$

$4x - (3x - 2) = 4$

$4x - 3x + 2 = 4$

$x + 2 = 4$

$x = 2$

$y = 3(2) - 2 = 6 - 2 = 4$

The solution is $\{(2,4)\}$.

c. $-3x + y = -2$

$4x - y = 4$

Add the equations and solve for x:

$-3x + y = -2$

$\underline{4x - y = 4}$

$x \quad\; = 2$

Substitute into the first equation and solve for y:

$-3(2) + y = -2$

$-6 + y = -2$

$y = 4$

The solution is $\{(2,4)\}$.

5. $y = -4x - 9$

$8x + 3y = -29$

Substitute the first equation into the second and solve for x:

$8x + 3(-4x - 9) = -29$

$8x - 12x - 27 = -29$

$-4x - 27 = -29$

$-4x = -2$

$x = \dfrac{1}{2}$

$y = -4x - 9 = -4\left(\dfrac{1}{2}\right) - 9 = -2 - 9 = -11$

The solution is $\left\{\left(\dfrac{1}{2}, -11\right)\right\}$.

No solution; $\{\ \}$; inconsistent system

b. $\dfrac{5}{2}x = \dfrac{5}{2}x + 1$

$0 = 1$

No solution; $\{\ \}$; inconsistent system

Write in standard form:

c. $5x = 2y \qquad\qquad y = \dfrac{5}{2}x + 1$

$5x - 2y = 0 \qquad\quad 2y = 5x + 2$

$-5x + 2y = 2$

Add the equations to solve for x:

$5x - 2y = 0$

$\underline{-5x + 2y = 2}$

$0 = 2$

No solution; $\{\ \}$; inconsistent system.

7. $5x - 3y = 2$

$7x + 4y = -30$

Multiply the first equation by 4 and the second equation by 3, add the results and solve for x:

$5x - 3y = 2 \xrightarrow{\times 4} 20x - 12y = 8$

$7x + 4y = -30 \xrightarrow{\times 3} \underline{21x + 12y = -90}$

$41x = -82$

$x = -2$

Substitute into the first equation and solve for y:

$5(-2) - 3y = 2$

$-10 - 3y = 2$

$-3y = 12$

$y = -4$

The solution is $\{(-2, -4)\}$.

Section 3.4   Applications of Systems of Linear Equations in Two Variables

1. **a.** $5 \cdot \$12 = \$60$

$x \cdot 12 = 12x$

b. $0.10 \cdot 20 = 2L$

$0.10 \cdot x = 0.10x$

c. $0.04 \cdot \$5000 = \$200$

$0.04 \cdot y = 0.04y$

d. $b - c$; $b + c$

e. $180°$

f. $180°$

g. $90°$

h. $90°$

3. Substitution:

$$y = 9 - 2x$$
$$3x - y = 16$$
$$3x - (9 - 2x) = 16$$
$$3x - 9 + 2x = 16$$
$$5x = 25$$
$$x = 5$$
$$y = 9 - 2x = 9 - 2(5) = 9 - 10 = -1$$

The solution is $\{(5, -1)\}$.

5. Let x = the number of premium tickets sold

y = the number of regular tickets sold

$30x$ = receipts from premium tickets

$20y$ = receipts from regular tickets

$$x + y = 1190 \rightarrow y = 1190 - x$$
$$30x + 20y = 30,180$$
$$30x + 20(1190 - x) = 30,180$$
$$30x + 23,800 - 20x = 30,180$$
$$10x = 6380$$
$$x = 638$$
$$y = 1190 - 638 = 552$$

There were 638 tickets sold at \$30 each and 552 tickets sold at \$20 each.

7. Let x = the cost of 1 hamburger

y = the cost of 1 fish sandwich

$$3x + 2y = 24.20$$
$$4x + y = 23.60 \rightarrow y = 23.60 - 4x$$
$$3x + 2(23.60 - 4x) = 24.20$$
$$3x + 47.20 - 8x = 24.20$$
$$-5x = -23$$
$$x = 4.60$$
$$y = 23.60 - 4(4.60)$$
$$= 23.60 - 18.40 = 5.20$$

Hamburgers cost \$4.60 and fish sandwiches cost \$5.20.

9. Let x = fat in 1 scoop of vanilla

y = fat in 1 scoop of mud pie

$$2x + y = 40 \rightarrow y = 40 - 2x$$
$$x + 2y = 44$$
$$x + 2(40 - 2x) = 44$$
$$x + 80 - 4x = 44$$
$$-3x = -36$$
$$x = 12$$
$$y = 40 - 2(12)$$
$$= 40 - 24 = 16$$

Vanilla has 12 g of fat per scoop and mud pie has 16 g of fat per scoop.

11. Let x = the amount of 18% moisturizer cream

y = the amount of 24% moisturizer cream

	18% Cr	24% Cr	22% Cr

13. Let x = the amount of 8% nitrogen fertilizer

y = the amount of 12% nitrogen fertilizer

	8% nit	12% nit	11% nit

oz cream	x	y	12
oz moist	$0.18x$	$0.24y$	
$\underline{0.22(12)}$			

$$x+\quad y=12$$
$$0.18x+0.24y=0.22(12)$$

Multiply the first equation by –0.18, add to the second equation and solve for y:
$$x+\quad y=12 \rightarrow -0.18x-0.18y=-2.16$$
$$0.18x+0.24y=2.64 \rightarrow \underline{0.18x+0.24y=2.64}$$
$$0.06y=0.48$$
$$y=8$$

Substitute into the first equation and solve for x:
$$x+8=12$$
$$x=4$$
The mixture contains 4 oz of 18% moisturizer and 8 oz of 24% moisturizer.

oz cream	x	y	8
oz moist	$0.08x$	$0.12y$	$0.11(8)$

$$x+\quad y=8$$
$$0.08x+0.12y=0.11(8)$$

Multiply the first equation by –0.08, add to the second equation and solve for y:
$$x+\quad y=8 \rightarrow -0.08x-0.08y=-0.64$$
$$0.08x+0.12y=0.88 \rightarrow \underline{0.08x+0.12y=0.88}$$
$$0.04y=0.24$$
$$y=6$$

Substitute into the first equation and solve for x:
$$x+6=8$$
$$x=2$$
The mixture contains 2 L of 8% nitrogen fertilizer and 6 L of 12% nitrogen fertilizer.

15. Let x = amount of pure (100%) bleach sol
y = the amount of 4% bleach solution

	100% bl	4% bl	12% bl
oz solution	x	y	12
oz bleach	$1.00x$	$0.04y$	$0.12(12)$

$$x+\quad y=12$$
$$1.00x+0.04y=0.12(12)$$

Multiply the first equation by –0.04, add to the second equation and solve for x:
$$x+\quad y=12 \rightarrow -0.04x-0.04y=-0.48$$
$$1x+0.04y=1.44 \rightarrow \underline{1.00x+0.04y=\quad1.44}$$
$$0.96x\quad\quad=0.96$$
$$x=1$$

Substitute into the first equation and solve for y:
$$1+y=12$$
$$y=11$$
The mixture contains 1 oz of pure bleach and 11 oz of 4% bleach solution.

17. Let x = the amount invested in 5% bonds
$3x$ = the amount invested in 8% stocks

	5% Acct	8% Acct	
Total			
Principal	x	$3x$	
Interest	$0.05x$	$0.08(3x)$	435

$$0.05x+0.08(3x)=435$$
$$0.05x+0.24x=435$$
$$0.29x=435$$
$$x=1500$$
$$3x=3(1500)=4500$$

He invested $1500 in the bond fund and $4500 in the stock fund.

19. Let x = the amount borrowed at 5.5%
y = the amount borrowed at 3.5%

	5.5% Acct	3.5% Acct	Total
Principal	x	y	
Interest	$0.055x$	$0.035y$	245

$$x=y+200$$
$$0.055x+0.035y=245$$
Substitute and solve for y:

21. Let x = the amount borrowed at 6%
y = the amount borrowed at 7%

	6% Acct	7% Acct	Total
Principal	x	y	15,000
Interest	$0.06x$	$0.07y$	4750/5

$$x+\quad y=15,000$$
$$0.06x+0.07y=\frac{4750}{5}=950$$

$$0.055(y+200)+0.035y=245$$
$$0.055y+11+0.035y=245$$
$$0.09y=234$$
$$y=2600$$

Substitute into the first equation and solve for x:
$$x=2600+200$$
$$x=2800$$
He borrowed $2800 at 5.5% and $2600 at 3.5%.

Multiply the first equation by –0.06, add to the second equation and solve for y:
$$x+\ y=15,000 \rightarrow -0.06x-0.06y=-900$$
$$0.06x+0.07y=950 \rightarrow \underline{0.06x+0.07y=\ 950}$$
$$0.01y=50$$
$$y=5000$$

Substitute into the first equation and solve for x:
$$x+5000=15,000$$
$$x=10,000$$
Alina borrowed $10,000 from the bank charging 6% interest and $5000 from the bank charging 7% interest.

23. Let b = the speed of the boat in still water
c = the speed of the current
$b+c$ = speed of boat with the current
$b-c$ = speed of boat against the current

	Distance	Rate	Time
With current	16	$b+c$	2
Against current	16	$b-c$	4

(rate)(time) = (distance)
$$(b+c)(2)=16$$
$$(b-c)(4)=16$$

Divide the first equation by 2, the second equation by 4, add the results, and solve:
$$(b+c)(2)=16 \xrightarrow{div\ 2} b+c=8$$
$$(b-c)(4)=16 \xrightarrow{div\ 4} \underline{b-c=4}$$
$$2b\ =12$$
$$b=6$$

Substitute and solve for c:
$$6+c=8$$
$$c=2$$
The speed of the boat is 6 mph and the speed of the current is 2 mph.

25. Let p = the speed of the plane in still air
Let w = the speed of the wind
$p+w$ = speed of the plane with the wind
$p-w$ = speed of plane against the wind

	Distance	Rate	Time
Tailwind	3200	$p+w$	4
Headwind	3200	$p-w$	5

(rate)(time) = (distance)
$$(p+w)(4)=3200$$
$$(p-w)(5)=3200$$

Divide the first equation by 4, the second equation by 5, add the results, and solve:
$$(p+w)(4)=3200 \xrightarrow{div\ 4} p+w=800$$
$$(p-w)(5)=3200 \xrightarrow{div\ 5} \underline{p-w=640}$$
$$2p\ =1440$$
$$p=720$$

Substitute and solve for w:
$$720+w=800$$
$$w=80$$
The speed of the plane is 720 km/hr in still air and the speed of the wind is 80 km/hr.

27. Let x = the walking speed
Let y = the speed of the moving sidewalk
$x+y$ = speed of walking with sidewalk
$x-y$ = speed of walking against sidewalk

	Distance	Rate	Time
With walk	100	$x+y$	20
Against walk	60	$x-y$	30

(rate)(time) = (distance)
$$(x+y)(20)=100$$
$$(x-y)(30)=60$$

Divide the first equation by 20, the second

29. Let x = one acute angle
Let y = the other acute angle
$$x=3y+6$$
$$x+y=90$$

Substitute and solve:
$$3y+6+y=90$$
$$4y=84$$
$$y=21$$
$$x=3(21)+6=63+6=69$$

The two acute angles measure 69° and 21°.

equation by 30, add the results, and solve:

$$(x+y)(20)=100 \xrightarrow{\ div\ 20\ } x+y=5$$

$$(x-y)(30)=60 \xrightarrow{\ div\ 30\ } \underline{x-y=2}$$

$$2x\ \ \ \ =7$$

$$x=3.5$$

Substitute and solve for y:

$$3.5+y=5$$

$$y=1.5$$

Stephen's speed on nonmoving ground is 3.5 ft/sec. The sidewalk moves at 1.5 ft/sec.

31. Let x = one angle
Let y = the other angle
$$y=3x-2$$
$$x+y=180$$
Substitute and solve:
$$x+3x-2=180$$
$$4x=182$$
$$x=45.5$$
$$y=3(45.5)-2=136.5-2=134.5$$
The two angles measure 45.5° and 134.5°.

33. Let x = one angle
Let y = the other angle
$$y=2x+6$$
$$x+y=90$$
Substitute and solve:
$$x+2x+6=90$$
$$3x=84$$
$$x=28$$
$$y=2(28)+6=56+6=62$$
The two angles measure 28° and 62°.

35. Let x = the amount of pure (100%) gold
y = the amount of 60% gold

	100% gold	60% gold	75% gold
g mix	x	y	20
g gold	$1.00x$	$0.60y$	$0.75(20)$

$$x+y=20$$
$$1.00x+0.60y=0.75(20)$$

Multiply the first equation by –0.60, add to the second equation and solve for x:

$$x+\ \ \ \ y=20 \rightarrow -0.60x-0.60y=-12$$
$$1.00x+0.60y=15 \rightarrow \ \underline{1.00x+0.60y=15}$$
$$0.40x=\ \ 3$$
$$x=7.5$$

7.5 g of pure gold must be used.

37. Let b = the speed of the boat in still water
Let c = the speed of the current
$b+c$ = speed of boat with the current
$b-c$ = speed of boat against the current

	Distance	Rate	Time
With current	16	$b+c$	2.5
Against current	10	$b-c$	2.5

(rate)(time) = (distance)
$$(b+c)(2.5)=16 \rightarrow 2.5b+2.5c=16$$
$$(b-c)(2.5)=10 \rightarrow 2.5b-2.5c=10$$

Add the two equations, and solve:
$$2.5b+2.5c=16$$
$$\underline{2.5b-2.5c=10}$$
$$5b\ \ \ \ \ \ =26$$
$$b=5.2$$

Substitute and solve for c:
$$2.5(5.2)+2.5c=16$$
$$13+2.5c=16$$
$$2.5c=3$$
$$c=1.2$$

The speed of the boat in still water is 5.2 mph and the speed of the current is 1.2 mph.

39. Let x = the cost of a grandstand ticket
y = the cost of a general admission ticket

41. Let x = the amount invested at 2%
y = the amount invested at 1.3%

$6x + 2y = 2330$

$4x + 4y = 2020$

Multiply the first equation by –2, add to the second equation and solve for x:

$6x + 2y = 2330 \xrightarrow{\times -2} -12x - 4y = -4660$

$4x + 4y = 2020 \longrightarrow \underline{\quad\;\; 4x + 4y = \;\; 2020}$

$\qquad\qquad\qquad\qquad -8x \qquad\;\; = -2640$

$\qquad\qquad\qquad\qquad\qquad\quad x = 330$

 Substitute and solve for y:

$6(330) + 2y = 2330$

$1980 + 2y = 2330$

$2y = 350$

$y = 175$

Grandstand tickets cost $330 and general admission tickets cost $175.

	2% Acct	1.3% Acct	Total
Principal	x	y	3,000
Interest	0.02x	0.013y	51.25

$x + \qquad y = 3,000$

$0.02x + 0.013y = 51.25$

Multiply the first equation by –0.02, add to the second equation and solve for y:

$x + \;\; y = 3,000 \rightarrow -.02x - .02y = -60$

$.02x + .013y = 51.25 \rightarrow \underline{.02x + .013y = 51.25}$

$\qquad\qquad\qquad\qquad\qquad -0.007y = -8.75$

$\qquad\qquad\qquad\qquad\qquad\qquad\quad y = 1250$

Substitute into the first equation and solve for x:

$x + 1250 = 3000$

$y = 1750$

Svetlana invested $1750 at 2% and $1250 at 1.3%.

43. Let w = the width of the rectangle
Let l = the length of the rectangle

$l = w + 1$

$2l + 2w = 42$

Substitute and solve:

$2(w + 1) + 2w = 42$

$2w + 2 + 2w = 42$

$4w + 2 = 42$

$4w = 40$

$w = 10$

$l = 10 + 1 = 11$

The width is 10 m and the length is 11 m.

45. Let d = the number of $1 coins
$\quad f$ = the number of 50 cent pieces

$d + f = 21 \;\rightarrow\; f = 21 - d$

$1d + 0.50f = 15.50$

$d + 0.50(21 - d) = 15.50$

$d + 10.50 - 0.50d = 15.50$

$0.50d = 5.00$

$d = 10$

$f = 21 - 10 = 11$

The collection contains 10 - $1 coins and 11 – 50 cent pieces.

47. a. $f(x) = 0.15x$

b. $g(x) = 0.10x + 4.95$

c. Substitute and solve:

$0.15x = 0.10x + 4.95$

$0.05x = 4.95$

$x = 99$

The cost of either plan would be the same for 99 text messages.

Section 3.5   Linear Inequalities and Systems of Linear Inequalities in Two Variables

1. **a.** linear
    b. is not; is
    c. dashed; is not
    d. solid; is;

3. $5 - x \leq 4$ and $6 > 3x - 3$

    $-x \leq -1$ and $9 > 3x$

    $x \geq 1$ and $x < 3$ $[1, 3)$

5. $-2x < 4$ or $3x - 1 \leq -13$

    $x > -2$ or $3x \leq -12$

    $x > -2$ or $x \leq -4$

    $(-\infty, -4] \cup (-2, \infty)$

7. $3y + x < 5$

    a. $3(7) + (-1) = 21 - 1 = 20 \not< 5$ No
    b. $3(0) + (5) = 0 + 5 = 5 \not< 5$ No
    c. $3(0) + (0) = 0 + 0 = 0 < 5$ Yes
    d. $3(-3) + (2) = -9 + 2 = -7 < 5$ Yes

9. $x \geq 5$

    a. $4 \not\geq 5$ No
    b. $5 \geq 5$ Yes
    c. $8 \geq 5$ Yes
    d. $0 \not\geq 5$ No

11. To choose the correct inequality symbol, three observations must be made. First, notice the shading occurs below the line. Second, since the coefficient of y is negative in the given statement, the direction of the inequality will change. Third, the boundary line is dashed indicating no equality. Thus use the symbol $>$ for the inequality: $x - y > 2$.

13. To choose the correct inequality symbol, three observations must be made. First, notice the shading occurs above the line. Second, since the coefficient of y is positive in the given statement, the direction of the inequality will not change. Third, the boundary line is solid indicating equality. Thus use the symbol $\geq$ for the inequality: $y \geq -4$.

15. The graph of $x \geq 0$ includes Quadrant I and Quadrant IV. The graph of $y \leq 0$ includes Quadrant III and Quadrant IV. The intersection of the graphs occurs in Quadrant IV. Thus, the statements are $x \geq 0$ and $y \leq 0$.

17. $x - 2y > 4$

    Graph the related equation $x - 2y = 4$ by using a dashed line.

    Test point above $(0, 0)$:   Test point below $(0, -3)$:

    $0 - 2(0) > 4$ $0 - 2(-3) > 4$

    $0 > 4$ $6 > 4$

    $(0, 0)$ is not a solution. $(0, -3)$ is a solution.

    Shade the region below the boundary line.

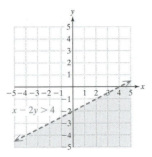

19. $5x - 2y < 10$

Graph the related equation $5x - 2y = 10$ by using a dashed line.

Test point above $(0,0)$:   Test point below $(2,-3)$:

$\quad\quad 5(0) - 2(0) < 10$ $\quad\quad 5(2) - 2(-3) < 10$

$\quad\quad\quad\quad 0 < 10$ $\quad\quad\quad\quad 16 < 10$

$(0,0)$ is a solution. $(2,-3)$ is not a solution.

Shade the region above the boundary line.

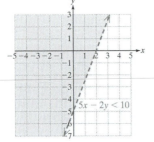

21. $2x \le -6y + 12$

Graph the related equation $2x = -6y + 12$ by using a solid line.

Test point above $(0,3)$:   Test point below $(0,0)$:

$\quad\quad 2(0) \le -6(3) + 12$ $\quad\quad 2(0) \le -6(0) + 12$

$\quad\quad\quad\quad 0 \le -6$ $\quad\quad\quad\quad 0 \le 12$

$(0,3)$ is not a solution. $(0,0)$ is a solution.

Shade the region below the boundary line.

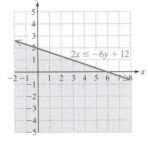

23. $2y \le 4x$

Graph the related equation $2y = 4x$ by using a solid line.

Test point above $(0,1)$:   Test point below $(0,-1)$:

$2(1) \le 4(0)$ $2(-1) \le 4(0)$

$\qquad 2 \le 0$ $\qquad -2 \le 0$

$(0,1)$ is not a solution. $(0,-1)$ is a solution.

Shade the region below the boundary line.

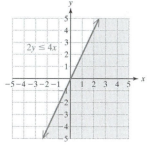

25. $y \ge -2$

Graph the related equation $y = -2$ by using a solid line.

Test point above $(0,0)$:   Test point below $(0,-3)$:

$\qquad 0 \ge -2$ $\qquad -3 \ge -2$

$(0,0)$ is a solution. $(0,-3)$ is not a solution.

Shade the region above the boundary line.

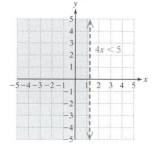

27. $4x < 5$ or $x < \dfrac{5}{4}$ represents all the points to the left of the vertical

line $x = \dfrac{5}{4}$. The boundary is a dashed line.

Shade the region to the left of the boundary line.

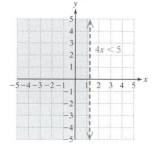

29. $y \geq \dfrac{2}{5}x - 4$

Graph the related equation $y = \dfrac{2}{5}x - 4$ by using a solid line.

Test point above $(0,0)$:   Test point below $(0,-5)$:

$$0 \geq \dfrac{2}{5}(0) - 4 \qquad\qquad -5 \geq \dfrac{2}{5}(0) - 4$$

$$0 \geq -4 \qquad\qquad\qquad\quad -5 \geq -4$$

$(0,0)$ is a solution. $(0,-5)$ is not a solution.

Shade the region above the boundary line.

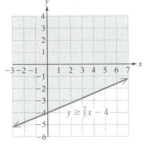

31. $y \leq \dfrac{1}{3}x + 6$

Graph the related equation $y = \dfrac{1}{3}x + 6$ by using a solid line.

Test point above $(0,7)$:   Test point below $(0,0)$:

$$7 \leq \dfrac{1}{3}(0) + 6 \qquad\qquad 0 \leq \dfrac{1}{3}(0) + 6$$

$$7 \leq 6 \qquad\qquad\qquad\quad 0 \leq 6$$

$(0,7)$ is not a solution. $(0,0)$ is a solution.

Shade the region below the boundary line.

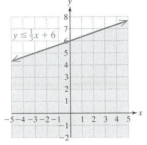

33. $y - 5x > 0$

Graph the related equation $y - 5x = 0$ by using a dashed line.

Test point above $(0,3)$:   Test point below $(0,-3)$:

$\quad\quad 3 - 5(0) > 0$ $\quad\quad -3 - 5(0) > 0$

$\quad\quad\quad 3 > 0$ $\quad\quad\quad -3 > 0$

$(0,3)$ is a solution. $(0,-3)$ is not a solution.

Shade the region above the boundary line.

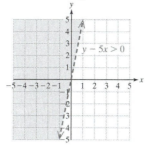

35. $\dfrac{x}{5} + \dfrac{y}{4} < 1$

Graph the related equation $\dfrac{x}{5} + \dfrac{y}{4} = 1$ by using a dashed line.

Test point above $(0,5)$:   Test point below $(0,0)$:

$\quad\quad \dfrac{0}{5} + \dfrac{5}{4} < 1$ $\quad\quad \dfrac{0}{5} + \dfrac{0}{4} < 1$

$\quad\quad\quad \dfrac{5}{4} < 1$ $\quad\quad\quad 0 < 1$

$(0,5)$ is not a solution. $(0,0)$ is a solution.

Shade the region below the boundary line.

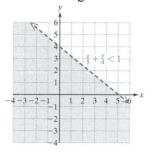

37. $0.1x + 0.2y \le 0.6$

Graph the related equation $0.1x + 0.2y = 0.6$ by using a solid line.

Test point above $(0,5)$:   Test point below $(0,0)$:

$\quad 0.1(0) + 0.2(5) \le 0.6$ $\quad 0.1(0) + 0.2(0) \le 0.6$

$\quad\quad\quad 1 \le 0.6$ $\quad\quad\quad 0 \le 0.6$

$(0,5)$ is not a solution. $(0,0)$ is a solution.

Shade the region below the boundary line.

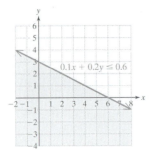

39. $x \le -\dfrac{2}{3}y$

Graph the related equation $x = -\dfrac{2}{3}y$ by using a solid line.

Test point above $(0,3)$:   Test point below $(0,-3)$:

$$0 \le -\frac{2}{3}(3)$$ $$0 \le -\frac{2}{3}(-3)$$

$$0 \le -2$$ $$0 \le 2$$

$(0,3)$ is not a solution. $(0,-3)$ is a solution.

Shade the region below the boundary line.

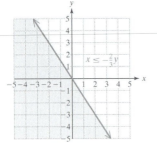

41. $y < 4$ and $y > -x + 2$

$y < 4$ represents the points below the horizontal line $y = 4$.

Shade the region below the boundary line using a dashed line border.

Graph the related equation $y = -x + 2$ by using a dashed line.

Test point above $(0,3)$:   Test point below $(0,0)$:

$$3 > -(0) + 2$$ $$0 > -(0) + 2$$

$$3 > 2$$ $$0 > 2$$

$(0,3)$ is a solution. $(0,0)$ is not a solution.

Shade the region above the boundary line.

The solution is the intersection of the graphs.

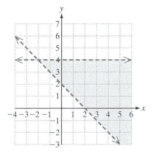

43. $2x + y \leq 5$ or $x \geq 3$

Graph the related equation $2x + y = 5$ by using a solid line.

Test point above $(0,6)$:   Test point below $(0,0)$:

$\quad\quad 2(0) + 6 \leq 5$ $\quad\quad 2(0) + 0 \leq 5$

$\quad\quad\quad\quad 6 \leq 5$ $\quad\quad\quad\quad 0 \leq 5$

$(0,6)$ is not a solution. $(0,0)$ is a solution.

Shade the region below the boundary line.

$x \geq 3$ represents the points to the right of the vertical line $x = 3$.

Shade the region to the right of the boundary line using a solid

line border. The solution is the union of the graphs.

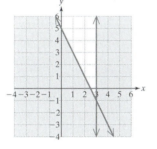

45. $x + y < 3$ and $4x + y < 6$

Graph the related equation $x + y = 3$ by using a dashed line.

Test point above $(0,4)$:   Test point below $(0,0)$:

$\quad\quad 0 + 4 < 3$ $\quad\quad 0 + 0 < 3$

$\quad\quad\quad\quad 4 < 3$ $\quad\quad\quad\quad 0 < 3$

$(0,4)$ is not a solution. $(0,0)$ is a solution.

Shade the region below the boundary line.

Graph the related equation $4x + y = 6$ by using a dashed line.

Test point above $(0,7)$:   Test point below $(0,0)$:

$\quad\quad 4(0) + 7 < 6$ $\quad\quad 4(0) + 0 < 6$

$\quad\quad\quad\quad 7 < 6$ $\quad\quad\quad\quad 0 < 6$

$(0,7)$ is not a solution. $(0,0)$ is a solution.

Shade the region below the boundary line.

The solution is the intersection of the graphs.

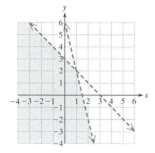

47. $2x - y \le 2$ or $2x + 3y \ge 6$

Graph the related equation $2x - y = 2$ by using a solid line.

Test point above $(0,0)$:   Test point below $(0,-3)$:

$$2(0) - 0 \le 2 \qquad\qquad 2(0) - (-3) \le 2$$

$$0 \le 2 \qquad\qquad\qquad 3 \le 2$$

$(0,0)$ is a solution. $(0,-3)$ is not a solution.

Shade the region above the boundary line.

Graph the related equation $2x + 3y = 6$ by using a solid line.

Test point above $(0,3)$:   Test point below $(0,0)$:

$$2(0) + 3(3) \ge 6 \qquad\qquad 2(0) + 3(0) \ge 6$$

$$9 \ge 6 \qquad\qquad\qquad 0 \ge 6$$

$(0,3)$ is a solution. $(0,0)$ is not a solution.

Shade the region above the boundary line.

The solution is the union of the graphs.

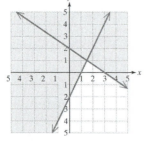

49. $x > 4$ and $y < 2$

$x > 4$ represents the points to the right of the vertical line $x = 4$.
Shade the region to the right of the boundary line using a dashed
line border.

$y < 2$ represents the points below the horizontal line $y = 2$.
Shade the region below the boundary line using a dashed
line border. The solution is the intersection of the graphs.

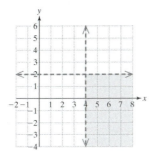

51. $x \leq -2$ or $y \leq 0$

$x \leq -2$ represents the points to the left of the vertical line $x = -2$.
Shade the region to the left of the boundary line using a solid line border.

$y \leq 0$ represents the points below the horizontal line $y = 0$.
Shade the region below the boundary line using a solid line border. The solution is the union of the graphs.

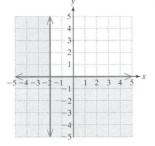

53. $x > 0$ and $x + y < 6$

$x > 0$ represents the points to the right of the vertical line $x = 0$.
Shade the region to the right of the boundary line using a dashed line border.

Graph the related equation $x + y = 6$ by using a dashed line.

Test point above $(0, 7)$:   Test point below $(0, 0)$:

$$0 + 7 < 6 \qquad\qquad\qquad 0 + 0 < 6$$

$$7 < 6 \qquad\qquad\qquad\qquad 0 < 6$$

$(0, 7)$ is not a solution. $(0, 0)$ is a solution.

Shade the region below the boundary line.

The solution is the intersection of the graphs.

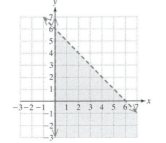

55. $y \le 0$ or $x - y \le -4$

$y \le 0$ represents the points below the horizontal line $y = 0$.

Shade the region below the boundary line using a solid line border.

Graph the related equation $x - y = -4$ by using a solid line.

Test point above $(0,5)$:   Test point below $(0,0)$:

$$0 - 5 \le -4 \qquad\qquad\qquad 0 - 0 \le -4$$
$$-5 \le -4 \qquad\qquad\qquad\quad 0 \le -4$$

$(0,5)$ is a solution. $(0,0)$ is not a solution.

Shade the region above the boundary line.

The solution is the union of the graphs.

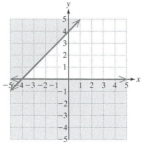

57. $x - y \le 2$ and $x \ge 0$ and $y \ge 0$

Graph the related equation $x - y = 2$ by using a solid line.

Test point above $(0,0)$:   Test point below $(0,-3)$:

$$0 - 0 \le 2 \qquad\qquad\qquad 0 - (-3) \le 2$$
$$0 \le 2 \qquad\qquad\qquad\qquad 3 \le 2$$

$(0,0)$ is a solution. $(0,-3)$ is not a solution.

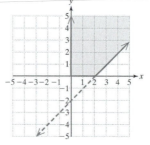

Shade the region above the boundary line.

$x \ge 0$ represents the points to the right of the vertical line $x = 0$.

Shade the region to the right of the boundary line using a solid line border.

$y \ge 0$ represents the points above the horizontal line $y = 0$.

Shade the region above the boundary line using a solid line border.

The solution is the intersection of the graphs.

59. $x \geq 0$ and $y \geq 0$ and $x + y \leq 5$ and $x + 2y \leq 6$

$x \geq 0$ represents the points to the right of the vertical line $x = 0$.
Shade the region to the right of the boundary line using a solid
line border.

$y \geq 0$ represents the points above the horizontal line $y = 0$.
Shade the region above the boundary line using a solid line border.

Graph the related equation $x + y = 5$ by using a solid line.

Test point above $(0,6)$:   Test point below $(0,0)$:

$$0 + 6 \leq 5 \qquad\qquad\qquad 0 + 0 \leq 5$$
$$6 \leq 5 \qquad\qquad\qquad\quad 0 \leq 5$$

$(0,6)$ is not a solution. $(0,0)$ is a solution.

Shade the region below the boundary line.

Graph the related equation $x + 2y = 6$ by using a solid line.

Test point above $(0,4)$:   Test point below $(0,0)$:

$$0 + 2(4) \leq 6 \qquad\qquad 0 + 2(0) \leq 6$$
$$8 \leq 6 \qquad\qquad\qquad\quad 0 \leq 6$$

$(0,4)$ is not a solution. $(0,0)$ is a solution.

Shade the region below the boundary line.

The solution is the intersection of the graphs.

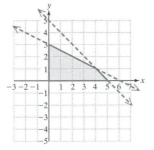

61. a. $2x + 2y \leq 40$

b. $x \geq 0$ and $y \geq 0$ and $2x + 2y \leq 40$

$x \geq 0$ represents the points to the right of the vertical line $x = 0$. Shade the region to the right
of the boundary line using a solid line border.

$y \geq 0$ represents the points above the horizontal line $y = 0$. Shade the region above the
boundary line using a solid line border.

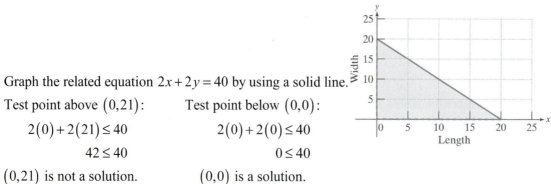

Graph the related equation $2x + 2y = 40$ by using a solid line.

Test point above $(0,21)$:

$2(0) + 2(21) \le 40$

$42 \le 40$

$(0,21)$ is not a solution.

Test point below $(0,0)$:

$2(0) + 2(0) \le 40$

$0 \le 40$

$(0,0)$ is a solution.

Shade the region below the boundary line.

63. **a.** $x \ge 0,\ y \ge 0$

 b. $x \le 40,\ y \le 40$

 c. $x + y \ge 65$

 d. $x \ge 0$ and $y \ge 0$ and $x \le 40$ and $y \le 40$ and $x + y \ge 65$

   $x \ge 0$ represents the points to the right of the vertical line $x = 0$. Shade the region to the right of the boundary line using a solid line border.

   $y \ge 0$ represents the points above the horizontal line $y = 0$. Shade the region above the boundary line using a solid line border.

   $x \le 40$ represents the points to the left of the vertical line $x = 40$. Shade the region to the left of the boundary line using a solid line border.

   $y \le 40$ represents the points below the horizontal line $y = 40$. Shade the region below the boundary line using a solid line border.

   Graph the related equation $x + y = 65$ by using a solid line.

   Test point above $(0,66)$:

   $0 + 66 \ge 65$

   $66 \ge 65$

   $(0,66)$ is a solution.

   Test point below $(0,0)$:

   $0 + 0 \ge 65$

   $0 \ge 65$

   $(0,0)$ is not a solution.

   Shade the region above the boundary line.

 e. The solution is the intersection of the graphs.

 f.

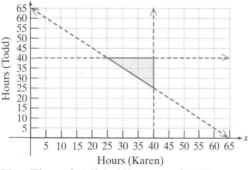

Yes. The point (35, 40) means that Karen works 35 hours and Todd works 40 hours.

No. The point (20, 40) means that Karen works 20 hours and Todd works 40 hours. This does not satisfy the constraint that there must be at least 65 hours total.

Section 3.6   Systems of Linear Equations in Three Variables and Applications

1. **a.** linear

    b. ordered triples

3. **a.** $3x + y = 4 \rightarrow y = -3x + 4$

      $4x + y = 5$

      $4x + (-3x + 4) = 5$

      $4x - 3x + 4 = 5$

      $x = 1$

      $y = -3x + 4 = -3(1) + 4 = -3 + 4 = 1$

      The solution is $\{(1,1)\}$.

      $3x + y = 4$

      $4x + y = 5$

    b. Multiply the first equation by –1, add to the second equation and solve for x:

$$3x + y = 4 \xrightarrow{\times -1} -3x - y = -4$$
$$4x + y = 5 \longrightarrow \underline{\quad 4x + y = \ 5}$$
$$x \quad = 1$$

      Substitute into the first equation and solve for y:

      $3(1) + y = 4$

      $y = 1$

      The solution is $\{(1,1)\}$.

5. Let b = the speed of the bike in still air

    Let w = the speed of the wind

      $b + w$ = speed of the bike with the wind

      $b - w$ = speed of bike against the wind

	Distance	Rate	Time
Tailwind	24	$b + w$	4/3
Headwind	24	$b - w$	2

(rate)(time) = (distance)

$$(b + w)\left(\frac{4}{3}\right) = 24$$

$$(b - w)(2) = 24$$

Divide the first equation by 4/3, the second equation by 2, add the results, and solve:

$$(b + w)\left(\frac{4}{3}\right) = 24 \xrightarrow{div\ 4/3} b + w = 18$$

$$(b - w)(2) = 24 \xrightarrow{div\ 2} \underline{b - w = 12}$$
$$2b \quad = 30$$
$$b = 15$$

Substitute and solve for w:

    $15 + w = 18$

    $w = 3$

Marge's speed is 15 mph in still air. The wind speed is 3 mph.

7.
$$2x - y + z = 10$$
$$4x + 2y - 3z = 10$$
$$x - 3y + 2z = 8$$

Substitute $(2,1,7)$:

$$2(2) - 1 + 7 = 4 - 1 + 7 = 10 = 10$$
$$4(2) + 2(1) - 3(7) = 8 + 2 - 21 = -11 \neq 10$$

Not a solution.

Substitute $(3,-10,-6)$:

$$2(3) - (-10) + (-6) = 6 + 10 - 6 = 10 = 10$$
$$4(3) + 2(-10) - 3(-6) = 12 - 20 + 18$$
$$= 10 = 10$$
$$3 - 3(-10) + 2(-6) = 3 + 30 - 12 = 21 \neq 8$$

Not a solution.

Substitute $(4,0,2)$:

$$2(4) - (0) + (2) = 8 - 0 + 2 = 10 = 10$$
$$4(4) + 2(0) - 3(2) = 16 + 0 - 6 = 10 = 10$$
$$4 - 3(0) + 2(2) = 4 - 0 + 4 = 8 = 8$$

$(4,0,2)$ is a solution.

9.
$$-x - y - 4z = -6$$
$$x - 3y + z = -1$$
$$4x + y - z = 4$$

Substitute $(12,2,-2)$:

$$-(12) - (2) - 4(-2) = -12 - 2 + 8 = -6 = -6$$
$$12 - 3(2) + (-2) = 12 - 6 - 2 = 4 \neq -1$$

Not a solution.

Substitute $(4,2,1)$:

$$-(4) - (2) - 4(1) = -4 - 2 - 4 = -10 \neq -6$$

Not a solution.

Substitute $(1,1,1)$:

$$-(1) - (1) - 4(1) = -1 - 1 - 4 = -6 = -6$$
$$1 - 3(1) + (1) = 1 - 3 + 1 = -1 = -1$$
$$4(1) + (1) - (1) = 4 + 1 - 1 = 4 = 4$$

$(1,1,1)$ is a solution.

11.
$$2x + y - 3z = -12$$
$$3x - 2y - z = 3$$
$$-x + 5y + 2z = -3$$

Multiply the first equation by 2 and add to the second equation to eliminate y:

$$2x + y - 3z = -12 \xrightarrow{\times 2} 4x + 2y - 6z = -24$$
$$3x - 2y - z = 3 \longrightarrow 3x - 2y - z = 3$$
$$\overline{\qquad 7x \qquad -7z = -21}$$
$$x - z = -3$$

Multiply the first equation by -5 and add to the third equation to eliminate y:

$$2x + y - 3z = -12 \rightarrow -10x - 5y + 15z = 60$$
$$-x + 5y + 2z = -3 \rightarrow -x + 5y + 2z = -3$$
$$\overline{\qquad -11x \qquad +17z = 57}$$

Multiply the first result by 11 and add to the second result to eliminate x:

$$x - z = -3 \xrightarrow{\times 11} 11x - 11z = -33$$
$$-11x + 17z = 57 \longrightarrow -11x + 17z = 57$$
$$\overline{\qquad 6z = 24}$$
$$z = 4$$

Substitute and solve for x and y:

13.
$$x - 3y - 4z = -7$$
$$5x + 2y + 2z = -1$$
$$4x - y - 5z = -6$$

Multiply the third equation by -3 and add to the first equation to eliminate y:

$$x - 3y - 4z = -7 \rightarrow x - 3y - 4z = -7$$
$$4x - y - 5z = -6 \rightarrow -12x + 3y + 15z = 18$$
$$\overline{\qquad -11x \qquad +11z = 11}$$
$$-x + z = 1$$

Multiply the third equation by 2 and add to the second equation to eliminate y:

$$5x + 2y + 2z = -1 \rightarrow 5x + 2y + 2z = -1$$
$$4x - y - 5z = -6 \rightarrow 8x - 2y - 10z = -12$$
$$\overline{\qquad 13x \qquad -8z = -13}$$

Multiply the first result by 8 and add to the second result to eliminate z:

$$-x + z = 1 \xrightarrow{\times 8} -8x + 8z = 8$$
$$13x - 8z = -13 \longrightarrow 13x - 8z = -13$$
$$\overline{\qquad 5x = -5}$$
$$x = -1$$

$x - z = -3$

$x - 4 = -3$

$x = 1$

The solution is $\{(1, -2, 4)\}$.

$2x + y - 3z = -12$

$2(1) + y - 3(4) = -12$

$2 + y - 12 = -12$

$y = -2$

Substitute and solve for y and z:

$4x - y - 5z = -6$

$-x + z = 1$ $4(-1) - y - 5(0) = -6$

$-(-1) + z = 1$ $-4 - y - 0 = -6$

$z = 0$ $-y = -2$

$y = 2$

The solution is $\{(-1, 2, 0)\}$.

15. $4x + 2z = 12 + 3y \rightarrow \quad 4x - 3y + 2z = 12$

$2y = 3x + 3z - 5 \rightarrow \quad -3x + 2y - 3z = -5$

$y = 2x + 7z + 8 \rightarrow \quad 2x - y + 7z = -8$

Multiply the third equation by -3 and add to the first equation to eliminate y:

$4x - 3y + 2z = 12 \longrightarrow \quad 4x - 3y + 2z = 12$

$2x - y + 7z = -8 \xrightarrow{\times -3} -6x + 3y - 21z = 24$

$\overline{\quad -2x \qquad\quad -19z = 36}$

Multiply the third equation by 2 and add to the second equation to eliminate y:

$-3x + 2y - 3z = -5 \longrightarrow -3x + 2y - 3z = -5$

$2x - y + 7z = -8 \xrightarrow{\times 2} 4x - 2y + 14z = -16$

$\overline{\quad x \qquad\quad +11z = -21}$

Multiply the second result by 2 and add to the first result to eliminate x:

$-2x - 19z = 36 \longrightarrow \quad -2x - 19z = \ 36$

$x + 11z = -21 \xrightarrow{\times 2} \quad 2x + 22z = -42$

$\overline{\qquad\qquad\qquad 3z = -6}$

$z = -2$

Substitute and solve for y and z:

$x + 11z = -21$

$x + 11(-2) = -21$

$x - 22 = -21$

$x = 1$

$2x - y + 7z = -8$

$2(1) - y + 7(-2) = -8$

$2 - y - 14 = -8$

$-y = 4$

$y = -4$

The solution is $\{(1, -4, -2)\}$.

17. $x + y + z = 6$

$-x + y - z = -2$

$2x + 3y + z = 11$

Add the first and second equations to eliminate z:

$x + y + z = 6$

$-x + y - z = -2$

$\overline{\quad 2y \quad\ = 4}$

$y = 2$

Add the second and third equations to eliminate z:

$-x + y - z = -2$

$2x + 3y + z = 11$

$\overline{\quad x + 4y \quad = 9}$

Substitute $y = 2$ and solve for x:

$x + 4(2) = 9$

$x + 8 = 9$

$x = 1$

Substitute into the first equation and solve for z:

$1 + 2 + z = 6$

$z = 3$

The solution is $\{(1, 2, 3)\}$.

19. $2x - 3y + 2z = -1$

$x + 2y \quad\ = -4$

$x \qquad +z = \ 1$

Multiply the third equation by -2 and add to the first equation to eliminate z:

21. $4x + 9y \quad\ = \ 8$

$8x \qquad +6z = -1$

$6y + 6z = -1$

Multiply the third equation by -1 and add to the second equation to eliminate z:

129

$$2x - 3y + 2z = -1 \longrightarrow 2x - 3y + 2z = -1$$
$$x \quad + z = 1 \xrightarrow{\times -2} \underline{-2x \quad\quad -2z = -2}$$
$$-3y \quad\quad = -3$$
$$y = 1$$

Substitute and solve for x and z:

$$x + 2y = -4$$
$$x + 2(1) = -4 \qquad x + z = 1$$
$$\qquad\qquad\qquad -6 + z = 1$$
$$x + 2 = -4$$
$$\qquad\qquad\qquad z = 7$$
$$x = -6$$

The solution is $\{(-6, 1, 7)\}$.

$$8x \quad\quad + 6z = -1 \longrightarrow \quad 8x \quad\quad + 6z = -1$$
$$6y + 6z = -1 \xrightarrow{\times -1} \quad \underline{-6y - 6z = 1}$$
$$8x - 6y \quad = 0$$

Multiply the first equation by -2 and add to this result to eliminate x:

$$4x + 9y = 8 \xrightarrow{\times -2} \quad -8x - 18y = -16$$
$$8x - 6y = 0 \longrightarrow \quad \underline{8x - 6y = 0}$$
$$-24y = -16$$
$$y = \frac{2}{3}$$

Substitute and solve for x and z:

$$4x + 9y = 8 \qquad\qquad 8x + 6z = -1$$
$$4x + 9\left(\frac{2}{3}\right) = 8 \qquad 8\left(\frac{1}{2}\right) + 6z = -1$$
$$4x + 6 = 8 \qquad\qquad 4 + 6z = -1$$
$$4x = 2 \qquad\qquad\qquad 6z = -5$$
$$x = \frac{1}{2} \qquad\qquad\qquad z = -\frac{5}{6}$$

The solution is $\left\{\left(\frac{1}{2}, \frac{2}{3}, -\frac{5}{6}\right)\right\}$.

---

23. Let $x =$ the first angle
Let $y =$ the second angle
Let $z =$ the third angle
$$x + y + z = 180$$
$$y = 2x + 5$$
$$z = 3x - 11$$
Substitute the second and third equations into the first and solve for x:
$$x + (2x + 5) + (3x - 11) = 180$$
$$6x - 6 = 180$$
$$6x = 186$$
$$x = 31$$
Substitute and solve for y and z:
$$y = 2x + 5 \qquad z = 3x - 11$$
$$y = 2(31) + 5 \qquad z = 3(31) - 11$$
$$y = 62 + 5 = 67 \qquad z = 93 - 11 = 82$$
The angles are $31°$, $67°$, and $82°$.

25. Let $x =$ the shortest side
Let $y =$ the middle side
Let $z =$ the longest side
$$x + y + z = 55 \qquad\rightarrow\qquad x + y + z = 55$$
$$x = y - 8 \qquad\rightarrow\qquad x - y \quad = -8$$
$$z = x + y - 1 \rightarrow -x - y + z = -1$$
Add the first and third equations to eliminate x:
$$x + y + z = 55$$
$$\underline{-x - y + z = -1}$$
$$2z = 54$$
$$z = 27$$
Add the second and third equations to eliminate x:
$$x - y \quad = -8$$
$$\underline{-x - y + z = -1}$$
$$-2y + z = -9$$
Substitute and solve for x and y:
$$-2y + z = -9$$
$$-2y + 27 = -9 \qquad\qquad x - y = -8$$
$$-2y = -36 \qquad\qquad x - 18 = -8$$
$$y = 18 \qquad\qquad\qquad x = 10$$

The lengths of the sides are 10 cm, 18 cm, and 27 cm.

27. Let x = the fiber in the supplement
Let y = the fiber in the oatmeal
Let z = the fiber in the cereal

$$3x + y + 4z = 19$$
$$2x + 4y + 2z = 25$$
$$5x + 3y + 2z = 30$$

Multiply the first equation by -4 and add to the second equation to eliminate y:

$$3x + y + 4z = 19 \xrightarrow{\times -4} -12x - 4y - 16z = -76$$
$$2x + 4y + 2z = 25 \xrightarrow{\hphantom{\times -4}} \underline{\hphantom{-1}2x + 4y + 2z = \hphantom{-}25}$$
$$-10x \hphantom{+4y} - 14z = -51$$

Multiply the first equation by -3 and add to the third equation to eliminate y:

$$3x + y + 4z = 19 \xrightarrow{\times -3} -9x - 3y - 12z = -57$$
$$5x + 3y + 2z = 30 \xrightarrow{\hphantom{\times -3}} \underline{\hphantom{-}5x + 3y + \hphantom{1}2z = \hphantom{-}30}$$
$$-4x \hphantom{+3y} - 10z = -27$$

Multiply the second new equation by -2.5 and add to the first new equation to eliminate x:

$$-10x - 14z = -51 \xrightarrow{\hphantom{\times -2.5}} -10x - 14z = -51$$
$$-4x - 10z = -27 \xrightarrow{\times -2.5} \underline{\hphantom{-}10x + 25z = 67.5}$$
$$11z = 16.5$$
$$z = 1.5$$

Substitute and solve for x and y:
$$-4x - 10z = -27$$
$$-4x - 10(1.5) = -27$$
$$-4x - 15 = -27$$
$$-4x = -12$$
$$x = 3$$
$$3x + y + 4z = 19$$
$$3(3) + y + 4(1.5) = 19$$
$$9 + y + 6 = 19$$
$$y + 15 = 19$$
$$y = 4$$

The fiber supplement has 3 g; the oatmeal has 4 g; and the cereal has 1.5 g.

29. Let x = the number of par 3 holes
Let y = the number of par 4 holes
Let z = the number of par 5 holes

$$x + y + z = 18$$
$$y = 3x$$
$$z = x + 3$$

Substitute the second and third equations into the first and solve for x:

$$x + (3x) + (x + 3) = 18$$
$$5x + 3 = 18$$
$$5x = 15$$
$$x = 3$$

Substitute and solve for y and z:
$$y = 3x \qquad\qquad z = x + 3$$
$$y = 3(3) \qquad\quad\; z = 3 + 3$$
$$y = 9 \qquad\qquad\; z = 6$$

There are three par 3 holes, nine par 4 holes, and six par 5 holes.

31. Let x = the price of a hat
Let y = the price of a T-shirt
Let z = the price of a jacket

33. Let x = the amount invested in small caps
Let y = the amount invested in global markets
Let z = the amount invested in the balanced fund

$3x + 2y + z = 140$

$2x + 2y + 2z = 170$

$x + 3y + 2z = 180$

Multiply the first equation by -2 and add to the second equation to eliminate z:

$3x + 2y + z = 140 \longrightarrow -6x - 4y - 2z = -280$

$2x + 2y + 2z = 170 \longrightarrow \underline{2x + 2y + 2z = 170}$

$$-4x - 2y \quad = -110$$

Multiply the third equation by -1 and add to the second equation to eliminate z:

$2x + 2y + 2z = 170 \longrightarrow 2x + 2y + 2z = 170$

$x + 3y + 2z = 180 \xrightarrow{\times -1} \underline{-x - 3y - 2z = -180}$

$$x - y \quad = -10$$

Multiply the second result by -2 and add to the first result to eliminate y:

$-4x - 2y = -110 \longrightarrow \quad -4x - 2y = -110$

$x - y = -10 \xrightarrow{\times -2} \underline{-2x + 2y = \quad 20}$

$$-6x \quad = -90$$

$$x = 15$$

Substitute and solve for y and z:

$x - y = -10 \qquad 3x + 2y + z = 140$

$15 - y = -10 \qquad 3(15) + 2(25) + z = 140$

$-y = -25 \qquad 45 + 50 + z = 140$

$y = 25 \qquad z = 45$

Hats cost \$15. T-shirts cost \$25, and jackets cost \$45.

35. $\quad 2x + y + 3z = 2$

$\qquad x - y + 2z = -4$

$-2x + 2y - 4z = 8$

Add the first and second equations to eliminate y:

$2x + y + 3z = 2$

$\underline{x - y + 2z = -4}$

$3x \quad + 5z = -2$

Multiply the second equation by 2 and add to the third equation to eliminate y:

$x - y + 2z = -4 \xrightarrow{\times 2} 2x - 2y + 4z = -8$

$-2x + 2y - 4z = 8 \longrightarrow \underline{-2x + 2y - 4z = 8}$

$$0 = 0$$

The equations are dependent.

$x + \quad y + \quad z = 25{,}000 \rightarrow x + 4y + 2z = \quad 1040$

$0.06x + 0.10y + 0.09z = \quad 2106 \rightarrow 6x + 10y + 9z = 216{,}000$

$\qquad y \quad = \quad 2z \rightarrow \qquad y \quad = \quad 2z$

Substitute into the first two equations to eliminate y:

$x + y + z = 25{,}000 \qquad 6x + 10y + 9z = 216{,}000$

$x + 2z + z = 25{,}000 \quad 6x + 10(2z) + 9z = 216{,}000$

$x + 3z = 25{,}000 \qquad 6x + 20z + 9z = 216{,}000$

$$6x + 29z = 216{,}000$$

Multiply the first new equation by -6 and add to the second new equation to eliminate x:

$x + 3z = 25{,}000 \xrightarrow{\times -6} -6x - 18z = -150{,}000$

$6x + 29z = 216{,}000 \longrightarrow \underline{6x + 29z = \quad 216{,}000}$

$$11z = \quad 66{,}000$$

$$z = \quad 6000$$

Substitute and solve for x:

$x + 3z = 25{,}000$

$x + 3(6000) = 25{,}000$

$x + 18{,}000 = 25{,}000$

$x = 7000$

Substitute and solve for y:

$x + y + z = 25{,}000$

$7000 + y + 6000 = 25{,}000$

$y + 13{,}000 = 25{,}000$

$y = 12{,}000$

Walter invested \$7000 in small caps, \$12,000 in global markets, and \$6000 in the balanced fund.

37. $\quad 6x - 2y + 2z = 2$

$\qquad 4x + 8y - 2z = 5$

$-2x - 4y + z = -2$

Multiply the third equation by 2 and add to the second equation to eliminate z:

$4x + 8y - 2z = 5 \longrightarrow \quad 4x + 8y - 2z = 5$

$-2x - 4y + z = -2 \xrightarrow{\times 2} \underline{-4x - 8y + 2z = -4}$

$$0 \neq 1$$

The system is inconsistent. There is no solution.

39. Multiply by the LCD of each equation:

$\frac{1}{2}x+\frac{2}{3}y = \frac{5}{2} \xrightarrow{\times 6} 3x+4y = 15$

$\frac{1}{5}x \quad -\frac{1}{2}z = -\frac{3}{10} \xrightarrow{\times 10} 2x \quad -5z = -3$

$\frac{1}{3}y-\frac{1}{4}z = \frac{3}{4} \xrightarrow{\times 12} 4y-3z = 9$

Multiply the first equation by –1 and add to the third equation to eliminate y:

$3x+4y = 15 \xrightarrow{\times -1} -3x-4y = -15$

$4y-3z = 9 \longrightarrow \underline{4y-3z = 9}$

$-3x \quad -3z = -6$

Multiply this result by 2/3 and add to the second equation to eliminate x:

$-3x-3z = -6 \xrightarrow{\times 2/3} -2x-2z = -4$

$2x-5z = -3 \longrightarrow \underline{2x-5z = -3}$

$-7z = -7$

$z = 1$

Substitute and solve for x and y:

$\begin{array}{ll} 2x-5z = -3 & 4y-3z = 9 \\ 2x-5(1) = -3 & 4y-3(1) = 9 \\ 2x-5 = -3 & 4y-3 = 9 \\ 2x = 2 & 4y = 12 \\ x = 1 & y = 3 \end{array}$

The solution is $\{(1,3,1)\}$.

41.
$-3x+y-z = 8$

$-4x+2y+3z = -3$

$2x+3y-2z = -1$

Multiply the first equation by 3 and add to the second equation to eliminate z:

$-3x+y-z = 8 \xrightarrow{\times 3} -9x+3y-3z = 24$

$-4x+2y+3z = -3 \longrightarrow \underline{-4x+2y+3z = -3}$

$-13x+5y = 21$

Multiply the first equation by –2 and add to the third equation to eliminate z:

$-3x+y-z = 8 \xrightarrow{\times -2} 6x-2y+2z = -16$

$2x+3y-2z = -1 \longrightarrow \underline{2x+3y-2z = -1}$

$8x+y = -17$

Multiply the second result by –5 and add to the first result to eliminate y:

$-13x+5y = 21 \longrightarrow -13x+5y = 21$

$8x+y = -17 \xrightarrow{\times -5} \underline{-40x-5y = 85}$

$-53x = 106$

$x = -2$

Substitute and solve for y and z:

$\begin{array}{ll} 8x+y = -17 & -3x+y-z = 8 \\ 8(-2)+y = -17 & -3(-2)+(-1)-z = 8 \\ -16+y = -17 & 6-1-z = 8 \\ y = -1 & -z = 3 \\ & z = -3 \end{array}$

43.
$2x+y = 3(z-1) \quad \rightarrow \quad 2x+y-3z = -3$

$3x-2(y-2z) = 1 \quad \rightarrow \quad 3x-2y+4z = 1$

$2(2x-3z) = -6-2y \rightarrow 4x+2y-6z = -6$

Multiply the first equation by –2 and add to the third equation to eliminate z:

$2x+y-3z = -3 \xrightarrow{\times -2} -4x-2y+6z = 6$

$4x+2y-6z = -6 \longrightarrow \underline{4x+2y-6z = -6}$

$0 = 0$

The equations are dependent.

The solution is $\{(-2,-1,-3)\}$.

45. Multiply each equation by 10:

$$-0.1y + 0.2z = 0.2 \rightarrow \quad -y + 2z = 2$$
$$0.1x + 0.1y + 0.1z = 0.2 \rightarrow \quad x + y + z = 2$$
$$-0.1x \quad\quad -0.3z = 0.2 \rightarrow \quad -x \quad -3z = 2$$

Add the first and second equations to eliminate y:

$$-y + 2z = 2$$
$$\underline{x + y + z = 2}$$
$$x \quad +3z = 4$$

Add this result to the third equation to eliminate x:

$$x + 3z = 4$$
$$\underline{-x - 3z = 2}$$
$$0 = 6$$

The system is inconsistent. There is no solution.

47.
$$2x - 4y + 8z = 0$$
$$-x - 3y + z = 0$$
$$x - 2y + 5z = 0$$

Add the second and third equations to eliminate x:

$$-x - 3y + z = 0$$
$$\underline{x - 2y + 5z = 0}$$
$$-5y + 6z = 0$$

Multiply the second equation by 2 and add to the first equation to eliminate x:

$$2x - 4y + 8z = 0 \longrightarrow \quad 2x - 4y + 8z = 0$$
$$-x - 3y + z = 0 \xrightarrow{\times 2} -2x - 6y + 2z = 0$$
$$\overline{\quad\quad\quad\quad -10y + 10z = 0}$$

Multiply the first result by -2 and add to the second result to eliminate y:

$$-5y + 6z = 0 \xrightarrow{\times -2} \quad 10y - 12z = 0$$
$$-10y + 10z = 0 \longrightarrow \underline{-10y + 10z = 0}$$
$$-2z = 0$$
$$z = 0$$

Substitute and solve for x and y:

$$-5y + 6z = 0 \quad\quad x - 2y + 5z = 0$$
$$-5y + 6(0) = 0 \quad x - 2(0) + 5(0) = 0$$
$$-5y = 0 \quad\quad\quad x - 0 + 0 = 0$$
$$y = 0 \quad\quad\quad\quad x = 0$$

The solution is $\{(0,0,0)\}$.

49.
$$4x - 2y - 3z = 0$$
$$-8x - y + z = 0$$
$$2x - y - \frac{3}{2}z = 0$$

Multiply the third equation by -2 and add to the first equation to eliminate y:

$$4x - 2y - 3z = 0 \longrightarrow \quad 4x - 2y - 3z = 0$$
$$2x - y - \frac{3}{2}z = 0 \xrightarrow{\times -2} -4x + 2y + 3z = 0$$
$$\overline{\quad\quad\quad\quad\quad\quad 0 = 0}$$

The equations are dependent.

Section 3.7   Solving Systems of Linear Equations by Using Matrices

1. **a.**   matrix; rows; columns
    b.   column; one; square
    c.   coefficient; augmented
    d.   row echelon

3. $x - 6y = 9$

$x + 2y = 13$

Multiply the first equation by –1, add to the second equation and solve for y:

$x - 6y = 9 \xrightarrow{\times -1} \quad -x + 6y = -9$

$x + 2y = 13 \longrightarrow \quad \underline{x + 2y = 13}$

$\qquad\qquad\qquad\qquad\quad 8y = 4$

$$y = \frac{1}{2}$$

Substitute into the first equation and solve for x:

$x - 6y = 9$

$x - 6\left(\dfrac{1}{2}\right) = 9$

$x - 3 = 9$

$x = 12$

The solution is $\left\{\left(12, \dfrac{1}{2}\right)\right\}$.

5. $2x - y + z = -4$

$-x + y + 3z = -7$

$x + 3y - 4z = 22$

Add the first and second equations to eliminate y:

$2x - y + z = -4$

$\underline{-x + y + 3z = -7}$

$x \qquad + 4z = -11$

Multiply the first equation by 3 and add to the third equation to eliminate y:

$2x - y + z = -4 \xrightarrow{\times 3} 6x - 3y + 3z = -12$

$x + 3y - 4z = 22 \longrightarrow \underline{x + 3y - 4z = 22}$

$\qquad\qquad\qquad\qquad\qquad 7x \qquad - z = 10$

Multiply the second result by 4 and add to the first result to eliminate z:

$x + 4z = -11 \longrightarrow \qquad\qquad x + 4z = -11$

$7x - z = 10 \xrightarrow{\times 4} \quad \underline{28x - 4z = 40}$

$\qquad\qquad\qquad\qquad\qquad 29x \qquad = 29$

$\qquad\qquad\qquad\qquad\qquad\qquad\quad x = 1$

Substitute and solve for y and z:

$x + 4z = -11 \qquad\qquad -x + y + 3z = -7$

$1 + 4z = -11 \qquad\qquad -1 + y + 3(-3) = -7$

$4z = -12 \qquad\qquad\qquad -1 + y - 9 = -7$

$z = -3 \qquad\qquad\qquad\qquad\quad y = 3$

The solution is $\{(1, 3, -3)\}$.

7. 3×1, column matrix

9. 2×2, square matrix

11. 1×4, row matrix

13. 2×3, none of these

15. $\begin{bmatrix} 1 & -2 & | & -1 \\ 2 & 1 & | & -7 \end{bmatrix}$

17. $\begin{bmatrix} 1 & -2 & 1 & | & 5 \\ 2 & 6 & 3 & | & -2 \\ 3 & -1 & -2 & | & 1 \end{bmatrix}$

19. $4x + 3y = 6$

$12x + 5y = -6$

21. $x = 4$

$y = -1$

$z = 7$

23. a. 7
  b. −2

25. $Z = \begin{bmatrix} 2 & 1 & | & 11 \\ 2 & -1 & | & 1 \end{bmatrix} \xrightarrow{\frac{1}{2}R_1 \Rightarrow R_1} \begin{bmatrix} 1 & \frac{1}{2} & | & \frac{11}{2} \\ 2 & -1 & | & 1 \end{bmatrix}$

27. $K = \begin{bmatrix} 5 & 2 & | & 1 \\ 1 & -4 & | & 3 \end{bmatrix} \xrightarrow{R_1 \Leftrightarrow R_2} \begin{bmatrix} 1 & -4 & | & 3 \\ 5 & 2 & | & 1 \end{bmatrix}$

29. $M = \begin{bmatrix} 1 & 5 & | & 2 \\ -3 & -4 & | & -1 \end{bmatrix} \xrightarrow{3R_1 + R_2 \Rightarrow R_2} \begin{bmatrix} 1 & 5 & | & 2 \\ 0 & 11 & | & 5 \end{bmatrix}$

31. a. $\begin{bmatrix} 1 & 3 & 0 & | & -1 \\ 4 & 1 & -5 & | & 6 \\ -2 & 0 & -3 & | & 10 \end{bmatrix}$

$\xrightarrow{-4R_1 + R_2 \Rightarrow R_2} \begin{bmatrix} 1 & 3 & 0 & | & -1 \\ 0 & -11 & -5 & | & 10 \\ -2 & 0 & -3 & | & 10 \end{bmatrix}$

  b. $\xrightarrow{2R_1 + R_3 \Rightarrow R_3} \begin{bmatrix} 1 & 3 & 0 & | & -1 \\ 0 & -11 & -5 & | & 10 \\ 0 & 6 & -3 & | & 8 \end{bmatrix}$

33. True

35. True

37. Interchange rows 1 and 2.

39. Multiply row 1 by −3 and add to row 2.
Replace row 2 with the result.

41. $x - 2y = -1$
$2x + y = -7$

$\begin{bmatrix} 1 & -2 & | & -1 \\ 2 & 1 & | & -7 \end{bmatrix} \xrightarrow{-2R_1 + R_2 \Rightarrow R_2} \begin{bmatrix} 1 & -2 & | & -1 \\ 0 & 5 & | & -5 \end{bmatrix} \xrightarrow{\frac{1}{5}R_2 \Rightarrow R_2} \begin{bmatrix} 1 & -2 & | & -1 \\ 0 & 1 & | & -1 \end{bmatrix} \xrightarrow{2R_2 + R_1 \Rightarrow R_1} \begin{bmatrix} 1 & 0 & | & -3 \\ 0 & 1 & | & -1 \end{bmatrix}$

The solution is $\{(-3, -1)\}$.

43. $x + 3y = 6$
$-4x - 9y = 3$

$\begin{bmatrix} 1 & 3 & | & 6 \\ -4 & -9 & | & 3 \end{bmatrix} \xrightarrow{4R_1 + R_2 \Rightarrow R_2} \begin{bmatrix} 1 & 3 & | & 6 \\ 0 & 3 & | & 27 \end{bmatrix} \xrightarrow{\frac{1}{3}R_2 \Rightarrow R_2} \begin{bmatrix} 1 & 3 & | & 6 \\ 0 & 1 & | & 9 \end{bmatrix} \xrightarrow{-3R_2 + R_1 \Rightarrow R_1} \begin{bmatrix} 1 & 0 & | & -21 \\ 0 & 1 & | & 9 \end{bmatrix}$

The solution is $\{(-21, 9)\}$.

45. $x + 3y = 3$
$4x + 12y = 12$

$\begin{bmatrix} 1 & 3 & | & 3 \\ 4 & 12 & | & 12 \end{bmatrix} \xrightarrow{-4R_1 + R_2 \Rightarrow R_2} \begin{bmatrix} 1 & 3 & | & 3 \\ 0 & 0 & | & 0 \end{bmatrix}$

Infinitely many solutions of the form $\{(x, y) | x + 3y = 3\}$. The equations are dependent.

47. $x - y = 4$

$2x + y = 5$

$$\begin{bmatrix} 1 & -1 & | & 4 \\ 2 & 1 & | & 5 \end{bmatrix} \xrightarrow{-2R_1 + R_2 \Rightarrow R_2} \begin{bmatrix} 1 & -1 & | & 4 \\ 0 & 3 & | & -3 \end{bmatrix} \xrightarrow{\frac{1}{3}R_2 \Rightarrow R_2} \begin{bmatrix} 1 & -1 & | & 4 \\ 0 & 1 & | & -1 \end{bmatrix} \xrightarrow{R_2 + R_1 \Rightarrow R_1} \begin{bmatrix} 1 & 0 & | & 3 \\ 0 & 1 & | & -1 \end{bmatrix}$$

The solution is $\{(3, -1)\}$.

49. $x + 3y = -1$

$-3x - 6y = 12$

$$\begin{bmatrix} 1 & 3 & | & -1 \\ -3 & -6 & | & 12 \end{bmatrix} \xrightarrow{3R_1 + R_2 \Rightarrow R_2} \begin{bmatrix} 1 & 3 & | & -1 \\ 0 & 3 & | & 9 \end{bmatrix} \xrightarrow{\frac{1}{3}R_2 \Rightarrow R_2} \begin{bmatrix} 1 & 3 & | & -1 \\ 0 & 1 & | & 3 \end{bmatrix} \xrightarrow{-3R_2 + R_1 \Rightarrow R_1} \begin{bmatrix} 1 & 0 & | & -10 \\ 0 & 1 & | & 3 \end{bmatrix}$$

The solution is $\{(-10, 3)\}$.

51. $3x + y = -4$

$-6x - 2y = 3$

$$\begin{bmatrix} 3 & 1 & | & -4 \\ -6 & -2 & | & 3 \end{bmatrix} \xrightarrow{2R_1 + R_2 \Rightarrow R_2} \begin{bmatrix} 3 & 1 & | & -4 \\ 0 & 0 & | & -5 \end{bmatrix}$$

There is no solution; $\{ \ \}$. The system is inconsistent.

53. $x + y + z = 6$

$x - y + z = 2$

$x + y - z = 0$

$$\begin{bmatrix} 1 & 1 & 1 & | & 6 \\ 1 & -1 & 1 & | & 2 \\ 1 & 1 & -1 & | & 0 \end{bmatrix} \xrightarrow[-R_1 + R_3 \Rightarrow R_3]{-R_1 + R_2 \Rightarrow R_2} \begin{bmatrix} 1 & 1 & 1 & | & 6 \\ 0 & -2 & 0 & | & -4 \\ 0 & 0 & -2 & | & -6 \end{bmatrix} \xrightarrow{-\frac{1}{2}R_2 \Rightarrow R_2} \begin{bmatrix} 1 & 1 & 1 & | & 6 \\ 0 & 1 & 0 & | & 2 \\ 0 & 0 & -2 & | & -6 \end{bmatrix}$$

$$\xrightarrow{-R_2 + R_1 \Rightarrow R_1} \begin{bmatrix} 1 & 0 & 1 & | & 4 \\ 0 & 1 & 0 & | & 2 \\ 0 & 0 & -2 & | & -6 \end{bmatrix} \xrightarrow{-\frac{1}{2}R_3 \Rightarrow R_3} \begin{bmatrix} 1 & 0 & 1 & | & 4 \\ 0 & 1 & 0 & | & 2 \\ 0 & 0 & 1 & | & 3 \end{bmatrix} \xrightarrow{-R_3 + R_1 \Rightarrow R_1} \begin{bmatrix} 1 & 0 & 0 & | & 1 \\ 0 & 1 & 0 & | & 2 \\ 0 & 0 & 1 & | & 3 \end{bmatrix}$$

The solution is $\{(1, 2, 3)\}$.

55. $x - 2y \quad = 5 - z \rightarrow \quad x - 2y + z = 5$

$2x + 6y + 3z = -10 \rightarrow 2x + 6y + 3z = -10$

$3x - y - 2z = 5 \rightarrow 3x - y - 2z = 5$

$$\begin{bmatrix} 1 & -2 & 1 & | & 5 \\ 2 & 6 & 3 & | & -10 \\ 3 & -1 & -2 & | & 5 \end{bmatrix} \xrightarrow[-3R_1 + R_3 \Rightarrow R_3]{-2R_1 + R_2 \Rightarrow R_2} \begin{bmatrix} 1 & -2 & 1 & | & 5 \\ 0 & 10 & 1 & | & -20 \\ 0 & 5 & -5 & | & -10 \end{bmatrix} \xrightarrow{\frac{1}{10}R_2 \Rightarrow R_2} \begin{bmatrix} 1 & -2 & 1 & | & 5 \\ 0 & 1 & \frac{1}{10} & | & -2 \\ 0 & 5 & -5 & | & -10 \end{bmatrix}$$

$$\xrightarrow[-5R_2 + R_3 \Rightarrow R_3]{2R_2 + R_1 \Rightarrow R_1} \begin{bmatrix} 1 & 0 & \frac{6}{5} & | & 1 \\ 0 & 1 & \frac{1}{10} & | & -2 \\ 0 & 0 & -\frac{11}{2} & | & 0 \end{bmatrix} \xrightarrow{-\frac{2}{11}R_3 \Rightarrow R_3} \begin{bmatrix} 1 & 0 & \frac{6}{5} & | & 1 \\ 0 & 1 & \frac{1}{10} & | & -2 \\ 0 & 0 & 1 & | & 0 \end{bmatrix} \xrightarrow[-\frac{1}{10}R_3 + R_2 \Rightarrow R_2]{-\frac{6}{5}R_3 + R_1 \Rightarrow R_1} \begin{bmatrix} 1 & 0 & 0 & | & 1 \\ 0 & 1 & 0 & | & -2 \\ 0 & 0 & 1 & | & 0 \end{bmatrix}$$

The solution is $\{(1, -2, 0)\}$.

Chapter 3 Group Activity

1.

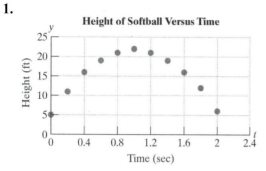

Height of Softball Versus Time

3. Answers will vary.

5. Answers will vary, but should be close to $y = -16t^2 + 32.5t + 5$. The equation represents the height of the ball t seconds after being thrown.

7. $y = -16(1.4)^2 + 32.5(1.4) + 5$

$\quad = -31.36 + 45.5 + 5$

$\quad = 19.14 \approx 19\,\text{ft}$

Answers will vary but should be close to the observed value of 19 ft.

Chapter 3 Review Exercises

Section 3.1

1.   a. $-5x - 7y = 4$

$y = -\dfrac{1}{2}x - 1$

Substitute $(2,2)$:

$-5(2) - 7(2) = -10 - 14 = -24 \neq 4$

$(2,2)$ is not a solution.

b.   Substitute $(2,-2)$:

$-5(2) - 7(-2) = -10 + 14 = 4 = 4$

$-2 = -\dfrac{1}{2}(2) - 1 = -1 - 1 = -2$

$(2,-2)$ is a solution.

3. True

5. $f(x) = x - 1$

$g(x) = 2x - 4$

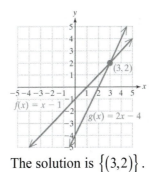

The solution is $\left\{(3,2)\right\}$.

7. $6x + 2y = 4$ $3x = -y + 2$

$2y = -6x + 4$ $y = -3x + 2$

$y = -3x + 2$

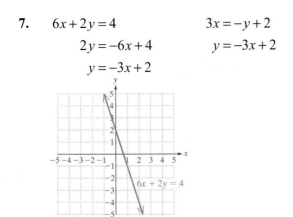

Infinitely many solutions of the form
$\left\{(x,y)\,\middle|\,6x + 2y = 4\right\}$; dependent equations.

Section 3.2

9.

$$y = \frac{3}{4}x - 4$$

$$-x + 2y = -6$$

$$-x + 2\left(\frac{3}{4}x - 4\right) = -6$$

$$-x + \frac{3}{2}x - 8 = -6$$

$$\frac{1}{2}x = 2$$

$$x = 4$$

$$y = \frac{3}{4}x - 4 = \frac{3}{4}(4) - 4 = 3 - 4 = -1$$

The solution is $\left\{(4,-1)\right\}$.

11. $2(x + y) = 10 - 3y \rightarrow 2x + 2y = 10 - 3y$

$$\rightarrow 2x + 5y = 10$$

$$0.4x + y = 1.2 \rightarrow y = 1.2 - 0.4x$$

$$2x + 5(1.2 - 0.4x) = 10$$

$$2x + 6 - 2x = 10$$

$$6 \neq 10$$

There is no solution; $\left\{\ \right\}$. The system is
inconsistent.

13. $60(5x - y) = 90 \rightarrow 300x - 60y = 90$

$$10x = 2y + 3 \rightarrow 2y = 10x - 3$$

$$\rightarrow y = 5x - \frac{3}{2}$$

15. $y = 105 + 45x$

$$y = 48.50x$$

$$105 + 45x = 48.50x$$

$$105 = 3.5x$$

$$30 = x$$

The cost would be the same for 30 months..

$$300x - 60\left(5x - \frac{3}{2}\right) = 90$$
$$300x - 300x + 90 = 90$$
$$90 = 90$$

Infinitely many solutions of the form
$\{(x,y) \mid 10x - 2y = 3\}$; dependent equations.

Section 3.3

17. Multiply each equation by its LCD:

$\frac{2}{5}x + \frac{3}{5}y = 1 \xrightarrow{\times 5} 2x + 3y = 5$

$x - \frac{2}{3}y = \frac{1}{3} \xrightarrow{\times 3} 3x - 2y = 1$

Multiply the first equation by 2 and the second equation by 3, add the results and solve for x:

$2x + 3y = 5 \xrightarrow{\times 2} 4x + 6y = 10$

$3x - 2y = 1 \xrightarrow{\times 3} 9x - 6y = 3$

$$\overline{13x = 13}$$
$$x = 1$$

Substitute into the first equation and solve for y:

$2(1) + 3y = 5$
$2 + 3y = 5$
$3y = 3$
$y = 1$ The solution is $\{(1,1)\}$.

19. $3x + 4y = 2$
$2x + 5y = -1$

Multiply the first equation by 5 and the second equation by –4, add the results and solve for x:

$3x + 4y = 2 \xrightarrow{\times 5} 15x + 20y = 10$

$2x + 5y = -1 \xrightarrow{\times -4} -8x - 20y = 4$

$$\overline{7x = 14}$$
$$x = 2$$

Substitute into the first equation and solve for y:

$3(2) + 4y = 2$
$6 + 4y = 2$
$4y = -4$
$y = -1$ The solution is $\{(2, -1)\}$.

21. Write in standard form:

$2y = 3x - 8 \;\rightarrow\; -3x + 2y = -8$

$-6x = -4y + 4 \;\rightarrow\; -6x + 4y = 4$

Multiply the first equation by –2, add to the second equation and solve for x:

$-3x + 2y = -8 \xrightarrow{\times -2} 6x - 4y = 16$

$-6x + 4y = 4 \xrightarrow{} -6x + 4y = 4$

$$\overline{0 \neq 20}$$

There is no solution; $\{\ \}$. The system is inconsistent.

23. Write in standard form:

$-(y + 4x) = 2x - 9 \rightarrow -6x - y = -9$

$-2x + 2y = -10 \quad \rightarrow -2x + 2y = -10$

Multiply the first equation by 2, add to the second equation and solve for x:

$-6x - y = -9 \xrightarrow{\times 2} -12x - 2y = -18$

$-2x + 2y = -10 \xrightarrow{} -2x + 2y = -10$

$$\overline{-14x = -28}$$
$$x = 2$$

Substitute into the first equation and solve for y:

$-6(2) - y = -9$
$-12 - y = -9$
$y = -3$

The solution is $\{(2, -3)\}$.

25. Multiply each equation by 10:

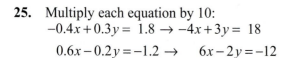

$-0.4x + 0.3y = 1.8 \rightarrow -4x + 3y = 18$

$0.6x - 0.2y = -1.2 \rightarrow 6x - 2y = -12$

Multiply the first equation by 2 and the second equation by 3, add the results and solve for x:

$$-4x + 3y = 18 \xrightarrow{\times 2} \quad -8x + 6y = 36$$
$$6x - 2y = -12 \xrightarrow{\times 3} \quad \underline{18x - 6y = -36}$$
$$10x \qquad = 0$$
$$x = 0$$

Substitute into the first equation and solve for y:

$$-4(0) + 3y = 18$$
$$0 + 3y = 18$$
$$y = 6$$

The solution is $\{(0,6)\}$.

Section 3.4

27. Let x = the amount invested at 5%
y = the amount invested at 3.5%

	5% Acct	3.5% Acct	Total
Principal	x	y	
Interest	$0.05x$	$0.035y$	303.75

$$x = 2y$$
$$0.05x + 0.035y = 303.75$$

Substitute and solve for y:

$$0.05(2y) + 0.035y = 303.75$$
$$0.10y + 0.035y = 303.75$$
$$0.135y = 303.75$$
$$y = 2250$$

Substitute into the first equation and solve for x: $x = 2y = 2(2250) = 4500$

$4500 was invested at 5%.

31. a. $f(x) = 9.95 + 0.10x$

b. $g(x) = 12.95 + 0.08x$

c.   Substitute and solve:
$$9.95 + 0.10x = 12.95 + 0.08x$$
$$0.02x = 3$$
$$x = 150$$
Both offers are the same when 150 min are used.

29. Let x = the amount of 20% saline solution
y = the amount of 50% saline solution

	20% sal	50% sal	31.25% sal
L solution	x	y	16
L saline	$0.20x$	$0.50y$	0.3125(16)

$$x + y = 16$$
$$0.20x + 0.50y = 0.3125(16)$$

Multiply the first equation by -0.20, add to the second equation and solve for y:

$$x + y = 16 \rightarrow -0.20x - 0.20y = -3.2$$
$$0.20x + 0.50y = 5 \rightarrow \underline{0.20x + 0.50y = 5.0}$$
$$0.30y = 1.8$$
$$y = 6$$

Substitute into the first equation and solve for x:

$$x + 6 = 16$$
$$x = 10$$

The mixture contains 10 L of 20% saline solution and 6 L of 50% saline solution.

Section 3.5

33. $2x > -y + 5$

Graph the related equation $2x = -y + 5$ by using a dashed line.

Test point above $(0,6)$:   Test point below $(0,0)$:

$\qquad 2(0) > -(6) + 5 \qquad\qquad 2(0) > -(0) + 5$

$\qquad\quad 0 > -1 \qquad\qquad\qquad\quad 0 > 5$

$(0,6)$ is a solution. $(0,0)$ is not a solution.

Shade the region above the boundary line.

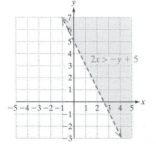

35. $x > -3$ represents all the points to the right of the vertical
line $x = -3$. The boundary is a dashed line.

Shade the region to the right of the boundary line.

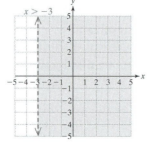

37. $x \ge \dfrac{1}{2} y$

Graph the related equation $x = \dfrac{1}{2} y$ by using a solid line.

Test point above $(0,2)$:   Test point below $(0,-2)$:

$\qquad 0 \ge \dfrac{1}{2}(2) \qquad\qquad\quad 0 \ge \dfrac{1}{2}(-2)$

$\qquad\quad 0 \ge 1 \qquad\qquad\qquad\quad 0 \ge -1$

$(0,2)$ is not a solution. $(0,-2)$ is a solution.

Shade the region below the boundary line.

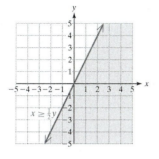

39. $2x - y > -2$ and $2x - y \le 2$

Graph the related equation $2x - y = -2$ by using a dashed line.

Test point above $(0,3)$:   Test point below $(0,0)$:

$$2(0) - 3 > -2$$ $$2(0) - 0 > -2$$

$$-3 > -2$$ $$0 > -2$$

$(0,3)$ is not a solution. $(0,0)$ is a solution.

Shade the region below the boundary line.

Graph the related equation $2x - y = 2$ by using a solid line.

Test point above $(0,0)$:   Test point below $(0,-3)$:

$$2(0) - 0 \le 2$$ $$2(0) - (-3) \le 2$$

$$0 \le 2$$ $$3 \le -2$$

$(0,0)$ is a solution. $(0,-3)$ is not a solution.

Shade the region above the boundary line.

The solution is the intersection of the graphs.

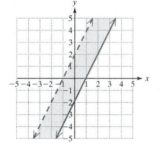

41. $y \ge x$ or $y \le -x$

Graph the related equation $y = x$ by using a solid line.

Test point above $(0,5)$: Test point below $(0,-5)$:

$5 \ge 0$ $-5 \ge 0$

$(0,5)$ is a solution. $(0,-5)$ is not a solution.

Shade the region above the boundary line.

Graph the related equation $y = -x$ by using a solid line.

Test point above $(1,5)$: Test point below $(1,-5)$:

$5 \le -1$ $-5 \le -1$

$(1,5)$ is not a solution. $(1,-5)$ is a solution.

Shade the region below the boundary line.

The solution is the union of the graphs.

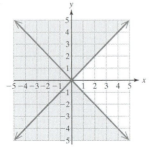

43. **a.** $x \ge 0,\ y \ge 0$

  b. $x + y \le 100$

  c. $x \ge 4y$

  $x \ge 0$ and $y \ge 0$ and $x + y \le 100$ and $x \ge 4y$

  $x \ge 0$ represents the points to the right of the vertical line $x = 0$. Shade the region to the right of the boundary line using a solid line border.

  $y \ge 0$ represents the points above the horizontal line $y = 0$. Shade the region above the boundary line using a solid line border.

  d. Graph the related equation $x + y = 100$ by using a solid line.

  Test point above $(0,101)$: Test point below $(0,0)$:

  $0 + 101 \le 100$ $0 + 0 \le 100$

  $101 \le 100$ $0 \le 100$

  $(0,101)$ is not a solution. $(0,0)$ is a solution.

  Shade the region below the boundary line.

  Graph the related equation $x = 4y$ by using a solid line.

  Test point above $(0,1)$: Test point below $(0,-1)$:

  $0 \ge 4(1)$ $0 \ge 4(-1)$

  $0 \ge 4$ $0 \ge -4$

  $(0,1)$ is not a solution. $(0,-1)$ is a solution.

  Shade the region below the boundary line.

  The solution is the intersection of the graphs.

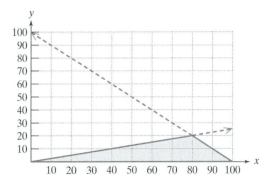

Section 3.6

45.
$$5x + 3y - z = 5$$
$$x + 2y + z = 6$$
$$-x - 2y - z = 8$$
Add the second and third equations to eliminate z:
$$x + 2y + z = 6$$
$$\underline{-x - 2y - z = 8}$$
$$0 \neq 14$$
The system is inconsistent. There is no solution.

47.
$$3x \quad + 4z = 5$$
$$2y + 3z = 2$$
$$2x - 5y \quad = 8$$
Multiply the first equation by 3 and the second equation by –4, and add the results to eliminate z:
$$3x \quad + 4z = 5 \xrightarrow{\times 3} \quad 9x \quad + 12z = 15$$
$$2y + 3z = 2 \xrightarrow{\times -4} \quad \underline{-8y - 12z = -8}$$
$$9x - 8y \quad = 7$$
Multiply the third equation by –9 and this result by 2, and add to eliminate x:
$$2x - 5y = 8 \xrightarrow{\times -9} -18x + 45y = -72$$
$$9x - 8y = 7 \xrightarrow{\times 2} \underline{18x - 16y = 14}$$
$$29y = -58$$
$$y = -2$$
Substitute and solve for x and z:
$$2x - 5y = 8 \qquad 2y + 3z = 2$$
$$2x - 5(-2) = 8 \qquad 2(-2) + 3z = 2$$
$$2x + 10 = 8 \qquad -4 + 3z = 2$$
$$2x = -2 \qquad 3z = 6$$
$$x = -1 \qquad z = 2$$
The solution is $\{(-1, -2, 2)\}$.

49. Let x = the rate of the slowest pump
Let y = the rate of the middle rate pump
Let z = the rate of the fastest pump
$$x + y + z = 950 \qquad \rightarrow \quad x + y + z = 950$$
$$x = z - 150 \qquad \rightarrow \quad x \quad - z = -150$$
$$z = x + y - 150 \rightarrow -x - y + z = -150$$
Add the first and third equations to eliminate x:

$$x + y + z = 950$$
$$\underline{-x - y + z = -150}$$
$$2z = 800$$
$$z = 400$$

Substitute and solve for x and y:

$x = z - 150$

$x = 400 - 150$

$x = 250$

$x + y + z = 950$

$250 + y + 400 = 950$

$y + 650 = 950$

$y = 300$

The pumps can drain 250, 300, and 400 gal/hr.

Section 3.7

51. 3×3

53. 1×4

55. $\begin{bmatrix} 1 & 1 & | & 3 \\ 1 & -1 & | & -1 \end{bmatrix}$

57. $x = 9$
$y = -3$

59. a. 4

b. $\begin{bmatrix} 1 & 3 & | & 1 \\ 4 & -1 & | & 6 \end{bmatrix} \xrightarrow{-4R_1 + R_2 \Rightarrow R_2} \begin{bmatrix} 1 & 3 & | & 1 \\ 0 & -13 & | & 2 \end{bmatrix}$

61. $x + y = 3$

$x - y = -1$

$\begin{bmatrix} 1 & 1 & | & 3 \\ 1 & -1 & | & -1 \end{bmatrix} \xrightarrow{-1R_1 + R_2 \Rightarrow R_2} \begin{bmatrix} 1 & 1 & | & 3 \\ 0 & -2 & | & -4 \end{bmatrix} \xrightarrow{-\frac{1}{2}R_2 \Rightarrow R_2} \begin{bmatrix} 1 & 1 & | & 3 \\ 0 & 1 & | & 2 \end{bmatrix} \xrightarrow{-1R_2 + R_1 \Rightarrow R_1} \begin{bmatrix} 1 & 0 & | & 1 \\ 0 & 1 & | & 2 \end{bmatrix}$

The solution is $\{(1,2)\}$.

63. $x - y + z = -4$

$2x + y - 2z = 9$

$x + 2y + z = 5$

$\begin{bmatrix} 1 & -1 & 1 & | & -4 \\ 2 & 1 & -2 & | & 9 \\ 1 & 2 & 1 & | & 5 \end{bmatrix} \xrightarrow[-R_1 + R_3 \Rightarrow R_3]{-2R_1 + R_2 \Rightarrow R_2} \begin{bmatrix} 1 & -1 & 1 & | & -4 \\ 0 & 3 & -4 & | & 17 \\ 0 & 3 & 0 & | & 9 \end{bmatrix} \xrightarrow{R_2 \Leftrightarrow R_3} \begin{bmatrix} 1 & -1 & 1 & | & -4 \\ 0 & 3 & 0 & | & 9 \\ 0 & 3 & -4 & | & 17 \end{bmatrix}$

$\xrightarrow{\frac{1}{3}R_2 \Rightarrow R_2} \begin{bmatrix} 1 & -1 & 1 & | & -4 \\ 0 & 1 & 0 & | & 3 \\ 0 & 3 & -4 & | & 17 \end{bmatrix} \xrightarrow[-3R_2 + R_3 \Rightarrow R_3]{R_2 + R_1 \Rightarrow R_1} \begin{bmatrix} 1 & 0 & 1 & | & -1 \\ 0 & 1 & 0 & | & 3 \\ 0 & 0 & -4 & | & 8 \end{bmatrix}$

$\xrightarrow{-\frac{1}{4}R_3 \Rightarrow R_3} \begin{bmatrix} 1 & 0 & 1 & | & -1 \\ 0 & 1 & 0 & | & 3 \\ 0 & 0 & 1 & | & -2 \end{bmatrix} \xrightarrow{-R_3 + R_1 \Rightarrow R_1} \begin{bmatrix} 1 & 0 & 0 & | & 1 \\ 0 & 1 & 0 & | & 3 \\ 0 & 0 & 1 & | & -2 \end{bmatrix}$

The solution is $\{(1,3,-2)\}$.

Chapter 3 Test

1. $4x - 3y = -5$

$12x + 2y = 7$

Substitute $\left(\frac{1}{4}, 2\right)$:

$4\left(\dfrac{1}{4}\right) - 3(2) = 1 - 6 = -5 = -5$

$12\left(\dfrac{1}{4}\right) + 2(2) = 3 + 4 = 7 = 7$

$\left(\frac{1}{4}, 2\right)$ is a solution.

3. c. The system is inconsistent and independent. There are no solutions.

5. $4x - 2y = -4$

$3x + y = 7$

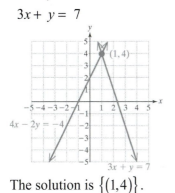

The solution is $\{(1, 4)\}$.

7. $3x + 5y = 13$

$y = x + 9$

$3x + 5(x + 9) = 13$

$3x + 5x + 45 = 13$

$8x = -32$

$x = -4$

$y = x + 9 = -4 + 9 = 5$

The solution is $\{(-4, 5)\}$.

9. Write in standard form:

$7y = 5x - 21 \rightarrow -5x + 7y = -21$

$9y + 2x = -27 \quad \rightarrow \quad 2x + 9y = -27$

Multiply the first equation by 2 and the second equation by 5, add the results and solve for y:

$-5x + 7y = -21 \xrightarrow{\times 2} -10x + 14y = -42$

$2x + 9y = -27 \xrightarrow{\times 5} \underline{10x + 45y = -135}$

$59y = -177$

$y = -3$

Substitute into the first equation and solve for x:

$2x + 9(-3) = -27$

$2x - 27 = -27$

$2x = 0$

$x = 0$

The solution is $\{(0, -3)\}$.

11. Multiply each equation by the LCD:

$\dfrac{1}{5}x = \dfrac{1}{2}y + \dfrac{17}{5} \rightarrow \quad 2x = 5y + 34$

$\rightarrow 2x - 5y = 34$

$\dfrac{1}{4}(x + 2) = -\dfrac{1}{6}y \quad \rightarrow \quad 3x + 6 = -2y$

$\rightarrow 3x + 2y = -6$

Multiply the first equation by 2 and the second equation by 5, add the results and solve for x:

$2x - 5y = 34 \xrightarrow{\times 2} \quad 4x - 10y = 68$

$3x + 2y = -6 \xrightarrow{\times 5} \underline{15x + 10y = -30}$

$19x = 38$

$x = 2$

Substitute into the first equation and solve for y:

$2(2) - 5y = 34$

$4 - 5y = 34$

$-5y = 30$

$y = -6$

The solution is $\{(2,-6)\}$.

13. Multiply each equation by the LCD:
$$-0.03y + 0.06x = 0.3 \quad \rightarrow -3y + 6x = 30$$
$$\rightarrow 6x - 3y = 30$$
$$0.4x - 2 = -0.5y \rightarrow 4x - 20 = -5y$$
$$\rightarrow 4x + 5y = 20$$

Multiply the first equation by 5 and the second equation by 3, add the results and solve for x:
$$6x - 3y = 30 \xrightarrow{\times 5} 30x - 15y = 150$$
$$4x + 5y = 20 \xrightarrow{\times 3} 12x + 15y = 60$$
$$42x = 210$$
$$x = 5$$

Substitute into the first equation and solve for y:
$$6(5) - 3y = 30$$
$$30 - 3y = 30$$
$$-3y = 0$$
$$y = 0$$

The solution is $\{(5,0)\}$.

15. $x + y < 3$ and $3x - 2y > -6$
Graph the related equation $x + y = 3$ by using a dashed line.

Test point above $(0,4)$:   Test point below $(0,0)$:
$$0 + 4 < 3 \qquad\qquad\qquad 0 + 0 < 3$$
$$4 < 3 \qquad\qquad\qquad\qquad 0 < 3$$

$(0,3)$ is not a solution. $(0,0)$ is a solution.

Shade the region below the boundary line.
Graph the related equation $3x - 2y = -6$ by using a dashed line.

Test point above $(0,4)$:   Test point below $(0,0)$:
$$3(0) - 2(4) > -6 \qquad\qquad 3(0) - 2(0) > -6$$
$$-8 > -6 \qquad\qquad\qquad\qquad 0 > -6$$

$(0,4)$ is not a solution. $(0,0)$ is a solution.

Shade the region below the boundary line.
The solution is the intersection of the graphs.

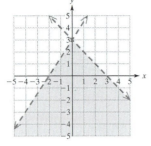

17. a. $x \geq 0,\ y \geq 0$
   b. $300x + 400y \geq 1200$

c. $x \geq 0$ and $y \geq 0$ and $300x + 400y \geq 1200$

$x \geq 0$ represents the points to the right of the vertical line $x = 0$. Shade the region to the right of the boundary line using a solid line border.

$y \geq 0$ represents the points above the horizontal line $y = 0$. Shade the region above the boundary line using a solid line border.

Graph the related equation $300x + 400y = 1200$ by using a solid line.

Test point above $(0,4)$:
$$300(0) + 400(4) \geq 1200$$
$$1600 \geq 1200$$
$(0,4)$ is a solution.

Test point below $(0,0)$:
$$300(0) + 400(0) \geq 1200$$
$$0 \geq 1200$$
$(0,0)$ is not a solution.

Shade the region above the boundary line.

The solution is the intersection of the graphs.

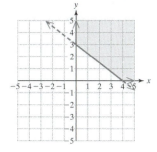

19. Write each equation in standard form:
$$2(x+z) = 6 + x - 3y \rightarrow \quad x + 3y + 2z = 6$$
$$2x = 11 + y - z \rightarrow \quad 2x - y + z = 11$$
$$x + 2(y+z) = 8 \quad \rightarrow \quad x + 2y + 2z = 8$$

Multiply the third equation by -1 and add to the first equation to eliminate x:
$$x + 3y + 2z = 6 \longrightarrow \quad x + 3y + 2z = 6$$
$$x + 2y + 2z = 8 \xrightarrow{\times -1} -x - 2y - 2z = -8$$
$$\overline{\qquad\qquad y = -2}$$

Multiply the second equation by -2 and add to the first equation to eliminate z:
$$x + 3y + 2z = 6 \longrightarrow \quad x + 3y + 2z = 6$$
$$2x - y + z = 11 \xrightarrow{\times -2} -4x + 2y - 2z = -22$$
$$\overline{\qquad -3x + 5y \qquad = -16}$$

Substitute and solve for x and z:
$$-3x + 5y = -16$$
$$-3x + 5(-2) = -16$$
$$-3x - 10 = -16$$
$$-3x = -6$$
$$x = 2$$

$$2x - y + z = 11$$
$$2(2) - (-2) + z = 11$$
$$4 + 2 + z = 11$$
$$z = 5$$

The solution is $\{(2, -2, 5)\}$.

21. Let x = the amount of 20% acid solution
Let y = the amount of 60% acid solution

	20% acid	60% acid	44% acid
L solution	x	y	200
L acid	$0.20x$	$0.60y$	$0.44(200)$

$$x + y = 200$$

$$0.20x + 0.60y = 0.44(200)$$

Multiply the first equation by -0.20, add to the second equation and solve for y:
$$x + y = 200 \rightarrow -0.20x - 0.20y = -40$$
$$0.20x + 0.60y = 88 \rightarrow \underline{0.20x + 0.60y = 88}$$
$$0.40y = 48$$
$$y = 120$$

Substitute into the first equation and solve for x:
$$x + 120 = 200$$
$$x = 80$$

The mixture contains 80 L of 20% acid solution and 120 L of 60% acid solution.

23. Let x = number of orders Joanne can process
Let y = number of orders Kent can process
Let z = number of orders Geoff can process
$$\begin{aligned} x+y+z &= 504 &\rightarrow&& x+y+z &= 504 \\ y &= x+20 &\rightarrow&& -x+y &= 20 \\ z &= x+y-104 &\rightarrow&& -x-y+z &= -104 \end{aligned}$$

Add the first and third equations to eliminate x:
$$\begin{aligned} x+y+z &= 504 \\ \underline{-x-y+z} &= \underline{-104} \\ 2z &= 400 \\ z &= 200 \end{aligned}$$

Add the first and second equations to eliminate x:
$$\begin{aligned} x+y+z &= 504 \\ \underline{-x+y} &= \underline{20} \\ 2y+z &= 524 \end{aligned}$$

Substitute and solve for x and y:
$$\begin{aligned} 2y+z &= 524 & -x+y &= 20 \\ 2y+200 &= 524 & -x+162 &= 20 \\ 2y &= 324 & -x &= -142 \\ y &= 162 & x &= 142 \end{aligned}$$

Joanne processes 142 orders, Kent processes 162 orders, and Geoff processes 200 orders.

25. a.
$$\left[\begin{array}{ccc|c} 1 & 2 & 1 & -3 \\ 4 & 0 & 1 & -2 \\ -5 & -6 & 3 & 0 \end{array}\right]$$

$$\xrightarrow{-4R_1+R_2 \Rightarrow R_2} \left[\begin{array}{ccc|c} 1 & 2 & 1 & -3 \\ 0 & -8 & -3 & 10 \\ -5 & -6 & 3 & 0 \end{array}\right]$$

b.
$$\xrightarrow{5R_1+R_3 \Rightarrow R_3} \left[\begin{array}{ccc|c} 1 & 2 & 1 & -3 \\ 0 & -8 & -3 & 10 \\ 0 & 4 & 8 & -15 \end{array}\right]$$

27.
$$\begin{aligned} x+y+z &= 1 \\ 2x+y &= 0 \\ -2y-z &= 5 \end{aligned}$$

$$\left[\begin{array}{ccc|c} 1 & 1 & 1 & 1 \\ 2 & 1 & 0 & 0 \\ 0 & -2 & -1 & 5 \end{array}\right] \xrightarrow{-2R_1+R_2 \Rightarrow R_2} \left[\begin{array}{ccc|c} 1 & 1 & 1 & 1 \\ 0 & -1 & -2 & -2 \\ 0 & -2 & -1 & 5 \end{array}\right] \xrightarrow{-R_2 \Rightarrow R_2} \left[\begin{array}{ccc|c} 1 & 1 & 1 & 1 \\ 0 & 1 & 2 & 2 \\ 0 & -2 & -1 & 5 \end{array}\right]$$

$$\xrightarrow[2R_2+R_3 \Rightarrow R_3]{-R_2+R_1 \Rightarrow R_1} \left[\begin{array}{ccc|c} 1 & 0 & -1 & -1 \\ 0 & 1 & 2 & 2 \\ 0 & 0 & 3 & 9 \end{array}\right] \xrightarrow{\frac{1}{3}R_3 \Rightarrow R_3} \left[\begin{array}{ccc|c} 1 & 0 & -1 & -1 \\ 0 & 1 & 2 & 2 \\ 0 & 0 & 1 & 3 \end{array}\right] \xrightarrow[-2R_3+R_2 \Rightarrow R_2]{R_3+R_1 \Rightarrow R_1} \left[\begin{array}{ccc|c} 1 & 0 & 0 & 2 \\ 0 & 1 & 0 & -4 \\ 0 & 0 & 1 & 3 \end{array}\right]$$

The solution is $\{(2,-4,3)\}$.

Chapters 1 – 3 Cumulative Review Exercises

1.
$$\begin{aligned} 2\left[3^2-8(7-5)\right] &= 2\left[3^2-8(2)\right] \\ &= 2\left[9-8(2)\right] = 2\left[9-16\right] \\ &= 2\left[-7\right] = -14 \end{aligned}$$

3.
$$\begin{aligned} -5(2x-1)-2(3x+1) &= 7-2(8x+1) \\ -10x+5-6x-2 &= 7-16x-2 \\ -16x+3 &= -16x+5 \\ 3 &\neq 5 \end{aligned}$$
There is no solution; $\{\ \}$.

5.
$$-4 = |2x-3| - 9$$
$$5 = |2x-3|$$
$$2x-3 = 5 \text{ or } 2x-3 = -5$$
$$2x = 8 \text{ or } \quad 2x = -2$$
$$x = 4 \text{ or } \quad x = -1 \quad \{4, -1\}$$

7.
$$4x > 16 \text{ or } -6x - 3 \geq 9$$
$$x > 4 \text{ or } \quad -6x \geq 12$$
$$x > 4 \text{ or } \quad x \leq -2 \quad (-\infty, -2] \cup (4, \infty)$$

9.
$$0 \leq \frac{3x-9}{6} \leq 5$$
$$0 \leq 3x - 9 \leq 30$$
$$9 \leq 3x \leq 39$$
$$3 \leq x \leq 13 \quad [3, 13]$$

11.
$$4 < |2x+4|$$
$$2x + 4 > 4 \text{ or } 2x + 4 < -4$$
$$2x > 0 \text{ or } \quad 2x < -8$$
$$x > 0 \text{ or } \quad x < -4$$
$$(-\infty, -4) \cup (0, \infty)$$

13.
$$5x - 2y = 15$$
$$-2y = -5x + 15$$
$$y = \frac{5}{2}x - \frac{15}{2}$$
Slope: $\frac{5}{2}$ y-intercept: $\left(0, -\frac{15}{2}\right)$
$$5x - 2(0) = 15$$
$$5x = 15$$
$$x = 3 \qquad x\text{-intercept: } (3, 0)$$

15. $x = -2$

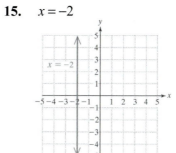

17.
$$m = \frac{-4 - (-8)}{2 - 3} = \frac{4}{-1} = -4$$
$$y - (-8) = -4(x - 3)$$
$$y + 8 = -4x + 12$$
$$y = -4x + 4$$

19.
$$2x + y = 4$$
$$y = 3x - 1$$
$$2x + (3x - 1) = 4$$
$$5x - 1 = 4$$
$$5x = 5$$
$$x = 1$$
$$y = 3x - 1 = 3(1) - 1 = 3 - 1 = 2$$
The solution is $\{(1, 2)\}$.

21. a. $f(x) = 2.50x + 25$

b. $g(x) = 3x + 10$

c.
$$3x + 10 = 2.50x + 25$$
$$0.50x = 15$$
$$x = 30$$
30 DVDs would need to be rented for the cost to be the same.

23. 2×3

25. $2x - 4y = -2$
$4x + y = 5$

$$\begin{bmatrix} 2 & -4 & \big| & -2 \\ 4 & 1 & \big| & 5 \end{bmatrix} \xrightarrow{\frac{1}{2}R_1 \Rightarrow R_1} \begin{bmatrix} 1 & -2 & \big| & -1 \\ 4 & 1 & \big| & 5 \end{bmatrix} \xrightarrow{-4R_1 + R_2 \Rightarrow R_2} \begin{bmatrix} 1 & -2 & \big| & -1 \\ 0 & 9 & \big| & 9 \end{bmatrix} \xrightarrow{\frac{1}{9}R_2 \Rightarrow R_2} \begin{bmatrix} 1 & -2 & \big| & -1 \\ 0 & 1 & \big| & 1 \end{bmatrix}$$

$$\xrightarrow{2R_2 + R_1 \Rightarrow R_1} \begin{bmatrix} 1 & 0 & \big| & 1 \\ 0 & 1 & \big| & 1 \end{bmatrix}$$

The solution is $\{(1,1)\}$.

Chapter 4   Polynomials

Are You Prepared?

Across

1. $(-2)^4 = (-2)(-2)(-2)(-2) = 16$ sixteen

6. distributive

8. $\left(\dfrac{1}{6}\right)^{-1} + \left(\dfrac{1}{3}\right)^{-2} = 6^1 + 3^2 = 6 + 9 = 15$

9. $y^0 = 1$ one

10. $\left(\dfrac{3r^4}{r^5}\right)^2 \cdot (2r)^3 = \left(\dfrac{3}{r}\right)^2 \cdot (2r)^3 = \dfrac{9}{r^2} \cdot 8r^3 = 72r$

Down

2. $3(x+4) - 2(x+2) = 3x + 12 - 2x - 4 = x + 8$

3. $-2^2 = -(2)(2) = -4$ negative four

4. true

5. $\dfrac{16d^4}{2d^3} = 8d$

7. $-4(5+t) + 5(t+3) + 10 = -20 - 4t + 5t + 15 + 10$
$$= t + 5$$

Section 4.1   Properties of Integer Exponents and Scientific Notation

1. a. exponent

   b. 1

   c. $\left(\dfrac{1}{b}\right)^n$ or $\dfrac{1}{b^n}$

   d. scientific notation

3. $ab^3 = a \cdot b \cdot b \cdot b$

$(ab)^3 = (ab) \cdot (ab) \cdot (ab)$

$\qquad = a \cdot a \cdot a \cdot b \cdot b \cdot b = a^3 \cdot b^3$

5. For example: $(5x)^2 = 5^2 x^2$

$\qquad (xy)^3 = x^3 y^3$

7.
For example: $\dfrac{x^5}{x^2} = x^3$

$\dfrac{8^4}{8^2} = 8^2$

9.   For example: $6^0 = 1$

$x^0 = 1 \ (x \neq 0)$

11. $\left(\dfrac{1}{3}\right)^{-1} = \left(\dfrac{3}{1}\right)^1 = 3$

13. $5^{-2} = \dfrac{1}{5^2} = \dfrac{1}{25}$

15. $-5^{-2} = -\dfrac{1}{5^2} = -\dfrac{1}{25}$

17. $(-5)^{-2} = \dfrac{1}{(-5)^2} = \dfrac{1}{25}$

19. $\left(-\dfrac{1}{4}\right)^{-3} = \left(-\dfrac{4}{1}\right)^3 = (-4)^3 = -64$

21. $\left(-\dfrac{3}{2}\right)^{-4} = \left(-\dfrac{2}{3}\right)^4 = \dfrac{(-2)^4}{3^4} = \dfrac{16}{81}$

23. $-\left(\dfrac{2}{5}\right)^{-3} = -\left(\dfrac{5}{2}\right)^3 = -\dfrac{5^3}{2^3} = -\dfrac{125}{8}$

25. $(10ab)^0 = 1$

27. $10ab^0 = 10a \cdot 1 = 10a$

29. $y^3 \cdot y^5 = y^{3+5} = y^8$

31. $\dfrac{13^8}{13^6} = 13^{8-6} = 13^2 = 169$

33. $\left(y^2\right)^4 = y^{2 \cdot 4} = y^8$

35. $\left(3x^2\right)^4 = 3^4 \left(x^2\right)^4 = 3^4 x^{2 \cdot 4} = 81x^8$

37. $p^{-3} = \dfrac{1}{p^3}$

39. $7^{10} \cdot 7^{-13} = 7^{10+(-13)} = 7^{-3} = \dfrac{1}{7^3} = \dfrac{1}{343}$

41. $\dfrac{w^3}{w^5} = w^{3-5} = w^{-2} = \dfrac{1}{w^2}$

43. $a^{-2}a^{-5} = a^{-2+(-5)} = a^{-7} = \dfrac{1}{a^7}$

45. $\dfrac{r}{r^{-1}} = r^{1-(-1)} = r^2$

47. $\dfrac{z^{-6}}{z^{-2}} = z^{-6-(-2)} = z^{-4} = \dfrac{1}{z^4}$

49. $\dfrac{a^3}{b^{-2}} = a^3 \cdot \dfrac{1}{b^{-2}} = a^3 b^2$

51. $\left(6xyz^2\right)^0 = 1$

53. $2^4 + 2^{-2} = 2^4 + \dfrac{1}{2^2} = 16 + \dfrac{1}{4} = 16\tfrac{1}{4}$ or $\dfrac{65}{4}$

55. $1^{-2} + 5^{-2} = \dfrac{1}{1^2} + \dfrac{1}{5^2} = \dfrac{1}{1} + \dfrac{1}{25} = 1\tfrac{1}{25}$ or $\dfrac{26}{25}$

57. $\left(\dfrac{2}{3}\right)^{-2} - \left(\dfrac{1}{2}\right)^2 + \left(\dfrac{1}{3}\right)^0 = \left(\dfrac{3}{2}\right)^2 - \dfrac{1}{4} + 1$

$= \dfrac{9}{4} - \dfrac{1}{4} + \dfrac{4}{4} = \dfrac{12}{4} = 3$

59. $\left(\dfrac{4}{5}\right)^{-1} + \left(\dfrac{3}{2}\right)^{2} - \left(\dfrac{2}{7}\right)^{0} = \dfrac{5}{4} + \dfrac{9}{4} - 1$

$$= \dfrac{5}{4} + \dfrac{9}{4} - \dfrac{4}{4} = \dfrac{10}{4} = \dfrac{5}{2}$$

61. $\dfrac{p^2 q}{p^5 q^{-1}} = p^{2-5}q^{1-(-1)} = p^{-3}q^2 = \dfrac{1}{p^3} \cdot q^2 = \dfrac{q^2}{p^3}$

63. $\dfrac{-48ab^{10}}{32a^4b^3} = -\dfrac{48}{32}a^{1-4}b^{10-3} = -\dfrac{3}{2}a^{-3}b^7$

$$= -\dfrac{3}{2} \cdot \dfrac{1}{a^3} \cdot b^7 = -\dfrac{3b^7}{2a^3}$$

65. $\left(-3x^{-4}y^5z^2\right)^{-4} = (-3)^{-4}\left(x^{-4}\right)^{-4}\left(y^5\right)^{-4}\left(z^2\right)^{-4}$

$$= \left(-\dfrac{1}{3}\right)^4 x^{16}y^{-20}z^{-8}$$

$$= \dfrac{1}{81} \cdot x^{16} \cdot \dfrac{1}{y^{20}} \cdot \dfrac{1}{z^8} = \dfrac{x^{16}}{81y^{20}z^8}$$

67. $\left(4m^{-2}n\right)\left(-m^6n^{-3}\right) = -4m^{-2+6}n^{1+(-3)}$

$$= -4m^4n^{-2} = -4m^4 \cdot \dfrac{1}{n^2}$$

$$= -\dfrac{4m^4}{n^2}$$

69. $\left(p^{-2}q\right)^3\left(2pq^4\right)^2 = \left(p^{-2}\right)^3 q^3 \cdot 2^2 p^2 \left(q^4\right)^2$

$$= p^{-6}q^3 \cdot 4p^2q^8 = 4p^{-6+2}q^{3+8}$$

$$= 4p^{-4}q^{11} = 4 \cdot \dfrac{1}{p^4} \cdot q^{11} = \dfrac{4q^{11}}{p^4}$$

71. $\left(\dfrac{x^2}{y}\right)^3\left(5x^2y\right) = \dfrac{x^6}{y^3}\left(5x^2y\right) = 5x^{6+2}y^{1-3}$

$$= 5x^8y^{-2} = 5x^8\dfrac{1}{y^2} = \dfrac{5x^8}{y^2}$$

73. $\dfrac{\left(-8a^2b^2\right)^4}{\left(16a^3b^7\right)^2} = \dfrac{(-8)^4\left(a^2\right)^4\left(b^2\right)^4}{(16)^2\left(a^3\right)^2\left(b^7\right)^2} = \dfrac{4096a^8b^8}{256a^6b^{14}}$

$$= 16a^{8-6}b^{8-14} = 16a^2b^{-6}$$

$$= 16a^2 \cdot \dfrac{1}{b^6} = \dfrac{16a^2}{b^6}$$

75. $\left(\dfrac{-2x^6y^{-5}}{3x^{-2}y^4}\right)^{-3} = \left(-\dfrac{2}{3}x^{6-(-2)}y^{-5-4}\right)^{-3}$

$$= \left(-\dfrac{2}{3}x^8y^{-9}\right)^{-3} = \left(-\dfrac{2}{3}\right)^{-3}\left(x^8\right)^{-3}\left(y^{-9}\right)^{-3}$$

$$= \left(-\dfrac{3}{2}\right)^3 x^{-24}y^{27} = -\dfrac{27}{8} \cdot \dfrac{1}{x^{24}} \cdot y^{27}$$

$$= -\dfrac{27y^{27}}{8x^{24}}$$

77. $\left(\dfrac{2x^{-3}y^0}{4x^6y^{-5}}\right)^{-2} = \left(\dfrac{1}{2}x^{-3-6}y^{0-(-5)}\right)^{-2} = \left(\dfrac{1}{2}x^{-9}y^5\right)^{-2}$

$$= \left(\dfrac{1}{2}\right)^{-2}\left(x^{-9}\right)^{-2}\left(y^5\right)^{-2}$$

$$= (2)^2 x^{18}y^{-10} = 4x^{18} \cdot \dfrac{1}{y^{10}} = \dfrac{4x^{18}}{y^{10}}$$

79.
$$3xy^5 \left(\frac{2x^4 y}{6x^5 y^3} \right)^{-2} = 3xy^5 \left(\frac{1}{3} x^{4-5} y^{1-3} \right)^{-2}$$
$$= 3xy^5 \left(\frac{1}{3} x^{-1} y^{-2} \right)^{-2}$$
$$= 3xy^5 \left(\frac{1}{3} \right)^{-2} \left(x^{-1} \right)^{-2} \left(y^{-2} \right)^{-2}$$
$$= 3xy^5 (3)^2 x^2 y^4 = 3 \cdot 9 x^{1+2} y^{5+4} = 27 x^3 y^9$$

81.
a. $\$8,000,000,000 = \$8 \times 10^9$
b. $3,000,000 = 3 \times 10^6$ DVDs
c. $14,000,000,000,000 = 1.4 \times 10^{13}$ eV
d. $0.000\,000\,000\,000\,000\,0001602$
$$= 1.602 \times 10^{-19} \text{ J}$$

83.
a. $2 \times 10^{11} = 200,000,000,000$
b. $4 \times 10^{-6} = 0.000004$
c. $1.082 \times 10^{11} = 108,200,000,000$

85.
$$35 \times 10^4 = 3.5 \times 10^1 \times 10^4 = 3.5 \times 10^5$$

87. 7.0×10^0 Proper

89. 9×10^1 Proper

91.
$$\left(6.5 \times 10^3 \right) \left(5.2 \times 10^{-8} \right) = 33.8 \times 10^{3+(-8)}$$
$$= 3.38 \times 10^1 \times 10^{-5} = 3.38 \times 10^{-4}$$

93.
$$(0.0000024)(6,700,000,000)$$
$$= \left(2.4 \times 10^{-6} \right) \left(6.7 \times 10^9 \right) = 16.08 \times 10^{-6+9}$$
$$= 1.608 \times 10^1 \times 10^3 = 1.608 \times 10^4$$

95.
$$\left(8.5 \times 10^{-2} \right) \div \left(2.5 \times 10^{-15} \right) = 3.4 \times 10^{-2-(-15)}$$
$$= 3.4 \times 10^{13}$$

97.
$$(900000000) \div (360000)$$
$$= \left(9 \times 10^8 \right) \div \left(3.6 \times 10^5 \right)$$
$$= 2.5 \times 10^{8-5} = 2.5 \times 10^3$$

99.
$$2 \cdot \left(6.02 \times 10^{23} \right) = 12.04 \times 10^{23}$$
$$= 1.204 \times 10^1 \times 10^{23}$$
$$= 1.204 \times 10^{24} \text{ hydrogen atoms}$$
$$1 \cdot \left(6.02 \times 10^{23} \right) = 6.02 \times 10^{23} \text{ oxygen atoms}$$

101.
$$2,200,000 \div 110 = \left(2.2 \times 10^6 \right) \div \left(1.1 \times 10^2 \right)$$
$$= 2 \times 10^4 \text{ or } 20,000 \text{ people per mi}^2$$

103.
$$\left(\$3.5 \times 10^9 \right) (15) = \$52.5 \times 10^9 = \$5.25 \times 10^{10}$$

105.
a. $45 \cdot 12 = 540$ months $\$20(540) = \$10,800$
b. $A = \$20 \left[\left(1 + \frac{0.06}{12} \right)^{540} - 1 \right] \left(1 + \frac{12}{0.06} \right)$
c. $= \$55,395.45$

107. $y^{a-5} y^{a+7} = y^{a-5+a+7} = y^{2a+2}$

109. $\dfrac{x^{3a-3}}{x^{a+1}} = x^{(3a-3)-(a+1)} = x^{3a-3-a-1} = x^{2a-4}$

111.
$$\frac{x^{2a-2} y^{a+3}}{x^{a+4} y^{a-3}} = x^{(2a-2)-(a+4)} y^{(a+3)-(a-3)}$$
$$= x^{2a-2-a-4} y^{a+3-a+3} = x^{a-6} y^6$$

Section 4.2   Addition and Subtraction of Polynomials and Polynomial Functions

1. **a.** polynomial

   b. coefficient; n

   c. 1; 1

   d. one

   e. binomial

   f. trinomial

   g. leading; leading coefficient

   h. greatest

   i. zero

   j. exponents

   k. polynomial

3. $\left(2ac^{-2}\right)\left(5a^{-1}c^4\right)=10a^{1+(-1)}c^{-2+4}=10a^0c^2=10c^2$

5. $\left(3.4\times10^5\right)\left(5.0\times10^{-2}\right)=17\times10^3=1.7\times10^4$

7. $-6a^3+a^2-a$
   leading coefficient: -6
   degree: 3

9. $3x^4+6x^2-x-1$
   leading coefficient: 3
   degree: 4

11. $-t^2+100$
    leading coefficient: -1
    degree: 2

13. For example: $3x^5$

15. For example: x^2+2x+1

17. For example: $6x^4-x^2$

19. $\left(-4m^2+4m\right)+\left(5m^2+6m\right)$

    $\quad=-4m^2+5m^2+4m+6m$

    $\quad=m^2+10m$

21. $\left(3x^4-x^3-x^2\right)+\left(3x^3-7x^2+2x\right)$

    $\quad=3x^4+\left(-x^3\right)+3x^3+\left(-x^2\right)+\left(-7x^2\right)+2x$

    $\quad=3x^4+2x^3-8x^2+2x$

23. $\left(\dfrac{1}{2}w^3 + \dfrac{2}{9}w^2 - 1.8w\right) + \left(\dfrac{3}{2}w^3 - \dfrac{1}{9}w^2 + 2.7w\right)$

$\quad = \dfrac{1}{2}w^3 + \dfrac{3}{2}w^3 + \dfrac{2}{9}w^2 - \dfrac{1}{9}w^2 - 1.8w + 2.7w$

$\quad = 2w^3 + \dfrac{1}{9}w^2 + 0.9w$

25. $\left(9x^2y - 5xy + 1\right) + \left(8x^2y + xy - 15\right)$

$\quad = 9x^2y + 8x^2y - 5xy + xy + 1 - 15$

$\quad = 17x^2y - 4xy - 14$

27. $\left(-7a + 6a^2 + 1\right) + \left(-8 - 4a - 2a^2\right)$

$\quad = 6a^2 - 2a^2 - 7a - 4a + 1 - 8$

$\quad = 4a^2 - 11a - 7$

29. $\left(-7a + 6a^2 + 1\right) + \left(-8 - 4a - 2a^2\right)$

$\quad = 6a^2 - 2a^2 - 7a - 4a + 1 - 8$

$\quad = 4a^2 - 11a - 7$

31. $-\left(-30y^3\right) = 30y^3$

33. $-\left(4p^3 + 2p - 12\right) = -4p^3 - 2p + 12$

35. $-\left(-11ab^2 + a^2b\right) = 11ab^2 - a^2b$

37. $\left(13z^5 - z^2\right) - \left(7z^5 + 5z^2\right)$

$\quad = \left(13z^5 - z^2\right) + \left(-7z^5 - 5z^2\right)$

$\quad = 13z^5 - 7z^5 - z^2 - 5z^2$

$\quad = 6z^5 - 6z^2$

39. $\left(-3x^3 + 3x^2 - x + 6\right) - \left(1 - x - x^2 - x^3\right)$

$\quad = \left(-3x^3 + 3x^2 - x + 6\right) + \left(-1 + x + x^2 + x^3\right)$

$\quad = \left(-3x^3 + 3x^2 - x + 6\right) + \left(x^3 + x^2 + x - 1\right)$

$\quad = -3x^3 + x^3 + 3x^2 + x^2 - x + x + 6 - 1$

$\quad = -2x^3 + 4x^2 + 5$

41. $\left(-3xy^3 + 3x^2y - x + 6\right) - \left(-xy^3 - xy - x + 1\right)$

$\quad = \left(-3xy^3 + 3x^2y - x + 6\right) + \left(xy^3 + xy + x - 1\right)$

$\quad = -3xy^3 + xy^3 + 3x^2y + xy - x + x + 6 - 1$

$\quad = -2xy^3 + 3x^2y + xy + 5$

43.

$\begin{array}{l} \quad 4t^3 - 6t^2 \quad\;\; -18 \rightarrow \\ -\left(3t^3 + 7t^2 + 9t - 5\right) \rightarrow \end{array}$ $\begin{array}{l} \quad 4t^3 - 6t^2 \quad\;\; -18 \\ +\left(-3t^3 - 7t^2 - 9t + 5\right) \\ \hline \quad t^3 - 13t^2 - 9t - 13 \end{array}$

45. $\left(\dfrac{1}{5}a^2 - \dfrac{1}{2}ab + \dfrac{1}{10}b^2 + 3\right) - \left(-\dfrac{3}{10}a^2 + \dfrac{2}{5}ab - \dfrac{1}{2}b^2 - 5\right)$

$\quad = \left(\dfrac{1}{5}a^2 - \dfrac{1}{2}ab + \dfrac{1}{10}b^2 + 3\right) + \left(\dfrac{3}{10}a^2 - \dfrac{2}{5}ab + \dfrac{1}{2}b^2 + 5\right)$

$\quad = \dfrac{1}{5}a^2 + \dfrac{3}{10}a^2 - \dfrac{1}{2}ab - \dfrac{2}{5}ab + \dfrac{1}{10}b^2 + \dfrac{1}{2}b^2 + 3 + 5$

$\quad = \dfrac{2}{10}a^2 + \dfrac{3}{10}a^2 - \dfrac{5}{10}ab - \dfrac{4}{10}ab + \dfrac{1}{10}b^2 + \dfrac{5}{10}b^2 + 3 + 5$

$\quad = \dfrac{1}{2}a^2 - \dfrac{9}{10}ab + \dfrac{3}{5}b^2 + 8$

47. $\left(8x^2 + x - 15\right) - \left(9x^2 - 5x + 1\right)$

$= \left(8x^2 + x - 15\right) + \left(-9x^2 + 5x - 1\right)$

$= 8x^2 - 9x^2 + x + 5x - 15 - 1$

$= -x^2 + 6x - 16$

49. $\left(3x^5 - 2x^3 + 4\right) - \left(x^4 + 2x^3 - 7\right)$

$= \left(3x^5 - 2x^3 + 4\right) + \left(-x^4 - 2x^3 + 7\right)$

$= 3x^5 - x^4 - 2x^3 - 2x^3 + 4 + 7$

$= 3x^5 - x^4 - 4x^3 + 11$

51. $\left(8y^2 - 4y^3\right) - \left(3y^2 - 8y^3\right)$

$= \left(8y^2 - 4y^3\right) + \left(-3y^2 + 8y^3\right)$

$= -4y^3 + 8y^3 + 8y^2 - 3y^2$

$= 4y^3 + 5y^2$

53. $\left(-2r - 6r^4\right) + \left(-r^4 - 9r\right)$

$= -6r^4 - r^4 - 2r - 9r$

$= -7r^4 - 11r$

55. $\left(5xy + 13x^2 + 3y\right) - \left(4x^2 - 8y\right)$

$= \left(5xy + 13x^2 + 3y\right) + \left(-4x^2 + 8y\right)$

$= 13x^2 - 4x^2 + 5xy + 3y + 8y$

$= 9x^2 + 5xy + 11y$

57. $\left(11ab - 23b^2\right) + \left(7ab - 19b^2\right)$

$= 11ab + 7ab - 23b^2 - 19b^2$

$= 18ab - 42b^2$

59. $\left[2p - \left(3p + 5\right)\right] + \left(4p - 6\right) + 2$

$= \left[2p - 3p - 5\right] + \left(4p - 6\right) + 2$

$= -p - 5 + 4p - 6 + 2$

$= -p + 4p - 5 - 6 + 2$

$= 3p - 9$

61. $5 - \left[2m^2 - \left(4m^2 + 1\right)\right] = 5 - \left[2m^2 - 4m^2 - 1\right]$

$= 5 - \left[-2m^2 - 1\right]$

$= 5 + 2m^2 + 1$

$= 2m^2 + 6$

63. $\left(6x^3 - 5\right) - \left(-3x^3 + 2x\right) - \left(2x^3 - 6x\right)$

$= 6x^3 - 5 + 3x^3 - 2x - 2x^3 + 6x$

$= 7x^3 + 4x - 5$

65. $\left(-ab + 5a^2b\right) - \left[7ab^2 - 2ab - \left(7a^2b + 2ab^2\right)\right] = -ab + 5a^2b - \left[7ab^2 - 2ab - 7a^2b - 2ab^2\right]$

$= -ab + 5a^2b - \left[5ab^2 - 2ab - 7a^2b\right]$

$= -ab + 5a^2b - 5ab^2 + 2ab + 7a^2b$

$= 12a^2b + ab - 5ab^2$

67. $\left(8x^3 - x^2 + 3\right) - \left[5x^2 + x - \left(4x^3 + x - 2\right)\right]$

$= \left(8x^3 - x^2 + 3\right) - \left[5x^2 + x - 4x^3 - x + 2\right]$

$= \left(8x^3 - x^2 + 3\right) - \left(-4x^3 + 5x^2 + 2\right)$

$= 8x^3 - x^2 + 3 + 4x^3 - 5x^2 - 2$

$= 8x^3 + 4x^3 - x^2 - 5x^2 + 3 - 2$

$= 12x^3 - 6x^2 + 1$

69.

$12a^2b - 4ab^2 - ab \rightarrow \qquad 12a^2b - 4ab^2 - ab$

$-\left(4a^2b + ab^2 - 5ab\right) \rightarrow +\left(-4a^2b - ab^2 + 5ab\right)$

$\overline{} \qquad \overline{8a^2b - 5ab^2 + 4ab}$

71.

$$-5x^4 \qquad -11x^2 \qquad +6 \quad \rightarrow \quad -5x^4 \qquad -11x^2 \qquad +6$$

$$\underline{-\left(-5x^4+3x^3+5x^2-10x+5\right)} \rightarrow \underline{+\left(5x^4-3x^3-5x^2+10x-5\right)}$$

$$-3x^3-16x^2+10x+1$$

73.

$$-2.2p^5-9.1p^4 \qquad +5.3p^2-7.9p$$

$$\underline{+\left(\qquad -6.4p^4-8.5p^3-10.3p^2 \qquad \right)}$$

$$-2.2p^5-15.5p^4-8.5p^3-\ 5p^2-7.9p$$

75.

$$P=\left(2x^3+6x\right)+\left(4x^3-5x\right)+\left(6x^3+x\right)$$

$$=2x^3+6x+4x^3-5x+6x^3+x$$

$$=12x^3+2x$$

77.

$$h(x)=\frac{2}{3}x^2-5$$

It is a polynomial function. The degree is 2.

79.

$$p(x)=8x^3+2x^2-\frac{3}{x}$$

It is not a polynomial function. The term $-\frac{3}{x}=-3x^{-1}$ and -1 is not a whole number.

81. $g(x)=-7$

It is a polynomial function. The degree is 0.

83. $M(x)=|x|+5x$

It is not a polynomial function. The term $|x|$ is not of the form ax^n.

85.

$$P(x)=-x^4+2x-5$$

a.
$$P(2)=-(2)^4+2(2)-5$$
$$=-16+4-5=-17$$

b.
$$P(-1)=-(-1)^4+2(-1)-5$$
$$=-1-2-5=-8$$

c.
$$P(0)=-(0)^4+2(0)-5$$
$$=0+0-5=-5$$

d.
$$P(1)=-(1)^4+2(1)-5$$
$$=-1+2-5=-4$$

87.

$$H(x)=\tfrac{1}{2}x^3-x+\tfrac{1}{4}$$

a.
$$H(0)=\frac{1}{2}(0)^3-(0)+\frac{1}{4}=0-0+\frac{1}{4}=\frac{1}{4}$$

b.
$$H(2)=\frac{1}{2}(2)^3-(2)+\frac{1}{4}$$
$$=4-2+\frac{1}{4}=2+\frac{1}{4}=\frac{9}{4}$$

c.
$$H(-2)=\frac{1}{2}(-2)^3-(-2)+\frac{1}{4}$$
$$=-4+2+\frac{1}{4}=-2+\frac{1}{4}=-\frac{7}{4}$$

d.
$$H(-1)=\frac{1}{2}(-1)^3-(-1)+\frac{1}{4}$$
$$=-\frac{1}{2}+1+\frac{1}{4}=\frac{3}{4}$$

89. Let $x =$ the width of the garden
$x + 3 =$ the length of the garden
$$P(x)=2x+2(x+3)=2x+2x+6=4x+6$$

91. a.
$$P(x)=R(x)-C(x)$$
$$=(12x)-(5.40x+99)$$
$$=12x-5.40x-99$$
$$=6.6x-99$$

b.
$$P(50)=6.6(50)-99$$
$$=330-99$$
$$=231$$
The profit will be $231.

93. a.
$$D(x) = 5.2x^2 + 40.4x + 1636$$
$$D(0) = 5.2(0)^2 + 40.4(0) + 1636$$
$$= 0 + 0 + 1636 = 1636$$
In 1990 the annual dormitory charge was $1636.
$$D(18) = 5.2(18)^2 + 40.4(18) + 1636$$
$$= 1684.8 + 727.2 + 1636 = 4048$$
In 2008 the annual dormitory charge was $4048.

b.
$$D(25) = 5.2(25)^2 + 40.4(25) + 1636$$
$$= 3250 + 1010 + 1636 = 5896$$
In 2015 the annual dormitory charge will be $5896.

95.
$$W(t) = 143t + 6580$$

a.
$$W(0) = 143(0) + 6580 = 6580$$
$$W(5) = 143(5) + 6580$$
$$= 715 + 6580 = 7295$$
$$W(10) = 143(10) + 6580$$
$$= 1430 + 6580 = 8010$$

b. In 2010, 8010 thousand (8,010,000) women will be due child support.

97. a.
$$x(t) = 25t$$
$$y(t) = -16t^2 + 43.3t$$
$$x(0) = 25(0) = 0$$
$$y(0) = -16(0)^2 + 43.3(0) = 0 + 0 = 0$$

b. (0, 0) At $t = 0$, the position of the rocket is at the origin.

c.
$$x(2) = 25(2) = 50$$
$$y(2) = -16(2)^2 + 43.3(2)$$
$$= -64 + 86.6 = 22.6$$
(50, 22.6) At $t = 2$ sec, the position of the rocket is (50, 22.6).
$$x(1) = 25(1) = 25$$
$$y(1) = -16(1)^2 + 43.3(1)$$
$$= -16 + 43.3 = 27.3$$
(25, 27.3) At $t = 1$ sec, the position of the rocket is (25, 27.3).

Section 4.3   Multiplication of Polynomials

1. a. distributive

b. $4x - 7$

c. squares; $a^2 - b^2$

d. perfect; $a^2 + 2ab + b^2$

3.
$$(-2 - 3x) - \left[5 - (6x^2 + 4x + 1)\right] = -2 - 3x - \left[5 - 6x^2 - 4x - 1\right]$$
$$= -2 - 3x - \left[-6x^2 - 4x + 4\right]$$
$$= -2 - 3x + 6x^2 + 4x - 4$$
$$= 6x^2 + x - 6$$

5.

$$g(x) = x^4 - x^2 - 3$$

a.

$$g(-1) = (-1)^4 - (-1)^2 - 3 = 1 - 1 - 3 = -3$$

b.

c.

$$g(2) = (2)^4 - (2)^2 - 3 = 16 - 4 - 3 = 9$$

$$g(0) = (0)^4 - (0)^2 - 3 = 0 - 0 - 3 = -3$$

7. $(7x^4 y)(-6xy^5) = 7(-6)(x^4 \cdot x)(y \cdot y^5) = -42x^5 y^6$

9. $(2.2a^6 b^4 c^7)(5ab^4 c^3) = 11a^7 b^8 c^{10}$

11. $\dfrac{1}{5}(2a - 3) = \dfrac{1}{5}(2a) + \dfrac{1}{5}(-3) = \dfrac{2}{5}a - \dfrac{3}{5}$

13. $2m^3 n^2 (m^2 n^3 - 3mn^2 + 4n)$

$$= 2m^3 n^2 (m^2 n^3) - 2m^3 n^2 (3mn^2) + 2m^3 n^2 (4n)$$

$$= 2m^5 n^5 - 6m^4 n^4 + 8m^3 n^3$$

15. $6xy^2 \left(\dfrac{1}{2}x - \dfrac{2}{3}xy \right) = 6xy^2 \left(\dfrac{1}{2}x \right) - 6xy^2 \left(\dfrac{2}{3}xy \right)$

$$= 3x^2 y^2 - 4x^2 y^3$$

17. $(x + y)(x - 2y) = x(x) - x(2y) + y(x) - y(2y)$

$$= x^2 - 2xy + xy - 2y^2$$

$$= x^2 - xy - 2y^2$$

19. $(6x - 1)(5 + 2x)$

$$= 6x(5) + 6x(2x) - 1(5) - 1(2x)$$

$$= 30x + 12x^2 - 5 - 2x$$

$$= 12x^2 + 28x - 5$$

21. $(y^2 - 12)(2y^2 + 3)$

$$= y^2 (2y^2) + y^2 (3) - 12(2y^2) - 12(3)$$

$$= 2y^4 + 3y^2 - 24y^2 - 36$$

$$= 2y^4 - 21y^2 - 36$$

23. $(5s + 3t)(5s - 2t)$

$$= 5s(5s) - 5s(2t) + 3t(5s) - 3t(2t)$$

$$= 25s^2 - 10st + 15st - 6t^2$$

$$= 25s^2 + 5st - 6t^2$$

25. $(n^2 + 10)(5n + 3)$

$$= n^2 (5n) + n^2 (3) + 10(5n) + 10(3)$$

$$= 5n^3 + 3n^2 + 50n + 30$$

27. $(1.3a - 4b)(2.5a + 7b)$

$$= 1.3a(2.5a) + 1.3a(7b) - 4b(2.5a) - 4b(7b)$$

$$= 3.25a^2 + 9.1ab - 10ab - 28b^2$$

$$= 3.25a^2 - 0.9ab - 28b^2$$

29. $(2x + y)(3x^2 + 2xy + y^2) = 2x(3x^2) + 2x(2xy) + 2x(y^2) + y(3x^2) + y(2xy) + y(y^2)$

$$= 6x^3 + 4x^2 y + 2xy^2 + 3x^2 y + 2xy^2 + y^3$$

$$= 6x^3 + 7x^2 y + 4xy^2 + y^3$$

31. $(x - 7)(x^2 + 7x + 49) = x(x^2) + x(7x) + x(49) - 7(x^2) - 7(7x) - 7(49)$

$$= x^3 + 7x^2 + 49x - 7x^2 - 49x - 343$$

$$= x^3 - 343$$

33. $(4a-b)(a^3-4a^2b+ab^2-b^3)$

$$= 4a(a^3)-4a(4a^2b)+4a(ab^2)-4a(b^3)-b(a^3)+b(4a^2b)-b(ab^2)+b(b^3)$$
$$= 4a^4-16a^3b+4a^2b^2-4ab^3-a^3b+4a^2b^2-ab^3+b^4$$
$$= 4a^4-17a^3b+8a^2b^2-5ab^3+b^4$$

35. $\left(\dfrac{1}{2}a-2b+c\right)(a+6b-c)$

$$= \dfrac{1}{2}a(a)+\dfrac{1}{2}a(6b)-\dfrac{1}{2}a(c)-2b(a)-2b(6b)+2b(c)+c(a)+c(6b)-c(c)$$
$$= \dfrac{1}{2}a^2+3ab-\dfrac{1}{2}ac-2ab-12b^2+2bc+ac+6bc-c^2$$
$$= \dfrac{1}{2}a^2+ab+\dfrac{1}{2}ac-12b^2+8bc-c^2$$

37. $(-x^2+2x+1)(3x-5)=-x^2(3x)+x^2(5)+2x(3x)-2x(5)+1(3x)-1(5)$

$$= -3x^3+5x^2+6x^2-10x+3x-5$$
$$= -3x^3+11x^2-7x-5$$

39. $\left(\dfrac{1}{5}y-10\right)\left(\dfrac{1}{2}y-15\right)$

$$= \dfrac{1}{5}y\left(\dfrac{1}{2}y\right)+\dfrac{1}{5}y(-15)-10\left(\dfrac{1}{2}y\right)-10(-15)$$
$$= \dfrac{1}{10}y^2-3y-5y+150$$
$$= \dfrac{1}{10}y^2-8y+150$$

41. $(a-8)(a+8)=a^2-8^2=a^2-64$

43. $(3p+1)(3p-1)=(3p)^2-1^2=9p^2-1$

45. $\left(x-\dfrac{1}{3}\right)\left(x+\dfrac{1}{3}\right)=x^2-\left(\dfrac{1}{3}\right)^2=x^2-\dfrac{1}{9}$

47. $(3h-k)(3h+k)=(3h)^2-k^2=9h^2-k^2$

49. $(3h-k)^2=(3h)^2-2(3h)(k)+k^2$
$$= 9h^2-6hk+k^2$$

51. $(t-7)^2=t^2-2(t)(7)+7^2=t^2-14t+49$

53. $(u+3v)^2=u^2+2(u)(3v)+(3v)^2$
$$= u^2+6uv+9v^2$$

55. $\left(h+\dfrac{1}{6}k\right)^2=h^2+2(h)\left(\dfrac{1}{6}k\right)+\left(\dfrac{1}{6}k\right)^2$
$$= h^2+\dfrac{1}{3}hk+\dfrac{1}{36}k^2$$

57. $(2z^2-w^3)(2z^2+w^3)=(2z^2)^2-(w^3)^2$
$$= 4z^4-w^6$$

59. $\left(5x^2 - 3y\right)^2 = \left(5x^2\right)^2 - 2\left(5x^2\right)\left(3y\right) + \left(3y\right)^2$

$$= 25x^4 - 30x^2y + 9y^2$$

61. a. When two conjugates are multiplied, the resulting binomial is a difference of squares.

$$(-5x + 4)(5x + 4)$$
$$= -25x^2 - 20x + 20x + 16$$
$$= 16 - 25x^2$$

Since $(-5x + 4)(5x + 4) = 16 - 25x^2$ is a difference of squares, the binomials are conjugates.

b. When two conjugates are multiplied, the resulting binomial is a difference of squares.

$$(-5x + 4)(5x - 4)$$
$$= -25x^2 + 20x + 20x + 16$$
$$= -25x^2 + 40x + 16$$

Since

$(-5x + 4)(5x - 4) = -25x^2 + 40x + 16$ is not a difference of squares, the binomials are not conjugates.

63. a. $\left(A - B\right)\left(A + B\right) = A^2 - B^2$

b. $\left[(x + y) - B\right]\left[(x + y) + B\right]$
$$= (x + y)^2 - B^2$$
$$= x^2 + 2xy + y^2 - B^2$$

Both are examples of multiplying conjugates to get a difference of squares.

65. $\left[(w + v) - 2\right]\left[(w + v) + 2\right] = (w + v)^2 - 2^2$
$$= w^2 + 2wv + v^2 - 4$$

67. $\left[2 - (x + y)\right]\left[2 + (x + y)\right] = 2^2 - (x + y)^2$
$$= 4 - \left(x^2 + 2xy + y^2\right)$$
$$= 4 - x^2 - 2xy - y^2$$

69. $\left[(3a - 4) + b\right]\left[(3a - 4) - b\right] = (3a - 4)^2 - b^2$
$$= (3a)^2 - 2(3a)(4) + 4^2 - b^2$$
$$= 9a^2 - 24a + 16 - b^2$$

71. Write $(x + y)^3$ as $(x + y)^2(x + y)$. Square the binomial and then use the distributive property to multiply the resulting trinomial by the remaining factor of $x + y$.

73. $(2x + y)^3 = (2x + y)^2(2x + y)$

$$= \left(4x^2 + 4xy + y^2\right)(2x + y)$$
$$= 4x^2(2x) + 4x^2(y) + 4xy(2x) + 4xy(y) + y^2(2x) + y^2(y)$$
$$= 8x^3 + 4x^2y + 8x^2y + 4xy^2 + 2xy^2 + y^3$$
$$= 8x^3 + 12x^2y + 6xy^2 + y^3$$

75. $(4a-b)^3 = (4a-b)^2(4a-b)$

$$= (16a^2 - 8ab + b^2)(4a-b)$$

$$= 16a^2(4a) - 16a^2(b) - 8ab(4a) + 8ab(b) + b^2(4a) - b^2(b)$$

$$= 64a^3 - 16a^2b - 32a^2b + 8ab^2 + 4ab^2 - b^3$$

$$= 64a^3 - 48a^2b + 12ab^2 - b^3$$

77. Multiply the first two binomials and simplify. Then multiply the resulting trinomial and the third binomial, using the distributive property.

79. $2a^2(a+5)(3a+1)$

$$= 2a^2[a(3a)+a(1)+5(3a)+5(1)]$$

$$= 2a^2[3a^2+a+15a+5]$$

$$= 2a^2(3a^2+16a+5)$$

$$= 2a^2(3a^2)+2a^2(16a)+2a^2(5)$$

$$= 6a^4 + 32a^3 + 10a^2$$

81. $(x+3)(x-3)(x+5) = (x^2-9)(x+5)$

$$= x^2(x)+x^2(5)-9(x)-9(5)$$

$$= x^3 + 5x^2 - 9x - 45$$

83. $-3(2x+7)-(4x-1)^2$

$$= -3(2x)-3(7)-(16x^2-8x+1)$$

$$= -6x-21-16x^2+8x-1$$

$$= -16x^2 + 2x - 22$$

85. $(y+1)^2 - (2y+3)^2$

$$= (y^2+2y+1)-(4y^2+12y+9)$$

$$= y^2 + 2y + 1 - 4y^2 - 12y - 9$$

$$= -3y^2 - 10y - 8$$

87. $(r+t)^2$

89. $x^2 - y^3$

91. The sum of the cube of p and the square of q.

93. The product of x and the square of y.

95. Let x = the width of the walk

$2x + 20$ = length of garden and walk

$2x + 15$ = width of garden and walk

$$A(x) = (2x+20)(2x+15)$$

$$= 2x(2x)+2x(15)+20(2x)+20(15)$$

$$= 4x^2 + 30x + 40x + 300$$

$$= 4x^2 + 70x + 300$$

97. Let x = the length of a side of the square

$8 - 2x$ = length and width of base

x = the height of the box

a. $V(x) = (8-2x)(8-2x)x$

$$= (64-32x+4x^2)x$$

$$= 4x^3 - 32x^2 + 64x$$

b.

$$V(1) = 4(1)^3 - 32(1)^2 + 64(1)$$
$$= 4 - 32 + 64$$
$$= 36 \text{ in}^3$$

99. $(x-2)^2 = x^2 - 2(x)(2) + 2^2 = x^2 - 4x + 4$

101. $(x-2)(x+2) = x^2 - 2^2 = x^2 - 4$

103. $\dfrac{1}{2}(2x-6)(x+3) = (x-3)(x+3)$
$$= x^2 - 3^2 = x^2 - 9$$

105. $x(3x)(3x+10) = 3x^2(3x+10)$
$$= 3x^2(3x) + 3x^2(10)$$
$$= 9x^3 + 30x^2$$

107. $\dfrac{\left[(x+h)^2 - 3(x+h) - 5\right] - \left(x^2 - 3x - 5\right)}{h}$

$$= \frac{x^2 + 2xh + h^2 - 3x - 3h - 5 - x^2 + 3x + 5}{h}$$

$$= \frac{x^2 - x^2 + 2xh + h^2 - 3x + 3x - 3h - 5 + 5}{h}$$

$$= \frac{2xh + h^2 - 3h}{h}$$

$$= \frac{h(2x + h - 3)}{h}$$

$$= 2x + h - 3$$

109. Multiply $(x+2)^2(x+2)^2$ by squaring the binomials. Then multiply the resulting trinomials using the distributive property.

111. $(5x-6)$
Check:
$(2x-3)(5x-6)$
$$= 2x(5x) - 2x(6) - 3(5x) + 3(6)$$
$$= 10x^2 - 12x - 15x + 18$$
$$= 10x^2 - 27x + 18$$

113. $(2y-1)$
Check:
$(4y+3)(2y-1)$
$$= 4y(2y) - 4y(1) + 3(2y) - 3(1)$$
$$= 8y^2 - 4y + 6y - 3$$
$$= 8y^2 + 2y - 3$$

Section 4.4   Division of Polynomials

1. a. division; quotient; remainder

    b. Synthetic

3. **a.** $(a-10b)-(5a+b)=a-10b-5a-b$
$$=-4a-11b$$

b. $(a-10b)(5a+b)$
$$=a(5a)+a(b)-10b(5a)-10b(b)$$
$$=5a^2+ab-50ab-10b^2$$
$$=5a^2-49ab-10b^2$$

5. **a.** $(x^2-x)+(6x^2+x+2)$
$$=x^2+6x^2-x+x+2$$
$$=7x^2+2$$

b. $(x^2-x)(6x^2+x+2)$
$$=x^2(6x^2)+x^2(x)+x^2(2)$$
$$-x(6x^2)-x(x)-x(2)$$
$$=6x^4+x^3+2x^2-6x^3-x^2-2x$$
$$=6x^4-5x^3+x^2-2x$$

7. For example:
$$(5y+1)^2=(5y)^2+2(5y)(1)+1^2$$
$$=25y^2+10y+1$$

9. $\dfrac{16t^4-4t^2+20t}{-4t}=\dfrac{16t^4}{-4t}-\dfrac{4t^2}{-4t}+\dfrac{20t}{-4t}$
$$=-4t^3+t-5$$

11. $(36y+24y^2+6y^3)\div(3y)$
$$=\dfrac{36y}{3y}+\dfrac{24y^2}{3y}+\dfrac{6y^3}{3y}=12+8y+2y^2$$

13. $(4x^3y+12x^2y^2-4xy^3)\div(4xy)$
$$=\dfrac{4x^3y}{4xy}+\dfrac{12x^2y^2}{4xy}-\dfrac{4xy^3}{4xy}=x^2+3xy-y^2$$

15. $(-8y^4-12y^3+32y^2)\div(-4y^2)$
$$=\dfrac{-8y^4}{-4y^2}-\dfrac{12y^3}{-4y^2}+\dfrac{32y^2}{-4y^2}=2y^2+3y-8$$

17. $(3p^4-6p^3+2p^2-p)\div(-6p)$
$$=\dfrac{3p^4}{-6p}-\dfrac{6p^3}{-6p}+\dfrac{2p^2}{-6p}-\dfrac{p}{-6p}$$
$$=-\dfrac{1}{2}p^3+p^2-\dfrac{1}{3}p+\dfrac{1}{6}$$

19. $(a^3+5a^2+a-5)\div(a)$
$$=\dfrac{a^3}{a}+\dfrac{5a^2}{a}+\dfrac{a}{a}-\dfrac{5}{a}$$
$$=a^2+5a+1-\dfrac{5}{a}$$

21. $\dfrac{6s^3t^5-8s^2t^4+10st^2}{-2st^4}$
$$=\dfrac{6s^3t^5}{-2st^4}-\dfrac{8s^2t^4}{-2st^4}+\dfrac{10st^2}{-2st^4}$$
$$=-3s^2t+4s-\dfrac{5}{t^2}$$

23. $(8p^4q^7-9p^5q^6-11p^3q-4)\div(p^2q)$
$$=\dfrac{8p^4q^7}{p^2q}-\dfrac{9p^5q^6}{p^2q}-\dfrac{11p^3q}{p^2q}-\dfrac{4}{p^2q}$$
$$=8p^2q^6-9p^3q^5-11p-\dfrac{4}{p^2q}$$

25. **a.**
$$\begin{array}{r}
2x^2-3x-1 \\
x-2\overline{)\,2x^3-7x^2+5x-1} \\
\underline{-(2x^3-4x^2)} \\
-3x^2+5x \\
\underline{-(-3x^2+6x)} \\
-x-1 \\
\underline{-(-x+2)} \\
-3
\end{array}$$

Divisior: $(x-2)$ Quotient: $(2x^2-3x-1)$
Remainder: (-3)

b.   Multiply the quotient and divisor; then add the remainder. The result should equal the dividend.

27.

$$
\begin{array}{r}
x+7 \\
x+4{\overline{\smash{\big)}\,x^2+11x+19}} \\
\underline{-\left(x^2+4x\right)} \\
7x+19 \\
\underline{-\left(7x+28\right)} \\
-9
\end{array}
$$

Solution: $x+7-\dfrac{9}{x+4}$

Check:

$(x+4)(x+7)+(-9)=x^2+11x+28-9$

$\qquad\qquad\qquad\quad =x^2+11x+19$

29.

$$
\begin{array}{r}
3y^2+2y+2 \\
y-3{\overline{\smash{\big)}\,3y^3-7y^2-4y+3}} \\
\underline{-\left(3y^3-9y^2\right)} \\
2y^2-4y \\
\underline{-\left(2y^2-6y\right)} \\
2y+3 \\
\underline{-\left(2y-6\right)} \\
9
\end{array}
$$

Solution: $3y^2+2y+2+\dfrac{9}{y-3}$

Check:

$(y-3)(3y^2+2y+2)+(9)$

$\quad =3y^3+2y^2+2y-9y^2-6y-6+9$

$\quad =3y^3-7y^2-4y+3$

31.

$$
\begin{array}{r}
-4a+11 \\
3a-11{\overline{\smash{\big)}\,-12a^2+77a-121}} \\
\underline{-\left(-12a^2+44a\right)} \\
33a-121 \\
\underline{-\left(33a-121\right)} \\
0
\end{array}
$$

Solution: $-4a+11$

Check:

$(3a-11)(-4a+11)+(0)$

$\quad =-12a^2+33a+44a-121$

$\quad =-12a^2+77a-121$

33.

$$
\begin{array}{r}
6y-5 \\
3y+4{\overline{\smash{\big)}\,18y^2+9y-20}} \\
\underline{-\left(18y^2+24y\right)} \\
-15y-20 \\
\underline{-\left(-15y-20\right)} \\
0
\end{array}
$$

Solution: $6y-5$

Check:

$(3y+4)(6y-5)+(0)$

$\quad =18y^2-15y+24y-20$

$\quad =18y^2+9y-20$

35.

$$
\begin{array}{r}
6x^2+4x+5 \\
3x-2{\overline{\smash{\big)}\,18x^3\qquad\;\;+7x+12}} \\
\underline{-\left(18x^3-12x^2\right)} \\
12x^2+7x \\
\underline{-\left(12x^2-8x\right)} \\
15x+12 \\
\underline{-\left(15x-10\right)} \\
22
\end{array}
$$

37.

$$
\begin{array}{r}
4a^2-2a+1 \\
2a+1{\overline{\smash{\big)}\,8a^3\qquad\qquad+1}} \\
\underline{-\left(8a^3+4a^2\right)} \\
-4a^2 \\
\underline{-\left(-4a^2-2a\right)} \\
2a+1 \\
\underline{-\left(2a+1\right)} \\
0
\end{array}
$$

Solution: $6x^2 + 4x + 5 + \dfrac{22}{3x-2}$

Check:

$(3x-2)(6x^2+4x+5)+(22)$

$\quad = 18x^3 + 12x^2 + 15x - 12x^2 - 8x - 10 + 22$

$\quad = 18x^3 + 7x + 12$

Solution: $4a^2 - 2a + 1$

Check:

$(2a+1)(4a^2-2a+1)+(0)$

$\quad = 8a^3 - 4a^2 + 2a + 4a^2 - 2a + 1$

$\quad = 8a^3 + 1$

39.

$$x^2 + x - 1 \enclose{longdiv}{x^4 - x^3 - x^2 + 4x - 2}$$

Quotient: $x^2 - 2x + 2$

$\dfrac{-(x^4 + x^3 - x^2)}{}$

$-2x^3 \qquad + 4x$

$\dfrac{-(-2x^3 - 2x^2 + 2x)}{}$

$2x^2 + 2x - 2$

$\dfrac{-(2x^2 + 2x - 2)}{0}$

Solution: $x^2 - 2x + 2$

Check:

$(x^2+x-1)(x^2-2x+2)+(0)$

$\quad = x^4 - 2x^3 + 2x^2 + x^3 - 2x^2 + 2x$

$\qquad\qquad\qquad\qquad -x^2 + 2x - 2$

$\quad = x^4 - x^3 - x^2 + 4x - 2$

41.

$$x^2 - 5 \enclose{longdiv}{x^4 + 2x^3 \qquad -10x - 25}$$

Quotient: $x^2 + 2x + 5$

$\dfrac{-(x^4 \qquad -5x^2)}{}$

$2x^3 + 5x^2 - 10x$

$\dfrac{-(2x^3 \qquad -10x)}{}$

$5x^2 \qquad -25$

$\dfrac{-(5x^2 \qquad -25)}{0}$

Solution: $x^2 + 2x + 5$

Check:

$(x^2-5)(x^2+2x+5)+(0)$

$\quad = x^4 + 2x^3 + 5x^2 - 5x^2 - 10x - 25$

$\quad = x^4 + 2x^3 - 10x - 25$

43.

$$x^2 - 2 \enclose{longdiv}{x^4 - 3x^2 + 10}$$

Quotient: $x^2 - 1$

$\dfrac{-(x^4 - 2x^2)}{}$

$-x^2 + 10$

$\dfrac{-(-x^2 + 2)}{8}$

Solution: $x^2 - 1 + \dfrac{8}{x^2 - 2}$

Check: $(x^2-2)(x^2-1)+(8)$

$\qquad\qquad = x^4 - x^2 - 2x^2 + 2 + 8$

$\qquad\qquad = x^4 - 3x^2 + 10$

45.

$$n - 2 \enclose{longdiv}{n^4 \qquad\qquad -16}$$

Quotient: $n^3 + 2n^2 + 4n + 8$

$\dfrac{-(n^4 - 2n^3)}{}$

$2n^3$

$\dfrac{-(2n^3 - 4n^2)}{}$

$4n^2$

$\dfrac{-(4n^2 - 8n)}{}$

$8n - 16$

$\dfrac{-(8n - 16)}{0}$

Solution: $n^3 + 2n^2 + 4n + 8$

Check: $(n-2)(n^3+2n^2+4n+8)+(0)$

$\qquad\qquad = n^4 + 2n^3 + 4n^2 + 8n - 2n^3 - 4n^2$

$\qquad\qquad\qquad\qquad\qquad\qquad -8n - 16$

$\qquad\qquad = n^4 - 16$

47. The divisor must be of the form $x - r$.

49. No, the divisor is not of the form $x - r$.

51. **a.** Divisor: $x - 5$

b. Quotient: $x^2 + 3x + 11$

c. Remainder: 58

53.
$$
\begin{array}{r|rrr}
8 & 1 & -2 & -48 \\
  &   & 8 & 48 \\
\hline
  & 1 & 6 & \underline{\hspace{1mm}}0
\end{array}
$$

Quotient: $x + 6$

Check:

$(x - 8)(x + 6) + (0) = x^2 + 6x - 8x - 48$

$ = x^2 - 2x - 48$

55.
$$
\begin{array}{r|rrr}
-1 & 1 & -3 & -4 \\
   &   & -1 & 4 \\
\hline
   & 1 & -4 & \underline{\hspace{1mm}}0
\end{array}
$$

Quotient: $t - 4$

Check:

$(t + 1)(t - 4) + (0) = t^2 - 4t + t - 4$

$ = t^2 - 3t - 4$

57.
$$
\begin{array}{r|rrr}
1 & 5 & 5 & 1 \\
  &   & 5 & 10 \\
\hline
  & 5 & 10 & \underline{\hspace{1mm}}11
\end{array}
$$

Quotient: $5y + 10 + \dfrac{11}{y - 1}$

Check:

$(y - 1)(5y + 10) + (11)$

$ = 5y^2 + 10y - 5y - 10 + 11$

$ = 5y^2 + 5y + 1$

59.
$$
\begin{array}{r|rrrr}
-3 & 3 & 7 & -4 & 3 \\
   &   & -9 & 6 & -6 \\
\hline
   & 3 & -2 & 2 & \underline{\hspace{1mm}}-3
\end{array}
$$

Quotient: $3y^2 - 2y + 2 + \dfrac{-3}{y + 3}$

Check:

$(y + 3)(3y^2 - 2y + 2) + (-3)$

$ = 3y^3 - 2y^2 + 2y + 9y^2 - 6y + 6 - 3$

$ = 3y^3 + 7y^2 - 4y + 3$

61.
$$
\begin{array}{r|rrrr}
2 & 1 & -3 & 0 & 4 \\
  &   & 2 & -2 & -4 \\
\hline
  & 1 & -1 & -2 & \underline{\hspace{1mm}}0
\end{array}
$$

Quotient: $x^2 - x - 2$

Check:

$(x - 2)(x^2 - x - 2) + (0)$

$ = x^3 - x^2 - 2x - 2x^2 + 2x + 4$

$ = x^3 - 3x^2 + 4$

63.
$$
\begin{array}{r|rrrrrr}
2 & 1 & 0 & 0 & 0 & 0 & -32 \\
  &   & 2 & 4 & 8 & 16 & 32 \\
\hline
  & 1 & 2 & 4 & 8 & 16 & \underline{\hspace{1mm}}0
\end{array}
$$

Quotient: $a^4 + 2a^3 + 4a^2 + 8a + 16$

Check:

$(a - 2)(a^4 + 2a^3 + 4a^2 + 8a + 16) + (0)$

$ = a^5 + 2a^4 + 4a^3 + 8a^2 + 16a$

$ -2a^4 - 4a^3 - 8a^2 - 16a - 32$

$ = a^5 - 32$

65.
$$
\begin{array}{r|rrrr}
6 & 1 & 0 & 0 & -216 \\
  &   & 6 & 36 & 216 \\
\hline
  & 1 & 6 & 36 & \underline{\hspace{1mm}}0
\end{array}
$$

Quotient: $x^2 + 6x + 36$

Check:

$(x - 6)(x^2 + 6x + 36) + (0)$

$ = x^3 + 6x^2 + 36x - 6x^2 - 36x - 216$

$ = x^3 - 216$

67.

$$-\frac{2}{3}\bigg|\;6\quad 7\quad -1\quad 3$$

$$\underline{\qquad -4\quad -2\quad 2\quad}$$

$$6\quad 3\quad -3\quad \lfloor 5$$

Quotient: $6t^2 + 3t - 3 + \dfrac{5}{t + \frac{2}{3}}$

Check:

$$\left(t + \frac{2}{3}\right)\left[(6t^2 + 3t - 3) + \frac{5}{t + \frac{2}{3}}\right]$$

$$= \left(t + \frac{2}{3}\right)(6t^2 + 3t - 3) + \left(t + \frac{2}{3}\right)\left(\frac{5}{t + \frac{2}{3}}\right)$$

$$= 6t^3 + 3t^2 - 3t + 4t^2 + 2t - 2 + 5$$

$$= 6t^3 + 7t^2 - t + 3$$

69.

$$\frac{1}{2}\bigg|\;4\quad 0\quad -1\quad 6\quad -3$$

$$\underline{\qquad 2\quad 1\quad 0\quad 3\quad}$$

$$4\quad 2\quad 0\quad 6\quad \lfloor 0$$

Quotient: $4w^3 + 2w^2 + 6$

Check:

$$\left(w - \frac{1}{2}\right)\left(4w^3 + 2w^2 + 6\right) + (0)$$

$$= 4w^4 + 2w^3 + 6w - 2w^3 - w^2 - 3$$

$$= 4w^4 - w^2 + 6w - 3$$

71.

$$-4\bigg|\quad -1\quad -8\quad -3\quad -2$$

$$\underline{\qquad 4\quad 16\quad -52\quad}$$

$$-1\quad -4\quad 13\quad \lfloor -54$$

Quotient: $-x^2 - 4x + 13 + \dfrac{-54}{x + 4}$

73. $\left(22x^2 - 11x + 33\right) \div \left(11x\right)$

$$= \frac{22x^2}{11x} - \frac{11x}{11x} + \frac{33}{11x} = 2x - 1 + \frac{3}{x}$$

75.

$$\begin{array}{r} 4y - 3 \\ 3y^2 - 2y + 5 \overline{)\; 12y^3 - 17y^2 + 30y - 10} \end{array}$$

$$\underline{-\left(12y^3 - 8y^2 + 20y\right)}$$

$$-9y^2 + 10y - 10$$

$$\underline{-\left(-9y^2 + 6y - 15\right)}$$

$$4y + 5$$

Quotient: $4y - 3 + \dfrac{4y + 5}{3y^2 - 2y + 5}$

77.

$$\begin{array}{r} 2x^2 + 3x - 1 \\ 2x^2 + 1 \overline{)\; 4x^4 + 6x^3 \qquad + 3x - 1} \end{array}$$

$$\underline{-\left(4x^4 \qquad + 2x^2\right)}$$

$$6x^3 - 2x^2 + 3x$$

$$\underline{-\left(6x^3 \qquad + 3x\right)}$$

$$-2x^2 \qquad -1$$

$$\underline{-\left(-2x^2 \qquad -1\right)}$$

$$0$$

Quotient: $2x^2 + 3x - 1$

79. $\left(16k^{11} - 32k^{10} + 8k^8 - 40k^4\right) \div \left(8k^8\right)$

$$= \frac{16k^{11}}{8k^8} - \frac{32k^{10}}{8k^8} + \frac{8k^8}{8k^8} - \frac{40k^4}{8k^8}$$

$$= 2k^3 - 4k^2 + 1 - \frac{5}{k^4}$$

81. $\left(5x^3 + 9x^2 + 10x\right) \div \left(5x^2\right)$

$$= \frac{5x^3}{5x^2} + \frac{9x^2}{5x^2} + \frac{10x}{5x^2}$$

$$= x + \frac{9}{5} + \frac{2}{x}$$

83. a. $P(-4) = 4(-4)^3 + 10(-4)^2 - 8(-4) - 20$

$$= 4(-64) + 10(16) + 32 - 20$$

$$= -256 + 160 + 32 - 20 = -84$$

85. $P(r)$ equals the remainder of $P(x) \div (x - r)$.

b.

$$\begin{array}{r} -4\lfloor \quad 4 \quad 10 \quad -8 \quad -20 \\ \underline{\quad\quad -16 \quad 24 \quad -64} \\ 4 \quad -6 \quad 16 \;\lfloor -84 \end{array}$$

Quotient: $4x^2 - 6x + 16 + \dfrac{-84}{x+4}$

c. The values are the same.

87. a.

$$\begin{array}{r} -1\lfloor \quad 8 \quad 13 \quad 5 \\ \underline{\quad\quad -8 \quad -5} \\ 8 \quad 5 \;\lfloor 0 \end{array}$$

b. Quotient: $8x + 5$
Yes

Problem Recognition Exercises

1. a. $(3x+1)^2 = (3x)^2 + 2(3x)(1) + 1^2$
$\qquad = 9x^2 + 6x + 1$

b. $(3x+1)(3x-1) = (3x)^2 - 1^2 = 9x^2 - 1$

c. $(3x+1) - (3x-1) = 3x + 1 - 3x + 1 = 2$

3. a. $\dfrac{4x^2 + 8x - 10}{2x} = \dfrac{4x^2}{2x} + \dfrac{8x}{2x} - \dfrac{10}{2x}$
$\qquad\qquad = 2x + 4 - \dfrac{5}{x}$

b.

$$\begin{array}{r} 2x+5 \\ 2x-1{\overline{\smash{\big)}\,4x^2 + 8x - 10}} \\ \underline{-(4x^2 - 2x)} \\ 10x - 10 \\ \underline{-(10x - 5)} \\ -5 \end{array}$$

Solution: $2x + 5 + \dfrac{-5}{2x-1}$

$$\begin{array}{r} 1\lfloor \quad 4 \quad 8 \quad -10 \\ \underline{\quad\quad 4 \quad 12} \\ 4 \quad 12 \;\lfloor 2 \end{array}$$

c. Quotient: $4x + 12 + \dfrac{2}{x-1}$

5. a. $(p-5)(p+5) - (p^2+5)$
$\qquad\qquad = p^2 - 25 - p^2 - 5 = -30$

b. $(p-5)(p+5) - (p+5)^2$
$\qquad\qquad = p^2 - 25 - p^2 - 10p - 25$
$\qquad\qquad = -10p - 50$

c. $(p-5)(p+5) - (p^2-25)$
$\qquad\qquad = p^2 - 25 - p^2 + 25 = 0$

7. $\left(5t^2 - 6t + 2\right) - \left(3t^2 - 7t + 3\right)$

$\qquad = 5t^2 - 6t + 2 - 3t^2 + 7t - 3$

$\qquad = 2t^2 + t - 1$

9. $\left(6z + 5\right)\left(6z - 5\right) = \left(6z\right)^2 - 5^2 = 36z^2 - 25$

11. $\left(3b - 4\right)\left(2b - 1\right)$

$\qquad = 3b\left(2b\right) - 3b\left(1\right) - 4\left(2b\right) + 4\left(1\right)$

$\qquad = 6b^2 - 3b - 8b + 4$

$\qquad = 6b^2 - 11b + 4$

13. $\left(t^3 - 4t^2 + t - 9\right) + \left(t + 12\right) - \left(2t^2 - 6t\right)$

$\qquad = t^3 - 4t^2 + t - 9 + t + 12 - 2t^2 + 6t$

$\qquad = t^3 - 6t^2 + 8t + 3$

15. $\left(k + 4\right)^2 + \left(-4k + 9\right)$

$\qquad = k^2 + 2\left(k\right)\left(4\right) + 4^2 - 4k + 9$

$\qquad = k^2 + 8k + 16 - 4k + 9$

$\qquad = k^2 + 4k + 25$

17. $-2t\left(t^2 + 6t - 3\right) + t\left(3t + 2\right)\left(3t - 2\right)$

$\qquad = -2t^3 - 12t^2 + 6t + t\left(9t^2 - 4\right)$

$\qquad = -2t^3 - 12t^2 + 6t + 9t^3 - 4t$

$\qquad = 7t^3 - 12t^2 + 2t$

19. $\left(\dfrac{1}{4}p^3 - \dfrac{1}{6}p^2 + 5\right) - \left(-\dfrac{2}{3}p^3 + \dfrac{1}{3}p^2 - \dfrac{1}{5}p\right)$

$\qquad = \dfrac{3}{12}p^3 - \dfrac{1}{6}p^2 + 5 + \dfrac{8}{12}p^3 - \dfrac{2}{6}p^2 + \dfrac{1}{5}p$

$\qquad = \dfrac{11}{12}p^3 - \dfrac{1}{2}p^2 + \dfrac{1}{5}p + 5$

21. $\left(6a^2 - 4b\right)^2 = \left(6a^2\right)^2 - 2\left(6a^2\right)\left(4b\right) + \left(4b\right)^2$

$\qquad = 36a^4 - 48a^2b + 16b^2$

23. $\left(m - 3\right)^2 - 2\left(m + 8\right) = m^2 - 6m + 9 - 2m - 16$

$\qquad = m^2 - 8m - 7$

25. $\left(m^2 - 6m + 7\right)\left(2m^2 + 4m - 3\right) = m^2\left(2m^2 + 4m - 3\right) - 6m\left(2m^2 + 4m - 3\right) + 7\left(2m^2 + 4m - 3\right)$

$\qquad = 2m^4 + 4m^3 - 3m^2 - 12m^3 - 24m^2 + 18m + 14m^2 + 28m - 21$

$\qquad = 2m^4 - 8m^3 - 13m^2 + 46m - 21$

27. $\left[5 - \left(a + b\right)\right]^2 = 5^2 - 2\left(5\right)\left(a + b\right) + \left(a + b\right)^2$

$\qquad = 25 - 10a - 10b + a^2 + 2ab + b^2$

29. $\left(x + y\right)^2 - \left(x - y\right)^2$

$\qquad = x^2 + 2xy + y^2 - \left(x^2 - 2xy + y^2\right)$

$\qquad = x^2 + 2xy + y^2 - x^2 + 2xy - y^2$

$\qquad = 4xy$

31. $\left(-\dfrac{1}{2}x + \dfrac{1}{3}\right)\left(\dfrac{1}{4}x - \dfrac{1}{2}\right) = -\dfrac{1}{8}x^2 + \dfrac{1}{4}x + \dfrac{1}{12}x - \dfrac{1}{6}$

$\qquad = -\dfrac{1}{8}x^2 + \dfrac{1}{3}x - \dfrac{1}{6}$

Section 4.5   Greatest Common Factor and Factoring by Grouping

1.   **a.** product

   b. greatest common factor

   c. greatest common factor

   d. grouping

3. $\left(7t^4+5t^3-9t\right)-\left(-2t^4+6t^2-3t\right)$

$$=7t^4+5t^3-9t+2t^4-6t^2+3t$$
$$=9t^4+5t^3-6t^2-6t$$

5. $\left(5y^2-3\right)\left(y^2+y+2\right)$

$$=5y^4+5y^3+10y^2-3y^2-3y-6$$
$$=5y^4+5y^3+7y^2-3y-6$$

7. $\dfrac{6v^3-12v^2+2v}{-2v}=\dfrac{6v^3}{-2v}-\dfrac{12v^2}{-2v}+\dfrac{2v}{-2v}$

$$=-3v^2+6v-1$$

9. $3x+12=3(x)+3(4)=3(x+4)$

11. $6z^2+4z=2z(3z)+2z(2)=2z(3z+2)$

13. $4p^6-4p=4p\left(p^5\right)-4p(1)=4p\left(p^5-1\right)$

15. $12x^4-36x^2=12x^2\left(x^2\right)-12x^2(3)$

$$=12x^2\left(x^2-3\right)$$

17. $9st^2+27t=9t(st)+9t(3)=9t(st+3)$

19. $9a^4b^3+27a^3b^4-18a^2b^5$

$$=9a^2b^3\left(a^2\right)+9a^2b^3(3ab)-9a^2b^3\left(2b^2\right)$$
$$=9a^2b^3\left(a^2+3ab-2b^2\right)$$

21. $10x^2y+15xy^2-5xy$

$$=5xy(2x)+5xy(3y)-5xy(1)$$
$$=5xy(2x+3y-1)$$

23. $13b^2-11a^2b-12ab$

$$=b(13b)-b\left(11a^2\right)-b(12a)$$
$$=b\left(13b-11a^2-12a\right)$$

25. $-x^2-10x+7=-1\left(x^2+10x-7\right)$

27. $-12x^3y-6x^2y-3xy$

$$=-3xy\left(4x^2\right)-3xy(2x)-3xy(1)$$
$$=-3xy\left(4x^2+2x+1\right)$$

29. $-2t^3+11t^2-3t=-t\left(2t^2\right)-t(-11t)-t(3)$

$$=-t\left(2t^2-11t+3\right)$$

31. $2a(3z-2b)-5(3z-2b)=(3z-2b)(2a-5)$

33. $2x^2(2x-3)+(2x-3)=(2x-3)\left(2x^2+1\right)$

35. $y(2x+1)^2-3(2x+1)^2=(2x+1)^2(y-3)$

37. $3y(x-2)^2 + 6(x-2)^2$

$$= 3\left[y(x-2)^2 + 2(x-2)^2\right]$$

$$= 3(x-2)^2(y+2)$$

39. For example: $3x^3 + 6x^2 + 12x^4$

41. For example: $6(c+d) + y(c+d)$

43. **a.** $2ax - ay + 6bx - 3by$

$$= a(2x-y) + 3b(2x-y)$$

$$= (2x-y)(a+3b)$$

b. $10w^2 - 5w - 6bw + 3b$

$$= 5w(2w-1) - 3b(2w-1)$$

$$= (2w-1)(5w-3b)$$

c.   In part (b), $-3b$ was factored out so that the signs in the last two terms were changed. The resulting binomial factor matches the binomial factor in the first two terms.

45. $y^3 + 4y^2 + 3y + 12 = y^2(y+4) + 3(y+4)$

$$= (y+4)(y^2+3)$$

47. $6p - 42 + pq - 7q = 6(p-7) + q(p-7)$

$$= (p-7)(6+q)$$

49. $2mx + 2nx + 3my + 3ny$

$$= 2x(m+n) + 3y(m+n)$$

$$= (m+n)(2x+3y)$$

51. $10ax - 15ay - 8bx + 12by$

$$= 5a(2x-3y) - 4b(2x-3y)$$

$$= (2x-3y)(5a-4b)$$

53. $x^3 - x^2 - 3x + 3 = x^2(x-1) - 3(x-1)$

$$= (x-1)(x^2-3)$$

55. $6p^2q + 18pq - 30p^2 - 90p$

$$= 6p\left[pq + 3q - 5p - 15\right]$$

$$= 6p\left[q(p+3) - 5(p+3)\right]$$

$$= 6p(p+3)(q-5)$$

57. $100x^3 - 300x^2 + 200x - 600$

$$= 100\left[x^3 - 3x^2 + 2x - 6\right]$$

$$= 100\left[x^2(x-3) + 2(x-3)\right]$$

$$= 100(x-3)(x^2+2)$$

59. $6ax - by + 2bx - 3ay = 6ax + 2bx - 3ay - by$

$$= 2x(3a+b) - y(3a+b)$$

$$= (3a+b)(2x-y)$$

61. $4a - 3b - ab + 12 = 4a - ab + 12 - 3b$

$$= a(4-b) + 3(4-b)$$

$$= (4-b)(a+3)$$

63. $7y^3 - 21y^2 + 5y - 10$ cannot be factored.

65. It is not possible to get a common binomial factor regardless of the order of the terms.

67.
$$U = Av + Acw$$
$$U = A(v + cw)$$
$$\frac{U}{v + cw} = A$$

69.
$$ay + bx = cy$$
$$bx = cy - ay$$
$$bx = y(c - a)$$
$$y = \frac{bx}{c - a} \text{ or } y = \frac{-bx}{a - c}$$

71.
$$A = 2w^2 + w$$
$$A = w(2w + 1)$$
The length of the rectangle is $2w + 1$.

73.
$$(a+3)^4 + 6(a+3)^5 = (a+3)^4\left[1 + 6(a+3)\right]$$
$$= (a+3)^4\left[1 + 6a + 18\right]$$
$$= (a+3)^4(6a + 19)$$

75.
$$24(3x+5)^3 - 30(3x+5)^2$$
$$= 6(3x+5)^2\left[4(3x+5) - 5\right]$$
$$= 6(3x+5)^2\left[12x + 20 - 5\right]$$
$$= 6(3x+5)^2(12x + 15)$$
$$= 6(3x+5)^2 \, 3(4x + 5)$$
$$= 18(3x+5)^2(4x + 5)$$

77.
$$(t+4)^2 - (t+4) = (t+4)\left[(t+4) - 1\right]$$
$$= (t+4)(t+3)$$

79.
$$15w^2(2w-1)^3 + 5w^3(2w-1)^2$$
$$= 5w^2(2w-1)^2\left[3(2w-1) + w\right]$$
$$= 5w^2(2w-1)^2\left[6w - 3 + w\right]$$
$$= 5w^2(2w-1)^2(7w - 3)$$

Section 4.6   Factoring Trinomials

1. a. positive

b. opposite

c.
$$(2x+3)(x-4) = 2x^2 - 8x + 3x - 12$$
$$= 2x^2 - 5x - 12$$
$$(x-4)(2x+3) = 2x^2 + 3x - 8x - 12$$
$$= 2x^2 - 5x - 12$$
Both are correct.

d.
$$6x^2 - 4x - 10 = 2(3x^2 - 2x - 5)$$
$$= 2(3x^2 + 3x - 5x - 5)$$
$$= 2[3x(x+1) - 5(x+1)]$$
$$= 2(3x-5)(x+1)$$

e. $(a+b)^2$; $(a-b)^2$

3. $36c^2d^7e^{11} + 12c^3d^5e^{15} - 6c^2d^4e^7$
$$= 6c^2d^4e^7(6d^3e^4 + 2cde^8 - 1)$$

5. $2x(3a-b) - (3a-b) = (3a-b)(2x-1)$

7. $wz^2 + 2wz - 33az - 66a$
$$= wz(z+2) - 33a(z+2)$$
$$= (z+2)(wz - 33a)$$

9. $b^2 - 12b + 32 = b^2 - 4b - 8b + 32$
$$= b(b-4) - 8(b-4)$$
$$= (b-4)(b-8)$$

11. $y^2 + 10y - 24 = y^2 + 12y - 2y - 24$
$$= y(y+12) - 2(y+12)$$
$$= (y+12)(y-2)$$

13. $x^2 + 13x + 30 = x^2 + 10x + 3x + 30$
$$= x(x+10) + 3(x+10)$$
$$= (x+10)(x+3)$$

15. $c^2 - 6c - 16 = c^2 - 8c + 2c - 16$
$$= c(c-8) + 2(c-8)$$
$$= (c-8)(c+2)$$

17 $2x^2 - 7x - 15 = 2x^2 - 10x + 3x - 15$
$$= 2x(x-5) + 3(x-5)$$
$$= (x-5)(2x+3)$$

19. $a + 6a^2 - 5 = 6a^2 + a - 5$
$$= 6a^2 + 6a - 5a - 5$$
$$= 6a(a+1) - 5(a+1)$$
$$= (a+1)(6a-5)$$

21. $s^2 + st - 6t^2 = s^2 + 3st - 2st - 6t^2$
$$= s(s+3t) - 2t(s+3t)$$
$$= (s+3t)(s-2t)$$

23. $3x^2 - 60x + 108 = 3(x^2 - 20x + 36)$
$$= 3(x^2 - 18x - 2x + 36)$$
$$= 3[x(x-18) - 2(x-18)]$$
$$= 3(x-18)(x-2)$$

25. $2c^2 - 2c - 24 = 2(c^2 - c - 12)$
$$= 2(c^2 - 4c + 3c - 12)$$
$$= 2[c(c-4) + 3(c-4)]$$
$$= 2(c-4)(c+3)$$

27. $2x^2 + 8xy - 10y^2 = 2(x^2 + 4xy - 5y^2)$
$$= 2(x^2 + 5xy - xy - 5y^2)$$
$$= 2[x(x+5y) - y(x+5y)]$$
$$= 2(x+5y)(x-y)$$

29. $33t^2 - 18t + 2$
Since there are not two factors of 66 whose sum is -18, the polynomial is prime.

31. $3x^2 + 14xy + 15y^2 = 3x^2 + 9xy + 5xy + 15y^2$
$$= 3x(x + 3y) + 5y(x + 3y)$$
$$= (x + 3y)(3x + 5y)$$

33. $5u^3v - 30u^2v^2 + 45uv^3 = 5uv(u^2 - 6uv + 9v^2)$
$$= 5uv(u^2 - 3uv - 3uv + 9v^2)$$
$$= 5uv[u(u - 3v) - 3v(u - 3v)]$$
$$= 5uv(u - 3v)(u - 3v)$$
$$= 5uv(u - 3v)^2$$

35. $x^3 - 5x^2 - 14x = x(x^2 - 5x - 14)$
$$= x(x^2 - 7x + 2x - 14)$$
$$= x[x(x - 7) + 2(x - 7)]$$
$$= x(x - 7)(x + 2)$$

37. $-23z - 5 + 10z^2 = 10z^2 - 23z - 5$
$$= 10z^2 - 25z + 2z - 5$$
$$= 5z(2z - 5) + (2z - 5)$$
$$= (2z - 5)(5z + 1)$$

39. $b^2 + 2b + 15$
Since there are not two factors of 15 whose sum is 2, the polynomial is prime.

41. $-2t^2 + 12t + 80 = -2(t^2 - 6t - 40)$
$$= -2(t^2 - 10t + 4t - 40)$$
$$= -2[t(t - 10) + 4(t - 10)]$$
$$= -2(t - 10)(t + 4)$$

43. $14a^2 + 13a - 12 = 14a^2 + 21a - 8a - 12$
$$= 7a(2a + 3) - 4(2a + 3)$$
$$= (2a + 3)(7a - 4)$$

45. $6a^2b + 22ab + 12b = 2b(3a^2 + 11a + 6)$
$$= 2b(3a^2 + 9a + 2a + 6)$$
$$= 2b[3a(a + 3) + 2(a + 3)]$$
$$= 2b(a + 3)(3a + 2)$$

47. a. $(x + 5)(x + 5) = x^2 + 5x + 5x + 25$
$$= x^2 + 10x + 25$$
b. $x^2 + 10x + 25 = (x + 5)^2$

49. a. $(3x - 2y)(3x - 2y)$
$$= 9x^2 - 6xy - 6xy + 4y^2$$
$$= 9x^2 - 12xy + 4y^2$$
b.

51. $9x^2 + (\_\_) + 25 = (3x)^2 + 2(3x)(5) + 5^2$
$$= 9x^2 + (\underline{30x}) + 25$$

53. $64z^4 + (\_\_) + t^2 = (8z^2)^2 + 2(8z^2)(t) + t^2$
$$= 64z^4 + (\underline{16z^2t}) + t^2$$

55. $y^2 - 8y + 16 = y^2 - 2(y)(4) + 4^2 = (y - 4)^2$

57. $64m^2 + 80m + 25 = (8m)^2 + 2(8m)(5) + 5^2$
$$= (8m + 5)^2$$

59. $w^2 - 5w + 9 = w^2 - 2(w)(3) + 3^2$

Not a perfect square trinomial.

61. $9a^2 - 30ab + 25b^2$

$$= (3a)^2 - 2(3a)(5b) + (5b)^2$$
$$= (3a - 5b)^2$$

63. $16t^2 - 80tv + 20v^2 = 4(4t^2 - 20tv + 5v^2)$

Not a perfect square trinomial.

65. $5b^4 - 20b^2 + 20 = 5(b^4 - 4b^2 + 4)$

$$= 5((b^2)^2 - 2(b^2)(2) + 2^2)$$
$$= 5(b^2 - 2)^2$$

67. a. $u^2 - 10u + 25 = u^2 - 2(u)(5) + 5^2$

$$= (u - 5)^2$$

b. $x^4 - 10x^2 + 25 = (x^2)^2 - 10x^2 + 25$

Let $u = x^2$

$u^2 - 10u + 25 = (u - 5)^2$

$$= (x^2 - 5)^2$$

$(a+1)^2 - 10(a+1) + 25$

Let $u = a + 1$

$u^2 - 10u + 25 = (u - 5)^2$

c.
$$= ((a+1) - 5)^2$$
$$= (a - 4)^2$$

69. a. $u^2 + 11u - 26 = u^2 + 13u - 2u - 26$

$$= u(u + 13) - 2(u + 13)$$
$$= (u + 13)(u - 2)$$

b. $w^6 + 11w^3 - 26 = (w^3)^2 + 11w^3 - 26$

Let $u = w^3$

$u^2 + 11u - 26 = (u + 13)(u - 2)$

$$= (w^3 + 13)(w^3 - 2)$$

$(y - 4)^2 + 11(y - 4) - 26$

Let $u = y - 4$

c. $u^2 + 11u - 26 = (u + 13)(u - 2)$

$$= ((y - 4) + 13)((y - 4) - 2)$$
$$= (y + 9)(y - 6)$$

71. $(3x - 1)^2 - (3x - 1) - 6$

Let $u = 3x - 1$

$u^2 - u - 6 = u^2 - 3u + 2u - 6$

$$= u(u - 3) + 2(u - 3)$$
$$= (u - 3)(u + 2)$$
$$= ((3x - 1) - 3)((3x - 1) + 2)$$
$$= (3x - 4)(3x + 1)$$

73. $2(x - 5)^2 + 9(x - 5) + 4$

Let $u = x - 5$

$2u^2 + 9u + 4 = 2u^2 + 8u + u + 4$

$$= 2u(u + 4) + (u + 4)$$
$$= (u + 4)(2u + 1)$$
$$= ((x - 5) + 4)(2(x - 5) + 1)$$
$$= (x - 1)(2x - 10 + 1)$$
$$= (x - 1)(2x - 9)$$

75. $3(y+4)^2 + 5(y+4) - 2$

Let $u = y+4$

$3u^2 + 5u - 2 = 3u^2 + 6u - u - 2$
$= 3u(u+2) - (u+2)$
$= (u+2)(3u-1)$
$= ((y+4)+2)(3(y+4)-1)$
$= (y+6)(3y+12-1)$
$= (y+6)(3y+11)$

77. $3y^6 + 11y^3 + 6$

Let $u = y^3$

$3u^2 + 11u + 6 = 3u^2 + 9u + 2u + 6$
$= 3u(u+3) + 2(u+3)$
$= (u+3)(3u+2)$
$= (y^3+3)(3y^3+2)$

79. $4p^4 + 5p^2 + 1$

Let $u = p^2$

$4u^2 + 5u + 1 = 4u^2 + 4u + u + 1$
$= 4u(u+1) + (u+1)$
$= (u+1)(4u+1)$
$= (p^2+1)(4p^2+1)$

81. $x^4 + 15x^2 + 36$

Let $u = x^2$

$u^2 + 15u + 36 = u^2 + 12u + 3u + 36$
$= u(u+12) + 3(u+12)$
$= (u+12)(u+3)$
$= (x^2+12)(x^2+3)$

83. The factorization $(2y-1)(2y-4)$ is not factored completely because the factor $2y-4$ has a greatest common factor of 2.

85. $w^4 + 12w^2 + 36 = (w^2)^2 + 2(w^2)(6) + 6^2$
$= (w^2+6)^2$

87. $81w^2 + 90w + 25 = (9w)^2 + 2(9w)(5) + 5^2$
$= (9w+5)^2$

89. $3x(a+b) - 6(a+b) = (a+b)(3x-6)$
$= 3(a+b)(x-2)$

91. $12a^2bc^2 + 4ab^2c^2 - 6abc^3$
$= 2abc^2(6a+2b-3c)$

93. $-20x^3 + 74x^2 - 60x = -2x(10x^2 - 37x + 30)$
$= -2x(10x^2 - 25x - 12x + 30)$
$= -2x[5x(2x-5) - 6(2x-5)]$
$= -2x(2x-5)(5x-6)$

95. $2y^2 - 9y - 4$

Since there are not two factors of –8 whose sum is –9, the polynomial is prime.

97. $2(w^2-5)^2 + (w^2-5) - 15$

Let $u = w^2 - 5$

$2u^2 + u - 15 = 2u^2 + 6u - 5u - 15$
$= 2u(u+3) - 5(u+3)$
$= (u+3)(2u-5)$
$= [(w^2-5)+3][2(w^2-5)-5]$
$= [w^2-5+3][2w^2-10-5]$
$= (w^2-2)(2w^2-15)$

99. $1 - 4d + 3d^2 = 1 - 3d - d + 3d^2$
$$= (1 - 3d) - d(1 - 3d)$$
$$= (1 - 3d)(1 - d) \text{ or } (3d - 1)(d - 1)$$

101. $ax - 5a^2 + 2bx - 10ab$
$$= a(x - 5a) + 2b(x - 5a)$$
$$= (x - 5a)(a + 2b)$$

103. $8z^2 + 24zw - 224w^2 = 8(z^2 + 3zw - 28w^2)$
$$= 8(z^2 + 7zw - 4zw - 28w^2)$$
$$= 8[z(z + 7w) - 4w(z + 7w)]$$
$$= 8(z + 7w)(z - 4w)$$

105. $ay + ax - 5cy - 5cx = a(y + x) - 5c(y + x)$
$$= (y + x)(a - 5c)$$

107. $g(x) = 3x^2 + 14x + 8$
$$= 3x^2 + 12x + 2x + 8$$
$$= 3x(x + 4) + 2(x + 4)$$
$$= (x + 4)(3x + 2)$$

109. $n(t) = t^2 + 20t + 100$
$$= t^2 + 2(t)(10) + 10^2$$
$$= (t + 10)^2$$

111. $n(t) = t^2 + 20t + 100$
$$= t^2 + 2(t)(10) + 10^2$$
$$= (t + 10)^2$$

113. $k(a) = a^3 - 4a^2 + 2a - 8$
$$= a^2(a - 4) + 2(a - 4)$$
$$= (a - 4)(a^2 + 2)$$

Section 4.7   Factoring Binomials

1. **a.** difference; $(a + b)(a - b)$

b. sum

c. is not

d. square

e. sum; cubes

f. difference; cubes

g. $a - b$; $a^2 + ab + b^2$

h. $a + b$; $a^2 - ab + b^2$

3. $4x^2 - 20x + 25 = (2x)^2 - 2(2x)(5) + 5^2$
$$= (2x - 5)^2$$

5. $10x + 6xy + 5 + 3y = 2x(5 + 3y) + (5 + 3y)$
$$= (5 + 3y)(2x + 1)$$

7. $32p^2 - 28p - 4 = 4(8p^2 - 7p - 1)$

$\qquad\qquad = 4(8p^2 - 8p + p - 1)$

$\qquad\qquad = 4[8p(p-1)+(p-1)]$

$\qquad\qquad = 4(p-1)(8p+1)$

9.   Look for a binomial of the form $a^2 - b^2$;

$\qquad a^2 - b^2 = (a+b)(a-b)$

11. $x^2 - 9 = x^2 - 3^2 = (x+3)(x-3)$

13. $16 - 49w^2 = 4^2 - (7w)^2 = (4+7w)(4-7w)$

15. $8a^2 - 162b^2 = 2(4a^2 - 81b^2)$

$\qquad\qquad = 2[(2a)^2 - (9b)^2]$

$\qquad\qquad = 2(2a+9b)(2a-9b)$

17. $25u^2 + 1$ Prime

19. $2a^4 - 32 = 2(a^4 - 16)$

$\qquad\qquad = 2(a^2+4)(a^2-4)$

$\qquad\qquad = 2(a^2+4)(a+2)(a-2)$

21. $49 - k^6 = 7^2 - (k^3)^2 = (7+k^3)(7-k^3)$

23. $x^3 - x^2 - 16x + 16 = x^2(x-1) - 16(x-1)$

$\qquad\qquad\qquad = (x-1)(x^2-16)$

$\qquad\qquad\qquad = (x-1)(x^2-4^2)$

$\qquad\qquad\qquad = (x-1)(x+4)(x-4)$

25. $4x^3 + 12x^2 - x - 3 = 4x^2(x+3) - (x+3)$

$\qquad\qquad\qquad = (x+3)(4x^2-1)$

$\qquad\qquad\qquad = (x+3)((2x)^2-1^2)$

$\qquad\qquad\qquad = (x+3)(2x+1)(2x-1)$

27. $9y^3 + 7y^2 - 36y - 28$

$\qquad\qquad = y^2(9y+7) - 4(9y+7)$

$\qquad\qquad = (9y+7)(y^2-4)$

$\qquad\qquad = (9y+7)(y^2-2^2)$

$\qquad\qquad = (9y+7)(y+2)(y-2)$

29. $49x^2 + 28x + 4 - y^2 = (49x^2 + 28x + 4) - y^2$

$\qquad\qquad = (7x+2)^2 - y^2$

$\qquad\qquad = (7x+2+y)(7x+2-y)$

31. $w^2 - 9n^2 + 6n - 1 = w^2 - (9n^2 - 6n + 1)$

$\qquad\qquad = w^2 - (3n-1)^2$

$\qquad\qquad = [w+(3n-1)][w-(3n-1)]$

$\qquad\qquad = (w+3n-1)(w-3n+1)$

33. $p^4 - 10p^2 + 25 - t^4 = (p^4 - 10p^2 + 25) - t^4$

$\qquad\qquad = (p^2-5)^2 - (t^2)^2$

$\qquad\qquad = (p^2-5+t^2)(p^2-5-t^2)$

35. $9u^4 - 4v^4 + 20v^2 - 25$

$$= 9u^4 - \left(4v^4 - 20v^2 + 25\right)$$
$$= \left(3u^2\right)^2 - \left(2v^2 - 5\right)^2$$
$$= \left[3u^2 + \left(2v^2 - 5\right)\right]\left[3u^2 - \left(2v^2 - 5\right)\right]$$
$$= \left(3u^2 + 2v^2 - 5\right)\left(3u^2 - 2v^2 + 5\right)$$

37. Look for a binomial of the form $a^3 + b^3$;
$$a^3 + b^3 = (a + b)\left(a^2 - ab + b^2\right)$$

39. $8x^3 - 1 = \left(2x\right)^3 - 1^3$

$$= (2x - 1)\left[\left(2x\right)^2 + (2x)(1) + 1^2\right]$$
$$= (2x - 1)\left(4x^2 + 2x + 1\right)$$

Check:
$$(2x - 1)\left(4x^2 + 2x + 1\right)$$
$$= 8x^3 + 4x^2 + 2x - 4x^2 - 2x - 1$$
$$= 8x^3 - 1$$

41. $125c^3 + 27 = \left(5c\right)^3 + 3^3$

$$= (5c + 3)\left[\left(5c\right)^2 - (5c)(3) + 3^2\right]$$
$$= (5c + 3)\left(25c^2 - 15c + 9\right)$$

43. $x^3 - 1000 = x^3 - 10^3$

$$= (x - 10)\left[x^2 + (x)(10) + 10^2\right]$$
$$= (x - 10)\left(x^2 + 10x + 100\right)$$

45. $64t^6 + 1 = \left(4t^2\right)^3 + 1^3$

$$= \left(4t^2 + 1\right)\left[\left(4t^2\right)^2 - \left(4t^2\right)(1) + 1^2\right]$$
$$= \left(4t^2 + 1\right)\left(16t^4 - 4t^2 + 1\right)$$

47. $2000y^6 + 2x^3 = 2\left(1000y^6 + x^3\right)$

$$= 2\left[\left(10y^2\right)^3 + x^3\right]$$
$$= 2\left(10y^2 + x\right)\left[\left(10y^2\right)^2 - \left(10y^2\right)(x) + x^2\right]$$
$$= 2\left(10y^2 + x\right)\left(100y^4 - 10y^2x + x^2\right)$$

49. $16z^4 - 54z = 2z\left(8z^3 - 27\right)$

$$= 2z\left[\left(2z\right)^3 - 3^3\right]$$
$$= 2z(2z - 3)\left[\left(2z\right)^2 + (2z)(3) + 3^2\right]$$
$$= 2z(2z - 3)\left(4z^2 + 6z + 9\right)$$

51. $p^{12} - 125 = \left(p^4\right)^3 - 5^3$

$$= \left(p^4 - 5\right)\left[\left(p^4\right)^2 + p^4(5) + 5^2\right]$$
$$= \left(p^4 - 5\right)\left(p^8 + 5p^4 + 25\right)$$

53.
$$36y^2 - \frac{1}{25} = \left(6y\right)^2 - \left(\frac{1}{5}\right)^2$$
$$= \left(6y + \frac{1}{5}\right)\left(6y - \frac{1}{5}\right)$$

55. $18d^{12} - 32 = 2\left(9d^{12} - 16\right) = 2\left[\left(3d^6\right)^2 - 4^2\right]$

$$= 2\left(3d^6 + 4\right)\left(3d^6 - 4\right)$$

57. $242v^2 + 32 = 2\left(121v^2 + 16\right)$

59. $4x^2 - 16 = 4\left(x^2 - 4\right) = 4\left(x^2 - 2^2\right)$

$$= 4(x + 2)(x - 2)$$

61. $25 - 49q^2 = 5^2 - \left(7q\right)^2$

$$= (5 + 7q)(5 - 7q)$$

63. $(t+2s)^2 - 36 = (t+2s)^2 - 6^2$
$$= (t+2s+6)(t+2s-6)$$

65. $27 - t^3 = 3^3 - t^3$
$$= (3-t)\left[3^2 + (3)(t) + t^2\right]$$
$$= (3-t)\left(9 + 3t + t^2\right)$$

67.
$$27a^3 + \frac{1}{8} = (3a)^3 + \left(\frac{1}{2}\right)^3$$
$$= \left(3a + \frac{1}{2}\right)\left[(3a)^2 - (3a)\left(\frac{1}{2}\right) + \left(\frac{1}{2}\right)^2\right]$$
$$= \left(3a + \frac{1}{2}\right)\left(9a^2 - \frac{3}{2}a + \frac{1}{4}\right)$$

69. $2m^3 + 16 = 2\left(m^3 + 8\right) = 2\left(m^3 + 2^3\right)$
$$= 2(m+2)\left[m^2 - (m)(2) + 2^2\right]$$
$$= 2(m+2)\left(m^2 - 2m + 4\right)$$

71. $x^4 - y^4 = \left(x^2\right)^2 - \left(y^2\right)^2 = \left(x^2 + y^2\right)\left(x^2 - y^2\right)$
$$= \left(x^2 + y^2\right)(x+y)(x-y)$$

73. $a^9 + b^9 = \left(a^3\right)^3 + \left(b^3\right)^3$
$$= \left(a^3 + b^3\right)\left[\left(a^3\right)^2 - \left(a^3\right)\left(b^3\right) + \left(b^3\right)^2\right]$$
$$= \left(a^3 + b^3\right)\left(a^6 - a^3b^3 + b^6\right)$$
$$= (a+b)\left[a^2 - (a)(b) + b^2\right]\left(a^6 - a^3b^3 + b^6\right)$$
$$= (a+b)\left(a^2 - ab + b^2\right)\left(a^6 - a^3b^3 + b^6\right)$$

75.
$$\frac{1}{8}p^3 - \frac{1}{125} = \left(\frac{1}{2}p\right)^3 - \left(\frac{1}{5}\right)^3$$
$$= \left(\frac{1}{2}p - \frac{1}{5}\right)\left[\left(\frac{1}{2}p\right)^2 + \left(\frac{1}{2}p\right)\left(\frac{1}{5}\right) + \left(\frac{1}{5}\right)^2\right]$$
$$= \left(\frac{1}{2}p - \frac{1}{5}\right)\left(\frac{1}{4}p^2 + \frac{1}{10}p + \frac{1}{25}\right)$$

77. $4w^2 + 25$ Prime

79.
$$\frac{1}{25}x^2 - \frac{1}{4}y^2 = \left(\frac{1}{5}x\right)^2 - \left(\frac{1}{2}y\right)^2$$
$$= \left(\frac{1}{5}x + \frac{1}{2}y\right)\left(\frac{1}{5}x - \frac{1}{2}y\right)$$

81. $a^6 - b^6 = \left(a^3\right)^2 - \left(b^3\right)^2 = \left(a^3 + b^3\right)\left(a^3 - b^3\right)$
$$= (a+b)\left(a^2 - ab + b^2\right)(a-b)\left(a^2 + ab + b^2\right)$$

83. $64 - y^6 = 8^2 - \left(y^3\right)^2$
$$= \left(8 + y^3\right)\left(8 - y^3\right)$$
$$= \left[2^3 + y^3\right]\left[2^3 - y^3\right]$$
$$= (2+y)\left(4 - 2y + y^2\right)(2-y)\left(4 + 2y + y^2\right)$$

85. $h^6 + k^6 = \left(h^2\right)^3 + \left(k^2\right)^3$
$$= \left(h^2 + k^2\right)\left(h^4 - h^2k^2 + k^4\right)$$

87. $8x^6 + 125 = \left(2x^2\right)^3 + 5^3$

$= \left(2x^2 + 5\right)\left[\left(2x^2\right)^2 - \left(2x^2\right)(5) + 5^2\right]$

$= \left(2x^2 + 5\right)\left(4x^4 - 10x^2 + 25\right)$

89. $(2x+3)(2x-3) = (2x)^2 - 3^2 = 4x^2 - 9$

91. $\left(4a^2 + 6a + 9\right)(2a - 3) = (2a)^3 - 3^3$

$= 8a^3 - 27$

93. $\left(4x^2 + y\right)\left(16x^4 - 4x^2y + y^2\right) = \left(4x^2\right)^3 + y^3$

$= 64x^6 + y^3$

95. a. $A = x^2 - y^2$

   b. $x^2 - y^2 = (x+y)(x-y)$

   c. $A = x^2 - y^2$

$= 6^2 - 4^2 = 36 - 16 = 20 \,\text{in}^2$

97. $x^2 - y^2 + x + y = (x+y)(x-y) + (x+y)$

$= (x+y)(x-y+1)$

99. $x^3 + y^3 + x + y = (x+y)\left(x^2 - xy + y^2\right) + x + y$

$= (x+y)\left(x^2 - xy + y^2 + 1\right)$

101. $576a^5 - 9a^2 - 64a^3c^2 + c^2$

$= 9a^2\left(64a^3 - 1\right) - c^2\left(64a^3 - 1\right)$

$= \left(9a^2 - c^2\right)\left(64a^3 - 1\right)$

$= (3a-c)(3a+c)(4a-1)\left(16a^2 + 4a + 1\right)$

Problem Recognition Exercises

1. A prime factor is an expression whose only factors are 1 and itself.

3. When factoring binomials, look for:
Difference of squares: $a^2 - b^2$;
Difference of cubes: $a^3 - b^3$; or
Sums of cubes: $a^3 + b^3$.

5. Try factoring by grouping (2 terms and two terms) or grouping 3 terms and one term.

7. a. Trinomial

   b. $6x^2 - 21x - 45 = 3\left(2x^2 - 7x - 15\right)$

$= 3\left(2x^2 - 10x + 3x - 15\right)$

$= 3\left[2x(x-5) + 3(x-5)\right]$

$= 3(x-5)(2x+3)$

9. a. Difference of squares

   b. $8a^2 - 50 = 2\left(4a^2 - 25\right) = 2\left[(2a)^2 - 5^2\right]$

$= 2(2a+5)(2a-5)$

11. a. Trinomial

   b. $14u^2 - 11uv + 2v^2$

$= 14u^2 - 7uv - 4uv + 2v^2$

$= 7u(2u-v) - 2v(2u-v)$

$= (2u-v)(7u-2v)$

13. a. Difference of cubes

   b. $16x^3 - 2 = 2\left(8x^3 - 1\right) = 2\left[(2x)^3 - 1^3\right]$

$= 2(2x-1)\left(4x^2 + 2x + 1\right)$

15. a. Sum of cubes

b. $27y^3 + 125 = (3y)^3 + 5^3$
$$= (3y+5)(9y^2 - 15y + 25)$$

17. a. Sum of cubes

b. $128p^6 + 54q^3 = 2(64p^6 + 27q^3)$
$$= 2\left[(4p^2)^3 + (3q)^3\right]$$
$$= 2(4p^2 + 3q)(16p^4 - 12p^2q + 9q^2)$$

19. a. Difference of squares

b. $16a^4 - 1 = (4a^2) - 1^2$
$$= (4a^2 + 1)(4a^2 - 1)$$
$$= (4a^2 + 1)(2a + 1)(2a - 1)$$

21. a. Grouping

b. $p^2 - 12p + 36 - c^2 = (p-6)^2 - c^2$
$$= (p - 6 + c)(p - 6 - c)$$

23. a. Grouping

b. $12ax - 6ay + 4bx - 2by$
$$= 2(6ax - 3ay + 2bx - by)$$
$$= 2\left[3a(2x - y) + b(2x - y)\right]$$
$$= 2(2x - y)(3a + b)$$

25. a. Trinomial

b. $5y^2 + 14y - 3 = 5y^2 + 15y - y - 3$
$$= 5y(y + 3) - (y + 3)$$
$$= (y + 3)(5y - 1)$$

27. a. Difference of squares

b. $t^2 - 100 = t^2 - 10^2 = (t - 10)(t + 10)$

29. a. Sum of cubes

$y^3 + 27 = y^3 + 3^3 = (y + 3)(y^2 - 3y + 9)$

b.

31. a. Trinomial

b. $d^2 + 3d - 28 = (d + 7)(d - 4)$

33. a. Perfect square trinomial

b. $x^2 - 12x + 36 = x^2 - 2(x)(6) + (6)^2$
$$= (x - 6)^2$$

35. a. Grouping

b. $2ax^2 - 5ax + 2bx - 5b$
$$= ax(2x - 5) + b(2x - 5)$$
$$= (ax + b)(2x - 5)$$

37. a. Trinomial

b. $10y^2 + 3y - 4 = (2y - 1)(5y + 4)$

39. a. Difference of squares

b. $10p^2 - 640 = 10(p^2 - 64)$
$$= 10(p - 8)(p + 8)$$

41. a. Difference of cubes

b. $z^4 - 64z = z(z^3 - 64)$
$$= z(z - 4)(z^2 + 4z + 16)$$

43. a. Trinomial

b. $b^3 - 4b^2 - 45b = b(b^2 - 4b - 45)$
$$= b(b - 9)(b + 5)$$

45. a. Perfect square trinomial

b. $9w^2 + 24wx + 16x^2$
$$= (3w)^2 + 2(3w)(4x) + (4x)^2$$
$$= (3w + 4x)^2$$

47. a. Grouping

b.

49. a. Difference of squares

b. $w^4 - 16 = (w^2 - 4)(w^2 + 4)$
$$= (w - 2)(w + 2)(w^2 + 4)$$

$$60x^2 - 20x + 30ax - 10a$$
$$= 10(6x^2 - 2x + 3ax - a)$$
$$= 10[2x(3x-1) + a(3x-1)]$$
$$= 10(2x + a)(3x - 1)$$

51. a. Difference of cubes

b. $t^6 - 8 = (t^2)^3 - 2^3$
$$= (t^2 - 2)(t^4 + 2t^2 + 4)$$

53. a. Trinomial

b. $8p^2 - 22p + 5 = (4p - 1)(2p - 5)$

55. a. Perfect square trinomial

b. $36y^2 - 12y + 1 = (6y)^2 - 2(6y)(1) + (1)^2$
$$= (6y - 1)^2$$

57. a. Sum of squares

b. $2x^2 + 50 = 2(x^2 + 25)$

59. a. Trinomial

b. $12r^2s^2 + 7rs^2 - 10s^2$
$$= s^2(12r^2 + 7r - 10)$$
$$= s^2(4r + 5)(3r - 2)$$

61. a. Trinomial

b. $x^2 + 8xy - 33y^2 = (x - 3y)(x + 11y)$

63. a. Sum of cubes

b. $m^6 + n^3 = (m^2)^3 + n^3$
$$= (m^2 + n)(m^4 - m^2n + n^2)$$

65. a. None of these

b. $x^2 - 4x = x(x - 4)$

67. $x^2(x + y) - y^2(x + y) = (x + y)(x^2 - y^2)$
$$= (x + y)(x + y)(x - y)$$
$$= (x + y)^2(x - y)$$

69. $(a + 3)^4 + 6(a + 3)^5 = (a + 3)^4(1 + 6(a + 3))$
$$= (a + 3)^4(1 + 6a + 18)$$
$$= (a + 3)^4(6a + 19)$$

71. $24(3x + 5)^3 - 30(3x + 5)^2$
$$= 6(3x + 5)^2[4(3x + 5) - 5]$$
$$= 6(3x + 5)^2[12x + 15]$$
$$= 6(3x + 5)^2 3(4x + 5)$$
$$= 18(3x + 5)^2(4x + 5)$$

73. $\dfrac{1}{100}x^2 + \dfrac{1}{35}x + \dfrac{1}{49}$
$$= \left(\frac{1}{10}x\right)^2 + 2\left(\frac{1}{10}x\right)\left(\frac{1}{7}\right) + \left(\frac{1}{7}\right)^2$$
$$= \left(\frac{1}{10}x + \frac{1}{7}\right)^2$$

75. $\left(5x^2 - 1\right)^2 - 4\left(5x^2 - 1\right) - 5$

Let $u = 5x^2 - 1$

$u^2 - 4u - 5 = (u - 5)(u + 1)$
$$= \left(5x^2 - 1 - 5\right)\left(5x^2 - 1 + 1\right)$$
$$= \left(5x^2 - 6\right)\left(5x^2\right)$$

77. $16p^4 - q^4 = (4p^2)^2 - (q^2)^2$
$$= (4p^2 + q^2)(4p^2 - q^2)$$
$$= (4p^2 + q^2)(2p + q)(2p - q)$$

79.
$$y^3 + \frac{1}{64} = y^3 + \left(\frac{1}{4}\right)^3$$
$$= \left(y + \frac{1}{4}\right)\left(y^2 - \frac{1}{4}y + \frac{1}{16}\right)$$

81. $6a^3 + a^2b - 6ab^2 - b^3$
$$= a^2(6a + b) - b^2(6a + b)$$
$$= (6a + b)(a^2 - b^2)$$
$$= (6a + b)(a + b)(a - b)$$

83.
$$\frac{1}{9}t^2 + \frac{1}{6}t + \frac{1}{16} = \left(\frac{1}{3}t\right)^2 + 2\left(\frac{1}{3}t\right)\left(\frac{1}{4}\right) + \left(\frac{1}{4}\right)^2$$
$$= \left(\frac{1}{3}t + \frac{1}{4}\right)^2$$

85. $x^2 + 12x + 36 - a^2 = (x + 6)^2 - a^2$
$$= (x + 6 + a)(x + 6 - a)$$

87. $p^2 + 2pq + q^2 - 81 = (p + q)^2 - 9^2$
$$= (p + q + 9)(p + q - 9)$$

89. $b^2 - (x^2 + 4x + 4) = b^2 - (x + 2)^2$
$$= (b + (x + 2))(b - (x + 2))$$
$$= (b + x + 2)(b - x - 2)$$

91. $4 - u^2 + 2uv - v^2 = 4 - \left(u^2 - 2uv + v^2\right)$
$$= 4 - (u - v)^2$$
$$= (2 + (u - v))(2 - (u - v))$$
$$= (2 + u - v)(2 - u + v)$$

93. $6ax - by + 2bx - 3ay = 6ax + 2bx - by - 3ay$
$$= 2x(3a + b) - y(3a + b)$$
$$= (3a + b)(2x - y)$$

95. $u^6 - 64$
$$= (u^3)^2 - (8)^2$$
$$= (u^3 + 8)(u^3 - 8)$$
$$= (u + 2)(u^2 - 2u + 4)(u - 2)(u^2 + 2u + 4)$$
$$= (u + 2)(u - 2)(u^2 - 2u + 4)(u^2 + 2u + 4)$$

97. $x^8 - 1 = (x^4)^2 - 1^2$
$$= (x^4 + 1)(x^4 - 1)$$
$$= (x^4 + 1)(x^2 + 1)(x^2 - 1)$$
$$= (x^4 + 1)(x^2 + 1)(x + 1)(x - 1)$$

99. $a^2 - b^2 + a + b = (a + b)(a - b) + (a + b)$
$$= (a + b)(a - b + 1)$$

101. $5wx^3 + 5wy^3 - 2zx^3 - 2zy^3$
$$= 5w(x^3 + y^3) - 2z(x^3 + y^3)$$
$$= (x^3 + y^3)(5w - 2z)$$
$$= (x + y)(x^2 - xy + y^2)(5w - 2z)$$

Section 4.8   Solving Equations by Using the Zero Product Rule

1. a. quadratic

   b. 0; 0

   c. Pythagorean; c^2

   d. quadratic

   e. $f(x) = 0$; y

   f. $x + 1$; $x + 2$; $x + 2$

g. lw

h. $\dfrac{1}{2}bh$

3. $10x^2 + 3x = x(10x + 3)$

5. $2p^2 - 9p - 5 = 2p^2 - 10p + p - 5$
$$= 2p(p - 5) + (p - 5)$$
$$= (p - 5)(2p + 1)$$

7. $t^3 - 1 = t^3 - 1^3 = (t - 1)(t^2 + t + 1)$

9. The equation must be set equal to 0, and the polynomial must be factored.

11. $2x(x - 3) = 0$ Correct form.

13. $3p^2 - 7p + 4 = 0$ Incorrect form. The polynomial is not factored.

15. $a(a + 3)^2 = 5$ Incorrect form. The equation is not set equal to 0.

17. a. $w^2 - 81 = (w + 9)(w - 9)$
$$w^2 - 81 = 0$$

b. $(w + 9)(w - 9) = 0$
$$w + 9 = 0 \quad \text{or} \quad w - 9 = 0$$
$$w = -9 \text{ or} \quad w = 9 \ \{-9, 9\}$$

19. a. $3x^2 + 14x - 5 = (3x - 1)(x + 5)$

b. $3x^2 + 14x - 5 = 0$
$$(3x - 1)(x + 5) = 0$$
$$3x - 1 = 0 \quad \text{or} \quad x + 5 = 0$$
$$x = \frac{1}{3} \text{ or} \quad x = -5 \ \left\{\frac{1}{3}, -5\right\}$$

21. $(x + 3)(x + 5) = 0$
$$x + 3 = 0 \text{ or } x + 5 = 0$$
$$x = -3 \text{ or} \quad x = -5 \ \{-3, -5\}$$

23. $(2w + 9)(5w - 1) = 0$
$$2w + 9 = 0 \text{ or } 5w - 1 = 0$$
$$2w = -9 \text{ or} \quad 5w = 1$$
$$w = -\frac{9}{2} \text{ or } w = \frac{1}{5} \ \left\{-\frac{9}{2}, \frac{1}{5}\right\}$$

25. $x(x + 4)(10x - 3) = 0$
$$x = 0 \text{ or } x + 4 = 0 \text{ or } 10x - 3 = 0$$
$$x = 0 \text{ or} \quad x = -4 \text{ or} \quad 10x = 3$$
$$x = 0 \text{ or} \quad x = -4 \text{ or} \quad x = \frac{3}{10}$$
$$\left\{0, -4, \frac{3}{10}\right\}$$

27. $0 = 5(y - 0.4)(y + 2.1)$
$$5 = 0 \text{ or } \quad y - 0.4 = 0 \text{ or } y + 2.1 = 0$$
$$\text{no solution} \quad y = 0.4 \text{ or} \quad y = -2.1$$
$$\{0.4, -2.1\}$$

29.

$$x^2 + 6x - 27 = 0$$
$$(x+9)(x-3) = 0$$
$$x+9=0 \text{ or } x-3=0$$
$$x=-9 \text{ or } \quad x=3 \ \{-9,3\}$$

31.

$$2x^2 + 5x = 3$$
$$2x^2 + 5x - 3 = 0$$
$$2x^2 + 6x - x - 3 = 0$$
$$2x(x+3) - (x+3) = 0$$
$$(x+3)(2x-1) = 0$$
$$x+3=0 \text{ or } 2x-1=0$$
$$x=-3 \text{ or } \quad 2x=1$$
$$x=-3 \text{ or } \quad x=\frac{1}{2} \ \left\{-3,\frac{1}{2}\right\}$$

33.

$$10x^2 = 15x$$
$$10x^2 - 15x = 0$$
$$5x(2x-3) = 0$$
$$5x=0 \text{ or } 2x-3=0$$
$$x=0 \text{ or } \quad 2x=3$$
$$x=0 \text{ or } \quad x=\frac{3}{2} \ \left\{0,\frac{3}{2}\right\}$$

35.

$$6(y-2) - 3(y+1) = 8$$
$$6y - 12 - 3y - 3 = 8$$
$$3y - 15 = 8$$
$$3y = 23$$
$$y = \frac{23}{3} \ \left\{\frac{23}{3}\right\}$$

37.

$$-9 = y(y+6)$$
$$-9 = y^2 + 6y$$
$$y^2 + 6y + 9 = 0$$
$$(y+3)^2 = 0$$
$$y+3 = 0$$
$$y = -3 \ \{-3\}$$

39.

$$9p^2 - 15p - 6 = 0$$
$$3(3p^2 - 5p - 2) = 0$$
$$3(3p^2 - 6p + p - 2) = 0$$
$$3[3p(p-2) + (p-2)] = 0$$
$$3(p-2)(3p+1) = 0$$
$$3=0 \text{ or } \qquad p-2=0 \text{ or } 3p+1=0$$
$$p=2 \text{ or } \quad 3p=-1$$

no solution $\qquad p=2 \text{ or } \qquad p=-\frac{1}{3}$

$$\left\{2,-\frac{1}{3}\right\}$$

41. $(x+1)(2x-1)(x-3) = 0$

$$x+1=0 \text{ or } \quad 2x-1=0 \text{ or } x-3=0$$
$$x=-1 \text{ or } \quad 2x=1 \text{ or } \quad x=3$$
$$x=-1 \text{ or } \quad x=\frac{1}{2} \text{ or } \quad x=3$$
$$\left\{-1,\frac{1}{2},3\right\}$$

43. $(y-3)(y+4) = 8$

$$y^2 + y - 12 = 8$$
$$y^2 + y - 20 = 0$$
$$(y+5)(y-4) = 0$$
$$y+5=0 \text{ or } y-4=0$$
$$y=-5 \text{ or } \quad y=4 \ \{-5,4\}$$

45.
$$(2a-1)(a-1)=6$$
$$2a^2-3a+1=6$$
$$2a^2-3a-5=0$$
$$(2a-5)(a+1)=0$$
$$2a-5=0 \text{ or } a+1=0$$
$$2a=5 \text{ or } \quad a=-1$$
$$a=\frac{5}{2} \text{ or } \quad a=-1 \quad \left\{\frac{5}{2},-1\right\}$$

47.
$$p^2+(p+7)^2=169$$
$$p^2+p^2+14p+49=169$$
$$2p^2+14p-120=0$$
$$2(p^2+7p-60)=0$$
$$2(p+12)(p-5)=0$$
$$2\neq0 \text{ or } \quad p+12=0 \text{ or } p-5=0$$
$$p=-12 \text{ or } p=5 \quad \{-12,5\}$$

49.
$$3t(t+5)-t^2=2t^2+4t-1$$
$$3t^2+15t-t^2=2t^2+4t-1$$
$$11t=-1$$
$$t=-\frac{1}{11} \quad \left\{-\frac{1}{11}\right\}$$

51.
$$2x^3-8x^2-24x=0$$
$$2x(x^2-4x-12)=0$$
$$2x(x-6)(x+2)=0$$
$$2x=0 \text{ or } x-6=0 \text{ or } x+2=0$$
$$x=0 \text{ or } \quad x=6 \text{ or } x=-2 \quad \{0,6,-2\}$$

53.
$$w^3=16w$$
$$w^3-16w=0$$
$$w(w^2-16)=0$$
$$w(w+4)(w-4)=0$$
$$w=0 \text{ or } \quad w+4=0 \text{ or } w-4=0$$
$$w=0 \text{ or } \quad x=-4 \text{ or } x=4$$
$$\{0,-4,4\}$$

55.
$$0=2x^3+5x^2-18x-45$$
$$0=x^2(2x+5)-9(2x+5)$$
$$0=(2x+5)(x^2-9)$$
$$0=(2x+5)(x+3)(x-3)$$
$$2x+5=0 \text{ or } x+3=0 \text{ or } x-3=0$$
$$2x=-5 \text{ or } \quad x=-3 \text{ or } \quad x=3$$
$$x=-\frac{5}{2} \text{ or } \quad x=-3 \text{ or } \quad x=3$$
$$\left\{-\frac{5}{2},-3,3\right\}$$

57. Let x = the number
$$x^2+5=30$$
$$x^2-25=0$$
$$(x+5)(x-5)=0$$
$$x+5=0 \text{ or } x-5=0$$
$$x=-5 \text{ or } \quad x=5$$

59. Let x = the number
$$x^2=x+12$$
$$x^2-x-12=0$$
$$(x+3)(x-4)=0$$
$$x+3=0 \text{ or } x-4=0$$
$$x=-3 \text{ or } \quad x=4$$

61. Let x = the first consecutive integer
$x+1$ = the second consecutive integer

63. Let x = the first consecutive odd integer
$x+2$ = second consecutive odd integer

$$x(x+1)=42$$
$$x^2+x=42$$
$$x^2+x-42=0$$
$$(x+7)(x-6)=0$$
$$x+7=0 \text{ or } x-6=0$$
$$x=-7 \text{ or } x=6$$
$$x+1=-7+1=-6 \text{ or } x+1=6+1=7$$

The consecutive integers are –7 and –6 or 6 and 7.

$$x(x+2)=63$$
$$x^2+2x=63$$
$$x^2+2x-63=0$$
$$(x+9)(x-7)=0$$
$$x+9=0 \text{ or } x-7=0$$
$$x=-9 \text{ or } x=7$$
$$x+2=-9+2=-7 \text{ or } x+2=7+2=9$$

The consecutive odd integers are –9 and –7 or 7 and 9.

65. Let x = the length
$x-2$ = the width
$$x(x-2)=35$$
$$x^2-2x=35$$
$$x^2-2x-35=0$$
$$(x+5)(x-7)=0$$
$$x+5=0 \text{ or } x-7=0$$
$$x\neq-5 \text{ or } x=7$$
$$\text{or } x-2=7-2=5$$

The length is 7 ft and the width is 5 ft.

67. Let x = the width
$x+5$ = the length
$$x(x+5)=300$$
$$x^2+5x=300$$
$$x^2+5x-300=0$$
$$(x+20)(x-15)=0$$
$$x+20=0 \text{ or } x-15=0$$
$$x\neq-20 \text{ or } x=15$$
$$\text{or } x+5=15+5=20$$

The width is 15 yd and the length is 20 yd.

69. a. Let b = the base of the triangle
$b+1$ = the height of the triangle
$$\frac{1}{2}b(b+1+2)=20$$
$$b(b+3)=40$$
$$b^2+3b=40$$
$$b^2+3b-40=0$$
b.
$$(b+8)(b-5)=0$$
$$b+8=0 \text{ or } b-5=0$$
$$b\neq-8 \text{ or } b=5$$
$$b+1=5+1=6$$

The base is 5 in and the height is 6 in.
$$A=\frac{1}{2}(5)(6)=15 \text{ in}^2$$

The area is 15 in².

71. Let h = the height of the triangle
$2h$ = the base of the triangle
$$\frac{1}{2}(2h)(h)=25$$
$$h^2=25$$
$$h^2-25=0$$
$$(h+5)(h-5)=0$$
$$h+5=0 \text{ or } h-5=0$$
$$h\neq-5 \text{ or } h=5$$
$$2h=2(5)=10$$

The height is 5 ft and the base is 10 ft.15 in².

73. Let x = the first positive consecutive integer
$x+1$ = second pos consecutive integer

75. a. Let x = the northern leg
$x-2$ = the eastern leg
$$x^2+(x-2)^2=10^2$$
$$x^2+x^2-4x+4=100$$
$$2x^2-4x-96=0$$

$$x^2 + (x+1)^2 = 41$$
$$x^2 + x^2 + 2x + 1 = 41$$
$$2x^2 + 2x - 40 = 0$$
$$2(x^2 + x - 20) = 0$$
$$2(x+5)(x-4) = 0$$
$$x+5 = 0 \text{ or } x-4 = 0$$
$$x \neq -5 \text{ or } \quad x = 4$$
$$x+1 = 4+1 = 5$$

The consecutive positive integers are 4 and 5.

b.
$$2(x^2 - 2x - 48) = 0$$
$$2(x+6)(x-8) = 0$$
$$x+6 = 0 \text{ or } x-8 = 0$$
$$x \neq -6 \text{ or } \quad x = 8$$
$$x-2 = 8-2 = 6$$

The alternative route is 8 mi + 6 mi = 14 mi.

$$t = \frac{d}{r} = \frac{10}{40} = \frac{1}{4} = 0.25 \text{ hr}$$

$$t = \frac{d}{r} = \frac{14}{60} = \frac{7}{30} \approx 0.23 \text{ hr}$$

The alternative route using superhighways takes less time.

77. Let x = the first consecutive even integer
$x + 2$ = second consecutive even integer
$x + 4$ = third consecutive even integer

$$x^2 + (x+2)^2 = (x+4)^2$$
$$x^2 + x^2 + 4x + 4 = x^2 + 8x + 16$$
$$x^2 - 4x - 12 = 0$$
$$(x+2)(x-6) = 0$$
$$x+2 = 0 \text{ or } x-6 = 0$$
$$x \neq -2 \text{ or } \quad x = 6$$
$$x+2 = 6+2 = 8$$
$$x+4 = 6+4 = 10$$

The lengths of the sides are 6 m, 8 m, and 10 m.

79. Let r = the radius of the circle
$$\pi r^2 = 2\pi r$$
$$r^2 = 2r$$
$$r^2 - 2r = 0$$
$$r(r-2) = 0$$
$$r = 0 \text{ or } r-2 = 0$$
$$r = 0 \text{ or } r = 2$$
The radius is 2 units.

81. **a.**
$$f(x) = x^2 - 3x = 0$$
$$x(x-3) = 0$$
$$x = 0 \text{ or } x-3 = 0$$
$$x = 0 \text{ or } \quad x = 3$$

b. $f(0) = 0^2 - 3(0) = 0 - 0 = 0$

83. **a.**
$$f(x) = x^2 - 6x - 7 = 0$$
$$(x-7)(x+1) = 0$$
$$x-7 = 0 \text{ or } x+1 = 0$$
$$x = 7 \text{ or } \quad x = -1$$

b. $f(0) = 0^2 - 6(0) - 7 = 0 - 0 - 7 = -7$

85. $f(x)=\dfrac{1}{2}(x-2)(x+1)(2x)=0$

$\dfrac{1}{2}\neq 0$ or $x-2=0$ or $x+1=0$ or $2x=0$

$\qquad\qquad x=2$ or $x=-1$ or $x=0$

$f(0)=\dfrac{1}{2}(0-2)(0+1)(2\cdot 0)$

$\qquad =\dfrac{1}{2}(-2)(1)(0)=0$

x-intercepts: (2, 0), (−1, 0), (0, 0)
y-intercept: (0, 0

87. $f(x)=x^2-2x+1=0$

$\qquad (x-1)^2=0$

$\qquad\quad x-1=0$

$\qquad\qquad x=1$

$f(0)=0^2-2(0)+1=0-0+1=1$

x-intercepts: (1, 0)
y-intercept: (0, 1)

89. $g(x)=(x+3)(x-3)=0$

$\qquad (x+3)=0$ or $x-3=0$

$\qquad\quad x=-3$ or $\quad x=3$

x-intercepts: (−3, 0), (3,0)
Graph d)

91. $f(x)=4(x+1)=0$

$\qquad 4\neq 0$ or $x+1=0$

$\qquad\qquad\qquad x=-1$

x-intercepts: (−1, 0)
Graph a.

93. a. The function is in the form
$\qquad s(t)=at^2+bt+c$

b. $s(t)=-4.9t^2+490t=0$

$\qquad -4.9t(t-100)=0$

$\qquad -4.9t=0$ or $t-100=0$

$\qquad\quad t=0$ or $\quad t=100$

t-intercepts (0, 0), (100, 0)

c. At 0 sec and 100 sec, the rocket is at
ground level (height = 0).

d. $s(t)=-4.9t^2+490t=485.1$

$\quad -4.9t^2+490t-485.1=0$

$\quad -4.9(t^2-100t+99)=0$

$\quad -4.9(t-1)(t-99)=0$

$\quad -4.9\neq 0$ or $t-1=0$ or $t-99=0$

$\qquad\qquad\qquad t=1$ or $\quad t=99$

The height is 485.1 m at 1 sec and 99 sec.

95. $f(x)=x^2-7x+10=0$

$f(x)=(x-5)(x-2)=0$

$\qquad x-5=0$ or $x-2=0$

$\qquad\quad x=5$ or $\quad x=2$

$x=5$ and $x=2$ represent the x-intercepts.

97. $f(x)=x^2+2x+1=0$

$\quad f(x)=(x+1)^2=0$

$\qquad\qquad x+1=0$

$\qquad\qquad\quad x=-1$

$x=-1$ represents the x-intercept.

99.
$$f(x) = -x^2 - 6x - 5 = 0$$
$$f(x) = -(x^2 + 6x + 5) = 0$$
$$f(x) = -(x+1)(x+5) = 0$$
$$x + 1 = 0 \quad \text{or} \quad x + 5 = 0$$
$$x = -1 \quad \text{or} \quad x = -5$$
$x = -1$ and $x = -5$ represent the x-intercepts.

101.
$$SA = 2\pi r^2 + 2\pi rh$$
$$2\pi r^2 + 2\pi r(7) = 156\pi$$
$$r^2 + 7r = 78$$
$$r^2 + 7r - 78 = 0$$
$$(r+13)(r-6) = 0$$
$$r + 13 = 0 \quad \text{or } r - 6 = 0$$
$$r = -13 \text{ or} \quad r = 6$$
The radius is 6 ft.

103. Let l = the length
w = the width
$$2l + 2w = 28$$
$$2w = 28 - 2l$$
$$w = 14 - l$$

(continued at the right)

$$A = l(14 - l) = 48$$
$$14l - l^2 = 48$$
$$0 = l^2 - 14l + 48$$
$$0 = (l-8)(l-6)$$
$$l - 8 = 0 \quad \text{or} \quad l - 6 = 0$$
$$l = 8 \quad \text{or} \quad l = 6$$
$$w = 14 - 8 = 6$$
The length is 8 ft and the width is 6 ft.

105. $x = 2$ and $x = -2$
$$(x-2)(x+2) = 0$$
$$x^2 - 4 = 0$$

107. $x = 0$ and $x = -3$
$$(x-0)(x+3) = 0$$
$$x^2 + 3x = 0$$

Chapter 4 Group Activity

1. $(a+b)^0 = 1$

3.
  a. 3
  b. 4
  c. 5
  d. 6
  e. $n + 1$

5.
$(a+b)^4$ 1 4 6 4 1
$(a+b)^5$ 1 5 10 10 5 1
$(a+b)^6$ 1 6 15 20 15 6 1
$(a+b)^7$ 1 7 21 35 35 21 7 1

Chapter 4 Review Exercises

Section 4.1

1. $(3x)^3(3x)^2 = (3x)^{2+3} = (3x)^5 = 3^5x^5 = 243x^5$

3. $\dfrac{24x^5y^3}{-8x^4y} = -3x^{5-4}y^{3-1} = -3xy^2$

5. $\left(-2a^2b^{-5}\right)^{-3} = \left(-2\right)^{-3}\left(a^2\right)^{-3}\left(b^{-5}\right)^{-3}$

$$= \left(-\frac{1}{2}\right)^3 a^{-6}b^{15} = -\frac{1}{8}\cdot\frac{1}{a^6}\cdot b^{15} = -\frac{b^{15}}{8a^6}$$

7. $\left(\dfrac{-4x^4y^{-2}}{5x^{-1}y^4}\right)^{-4} = \left(-\dfrac{4}{5}x^{4-(-1)}y^{-2-4}\right)^{-4}$

$$= \left(-\frac{4}{5}x^5y^{-6}\right)^{-4} = \left(-\frac{4}{5}\right)^{-4}\left(x^5\right)^{-4}\left(y^{-6}\right)^{-4}$$

$$= \left(-\frac{5}{4}\right)^4 x^{-20}y^{24} = \frac{5^4}{4^4}\cdot\frac{1}{x^{20}}\cdot y^{24} = \frac{5^4y^{24}}{4^4x^{20}}$$

9.   **a.** $3,686,600,000 = 3.6866\times10^9$

      b. $0.000001 = 1.0\times10^{-6}$

11.   **a.** $1\times10^{-3} = 0.001$

      b $1\times10^{-9} = 0.000000001$

13. $\dfrac{2,500,000}{0.0004} = \dfrac{2.5\times10^6}{4\times10^{-4}} = 0.625\times10^{6-(-4)}$

$$= 6.25\times10^{-1}\times10^{10} = 6.25\times10^9$$

15. $\left(3.6\times10^8\right)\left(9.0\times10^{-2}\right) = 32.4\times10^{8+(-2)}$

$$= 3.24\times10^1\times10^6 = 3.24\times10^7$$

Section 4.2

17. $6x^4 + 10x - 1$ Trinomial; degree 4

19. $g(x) = 4x - 7$

   a. $g(0) = 4(0) - 7 = 0 - 7 = -7$

   b. $g(-4) = 4(-4) - 7 = -16 - 7 = -23$

   c. $g(3) = 4(3) - 7 = 12 - 7 = 5$

21.   **a.** $A(x) = 0.047x^2 + 1.46x + 16.8$

$$A(5) = 0.047(5^2) + 1.46(5) + 16.8$$

$$= 0.047(25) + 1.46(5) + 16.8$$

$$= 1.175 + 7.3 + 16.8 = 25.275$$

$A(5) \approx 25$ means that in the year 2005, each American on average consumed approximately 25 gal of bottled water.

   b. $A(15) = 0.047(15^2) + 1.46(15) + 16.8$

$$= 0.047(225) + 1.46(15) + 16.8$$

$$= 10.575 + 21.9 + 16.8 = 49.275$$

$A(15) \approx 49$ means that in the year 2015, each American will consume approximately 49 gal of bottled water if this trend continues.

23. $7xy - 3xz + 5yz$

     $\underline{+13xy - 15xz - 8yz}$

     $20xy - 18xz - 3yz$

25. $\left(3a^2 - 2a - a^3\right) - \left(5a^2 - a^3 - 8a\right)$

$$= 3a^2 - 2a - a^3 - 5a^2 + a^3 + 8a$$

$$= -2a^2 + 6a$$

27. $\left(\dfrac{5}{6}x^4 + \dfrac{1}{2}x^2 - \dfrac{1}{3}\right) - \left(-\dfrac{1}{6}x^4 - \dfrac{1}{4}x^2 - \dfrac{1}{3}\right)$

$$= \frac{5}{6}x^4 + \frac{1}{2}x^2 - \frac{1}{3} + \frac{1}{6}x^4 + \frac{1}{4}x^2 + \frac{1}{3}$$

$$= x^4 + \frac{3}{4}x^2$$

29. $-(4x - 4y) - \left[(4x + 2y) - (3x + 7y)\right]$

$$= -4x + 4y - \left[4x + 2y - 3x - 7y\right]$$

$$= -4x + 4y - 4x - 2y + 3x + 7y$$

$$= -5x + 9y$$

31. $\left(2x^2 - 4x\right) + \left(2x^2 - 7x\right) = 4x^2 - 11x$

33. $\left(2x^2 - 7x\right) - \left(2x^2 - 4x\right) = 2x^2 - 7x - 2x^2 + 4x$
$$= -3x$$

Section 4.3

35. $-3x\left(6x^2 - 5x + 4\right) = -18x^3 + 15x^2 - 12x$

37. $(x - 2)(x - 9) = x^2 - 9x - 2x + 18$
$$= x^2 - 11x + 18$$

39. $\left(-\dfrac{1}{5} + 2y\right)\left(\dfrac{1}{5} + y\right) = -\dfrac{1}{25} - \dfrac{1}{5}y + \dfrac{2}{5}y + 2y^2$
$$= 2y^2 + \dfrac{1}{5}y - \dfrac{1}{25}$$

41. $(x - y)\left(x^2 + xy + y^2\right)$
$$= x^3 + x^2y + xy^2 - x^2y - xy^2 - y^3$$
$$= x^3 - y^3$$

43. $\left(\dfrac{1}{2}x + 4\right)^2 = \left(\dfrac{1}{2}x\right)^2 + 2\left(\dfrac{1}{2}x\right)(4) + (4)^2$
$$= \dfrac{1}{4}x^2 + 4x + 16$$

45. $(6w - 1)(6w + 1) = (6w)^2 - (1)^2$
$$= 36w^2 - 1$$

47. $\left(z + \dfrac{1}{4}\right)\left(z - \dfrac{1}{4}\right) = (z)^2 - \left(\dfrac{1}{4}\right)^2$
$$= z^2 - \dfrac{1}{16}$$

49. $\left[c - (w + 3)\right]\left[c + (w + 3)\right] = c^2 - (w + 3)^2$
$$= c^2 - \left(w^2 + 6w + 9\right)$$
$$= c^2 - w^2 - 6w - 9$$

51. $\left(y^2 - 3\right)^3 = \left(y^2 - 3\right)\left(y^2 - 3\right)^2$
$$= \left(y^2 - 3\right)\left(y^4 - 6y^2 + 9\right)$$
$$= y^6 - 6y^4 + 9y^2 - 3y^4 + 18y^2 - 27$$
$$= y^6 - 9y^4 + 27y^2 - 27$$

53.

Let x = the width of the rectangle
$3x + 2$ = the length of the

a. rectangle
$$P(x) = 2(3x + 2) + 2x$$

b. $= 6x + 4 + 2x = 8x + 4$
$$A(x) = (3x + 2)(x) = 3x^2 + 2x$$

Section 4.4

55. $\left(6x^3y + 12x^2y^2 - 9xy^3\right) \div (3xy)$

$\dfrac{6x^3y}{3xy} + \dfrac{12x^2y^2}{3xy} - \dfrac{9xy^3}{3xy} = 2x^2 + 4xy - 3y^2$

57. **a.**

$$
\begin{array}{r}
3y^3 - 2y^2 + 6y - 4 \\
3y+2\overline{\smash{\big)}\ 9y^4 \qquad\quad +14y^2 \qquad -8} \\
\underline{-\left(9y^4 + 6y^3\right)} \\
-6y^3 + 14y^2 \\
\underline{-\left(-6y^3 - 4y^2\right)} \\
18y^2 \\
\underline{-\left(18y^2 + 12y\right)} \\
-12y - 8 \\
\underline{-(-12y - 8)} \\
0
\end{array}
$$

b.   Quotient: $3y^3 - 2y^2 + 6y - 4$
Remainder: 0

c.   Multiply the quotient and the divisor.

59.

$$
\begin{array}{r}
x + 4 \\
x+4\overline{\smash{\big)}\ x^2 + 8x - 16} \\
\underline{-\left(x^2 + 4x\right)} \\
4x - 16 \\
\underline{-\left(4x + 16\right)} \\
-32
\end{array}
$$

The quotient is $x + 4 + \dfrac{-32}{x+4}$.

61.

$$
\begin{array}{r}
2x^3 - 2x^2 + 5x - 4 \\
x^2+x\overline{\smash{\big)}\ 2x^5 \qquad +3x^3 + x^2 \qquad -4} \\
\underline{-\left(2x^5 + 2x^4\right)} \\
-2x^4 + 3x^3 \\
\underline{-\left(-2x^4 - 2x^3\right)} \\
5x^3 + x^2 \\
\underline{-\left(5x^3 + 5x^2\right)} \\
-4x^2 \\
\underline{-\left(-4x^2 - 4x\right)} \\
4x - 4
\end{array}
$$

The quotient is $2x^3 - 2x^2 + 5x - 4 + \dfrac{4x-4}{x^2+x}$.

63. **a.**   Divisor: $x - 3$
b.   Quotient: $2x^3 + 11x^2 + 31x + 99$
c.   Remainder: 298

65.

$$
\begin{array}{r|rrr}
-5 & 1 & 7 & 14 \\
   &   & -5 & -10 \\
\hline
   & 1 & 2 & \underline{\,4} \\
\end{array}
$$

Quotient: $x + 2 + \dfrac{4}{x+5}$

67.

$$
\begin{array}{r|rrrr}
3 & 1 & -6 & 0 & 8 \\
  &   & 3 & -9 & -27 \\
\hline
  & 1 & -3 & -9 & \underline{-19} \\
\end{array}
$$

Quotient: $w^2 - 3w - 9 + \dfrac{-19}{w-3}$

Section 4.5

69. $-x^3 - 4x^2 + 11x = -x(x^2 + 4x - 11)$

or $x(-x^2 - 4x + 11)$

71. $5x(x-7) - 2(x-7) = (x-7)(5x-2)$

73. $m^3 - 8m^2 + m - 8 = m^2(m-8) + (m-8)$
$= (m-8)(m^2+1)$

75. $4ax^2 + 2bx^2 - 6ax - 3xb$
$= x(4ax + 2bx - 6a - 3b)$
$= x[2x(2a+b) - 3(2a+b)]$
$= x(2a+b)(2x-3)$

Section 4.6

77. The trinomial must be of the form $a^2 + 2ab + b^2$ or $a^2 - 2ab + b^2$.

79. $3m^2 + mt - 10t^2 = 3m^2 + 6mt - 5mt - 10t^2$
$= 3m(m+2t) - 5t(m+2t)$
$= (m+2t)(3m-5t)$

81. $2k^2 + 7k^3 + 6k^4 = k^2(6k^2 + 7k + 2)$
$= k^2(6k^2 + 4k + 3k + 2)$
$= k^2[2k(3k+2) + (3k+2)]$
$= k^2(3k+2)(2k+1)$

83. $80z + 32 + 50z^2 = 50z^2 + 80z + 32$
$= 2(25z^2 + 40z + 16)$
$= 2[(5z)^2 + 2(5z)(4) + 4^2]$
$= 2(5z+4)^2$

85. $(4x+3)^2 - 12(4x+3) + 36$
Let $u = 4x + 3$
$u^2 - 12u + 36 = u^2 - 2(u)(6) + 6^2$
$= (u-6)^2$
$= (4x+3-6)^2$
$= (4x-3)^2$

87. $3w^4 - 2w^2 - 5 = 3w^4 - 5w^2 + 3w^2 - 5$
$= w^2(3w^2 - 5) + (3w^2 - 5)$
$= (3w^2 - 5)(w^2 + 1)$

Section 4.7

89. $x^3 - \dfrac{1}{27} = x^3 - \left(\dfrac{1}{3}\right)^3 = \left(x - \dfrac{1}{3}\right)\left(x^2 + \dfrac{1}{3}x + \dfrac{1}{9}\right)$

91. $h^3 + 9h = h(h^2 + 9)$

93. $k^4 - 16 = (k^2)^2 - 4^2 = (k^2 + 4)(k^2 - 4)$
$= (k^2 + 4)(k+2)(k-2)$

95. $x^2 - 8xy + 16y^2 - 9 = (x-4y)^2 - 3^2$
$= (x-4y+3)(x-4y-3)$

97. $t^2 + 16t + 64 - 25c^2 = (t+8)^2 - (5c)^2$
$= (t+8+5c)(t+8-5c)$

Section 4.8

99. A quadratic equation can be written in the form $ax^2 + bx + c = 0$ $(a \neq 0)$.

101. $x^2 + 6x = 7$

$x^2 + 6x - 7 = 0$ Quadratic

103. $2x - 5 = 3$

$2x - 8 = 0$ Linear

105.
  a. $5x^2 + 6x - 8 = (5x - 4)(x + 2)$

  b. $5x^2 + 6x - 8 = 0$

$(5x - 4)(x + 2) = 0$

$5x - 4 = 0$ or $x + 2 = 0$

$x = \dfrac{4}{5}$ or $x = -2$ $\left\{\dfrac{4}{5}, -2\right\}$

107.
$x^2 - 2x - 15 = 0$

$(x - 5)(x + 3) = 0$

$x - 5 = 0$ or $x + 3 = 0$

$x = 5$ or $x = -3$

$\{5, -3\}$

109.
$2t(t + 5) + 1 = 3t - 3 - t^2$

$2t^2 + 10t + 1 = 3t - 3 - t^2$

$3t^2 + 7t + 4 = 0$

$(3t + 4)(t + 1) = 0$

$3t + 4 = 0$ or $t + 1 = 0$

$3t = -4$ or $t = -1$

$t = -\dfrac{4}{3}$ or $t = -1$ $\left\{-\dfrac{4}{3}, -1\right\}$

111.
$f(x) = -4x^2 + 4 = 0$

$-4(x^2 - 1) = 0$

$-4(x + 1)(x - 1) = 0$

$-4 \neq 0$ or $(x + 1) = 0$ or $x - 1 = 0$

$x = -1$ or $x = 1$

$f(0) = -4(0)^2 + 4 = 0 + 4 = 4$

x-intercepts: $(-1, 0), (1, 0)$

y-intercept: $(0, 4)$

Graph b.

113.
$h(x) = 5x^3 - 10x^2 - 20x + 40 = 0$

$5(x^3 - 2x^2 - 4x + 8) = 0$

$5[x^2(x - 2) - 4(x - 2)] = 0$

$5(x - 2)(x^2 - 4) = 0$

$5(x - 2)(x - 2)(x + 2) = 0$

$5 \neq 0$ or $(x - 2) = 0$ or $x - 2 = 0$ or $x + 2 = 0$

$x = 2$ or $x = 2$ or $x = -2$

$h(0) = 5(0)^3 - 10(0)^2 - 20(0) + 40 = 40$

x-intercepts: $(-2, 0), (2, 0)$

y-intercept: $(0, 40)$

Graph c.

115. Let x = the width of the truck

$2x - 1$ = the length of the truck

$$V = 10(x)(2x-1) = 1200$$
$$x(2x-1) = 120$$
$$2x^2 - x - 120 = 0$$
$$2x^2 - 16x + 15x - 120 = 0$$
$$2x(x-8) + 15(x-8) = 0$$
$$(x-8)(2x+15) = 0$$
$$x - 8 = 0 \text{ or } 2x + 15 = 0$$
$$x = 8 \text{ or } \qquad 2x = -15$$
$$x = 8 \text{ or } \qquad x \neq -\frac{15}{2}$$
$$2x - 1 = 2(8) - 1 = 15$$

The width of the truck is 8 ft, the length of the truck is 15 ft, and the height is 10 ft.

Chapter 4   Test

1. $\dfrac{20a^7}{4a^{-6}} = 5a^{7-(-6)} = 5a^{13}$

3. $\left(\dfrac{-3x^6}{5y^7}\right)^2 = \dfrac{(-3)^2(x^6)^2}{(5)^2(y^7)^2} = \dfrac{9x^{12}}{25y^{14}}$

5. $(8.0 \times 10^{-6})(7.1 \times 10^5) = 56.8 \times 10^{-6+5}$
$\qquad = 5.68 \times 10^1 \times 10^{-1} = 5.68 \times 10^0 = 5.68$

7. $F(x) = 5x^3 - 2x^2 + 8$
$F(-1) = 5(-1)^3 - 2(-1)^2 + 8 = -5 - 2 + 8 = 1$
$F(2) = 5(2)^3 - 2(2)^2 + 8 = 40 - 8 + 8 = 40$
$F(0) = 5(0)^3 - 2(0)^2 + 8 = 0 - 0 + 8 = 8$

9. $(2a-5)(a^2 - 4a - 9)$
$\qquad = 2a^3 - 8a^2 - 18a - 5a^2 + 20a + 45$
$\qquad = 2a^3 - 13a^2 + 2a + 45$

11. $(5x - 4y^2)(5x + 4y^2) = (5x)^2 - (4y^2)^2$
$\qquad = 25x^2 - 16y^4$

13. $(7x - 4)^2 = (7x)^2 - 2(7x)(4) + 4^2$
$\qquad = 49x^2 - 56x + 16$

15.

$$
\begin{array}{r}
5p^2 - p + 1 \\
2p+3 \overline{) 10p^3 + 13p^2 - p + 3} \\
\underline{-(10p^3 + 15p^2)} \\
-2p^2 - p \\
\underline{-(-2p^2 - 3p)} \\
2p + 3 \\
\underline{-(2p + 3)} \\
0
\end{array}
$$

Quotient: $5p^2 - p + 1$

17. 1. Take out the GCF.
2. If there are more than three terms, try grouping.
3. If a trinomial, look for a perfect square trinomial. Otherwise, use the grouping or trial-and-error method.
4. If a binomial, look for a difference of squares, a difference of cubes, or a sum of cubes.

19.
$$3a^2 + 27ab + 54b^2 = 3(a^2 + 9ab + 18b^2)$$
$$= 3(a + 6b)(a + 3b)$$

21.
$$xy - 7x + 3y - 21 = x(y - 7) + 3(y - 7)$$
$$= (y - 7)(x + 3)$$

23.
$$-10u^2 + 30u - 20 = -10(u^2 - 3u + 2)$$
$$= -10(u - 2)(u - 1)$$

25.
$$5y^2 - 50y + 125 = 5(y^2 - 10y + 25)$$
$$= 5(y - 5)(y - 5)$$
$$= 5(y - 5)^2$$

27.
$$2x^3 + x^2 - 8x - 4 = x^2(2x + 1) - 4(2x + 1)$$
$$= (2x + 1)(x^2 - 4)$$
$$= (2x + 1)(x + 2)(x - 2)$$

29.
$$x^2 + 8x + 16 - y^2 = (x + 4)^2 - y^2$$
$$= (x + 4 + y)(x + 4 - y)$$

31.
$$\left(x^2 + 1\right)^2 + 3\left(x^2 + 1\right) + 2$$
Let $u = x^2 + 1$
$$u^2 + 3u + 2 = (u + 2)(u + 1)$$
$$= \left(x^2 + 1 + 2\right)\left(x^2 + 1 + 1\right)$$
$$= \left(x^2 + 3\right)\left(x^2 + 2\right)$$

33.
$$(2x - 3)(x + 5) = 0$$
$$2x - 3 = 0 \quad \text{or} \quad x + 5 = 0$$
$$x = \frac{3}{2} \qquad x = -5 \quad \left\{\frac{3}{2}, -5\right\}$$

35.
$$y^2 - 6y = 16$$
$$y^2 - 6y - 16 = 0$$
$$(y + 2)(y - 8) = 0$$
$$y + 2 = 0 \quad \text{or} \quad y - 8 = 0$$
$$y = -2 \qquad y = 8 \quad \{-2, 8\}$$

37.
$$4p - 64p^3 = 0$$
$$-4p\left(16p^2 - 1\right) = 0$$
$$-4p(4p + 1)(4p - 1) = 0$$
$$-4p = 0 \quad \text{or} \quad 4p + 1 = 0 \quad \text{or} \quad 4p - 1 = 0$$
$$p = 0 \quad \text{or} \qquad 4p = -1 \quad \text{or} \qquad 4p = 1$$
$$p = 0 \quad \text{or} \qquad p = -\frac{1}{4} \quad \text{or} \qquad p = \frac{1}{4}$$
$$\left\{0, -\frac{1}{4}, \frac{1}{4}\right\}$$

39.
$$f(x) = x^2 - 6x + 8 = 0$$
$$(x - 4)(x - 2) = 0$$
$$x - 4 = 0 \quad \text{or} \quad x - 2 = 0$$
$$x = 4 \quad \text{or} \qquad x = 2$$
$$f(0) = (0)^2 - 6(0) + 8 = 0 - 0 + 8 = 8$$

x-intercepts: (4, 0), (2, 0)
y-intercept: (0, 8)
Graph c.

41.
$$p(x) = -2x^2 - 8x - 6 = 0$$
$$-2(x^2 + 4x + 3) = 0$$
$$-2(x + 3)(x + 1) = 0$$
$$-2 \neq 0 \text{ or } x + 3 = 0 \text{ or } x + 1 = 0$$
$$x = -3 \text{ or } x = -1$$
$$p(0) = -2(0)^2 - 8(0) - 6 = 0 - 0 - 6 = -6$$
x-intercepts: $(-3, 0)$, $(-1, 0)$
y-intercept: $(0, -6)$
Graph d.

43.
$$h(x) = -\frac{x^2}{256} + x = 0$$
$$x^2 - 256x = 0$$
$$x(x - 256) = 0$$
$$x = 0 \text{ or } x - 256 = 0$$
$$x = 0 \text{ or } x = 256$$
The rocket hits the ground 256 ft from the launch pad

Chapters 1 – 4   Cumulative Review Exercises

1.
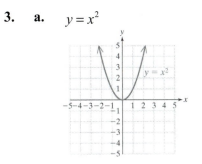
$[5, 12]$

3. a. $y = x^2$

b. $y = |x|$

5.
$$2(9.85 \times 10^7) = 19.7 \times 10^7$$
$$= 1.97 \times 10^1 \times 10^7$$
$$= 1.97 \times 10^8 \text{ people}$$

7.
$$x + (2x) + (2x - 5) = 180$$
$$5x - 5 = 180$$
$$5x = 185$$
$$x = 37$$
$$2x = 2(37) = 74$$
$$2x - 5 = 2(37) - 5 = 69$$
The angles are 37°, 74°, and 69°.

9.
$$4x - 3y = -9$$
$$-3y = -4x - 9$$
$$\frac{-3y}{-3} = \frac{-4x}{-3} - \frac{9}{-3}$$
$$y = \frac{4}{3}x + 3$$
Slope $= \frac{4}{3}$; y-intercept: $(0, 3)$

11.
$$\left(\frac{36a^{-2}b^4}{18b^{-6}}\right)^{-3} = \left(2a^{-2}b^{4-(-6)}\right)^{-3} = \left(2a^{-2}b^{10}\right)^{-3}$$
$$= 2^{-3}a^6b^{-30} = \frac{1}{2^3}a^6\frac{1}{b^{30}} = \frac{a^6}{8b^{30}}$$
Substitute this result into the first sum and solve for y:
$$-0 + 4y = 12$$
$$y = 3$$
Substitute the values of x and y into the original first equation and solve for z:

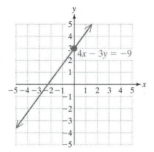

$$2(0)-3+2z=1$$
$$-3+2z=1$$
$$2z=4$$
$$z=2$$

The solution is $\{(0, 3, 2)\}$.

13. **a.**   Function
    b.   Not a function

15.
$$P(x)=\frac{1}{6}x^2+x-5$$
$$P(6)=\frac{1}{6}(6)^2+6-5=\frac{1}{6}(36)+6-5$$
$$=6+6-5=7$$

17. $3x-2y=5$
$$-2y=-3x+5$$
$$\frac{-2y}{-2}=\frac{-3x}{-2}+\frac{5}{-2}$$
$$y=\frac{3}{2}x-\frac{5}{2}$$

19.

	40% Solution	15% Solution	30% Solution
Amount of Solution	x	$25-x$	25
Amount of Alcohol	$0.40x$	$0.15(25-x)$	$0.30(25)$

(amt of 40%) + (amt of 15%) = (amt of 30%)
$$0.40x+0.15(25-x)=0.30(25)$$
$$0.40x+3.75-0.15x=7.5$$
$$0.25x+3.75=7.5$$
$$0.25x=3.75$$
$$x=15$$
$$25-x=25-15=10$$
15 L of 40% solution and 10 L of 15% solution must be mixed.

21. $\left(5a^2+3a-1\right)+\left(3a^3-5a+6\right)$
$$=3a^3+5a^2-2a+5$$

23.
$$y^2-5y=14$$
$$y^2-5y-14=0$$
$$(y-7)(y+2)=0$$
$$y-7=0 \ \text{ or } \ y+2=0$$
$$y=7 \ \text{ or } \quad y=-2 \ \ \{7,-2\}$$

25.
$$-5<-8+|2x-5|$$
$$3<|2x-5|$$
$$2x-5>3 \text{ or } 2x-5<-3$$
$$2x>8 \text{ or } \quad 2x<2$$
$$x>4 \text{ or } \quad x<1$$
$$(-\infty,1)\cup(4,\infty)$$

Chapter 5   Rational Expressions and Rational Equations

Are You Prepared?

1. $\dfrac{4}{9} \cdot \dfrac{15}{40} = \dfrac{\cancel{4}}{3 \cdot \cancel{3}} \cdot \dfrac{\cancel{3} \cdot \cancel{5}}{\cancel{4} \cdot 2 \cdot \cancel{5}} = \dfrac{1}{6}$

2. $\dfrac{6}{35} \div \dfrac{2}{7} = \dfrac{6}{35} \cdot \dfrac{7}{2} = \dfrac{3 \cdot \cancel{2}}{5 \cdot \cancel{7}} \cdot \dfrac{\cancel{7}}{\cancel{2}} = \dfrac{3}{5}$

3. $\dfrac{8}{3} - \dfrac{7}{6} = \dfrac{16}{6} - \dfrac{7}{6} = \dfrac{9}{6} = \dfrac{3}{2}$

4. $-\dfrac{6}{5} + \dfrac{2}{3} = -\dfrac{18}{15} + \dfrac{10}{15} = -\dfrac{8}{15}$

5. $-\dfrac{16}{3} \div 6 = -\dfrac{16}{3} \cdot \dfrac{1}{6} = -\dfrac{8 \cdot \cancel{2}}{3} \cdot \dfrac{1}{\cancel{2} \cdot 3} = -\dfrac{8}{9}$

6. $-4 \div \left(-\dfrac{2}{3}\right) = -\dfrac{4}{1} \cdot \left(-\dfrac{3}{2}\right) = -\dfrac{2 \cdot \cancel{2}}{1} \cdot \left(-\dfrac{3}{\cancel{2}}\right) = 6$

7.
$$\dfrac{x-5}{2} = \dfrac{x}{3}$$
$$6\left(\dfrac{x-5}{2}\right) = 6\left(\dfrac{x}{3}\right)$$
$$3(x-5) = 2x$$
$$3x - 15 = 2x$$
$$3x - 2x - 15 = 2x - 2x$$
$$x - 15 = 0$$
$$x - 15 + 15 = 0 + 15$$
$$x = 15 \quad \{15\}$$

8.
$$\dfrac{1}{2}x - \dfrac{2}{3} = 1 + \dfrac{1}{6}x$$
$$6\left(\dfrac{1}{2}x - \dfrac{2}{3}\right) = 6\left(1 + \dfrac{1}{6}x\right)$$
$$3x - 4 = 6 + x$$
$$3x - x - 4 = 6 + x - x$$
$$2x - 4 = 6$$
$$2x - 4 + 4 = 6 + 4$$
$$2x = 10$$
$$\dfrac{2x}{2} = \dfrac{10}{2}$$
$$x = 5 \quad \{5\}$$

To <u>**add**</u> or <u>**subtract**</u> <u>**rational**</u> <u>**expressions**</u> <u>**we**</u> <u>**need**</u>, a <u>**common**</u> <u>**denominator**</u>.
 2   4   8   3   1   6   5   7

Section 5.1   Rational Expressions and Rational Functions

1.
   a. rational
   b. denominator
   c. $\dfrac{p}{q}$
   d. $1; -1$

3.
$$k(x) = \dfrac{-3}{x+4}$$
$$k(0) = \dfrac{-3}{0+4} = -\dfrac{3}{4}$$
$$k(-1) = \dfrac{-3}{-1+4} = -\dfrac{3}{3} = -1$$
$$k(2) = \dfrac{-3}{2+4} = -\dfrac{3}{6} = -\dfrac{1}{2}$$
$$k(-4) = \dfrac{-3}{-4+4} = -\dfrac{3}{0} \text{ is undefined}$$

5.

$$n(a) = \frac{3a+1}{a^2+1}$$

$$n(1) = \frac{3(1)+1}{(1)^2+1} = \frac{3+1}{1+1} = \frac{4}{2} = 2$$

$$n(0) = \frac{3(0)+1}{(0)^2+1} = \frac{0+1}{0+1} = \frac{1}{1} = 1$$

$$n\left(-\frac{1}{3}\right) = \frac{3\left(-\frac{1}{3}\right)+1}{\left(-\frac{1}{3}\right)^2+1} = \frac{-1+1}{\frac{1}{9}+1} = \frac{0}{\frac{10}{9}} = 0$$

$$n(-1) = \frac{3(-1)+1}{(-1)^2+1} = \frac{-3+1}{1+1} = \frac{-2}{2} = -1$$

7.

$$f(x) = \frac{9}{x}$$

$$x \neq 0$$

a. $\{x \mid x \text{ is a real number and } x \neq 0\}$

b. $(-\infty, 0) \cup (0, \infty)$

9.

$$h(v) = \frac{v+1}{v-7}$$

$$v - 7 \neq 0$$

$$v \neq 7$$

a. $\{v \mid v \text{ is a real number and } v \neq 7\}$

b. $(-\infty, 7) \cup (7, \infty)$

11.

$$k(x) = \frac{3x-1}{2x-5}$$

$$2x - 5 \neq 0$$

$$2x \neq 5$$

$$x \neq \frac{5}{2}$$

a. $\left\{x \mid x \text{ is a real number and } x \neq \frac{5}{2}\right\}$

b. $\left(-\infty, \frac{5}{2}\right) \cup \left(\frac{5}{2}, \infty\right)$

13.

$$f(q) = \frac{q+1}{q^2+6q-27}$$

$$q^2 + 6q - 27 \neq 0$$

$$(q+9)(q-3) \neq 0$$

$$q + 9 \neq 0 \quad \text{or} \quad q - 3 \neq 0$$

$$q \neq -9 \quad \text{or} \quad q \neq 3$$

a. $\{q \mid q \text{ is a real number and } q \neq -9, q \neq 3\}$

b. $(-\infty, -9) \cup (-9, 3) \cup (3, \infty)$

15.

$$h(c) = \frac{c}{c^2+25}$$

Because c^2 is nonnegative for any real number c, the denominator c^2+25 cannot equal zero; therefore, no real numbers are excluded from the domain.

a. $\{c \mid c \text{ is a real number}\}$

b. $(-\infty, \infty)$

17.
$$f(x) = \frac{x+5}{x^2 - 25}$$
$$x^2 - 25 \neq 0$$
$$(x+5)(x-5) \neq 0$$
$$x+5 \neq 0 \quad \text{or} \quad x-5 \neq 0$$
$$x \neq -5 \quad \text{or} \qquad x \neq 5$$

a. $\{x \mid x \text{ is a real number and } x \neq -5, \ x \neq 5\}$

b. $(-\infty, -5) \cup (-5, 5) \cup (5, \infty)$

19.
$$p(x) = \frac{x-5}{3}$$

a. $\{x \mid x \text{ is a real number}\}$

b. $(-\infty, \infty)$

21.
$$m(x) = \frac{1}{x+4}$$
$$x + 4 = 0$$
$$x = -4$$
$$D: (-\infty, -4) \cup (-4, \infty)$$
Graph: b

23.
$$q(x) = \frac{1}{x-4}$$
$$x - 4 = 0$$
$$x = 4$$
$$D: (-\infty, 4) \cup (4, \infty)$$
Graph: d

25. a. $\dfrac{8x}{4y} = \dfrac{2x}{y}$

b. $\dfrac{8+x}{4+y}$ cannot be simplified

27.
$$\frac{x^2 + 6x + 8}{x^2 + 3x - 4}$$

a. $\dfrac{(x+4)(x+2)}{(x+4)(x-1)}$

b. $(x+4)(x-1) \neq 0$
$$x+4 \neq 0 \quad \text{or} \quad x-1 \neq 0$$
$$x \neq -4 \quad \text{or} \qquad x \neq 1$$

c. $\dfrac{\cancel{(x+4)}(x+2)}{\cancel{(x+4)}(x-1)} = \dfrac{x+2}{x-1}$

provided $x \neq -4, \ x \neq 1$

29.
$$\frac{x^2 - 18x + 81}{x^2 - 81}$$

a. $\dfrac{(x-9)(x-9)}{(x+9)(x-9)}$

b. $(x+9)(x-9) \neq 0$
$$x+9 \neq 0 \ \text{or} \ x-9 \neq 0$$
$$x \neq -9 \ \text{or} \ x \neq 9$$

c. $\dfrac{(x-9)\cancel{(x-9)}}{(x+9)\cancel{(x-9)}} = \dfrac{x-9}{x+9}$

provided $x \neq -9, \ x \neq 9$

31. $\dfrac{100x^3 y^5}{36xy^8} = \dfrac{25}{9} x^{3-1} y^{5-8} = \dfrac{25}{9} x^2 y^{-3}$

$\qquad = \dfrac{25x^2}{9y^3}$ provided $x \neq 0, \ y \neq 0$

33. $\dfrac{7w^{11}z^6}{14w^3z^3} = \dfrac{1}{2}w^{11-3}z^{6-3} = \dfrac{1}{2}w^8z^3$

$\qquad = \dfrac{w^8z^3}{2}$ provided $w \neq 0,\ z \neq 0$

35. $\dfrac{-3m^4n}{12m^6n^4} = -\dfrac{1}{4}m^{4-6}n^{1-4} = -\dfrac{1}{4}m^{-2}n^{-3}$

$\qquad = -\dfrac{1}{4m^2n^3}$ provided $m \neq 0,\ n \neq 0$

37. $\dfrac{6a+18}{9a+27} = \dfrac{6\cancel{(a+3)}}{9\cancel{(a+3)}} = \dfrac{2}{3}$ provided $a \neq -3$

39. $\dfrac{x-5}{x^2-25} = \dfrac{\cancel{x-5}}{(x+5)\cancel{(x-5)}} = \dfrac{1}{x+5}$

$\qquad$ provided $x \neq -5,\ x \neq 5$

41. $\dfrac{-7c}{21c^2-35c} = \dfrac{-1\cdot\cancel{7c}}{\cancel{7c}(3c-5)} = -\dfrac{1}{3c-5}$

$\qquad$ provided $c \neq 0,\ c \neq \dfrac{5}{3}$

43. $\dfrac{2t^2+7t-4}{-2t^2-5t+3} = \dfrac{(2t-1)(t+4)}{-(2t^2+5t-3)}$

$\qquad = \dfrac{\cancel{(2t-1)}(t+4)}{-\cancel{(2t-1)}(t+3)}$

$\qquad = -\dfrac{t+4}{t+3}$ provided $t \neq \dfrac{1}{2},\ t \neq -3$

45. $\dfrac{(p+1)(2p-1)^4}{(p+1)^2(2p-1)^2} = (p+1)^{1-2}(2p-1)^{4-2}$

$\qquad = (p+1)^{-1}(2p-1)^2$

$\qquad = \dfrac{(2p-1)^2}{p+1}$ provided $p \neq \dfrac{1}{2},\ p \neq -1$

47. $\dfrac{9-z^2}{2z^2+z-15} = \dfrac{\cancel{(3+z)}(3-z)}{(2z-5)\cancel{(z+3)}} = \dfrac{3-z}{2z-5}$

$\qquad$ provided $z \neq \dfrac{5}{2},\ z \neq -3$

49. $\dfrac{2z^3+128}{16+8z+z^2} = \dfrac{2(z^3+64)}{z^2+8z+16}$

$\qquad = \dfrac{2\cancel{(z+4)}(z^2-4z+16)}{(z+4)\cancel{(z+4)}} = \dfrac{2(z^2-4z+16)}{z+4}$

$\qquad$ provided $z \neq -4$

51. $\dfrac{10x^3-25x^2+4x-10}{-4-10x^2}$

$\qquad = \dfrac{5x^2(2x-5)+2(2x-5)}{-2(5x^2+2)}$

$\qquad = \dfrac{(2x-5)\cancel{(5x^2+2)}}{-2\cancel{(5x^2+2)}} = -\dfrac{2x-5}{2}$

53. $\dfrac{r+6}{6+r} = \dfrac{r+6}{r+6} = 1$ provided $r \neq -6$

55. $\dfrac{b+8}{-b-8} = \dfrac{\cancel{b+8}}{-\cancel{(b+8)}} = -1$ provided $b \neq -8$

57. $\dfrac{10-x}{x-10} = \dfrac{-\cancel{(x-10)}}{\cancel{x-10}} = -1$ provided $x \neq 10$

59. $\dfrac{2t-2}{1-t} = \dfrac{2\cancel{(t-1)}}{-\cancel{(t-1)}} = -2$ provided $t \neq 1$

61. $\dfrac{c+4}{c-4}$ cannot be simplified

63. $\dfrac{y-x}{12x^2-12y^2}=\dfrac{-(x-y)}{12\left(x^2-y^2\right)}$

$$=\dfrac{-\cancel{(x-y)}}{12(x+y)\cancel{(x-y)}}=-\dfrac{1}{12(x+y)}$$

provided $x \neq y,\ x \neq -y$

65. $\dfrac{t^2-1}{t^2+7t+6}=\dfrac{(t-1)\cancel{(t+1)}}{(t+6)\cancel{(t+1)}}=\dfrac{t-1}{t+6}$

provided $t \neq -6,\ t \neq -1$

67. $\dfrac{8p+8}{2p^2-4p-6}=\dfrac{8(p+1)}{2\left(p^2-2p-3\right)}$

$$=\dfrac{\cancel{2}\cdot 4\cancel{(p+1)}}{\cancel{2}(p-3)\cancel{(p+1)}}=\dfrac{4}{p-3}$$

provided $p \neq 3,\ p \neq -1$

69. $\dfrac{-16a^2bc^4}{8ab^2c^4}=-\dfrac{16}{8}a^{2-1}b^{1-2}c^{4-4}=-2a^1b^{-1}c^0$

$$=-\dfrac{2a}{b}\ \text{ provided } a \neq 0,\ b \neq 0,\ c \neq 0$$

71. $\dfrac{x^2-y^2}{8y-8x}=\dfrac{\cancel{(x-y)}(x+y)}{-8\cancel{(x-y)}}=-\dfrac{x+y}{8}$

provided $x \neq y$

73. $\dfrac{b+4}{2b^2+5b-12}=\dfrac{\cancel{b+4}}{(2b-3)\cancel{(b+4)}}=\dfrac{1}{2b-3}$

provided $b \neq \dfrac{3}{2},\ b \neq -4$

75. $\dfrac{-2x+34}{-4x+6}=\dfrac{\cancel{-2}(x-17)}{\cancel{-2}(2x-3)}=\dfrac{x-17}{2x-3}$

provided $x \neq \dfrac{3}{2}$

77. $\dfrac{(a-2)^2(a-5)^3}{(a-2)^3(a-5)}=(a-2)^{2-3}(a-5)^{3-1}$

$$=(a-2)^{-1}(a-5)^2=\dfrac{(a-5)^2}{a-2}$$

provided $a \neq 2,\ a \neq 5$

79. $\dfrac{4x-2x^2}{5x-10}=\dfrac{-2x\cancel{(x-2)}}{5\cancel{(x-2)}}=-\dfrac{2x}{5}$

provided $x \neq 2$

81. $\dfrac{x^3-2x^2-25x+50}{x^3+5x^2-4x-20}=\dfrac{x^2(x-2)-25(x-2)}{x^2(x+5)-4(x+5)}$

$$=\dfrac{(x-2)\left(x^2-25\right)}{(x+5)\left(x^2-4\right)}$$

$$=\dfrac{\cancel{(x-2)}\cancel{(x+5)}(x-5)}{\cancel{(x+5)}(x+2)\cancel{(x-2)}}=\dfrac{x-5}{x+2}$$

provided $x \neq -5,\ x \neq -2,\ x \neq 2$

83. $\dfrac{t^3+8}{3t^2+t-10}=\dfrac{\cancel{(t+2)}\left(t^2-2t+4\right)}{(3t-5)\cancel{(t+2)}}$

$$=\dfrac{t^2-2t+4}{3t-5}$$

provided $t \neq \dfrac{5}{3},\ t \neq -2$

85. For example: $\dfrac{1}{x-2}$

87. For example: $f(x)=\dfrac{1}{x+5}$

Section 5.2　Multiplication and Division of Rational Expressions

1.

　　a.　$\dfrac{pr}{qs}$

　　b.　$\dfrac{ps}{qr}$

3.　$\dfrac{t^2-5t-6}{t^2-7t+6}=\dfrac{(t-6)(t+1)}{(t-6)(t-1)}=\dfrac{t+1}{t-1}$

5.　$\dfrac{2-p}{p^2-p-2}=\dfrac{-1(p-2)}{(p-2)(p+1)}=-\dfrac{1}{p+1}$

7.　$\dfrac{7x+14}{7x^2-7x-42}=\dfrac{7(x+2)}{7(x^2-x-6)}$

$$=\dfrac{7(x+2)}{7(x-3)(x+2)}=\dfrac{1}{x-3}$$

9.　$\dfrac{a^3b^2c^5}{2a^3bc^2}=\dfrac{1}{2}a^{3-3}b^{2-1}c^{5-2}=\dfrac{1}{2}a^0bc^3=\dfrac{bc^3}{2}$

11.　$\dfrac{16}{z^7}\cdot\dfrac{z^4}{8}=\dfrac{16z^4}{8z^7}=2z^{4-7}=2z^{-3}=\dfrac{2}{z^3}$

13.　$\dfrac{27r^5}{7s}\cdot\dfrac{28rs^3}{9r^3s^2}=\dfrac{9\cdot3r^3\cdot r^2}{7s}\cdot\dfrac{7\cdot4rs^2\cdot s}{9r^3s^2}$

$$=3\cdot4\cdot r^2\cdot r=12r^{2+1}=12r^3$$

15.　$\dfrac{x^2y}{x^2-4x-5}\cdot\dfrac{2x^2-13x+15}{xy^3}$

$$=\dfrac{x\cdot x\cdot y}{(x-5)(x+1)}\cdot\dfrac{(2x-3)(x-5)}{x\cdot y\cdot y^2}=\dfrac{x(2x-3)}{y^2(x+1)}$$

17.　$\dfrac{10w-8}{w+2}\cdot\dfrac{3w^2-w-14}{25w^2-16}$

$$=\dfrac{2(5w-4)}{w+2}\cdot\dfrac{(3w-7)(w+2)}{(5w+4)(5w-4)}$$

$$=\dfrac{2(3w-7)}{5w+4}$$

19.　$\dfrac{3x-15}{4x^2-2x}\cdot\dfrac{10x-20x^2}{5-x}$

$$=\dfrac{3(x-5)}{2x(2x-1)}\cdot\dfrac{-5\cdot2x(2x-1)}{-(x-5)}=15$$

21.　$y(y^2-4)\cdot\dfrac{y}{y+2}=\dfrac{y^2(y+2)(y-2)}{y+2}$

$$=y^2(y-2)$$

23.　$\dfrac{2a}{7b^3}\div\dfrac{10a^5}{77}=\dfrac{2a}{7b^3}\cdot\dfrac{77}{10a^5}=\dfrac{2a}{7b^3}\cdot\dfrac{7\cdot11}{2a\cdot5a^4}$

$$=\dfrac{11}{5a^4b^3}$$

25.　$\dfrac{(r+3)^2}{4r^3s}\div\dfrac{r+3}{rs}=\dfrac{(r+3)^2}{4r^3s}\cdot\dfrac{rs}{r+3}$

$$=\dfrac{(r+3)(r+3)}{4r^2\cdot r\cdot s}\cdot\dfrac{r\cdot s}{r+3}=\dfrac{r+3}{4r^2}$$

27.　$\dfrac{6p+7}{p+2}\div(36p^2-49)=\dfrac{6p+7}{p+2}\cdot\dfrac{1}{36p^2-49}$

$$=\dfrac{6p+7}{p+2}\cdot\dfrac{1}{(6p+7)(6p-7)}$$

$$=\dfrac{1}{(p+2)(6p-7)}$$

29. $\dfrac{b^2-6b+9}{b^2-b-6} \div \dfrac{b^2-9}{4} = \dfrac{b^2-6b+9}{b^2-b-6} \cdot \dfrac{4}{b^2-9}$

$= \dfrac{(b-3)(b-3)}{(b-3)(b+2)} \cdot \dfrac{4}{(b+3)(b-3)}$

$= \dfrac{4}{(b+2)(b+3)}$

31. $\dfrac{6s^2+st-2t^2}{6s^2-5st+t^2} \div \dfrac{3s^2+17st+10t^2}{6s^2+13st-5t^2}$

$= \dfrac{6s^2+st-2t^2}{6s^2-5st+t^2} \cdot \dfrac{6s^2+13st-5t^2}{3s^2+17st+10t^2}$

$= \dfrac{(3s+2t)(2s-t)}{(3s-t)(2s-t)} \cdot \dfrac{(3s-t)(2s+5t)}{(3s+2t)(s+5t)}$

$= \dfrac{2s+5t}{s+5t}$

33. $\dfrac{a^3+a+a^2+1}{a^3+a^2+ab^2+b^2} \div \dfrac{a^3+a+a^2b+b}{2a^2+2ab+ab^2+b^3} = \dfrac{a^3+a+a^2+1}{a^3+a^2+ab^2+b^2} \cdot \dfrac{2a^2+2ab+ab^2+b^3}{a^3+a+a^2b+b}$

$= \dfrac{a(a^2+1)+(a^2+1)}{a^2(a+1)+b^2(a+1)} \cdot \dfrac{2a(a+b)+b^2(a+b)}{a(a^2+1)+b(a^2+1)} = \dfrac{(a^2+1)(a+1)}{(a+1)(a^2+b^2)} \cdot \dfrac{(a+b)(2a+b^2)}{(a^2+1)(a+b)} = \dfrac{2a+b^2}{a^2+b^2}$

35. $\dfrac{8x-4x^2}{xy-2y+3x-6} \div \dfrac{3x+6}{y+3}$

$= \dfrac{8x-4x^2}{y(x-2)+3(x-2)} \cdot \dfrac{y+3}{3x+6}$

$= \dfrac{-4x(x-2)}{(x-2)(y+3)} \cdot \dfrac{y+3}{3(x+2)} = \dfrac{-4x}{3(x+2)}$

37. $\dfrac{3x^5}{2x^2y^7} \div \dfrac{4x^3y}{6y^6} = \dfrac{3x^5}{2x^2y^7} \cdot \dfrac{6y^6}{4x^3y}$

$= \dfrac{3x \cdot x}{2 \cdot x \cdot y^6 \cdot y} \cdot \dfrac{2 \cdot 3 y^6}{4x^3 y} = \dfrac{9}{4y^2}$

39. $\dfrac{4y}{7} \div \dfrac{y^2}{14} \cdot \dfrac{3}{y} = \dfrac{4y}{7} \cdot \dfrac{14}{y^2} \cdot \dfrac{3}{y} = \dfrac{4y}{7} \cdot \dfrac{7 \cdot 2}{y^2} \cdot \dfrac{3}{y} = \dfrac{24}{y^2}$

41. $\dfrac{6a^2+ab-b^2}{10a^2+5ab} \cdot \dfrac{2a^3+4a^2b}{3a^2+5ab-2b^2}$

$= \dfrac{(3a-b)(2a+b)}{5a(2a+b)} \cdot \dfrac{2a^2(a+2b)}{(3a-b)(a+2b)}$

$= \dfrac{2a}{5}$

43. $(2x^2+8) \div \dfrac{x^4-16}{x^2+x-6} = \dfrac{2x^2+8}{1} \cdot \dfrac{x^2+x-6}{x^4-16}$

$= \dfrac{2(x^2+4)}{1} \cdot \dfrac{(x+3)(x-2)}{(x^2+4)(x^2-4)}$

$= \dfrac{2(x^2+4)}{1} \cdot \dfrac{(x+3)(x-2)}{(x^2+4)(x+2)(x-2)}$

$= \dfrac{2(x+3)}{x+2}$

45. $\dfrac{m^2-n^2}{(m-n)^2} \div \dfrac{m^2-2mn+n^2}{m^2-mn+n^2} \cdot \dfrac{(m-n)^4}{m^3+n^3} = \dfrac{m^2-n^2}{(m-n)^2} \cdot \dfrac{m^2-mn+n^2}{m^2-2mn+n^2} \cdot \dfrac{(m-n)^4}{m^3+n^3}$

$$= \dfrac{\cancel{(m+n)}\,(m-n)}{\cancel{(m-n)^2}} \cdot \dfrac{\cancel{m^2-mn+n^2}}{\cancel{(m-n)^2}} \cdot \dfrac{\cancel{(m-n)^2}\,\cancel{(m-n)^2}}{\cancel{(m+n)}\,\cancel{(m^2-mn+n^2)}} = m-n$$

47. $\dfrac{x^2-6xy+9y^2}{x^2-4y^2} \cdot \dfrac{x^2-5xy+6y^2}{3y-x} \div \dfrac{x^2-9y^2}{x+2y} = \dfrac{x^2-6xy+9y^2}{x^2-4y^2} \cdot \dfrac{x^2-5xy+6y^2}{3y-x} \cdot \dfrac{x+2y}{x^2-9y^2}$

$$= \dfrac{(x-3y)\cancel{(x-3y)}}{\cancel{(x+2y)}\cancel{(x-2y)}} \cdot \dfrac{\cancel{(x-3y)}\cancel{(x-2y)}}{-\cancel{(x-3y)}} \cdot \dfrac{\cancel{x+2y}}{(x+3y)\cancel{(x-3y)}} = -\dfrac{x-3y}{x+3y} \text{ or } \dfrac{3y-x}{x+3y}$$

49. $\dfrac{25m^2-1}{125m^3-1} \div \dfrac{5m+1}{25m^2+5m+1} = \dfrac{25m^2-1}{125m^3-1} \cdot \dfrac{25m^2+5m+1}{5m+1}$

$$= \dfrac{\cancel{(5m+1)}\,\cancel{(5m-1)}}{\cancel{(5m-1)}\,\cancel{(25m^2+5m+1)}} \cdot \dfrac{\cancel{25m^2+5m+1}}{\cancel{5m+1}} = 1$$

51. $\dfrac{2a^2+ab-8a-4b}{2a^2-6a+ab-3b} \cdot \dfrac{a^2-6a+9}{a^2-16} = \dfrac{a(2a+b)-4(2a+b)}{2a(a-3)+b(a-3)} \cdot \dfrac{(a-3)(a-3)}{(a+4)(a-4)}$

$$= \dfrac{\cancel{(2a+b)}\,\cancel{(a-4)}}{\cancel{(a-3)}\,\cancel{(2a+b)}} \cdot \dfrac{(a-3)\cancel{(a-3)}}{(a+4)\cancel{(a-4)}} = \dfrac{a-3}{a+4}$$

53. $\dfrac{45}{2x+1} \cdot (8x+4) \div \dfrac{27}{4x+2} = \dfrac{5\cdot\cancel{9}}{\cancel{2x+1}} \cdot \dfrac{4\cancel{(2x+1)}}{1} \cdot \dfrac{2(2x+1)}{\cancel{9}\cdot 3} = \dfrac{40(2x+1)}{3}$

55. $\dfrac{2x^2-11x-6}{3x-2} \div \dfrac{2x^2-5x-3}{3x^2-7x-6} = \dfrac{2x^2-11x-6}{3x-2} \cdot \dfrac{3x^2-7x-6}{2x^2-5x-3} = \dfrac{\cancel{(2x+1)}(x-6)}{3x-2} \cdot \dfrac{(3x+2)\cancel{(x-3)}}{\cancel{(2x+1)}\cancel{(x-3)}}$

$$= \dfrac{(x-6)(3x+2)}{3x-2}$$

57. $A = \dfrac{1}{2}\left(\dfrac{k^2}{2h^2}\right)\left(\dfrac{8}{hk}\right) = \dfrac{1}{\cancel{2}}\left(\dfrac{k\cdot\cancel{k}}{\cancel{2}h^2}\right)\left(\dfrac{\cancel{2}\cdot\cancel{2}\cdot 2}{h\cancel{k}}\right)$

$$= \dfrac{2k}{h^3} \text{ cm}^2$$

59. $A = \dfrac{x^2}{x-3} \cdot \dfrac{5x-15}{4x} = \dfrac{x\cdot\cancel{x}}{\cancel{x-3}} \cdot \dfrac{5\cancel{(x-3)}}{4\cancel{x}}$

$$= \dfrac{5x}{4} \text{ ft}^2$$

Section 5.3   Addition and Subtraction of Rational Expressions

1.

    a. $\dfrac{p+r}{q}$; $\dfrac{p-r}{q}$.

    b.   least common denominator

3.
$$\frac{9b+9}{4b+8}\cdot\frac{2b+4}{3b-3}=\frac{\cancel{3}\cdot3(b+1)}{\cancel{2}\cdot2\cancel{(b+2)}}\cdot\frac{\cancel{2}\cancel{(b+2)}}{\cancel{3}(b-1)}=\frac{3(b+1)}{2(b-1)}$$

5.
$$\frac{(5-a)^2}{10a-2}\cdot\frac{25a^2-1}{a^2-10a+25}=\frac{(-1)^2\cancel{(a-5)^2}}{2\cancel{(5a-1)}}\cdot\frac{(5a+1)\cancel{(5a-1)}}{\cancel{(a-5)^2}}=\frac{5a+1}{2}$$

7.
$$\frac{3}{5x}+\frac{7}{5x}=\frac{10}{5x}=\frac{\cancel{5}\cdot2}{\cancel{5}x}=\frac{2}{x}$$

9.
$$\frac{x}{x^2-2x-3}-\frac{3}{x^2-2x-3}=\frac{x-3}{x^2-2x-3}$$
$$=\frac{\cancel{x-3}}{\cancel{(x-3)}(x+1)}=\frac{1}{x+1}$$

11.
$$\frac{5x-1}{(2x+9)(x-6)}-\frac{3x-6}{(2x+9)(x-6)}$$
$$=\frac{5x-1-(3x-6)}{(2x+9)(x-6)}$$
$$=\frac{5x-1-3x+6}{(2x+9)(x-6)}$$
$$=\frac{2x+5}{(2x+9)(x-6)}$$

13.
$$\frac{x+2}{x-5}+\frac{x-12}{x-5}=\frac{2x-10}{x-5}=\frac{2\cancel{(x-5)}}{\cancel{x-5}}=2$$

15.
$$\frac{5}{8}=\frac{5}{2^3},\qquad\frac{3}{20x}=\frac{3}{5\cdot2^2x}$$
$$\text{LCD}=2^3\cdot5\cdot x=40x$$

17.
$$\frac{-5}{6m^4}=\frac{-5}{2\cdot3\cdot m^4},\qquad\frac{1}{15mn^7}=\frac{1}{3\cdot5\cdot mn^7}$$
$$\text{LCD}=2\cdot3\cdot5\cdot m^4n^7=30m^4n^7$$

19.
$$\frac{6}{(x-4)(x+2)},\qquad\frac{-8}{(x-4)(x-6)}$$
$$\text{LCD}=(x-4)(x+2)(x-6)$$

21.
$$\frac{3}{x(x-1)(x+7)^2},\qquad\frac{-1}{x^2(x+7)}$$
$$\text{LCD}=x^2(x-1)(x+7)^2$$

23.
$$\frac{5}{x-6},\qquad\frac{x-5}{x^2-8x+12}=\frac{x-5}{(x-6)(x-2)}$$
$$\text{LCD}=(x-6)(x-2)$$

25.
$$\frac{3a}{a-4},\qquad\frac{5}{4-a}=\frac{5(-1)}{(4-a)(-1)}=\frac{-5}{a-4}$$
$$\text{LCD}=a-4\ \text{ or }\ 4-a$$

27.
$$\frac{5}{3x}=\frac{}{9x^2y}$$
$$\frac{5}{3x}\cdot\frac{3xy}{3xy}=\frac{15xy}{9x^2y}$$

29.
$$\frac{2x}{x-1}=\frac{}{x(x-1)(x+2)}$$
$$\frac{2x}{x-1}\cdot\frac{x(x+2)}{x(x+2)}=\frac{2x^2(x+2)}{x(x-1)(x+2)}$$
$$=\frac{2x^3+4x^2}{x(x-1)(x+2)}$$

31.

$$\frac{y}{y+6} = \frac{y}{y^2+5y-6}$$

$$\frac{y}{y+6} = \frac{y}{(y+6)(y-1)}$$

$$\frac{y}{y+6} \cdot \frac{y-1}{y-1} = \frac{y^2-y}{(y+6)(y-1)}$$

33.

$$\frac{4}{3p} - \frac{5}{2p^2} \qquad \text{LCD} = 2 \cdot 3 \cdot p^2 = 6p^2$$

$$= \frac{4}{3p} \cdot \frac{2p}{2p} - \frac{5}{2p^2} \cdot \frac{3}{3} = \frac{8p}{6p^2} - \frac{15}{6p^2} = \frac{8p-15}{6p^2}$$

35.

$$\frac{s-1}{s} - \frac{t+1}{t} \qquad \text{LCD} = st$$

$$= \frac{s-1}{s} \cdot \frac{t}{t} - \frac{t+1}{t} \cdot \frac{s}{s} = \frac{st-t}{st} - \frac{st+s}{st}$$

$$= \frac{st-t-st-s}{st} = \frac{-t-s}{st}$$

37.

$$\frac{4a-2}{3a+12} - \frac{a-2}{a+4} = \frac{4a-2}{3(a+4)} - \frac{a-2}{a+4}$$

$$\text{LCD} = 3(a+4)$$

$$= \frac{4a-2}{3(a+4)} - \frac{a-2}{a+4} \cdot \frac{3}{3} = \frac{4a-2-3(a-2)}{3(a+4)}$$

$$= \frac{4a-2-3a+6}{3(a+4)} = \frac{\cancel{a+4}}{3\cancel{(a+4)}} = \frac{1}{3}$$

39.

$$\frac{10}{b(b+5)} + \frac{2}{b} \qquad \text{LCD} = b(b+5)$$

$$= \frac{10}{b(b+5)} + \frac{2}{b} \cdot \frac{b+5}{b+5} = \frac{10+2(b+5)}{b(b+5)}$$

$$= \frac{10+2b+10}{b(b+5)} = \frac{2b+20}{b(b+5)}$$

41.

$$\frac{x-2}{x-6} - \frac{x+2}{6-x} = \frac{x-2}{x-6} - \frac{x+2}{6-x} \cdot \frac{(-1)}{(-1)}$$

$$= \frac{x-2}{x-6} - \frac{-(x+2)}{x-6} = \frac{x-2+x+2}{x-6} = \frac{2x}{x-6}$$

43.

$$\frac{6b}{b-4} - \frac{1}{b+1} \qquad \text{LCD} = (b-4)(b+1)$$

$$= \frac{6b}{b-4} \cdot \frac{b+1}{b+1} - \frac{1}{b+1} \cdot \frac{b-4}{b-4}$$

$$= \frac{6b(b+1)-1(b-4)}{(b-4)(b+1)}$$

$$= \frac{6b^2+6b-b+4}{(b-4)(b+1)} = \frac{6b^2+5b+4}{(b-4)(b+1)}$$

45.

$$\frac{2}{2x+1} + \frac{4}{x-2}$$

$$\text{LCD} = (2x+1)(x-2)$$

$$= \frac{2}{2x+1} \cdot \frac{x-2}{x-2} + \frac{4}{x-2} \cdot \frac{2x+1}{2x+1}$$

$$= \frac{2x-4+8x+4}{(2x+1)(x-2)} = \frac{10x}{(2x+1)(x-2)}$$

47.

$$\frac{y-2}{y-4} + \frac{2y^2-15y+12}{y^2-16} = \frac{y-2}{y-4} + \frac{2y^2-15y+12}{(y+4)(y-4)} \qquad \text{LCD} = (y+4)(y-4)$$

$$= \frac{y-2}{y-4} \cdot \frac{y+4}{y+4} + \frac{2y^2-15y+12}{(y+4)(y-4)} = \frac{y^2+2y-8+2y^2-15y+12}{(y+4)(y-4)} = \frac{3y^2-13y+4}{(y+4)(y-4)}$$

$$= \frac{(3y-1)\cancel{(y-4)}}{(y+4)\cancel{(y-4)}} = \frac{3y-1}{y+4}$$

49. $\dfrac{x+2}{x^2-36}-\dfrac{x}{x^2+9x+18}=\dfrac{x+2}{(x+6)(x-6)}-\dfrac{x}{(x+6)(x+3)}$ $\quad LCD=(x+6)(x-6)(x+3)$

$\qquad =\dfrac{x+2}{(x+6)(x-6)}\cdot\dfrac{x+3}{x+3}-\dfrac{x}{(x+6)(x+3)}\cdot\dfrac{x-6}{x-6}=\dfrac{(x+2)(x+3)-x(x-6)}{(x+6)(x-6)(x+3)}=\dfrac{x^2+5x+6-x^2+6x}{(x+6)(x-6)(x+3)}$

$\qquad =\dfrac{11x+6}{(x+6)(x-6)(x+3)}$

51. $\dfrac{5}{w}+\dfrac{8}{-w}=\dfrac{5}{w}+\dfrac{8}{-w}\cdot\dfrac{(-1)}{(-1)}=\dfrac{5}{w}+\dfrac{-8}{w}=-\dfrac{3}{w}$

53. $\dfrac{n}{5-n}+\dfrac{2n-5}{n-5}=\dfrac{n}{5-n}\cdot\dfrac{(-1)}{(-1)}+\dfrac{2n-5}{n-5}$

$\qquad =\dfrac{-n}{n-5}+\dfrac{2n-5}{n-5}=\dfrac{n-5}{n-5}=1$

55. $\dfrac{2}{3x-15}+\dfrac{x}{25-x^2}=\dfrac{2}{3x-15}+\dfrac{x}{25-x^2}\cdot\dfrac{(-1)}{(-1)}=\dfrac{2}{3(x-5)}+\dfrac{-x}{(x+5)(x-5)}$ $\quad LCD=3(x+5)(x-5)$

$\qquad =\dfrac{2}{3(x-5)}\cdot\dfrac{x+5}{x+5}+\dfrac{-x}{(x+5)(x-5)}\cdot\dfrac{3}{3}=\dfrac{2x+10-3x}{3(x+5)(x-5)}=\dfrac{10-x}{3(x+5)(x-5)}$

57. $\dfrac{m}{20+9m+m^2}-\dfrac{4}{12+7m+m^2}=\dfrac{m}{(m+5)(m+4)}-\dfrac{4}{(m+4)(m+3)}$

$\quad LCD=(m+5)(m+4)(m+3)$

$\qquad =\dfrac{m}{(m+5)(m+4)}\cdot\dfrac{m+3}{m+3}-\dfrac{4}{(m+4)(m+3)}\cdot\dfrac{m+5}{m+5}=\dfrac{m^2+3m}{(m+5)(m+4)(m+3)}-\dfrac{4m+20}{(m+5)(m+4)(m+3)}$

$\qquad =\dfrac{m^2+3m-4m-20}{(m+5)(m+4)(m+3)}=\dfrac{m^2-m-20}{(m+5)(m+4)(m+3)}=\dfrac{(m-5)\cancel{(m+4)}}{(m+5)\cancel{(m+4)}(m+3)}=\dfrac{m-5}{(m+5)(m+3)}$

59. $\dfrac{x+3}{x^2}+\dfrac{x+5}{2x}$ $\quad LCD=2x^2$

$\qquad =\dfrac{x+3}{x^2}\cdot\dfrac{2}{2}+\dfrac{x+5}{2x}\cdot\dfrac{x}{x}=\dfrac{2x+6}{2x^2}+\dfrac{x^2+5x}{2x^2}$

$\qquad =\dfrac{x^2+7x+6}{2x^2}$

61. $\dfrac{n}{5-n}+\dfrac{2n-5}{n-5}=\dfrac{n}{5-n}\cdot\dfrac{(-1)}{(-1)}+\dfrac{2n-5}{n-5}$

$\qquad =\dfrac{-n}{n-5}+\dfrac{2n-5}{n-5}=\dfrac{n-5}{n-5}=1$

63. $\dfrac{9}{x^2-2x+1}-\dfrac{x-3}{x^2-x}=\dfrac{9}{(x-1)^2}-\dfrac{x-3}{x(x-1)}$ $\quad LCD=x(x-1)^2$

$\qquad =\dfrac{9}{(x-1)^2}\cdot\dfrac{x}{x}-\dfrac{x-3}{x(x-1)}\cdot\dfrac{x-1}{x-1}=\dfrac{9x-(x^2-4x+3)}{x(x-1)^2}=\dfrac{9x-x^2+4x-3}{x(x-1)^2}=\dfrac{-x^2+13x-3}{x(x-1)^2}$

65. $\dfrac{t+1}{t+3}-\dfrac{t-2}{t-3}+\dfrac{6}{t^2-9}=\dfrac{t+1}{t+3}-\dfrac{t-2}{t-3}+\dfrac{6}{(t+3)(t-3)}$ $\text{LCD}=(t+3)(t-3)$

$$=\dfrac{t+1}{t+3}\cdot\dfrac{t-3}{t-3}-\dfrac{t-2}{t-3}\cdot\dfrac{t+3}{t+3}+\dfrac{6}{(t+3)(t-3)}=\dfrac{t^2-2t-3-\left(t^2+t-6\right)+6}{(t+3)(t-3)}$$

$$=\dfrac{t^2-2t-3-t^2-t+6+6}{(t+3)(t-3)}=\dfrac{-3t+9}{(t+3)(t-3)}=\dfrac{-3(t-3)}{(t+3)(t-3)}=-\dfrac{3}{t+3}$$

67. $(x-1)\cdot\left[\dfrac{3}{x^2-1}+\dfrac{x}{2x-2}\right]=\dfrac{3(x-1)}{(x+1)(x-1)}+\dfrac{x(x-1)}{2(x-1)}=\dfrac{3}{x+1}+\dfrac{x}{2}$ $\text{LCD}=2(x+1)$

$$=\dfrac{3}{x+1}\cdot\dfrac{2}{2}+\dfrac{x}{2}\cdot\dfrac{x+1}{x+1}=\dfrac{6+x^2+x}{2(x+1)}=\dfrac{x^2+x+6}{2(x+1)}$$

69. $\dfrac{3z}{z-3}-\dfrac{z}{z+4}$ $\text{LCD}=(z-3)(z+4)$

$$=\dfrac{3z}{z-3}\cdot\dfrac{z+4}{z+4}-\dfrac{z}{z+4}\cdot\dfrac{z-3}{z-3}=\dfrac{3z^2+12z-\left(z^2-3z\right)}{(z-3)(z+4)}=\dfrac{3z^2+12z-z^2+3z}{(z-3)(z+4)}=\dfrac{2z^2+15z}{(z-3)(z+4)}$$

71. $\dfrac{2x}{x^2-y^2}-\dfrac{1}{x-y}+\dfrac{1}{y-x}=\dfrac{2x}{(x+y)(x-y)}-\dfrac{1}{x-y}+\dfrac{1}{y-x}$ $\text{LCD}=(x+y)(x-y)$

$$=\dfrac{2x}{(x+y)(x-y)}-\dfrac{1}{x-y}\cdot\dfrac{x+y}{x+y}+\dfrac{1}{y-x}\cdot\dfrac{(-1)}{(-1)}\cdot\dfrac{x+y}{x+y}=\dfrac{2x-1(x+y)-1(x+y)}{(x+y)(x-y)}=\dfrac{2x-x-y-x-y}{(x+y)(x-y)}$$

$$=\dfrac{-2y}{(x+y)(x-y)}$$

73. $(2p+1)\cdot\left[\dfrac{2p}{6p+3}-\dfrac{1}{p+4}\right]=\dfrac{2p(2p+1)}{3(2p+1)}-\dfrac{2p+1}{p+4}=\dfrac{2p}{3}-\dfrac{2p+1}{p+4}$ $\text{LCD}=3(p+4)$

$$=\dfrac{2p}{3}\cdot\dfrac{p+4}{p+4}-\dfrac{2p+1}{p+4}\cdot\dfrac{3}{3}=\dfrac{2p^2+8p-(6p+3)}{3(p+4)}=\dfrac{2p^2+8p-6p-3}{3(p+4)}=\dfrac{2p^2+2p-3}{3(p+4)}$$

75. $\dfrac{1}{x+5}+\dfrac{3}{(x+5)^2}-\dfrac{2}{(x+5)^3}$ $LCD=(x+5)^3$

$=\dfrac{1}{x+5}\cdot\dfrac{(x+5)^2}{(x+5)^2}+\dfrac{3}{(x+5)^2}\cdot\dfrac{x+5}{x+5}-\dfrac{2}{(x+5)^3}$

$=\dfrac{(x+5)^2}{(x+5)^3}+\dfrac{3x+15}{(x+5)^3}-\dfrac{2}{(x+5)^3}$

$=\dfrac{x^2+10x+25+3x+15-2}{(x+5)^3}$

$=\dfrac{x^2+13x+38}{(x+5)^3}$

77. $\dfrac{-10}{z^2-6z+5}+\dfrac{15}{z^2-4z-5}=\dfrac{-10}{(z-5)(z-1)}+\dfrac{15}{(z-5)(z+1)}$ $LCD=(z-5)(z-1)(z+1)$

$=\dfrac{-10}{(z-5)(z-1)}\cdot\dfrac{z+1}{z+1}+\dfrac{15}{(z-5)(z+1)}\cdot\dfrac{z-1}{z-1}=\dfrac{-10z-10}{(z-5)(z-1)(z+1)}+\dfrac{15z-15}{(z-5)(z-1)(z+1)}$

$=\dfrac{5z-25}{(z-5)(z-1)(z+1)}=\dfrac{5\cancel{(z-5)}}{\cancel{(z-5)}(z-1)(z+1)}=\dfrac{5}{(z-1)(z+1)}$

79. $\dfrac{5}{x^2-4}+\dfrac{2}{x^3-8}$

$=\dfrac{5}{(x-2)(x+2)}+\dfrac{2}{(x-2)\left(x^2+2x+4\right)}$

$LCD=(x-2)(x+2)\left(x^2+2x+4\right)$

$=\dfrac{5}{(x-2)(x+2)}\cdot\dfrac{x^2+2x+4}{x^2+2x+4}+\dfrac{2}{(x-2)\left(x^2+2x+4\right)}\cdot\dfrac{x+2}{x+2}$

$=\dfrac{5x^2+10x+20+2x+4}{(x-2)(x+2)\left(x^2+2x+4\right)}$

$=\dfrac{5x^2+12x+24}{(x-2)(x+2)\left(x^2+2x+4\right)}$

81. $\dfrac{2}{3x}+\dfrac{x+1}{x}+\dfrac{6}{x^2}$ $LCD=3x^2$

$=\dfrac{2}{3x}\cdot\dfrac{x}{x}+\dfrac{x+1}{x}\cdot\dfrac{3x}{3x}+\dfrac{6}{x^2}\cdot\dfrac{3}{3}$

$=\dfrac{2x+3x^2+3x+18}{3x^2}=\dfrac{3x^2+5x+18}{3x^2}$ cm

83. $2\left(\dfrac{5}{x-3}\right)+2\left(\dfrac{2x}{x+5}\right)$ $LCD=(x-3)(x+5)$

$=\dfrac{10}{x-3}\cdot\dfrac{x+5}{x+5}+\dfrac{4x}{x+5}\cdot\dfrac{x-3}{x-3}$

$=\dfrac{10x+50+4x^2-12x}{(x-3)(x+5)}=\dfrac{4x^2-2x+50}{(x-3)(x+5)}$ m

Section 5.4   Complex Fractions

1.   complex

3. $\dfrac{25a^3b^3c}{15a^4bc} = \dfrac{25}{15}a^{3-4}b^{3-1}c^{1-1} = \dfrac{5}{3}a^{-1}b^2c^0 = \dfrac{5b^2}{3a}$

5. $\dfrac{\dfrac{5}{x^2} + \dfrac{3}{2x}}{}$ $\text{LCD} = 2x^2$

$= \dfrac{5}{x^2} \cdot \dfrac{2}{2} + \dfrac{3}{2x} \cdot \dfrac{x}{x} = \dfrac{10 + 3x}{2x^2}$

7. $\dfrac{3}{a-5} - \dfrac{1}{a+1}$ $\text{LCD} = (a-5)(a+1)$

$= \dfrac{3}{a-5} \cdot \dfrac{a+1}{a+1} - \dfrac{1}{a+1} \cdot \dfrac{a-5}{a-5}$

$= \dfrac{3a+3-a+5}{(a-5)(a+1)} = \dfrac{2a+8}{(a-5)(a+1)}$

9. $\dfrac{\dfrac{5x^2}{9y^2}}{\dfrac{3x}{y^2x}} = \dfrac{5x^2}{9y^2} \cdot \dfrac{y^2 \cancel{x}}{3\cancel{x}} = \dfrac{5x^2}{27}$

11. $\dfrac{\dfrac{x-6}{3x}}{\dfrac{3x-18}{9}} = \dfrac{x-6}{3x} \cdot \dfrac{9}{3x-18}$

$= \dfrac{\cancel{x-6}}{\cancel{3}x} \cdot \dfrac{\cancel{3} \cdot \cancel{3}}{\cancel{3}(\cancel{x-6})} = \dfrac{1}{x}$

13. $\dfrac{\dfrac{2}{3} + \dfrac{1}{6}}{\dfrac{1}{2} - \dfrac{1}{4}} = \dfrac{\dfrac{4}{6} + \dfrac{1}{6}}{\dfrac{2}{4} - \dfrac{1}{4}} = \dfrac{\dfrac{5}{6}}{\dfrac{1}{4}} = \dfrac{5}{6} \cdot \dfrac{4}{1}$

$= \dfrac{5}{\cancel{2} \cdot 3} \cdot \dfrac{\cancel{2} \cdot 2}{1} = \dfrac{10}{3}$

15. $\dfrac{8 - \dfrac{5}{2x}}{\dfrac{5}{8x} - 2} = \dfrac{8 \cdot \dfrac{2x}{2x} - \dfrac{5}{2x}}{\dfrac{5}{8x} - 2 \cdot \dfrac{8x}{8x}} = \dfrac{\dfrac{16x-5}{2x}}{\dfrac{5-16x}{8x}}$

$= \dfrac{\cancel{16x-5}}{\cancel{2}x} \cdot \dfrac{\cancel{2} \cdot 4\cancel{x}}{-(\cancel{16x-5})} = \dfrac{4}{-1} = -4$

17. $\dfrac{\dfrac{7y}{y+3}}{\dfrac{1}{4y+12}} = \dfrac{\dfrac{7y}{y+3}}{\dfrac{1}{4(y+3)}}$ $\text{LCD} = 4(y+3)$

$\dfrac{4(\cancel{y+3})\left(\dfrac{7y}{\cancel{y+3}}\right)}{\cancel{4(y+3)}\left(\dfrac{1}{\cancel{4(y+3)}}\right)} = \dfrac{4(7y)}{1} = 28y$

19. $\dfrac{1 + \dfrac{1}{3}}{\dfrac{5}{6} - 1}$ $\text{LCD} = 6$

$\dfrac{6\left(1 + \dfrac{1}{3}\right)}{6\left(\dfrac{5}{6} - 1\right)} = \dfrac{6 \cdot 1 + 6\left(\dfrac{1}{3}\right)}{6\left(\dfrac{5}{6}\right) - 6 \cdot 1} = \dfrac{6+2}{5-6} = \dfrac{8}{-1} = -8$

21.

$$\frac{\dfrac{3q}{p}-q}{q-\dfrac{q}{p}} \qquad LCD = p$$

$$\frac{p\left(\dfrac{3q}{p}-q\right)}{p\left(q-\dfrac{q}{p}\right)} = \frac{\cancel{p}\left(\dfrac{3q}{\cancel{p}}\right)-pq}{pq-\cancel{p}\left(\dfrac{q}{\cancel{p}}\right)} = \frac{3q-pq}{pq-q}$$

$$= \frac{\cancel{q}(3-p)}{\cancel{q}(p-1)} = \frac{3-p}{p-1}$$

23.

$$\frac{\dfrac{2}{a}+\dfrac{3}{a^2}}{\dfrac{4}{a^2}-\dfrac{9}{a}} \qquad LCD = a^2$$

$$\frac{a^2\left(\dfrac{2}{a}+\dfrac{3}{a^2}\right)}{a^2\left(\dfrac{4}{a^2}-\dfrac{9}{a}\right)} = \frac{a^2\left(\dfrac{2}{a}\right)+a^2\left(\dfrac{3}{a^2}\right)}{a^2\left(\dfrac{4}{a^2}\right)-a^2\left(\dfrac{9}{a}\right)} = \frac{2a+3}{4-9a}$$

25.

$$\frac{t^{-1}-1}{1-t^{-2}} = \frac{\dfrac{1}{t}-1}{1-\dfrac{1}{t^2}} \qquad LCD = t^2$$

$$\frac{t^2\left(\dfrac{1}{t}-1\right)}{t^2\left(1-\dfrac{1}{t^2}\right)} = \frac{t^2\left(\dfrac{1}{t}\right)-t^2(1)}{t^2(1)-t^2\left(\dfrac{1}{t^2}\right)} = \frac{t-t^2}{t^2-1}$$

$$= \frac{-t(\cancel{t-1})}{(t+1)(\cancel{t-1})} = -\frac{t}{t+1}$$

27.

$$\frac{-8}{\dfrac{6w}{w-1}-4} \qquad LCD = w-1$$

$$\frac{(w-1)(-8)}{(w-1)\left(\dfrac{6w}{w-1}-4\right)}$$

$$= \frac{(w-1)(-8)}{(\cancel{w-1})\left(\dfrac{6w}{\cancel{w-1}}\right)-(w-1)4}$$

$$= \frac{-8w+8}{6w-4w+4} = \frac{-8w+8}{2w+4}$$

$$= \frac{-4\cdot\cancel{2}(w-1)}{\cancel{2}(w+2)} = -\frac{4(w-1)}{w+2}$$

29.

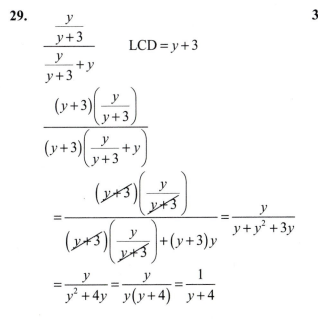

$$\frac{\dfrac{y}{y+3}}{\dfrac{y}{y+3}+y} \qquad LCD = y+3$$

$$\frac{(y+3)\left(\dfrac{y}{y+3}\right)}{(y+3)\left(\dfrac{y}{y+3}+y\right)}$$

$$= \frac{(\cancel{y+3})\left(\dfrac{y}{\cancel{y+3}}\right)}{(\cancel{y+3})\left(\dfrac{y}{\cancel{y+3}}\right)+(y+3)y} = \frac{y}{y+y^2+3y}$$

$$= \frac{y}{y^2+4y} = \frac{y}{y(y+4)} = \frac{1}{y+4}$$

31.

$$\frac{1-\dfrac{1}{x}-\dfrac{6}{x^2}}{1-\dfrac{4}{x}+\dfrac{3}{x^2}} \qquad LCD = x^2$$

$$\frac{x^2\left(1-\dfrac{1}{x}-\dfrac{6}{x^2}\right)}{x^2\left(1-\dfrac{4}{x}+\dfrac{3}{x^2}\right)} = \frac{x^2(1)-x^2\left(\dfrac{1}{x}\right)-x^2\left(\dfrac{6}{x^2}\right)}{x^2(1)-x^2\left(\dfrac{4}{x}\right)+x^2\left(\dfrac{3}{x^2}\right)}$$

$$= \frac{x^2-x-6}{x^2-4x+3} = \frac{(\cancel{x-3})(x+2)}{(\cancel{x-3})(x-1)} = \frac{x+2}{x-1}$$

33.

$$\dfrac{2-\dfrac{2}{t+1}}{2+\dfrac{2}{t}} \qquad \text{LCD}=t(t+1)$$

$$\dfrac{t(t+1)\left(2-\dfrac{2}{t+1}\right)}{t(t+1)\left(2+\dfrac{2}{t}\right)}=\dfrac{t(t+1)(2)-t\cancel{(t+1)}\left(\dfrac{2}{\cancel{t+1}}\right)}{t(t+1)(2)+\cancel{t}(t+1)\left(\dfrac{2}{\cancel{t}}\right)}=\dfrac{2t^2+2t-2t}{2t^2+2t+2t+2}=\dfrac{\cancel{2}t^2}{\cancel{2}\left(t^2+2t+1\right)}=\dfrac{t^2}{(t+1)^2}$$

35.

$$\dfrac{\dfrac{2}{a}-\dfrac{3}{a+1}}{\dfrac{2}{a+1}-\dfrac{3}{a}} \qquad \text{LCD}=a(a+1)$$

$$\dfrac{a(a+1)\left(\dfrac{2}{a}-\dfrac{3}{a+1}\right)}{a(a+1)\left(\dfrac{2}{a+1}-\dfrac{3}{a}\right)}=\dfrac{\cancel{a}(a+1)\left(\dfrac{2}{\cancel{a}}\right)-a\cancel{(a+1)}\left(\dfrac{3}{\cancel{a+1}}\right)}{a\cancel{(a+1)}\left(\dfrac{2}{\cancel{a+1}}\right)-\cancel{a}(a+1)\left(\dfrac{3}{\cancel{a}}\right)}=\dfrac{2a+2-3a}{2a-3a-3}=\dfrac{-a+2}{-a-3}$$

37.

$$\dfrac{\dfrac{1}{y+2}+\dfrac{4}{y-3}}{\dfrac{2}{y-3}-\dfrac{7}{y+2}} \qquad \text{LCD}=(y+2)(y-3)$$

$$\dfrac{(y+2)(y-3)\left(\dfrac{1}{y+2}+\dfrac{4}{y-3}\right)}{(y+2)(y-3)\left(\dfrac{2}{y-3}-\dfrac{7}{y+2}\right)}=\dfrac{\cancel{(y+2)}(y-3)\left(\dfrac{1}{\cancel{y+2}}\right)+(y+2)\cancel{(y-3)}\left(\dfrac{4}{\cancel{y-3}}\right)}{(y+2)\cancel{(y-3)}\left(\dfrac{2}{\cancel{y-3}}\right)-\cancel{(y+2)}(y-3)\left(\dfrac{7}{\cancel{y+2}}\right)}$$

$$=\dfrac{y-3+4y+8}{2y+4-7y+21}=\dfrac{5y+5}{-5y+25}=\dfrac{\cancel{5}(y+1)}{-\cancel{5}(y-5)}=-\dfrac{y+1}{y-5}$$

39.

$$\dfrac{\dfrac{2}{x+h}-\dfrac{2}{x}}{h} \qquad \text{LCD}=x(x+h)$$

$$\dfrac{x(x+h)\left(\dfrac{2}{x+h}-\dfrac{2}{x}\right)}{x(x+h)(h)}=\dfrac{x\cancel{(x+h)}\left(\dfrac{2}{\cancel{x+h}}\right)-\cancel{x}(x+h)\left(\dfrac{2}{\cancel{x}}\right)}{x(x+h)(h)}=\dfrac{2x-2x-2h}{x(x+h)(h)}=\dfrac{-2\cancel{h}}{x(x+h)\cancel{(h)}}$$

$$=-\dfrac{2}{x(x+h)}$$

41.

$$\frac{x^{-2}}{x+3x^{-1}} = \frac{\dfrac{1}{x^2}}{x+\dfrac{3}{x}} \qquad \text{LCD} = x^2$$

$$\frac{x^2\left(\dfrac{1}{x^2}\right)}{x^2\left(x+\dfrac{3}{x}\right)} = \frac{x^2\left(\dfrac{1}{x^2}\right)}{x^2(x)+x^2\left(\dfrac{3}{x}\right)} = \frac{1}{x^3+3x}$$

$$= \frac{1}{x\left(x^2+3\right)}$$

43.

$$\frac{2a^{-1}+3b^{-2}}{a^{-1}-b^{-1}} = \frac{\dfrac{2}{a}+\dfrac{3}{b^2}}{\dfrac{1}{a}-\dfrac{1}{b}} \qquad \text{LCD} = ab^2$$

$$\frac{ab^2\left(\dfrac{2}{a}+\dfrac{3}{b^2}\right)}{ab^2\left(\dfrac{1}{a}-\dfrac{1}{b}\right)} = \frac{ab^2\left(\dfrac{2}{a}\right)+ab^2\left(\dfrac{3}{b^2}\right)}{ab^2\left(\dfrac{1}{a}\right)-ab^2\left(\dfrac{1}{b}\right)}$$

$$= \frac{2b^2+3a}{b^2-ab} = \frac{2b^2+3a}{b(b-a)}$$

45.

$$\frac{\dfrac{1}{4+h}-\dfrac{1}{4}}{h} \qquad \text{LCD} = 4(4+h)$$

$$\frac{4(4+h)\left(\dfrac{1}{4+h}-\dfrac{1}{4}\right)}{4(4+h)h} = \frac{4-(4+h)}{4h(4+h)}$$

$$= \frac{-\cancel{h}}{4\cancel{h}(4+h)} = \frac{-1}{4(4+h)}$$

47.

$$\frac{\dfrac{6}{x+h}-\dfrac{6}{x}}{h} \qquad \text{LCD} = x(x+h)$$

$$\frac{x(x+h)\left(\dfrac{6}{x+h}-\dfrac{6}{x}\right)}{x(x+h)h} = \frac{6x-6(x+h)}{xh(x+h)}$$

$$= \frac{6x-6x-6h}{xh(x+h)} = \frac{-6\cancel{h}}{x\cancel{h}(x+h)} = \frac{-6}{x(x+h)}$$

49.

$$m = \frac{y_2-y_1}{x_2-x_1}$$

51.

$$m = \frac{-3-\dfrac{3}{5}}{-1-\left(-\dfrac{3}{7}\right)} = \frac{-3-\dfrac{3}{5}}{-1+\dfrac{3}{7}} \qquad \text{LCD} = 35$$

$$= \frac{35\left(-3-\dfrac{3}{5}\right)}{35\left(-1+\dfrac{3}{7}\right)} = \frac{-105-21}{-35+15} = \frac{-126}{-20} = \frac{63}{10}$$

53.

$$m = \frac{\dfrac{1}{6}-\dfrac{1}{3}}{\dfrac{1}{8}-\dfrac{1}{4}} \qquad \text{LCD} = 24$$

$$= \frac{24\left(\dfrac{1}{6}-\dfrac{1}{3}\right)}{24\left(\dfrac{1}{8}-\dfrac{1}{4}\right)} = \frac{4-8}{3-6} = \frac{-4}{-3} = \frac{4}{3}$$

55.

$$\left(x^{-1}+y^{-1}\right)^{-1} = \frac{1}{x^{-1}+y^{-1}} = \frac{1}{\dfrac{1}{x}+\dfrac{1}{y}} \qquad \text{LCD} = xy$$

$$= \frac{xy(1)}{xy\left(\dfrac{1}{x}+\dfrac{1}{y}\right)} = \frac{xy}{y+x} = \frac{xy}{x+y}$$

57.

$$\frac{x}{1-\left(1-\frac{1}{x}\right)^{-1}} = \frac{x}{1-\dfrac{1}{1-\dfrac{1}{x}}} = \frac{x}{1-\dfrac{1}{\dfrac{x}{x}-\dfrac{1}{x}}} = \frac{x}{1-\dfrac{1}{\dfrac{x-1}{x}}} = \frac{x}{1-1\cdot\dfrac{x}{x-1}} = \frac{x}{\dfrac{x-1}{x-1}-\dfrac{x}{x-1}} = \frac{x}{\dfrac{-1}{x-1}}$$

$$= x\cdot\frac{x-1}{-1} = -x(x-1)$$

Problem Recognition Exercises

1. $\dfrac{2}{2y-3}-\dfrac{3}{2y}+1 \qquad \text{LCD}=2y(2y-3)$

$$=\frac{2}{2y-3}\cdot\frac{2y}{2y}-\frac{3}{2y}\cdot\frac{2y-3}{2y-3}+1\cdot\frac{2y(2y-3)}{2y(2y-3)}$$

$$=\frac{4y-3(2y-3)+2y(2y-3)}{2y(2y-3)}$$

$$=\frac{4y-6y+9+4y^2-6y}{2y(2y-3)}=\frac{4y^2-8y+9}{2y(2y-3)}$$

3. $\dfrac{5x^2-6x+1}{x^2-1}\div\dfrac{16x^2-9}{4x^2+7x+3}-\dfrac{x}{4x-3}$

$$=\frac{5x^2-6x+1}{x^2-1}\cdot\frac{4x^2+7x+3}{16x^2-9}-\frac{x}{4x-3}$$

$$=\frac{(5x-1)\cancel{(x-1)}}{\cancel{(x+1)}\cancel{(x-1)}}\cdot\frac{\cancel{(4x+3)}\cancel{(x+1)}}{\cancel{(4x+3)}(4x-3)}-\frac{x}{4x-3}$$

$$=\frac{5x-1}{4x-3}-\frac{x}{4x-3}=\frac{4x-1}{4x-3}$$

5. $\dfrac{4}{y+1}+\dfrac{y+2}{y^2-1}-\dfrac{3}{y-1}=\dfrac{4}{y+1}+\dfrac{y+2}{(y+1)(y-1)}-\dfrac{3}{y-1} \qquad \text{LCD}=(y+1)(y-1)$

$$=\frac{4}{y+1}\cdot\frac{y-1}{y-1}+\frac{y+2}{(y+1)(y-1)}-\frac{3}{y-1}\cdot\frac{y+1}{y+1}=\frac{4(y-1)+y+2-3(y+1)}{(y+1)(y-1)}=\frac{4y-4+y+2-3y-3}{(y+1)(y-1)}$$

$$=\frac{2y-5}{(y+1)(y-1)}$$

7. $\dfrac{a^2-16}{2x+6}\cdot\dfrac{x+3}{a-4}=\dfrac{(a+4)\cancel{(a-4)}}{2\cancel{(x+3)}}\cdot\dfrac{\cancel{x+3}}{\cancel{a-4}}$

$$=\frac{a+4}{2}$$

9. $\dfrac{2+\dfrac{1}{a}}{4-\dfrac{1}{a^2}} \qquad \text{LCD}=a^2$

$$\frac{a^2\left(2+\dfrac{1}{a}\right)}{a^2\left(4-\dfrac{1}{a^2}\right)}=\frac{a^2(2)+a^2\left(\dfrac{1}{a}\right)}{a^2(4)-a^2\left(\dfrac{1}{a^2}\right)}=\frac{2a^2+a}{4a^2-1}$$

$$=\frac{a\cancel{(2a+1)}}{\cancel{(2a+1)}(2a-1)}=\frac{a}{2a-1}$$

11. $\dfrac{6xy}{x^2-y^2}+\dfrac{x+y}{y-x}=\dfrac{6xy}{(x+y)(x-y)}+\dfrac{x+y}{y-x}$

$\quad$ LCD $=(x+y)(x-y)$

$\qquad =\dfrac{6xy}{(x+y)(x-y)}+\dfrac{x+y}{y-x}\cdot\dfrac{(-1)}{(-1)}\cdot\dfrac{x+y}{x+y}$

$\qquad =\dfrac{6xy-\left(x^2+2xy+y^2\right)}{(x+y)(x-y)}$

$\qquad =\dfrac{6xy-x^2-2xy-y^2}{(x+y)(x-y)}=\dfrac{-x^2+4xy-y^2}{(x+y)(x-y)}$

13. $\dfrac{3}{x-2}-\dfrac{x-2}{6}\quad$ LCD $=6(x-2)$

$\qquad =\dfrac{3}{x-2}\cdot\dfrac{6}{6}-\dfrac{x-2}{6}\cdot\dfrac{x-2}{x-2}$

$\qquad =\dfrac{18-\left(x^2-4x+4\right)}{6(x-2)}=\dfrac{18-x^2+4x-4}{6(x-2)}$

$\qquad =\dfrac{-x^2+4x+14}{6(x-2)}$

15. $\dfrac{1}{w-1}-\dfrac{w+2}{3w-3}=\dfrac{1}{w-1}-\dfrac{w+2}{3(w-1)}$

$\quad$ LCD $=3(w-1)$

$\qquad =\dfrac{3}{3}\cdot\dfrac{1}{w-1}-\dfrac{w+2}{3(w-1)}=\dfrac{3-w-2}{3(w-1)}$

$\qquad =\dfrac{1-w}{3(w-1)}=\dfrac{-1\cancel{(w-1)}}{3\cancel{(w-1)}}=-\dfrac{1}{3}$

17. $\dfrac{y+\dfrac{2}{y}-3}{1-\dfrac{2}{y}}\qquad$ LCD $=y$

$\dfrac{y\left(y+\dfrac{2}{y}-3\right)}{y\left(1-\dfrac{2}{y}\right)}=\dfrac{y(y)+y\left(\dfrac{2}{y}\right)-y(3)}{y(1)-y\left(\dfrac{2}{y}\right)}=\dfrac{y^2+2-3y}{y-2}=\dfrac{\cancel{(y-2)}(y-1)}{\cancel{y-2}}=y-1$

19. $\dfrac{4x^2+22x+24}{4x+4}\cdot\dfrac{6x+6}{4x^2-9}=\dfrac{2\left(2x^2+11x+12\right)}{4(x+1)}\cdot\dfrac{6(x+1)}{(2x-3)(2x+3)}$

$\qquad =\dfrac{\cancel{2}\cancel{(2x+3)}(x+4)}{\cancel{2}\cdot\cancel{2}\cancel{(x+1)}}\cdot\dfrac{\cancel{2}\cdot3\cancel{(x+1)}}{(2x-3)\cancel{(2x+3)}}=\dfrac{3(x+4)}{2x-3}$

21. $\dfrac{3x-1}{4}+\dfrac{7}{6x-2}=\dfrac{3x-1}{4}+\dfrac{7}{2(3x-1)}$

$\quad$ LCD $=4(3x-1)$

$\qquad =\dfrac{3x-1}{4}\cdot\dfrac{3x-1}{3x-1}+\dfrac{7}{2(3x-1)}\cdot\dfrac{2}{2}$

$\qquad =\dfrac{9x^2-6x+1+14}{4(3x-1)}=\dfrac{9x^2-6x+15}{4(3x-1)}$

23.

$$(y+2) \cdot \frac{2y+1}{y^2-4} - \frac{y-2}{y+3} = \frac{\cancel{(y+2)}(2y+1)}{\cancel{(y+2)}(y-2)} - \frac{y-2}{y+3} = \frac{2y+1}{y-2} - \frac{y-2}{y+3}$$

$$\text{LCD} = (y-2)(y+3)$$

$$\left(\frac{y+3}{y+3}\right)\left(\frac{2y+1}{y-2}\right) - \left(\frac{y-2}{y-2}\right)\left(\frac{y-2}{y+3}\right) = \frac{2y^2+7y+3-\left(y^2-4y+4\right)}{(y-2)(y+3)}$$

$$= \frac{2y^2+7y+3-y^2+4y-4}{(y-2)(y+3)} = \frac{y^2+11y-1}{(y-2)(y+3)}$$

Section 5.5   Solving Rational Equations

1. a. rational.
  b. denominator.
  c. No.

3.

$$\frac{3}{y^2-1} - \frac{2}{y^2-2y+1} = \frac{3}{(y+1)(y-1)} - \frac{2}{(y-1)^2} \qquad \text{LCD} = (y-1)^2(y+1)$$

$$= \frac{3}{(y+1)(y-1)} \cdot \frac{y-1}{y-1} - \frac{2}{(y-1)^2} \cdot \frac{y+1}{y+1} = \frac{3y-3-2y-2}{(y-1)^2(y+1)} = \frac{y-5}{(y-1)^2(y+1)}$$

5.

$$\frac{2t^2+7t+3}{4t^2-1} \div (t+3) = \frac{2t^2+7t+3}{4t^2-1} \cdot \frac{1}{t+3} = \frac{\cancel{(2t+1)}\cancel{(t+3)}}{\cancel{(2t+1)}(2t-1)} \cdot \frac{1}{\cancel{t+3}} = \frac{1}{2t-1}$$

7.

$$\frac{x+y}{x^{-1}+y^{-1}} = \frac{x+y}{\dfrac{1}{x}+\dfrac{1}{y}} \qquad \text{LCD} = xy$$

$$\frac{xy(x+y)}{xy\left(\dfrac{1}{x}+\dfrac{1}{y}\right)} = \frac{xy(x)+xy(y)}{xy\left(\dfrac{1}{x}\right)+xy\left(\dfrac{1}{y}\right)} = \frac{x^2y+xy^2}{y+x}$$

$$= \frac{xy\cancel{(x+y)}}{\cancel{x+y}} = xy$$

9.

$$\frac{x+2}{3} - \frac{x-4}{4} = \frac{1}{2} \qquad \text{LCD} = 12$$

$$12\left(\frac{x+2}{3} - \frac{x-4}{4}\right) = 12\left(\frac{1}{2}\right)$$

$$12\left(\frac{x+2}{3}\right) - 12\left(\frac{x-4}{4}\right) = 12\left(\frac{1}{2}\right)$$

$$4(x+2) - 3(x-4) = 6$$

$$4x+8-3x+12 = 6$$

$$x+20 = 6$$

$$x = -14 \quad \{-14\}$$

11.

$$\frac{3y}{4} - 2 = \frac{5y}{6} \qquad LCD = 12$$

$$12\left(\frac{3y}{4} - 2\right) = 12\left(\frac{5y}{6}\right)$$

$$12\left(\frac{3y}{4}\right) - 12(2) = 12\left(\frac{5y}{6}\right)$$

$$9y - 24 = 10y$$

$$-24 = y \quad \{-24\}$$

13.

$$\frac{5}{4p} - \frac{7}{6} + 3 = 0$$

$$LCD = 12p \quad so\ p \neq 0$$

$$12p\left(\frac{5}{4p} - \frac{7}{6} + 3\right) = 12p(0)$$

$$12p\left(\frac{5}{4p}\right) - 12p\left(\frac{7}{6}\right) + 12p(3) = 0$$

$$15 - 14p + 36p = 0$$

$$15 + 22p = 0$$

$$22p = -15$$

$$p = -\frac{15}{22} \quad \left\{-\frac{15}{22}\right\}$$

15.

$$\frac{1}{2} - \frac{3}{2x} = \frac{4}{x} - \frac{5}{12}$$

$$LCD = 12x \quad so\ x \neq 0$$

$$12x\left(\frac{1}{2} - \frac{3}{2x}\right) = 12x\left(\frac{4}{x} - \frac{5}{12}\right)$$

$$12x\left(\frac{1}{2}\right) - 12x\left(\frac{3}{2x}\right) = 12x\left(\frac{4}{x}\right) - 12x\left(\frac{5}{12}\right)$$

$$6x - 18 = 48 - 5x$$

$$11x = 66$$

$$x = 6 \quad \{6\}$$

17.

$$\frac{3}{x-4} + 2 = \frac{5}{x-4}$$

$$LCD = x - 4 \quad so\ x \neq 4$$

$$(x-4)\left(\frac{3}{x-4} + 2\right) = (x-4)\left(\frac{5}{x-4}\right)$$

$$(x-4)\left(\frac{3}{x-4}\right) + (x-4)(2) = 5$$

$$3 + 2x - 8 = 5$$

$$2x - 5 = 5$$

$$2x = 10$$

$$x = 5 \quad \{5\}$$

19.

$$\frac{1}{3} + \frac{2}{w-3} = 1$$

$$LCD = 3(w-3) \quad so\ w \neq 3$$

$$3(w-3)\left(\frac{1}{3} + \frac{2}{w-3}\right) = 3(w-3)(1)$$

$$3(w-3)\left(\frac{1}{3}\right) + 3(w-3)\left(\frac{2}{w-3}\right) = 3w - 9$$

$$w - 3 + 6 = 3w - 9$$

$$w + 3 = 3w - 9$$

$$-2w = -12$$

$$w = 6 \quad \{6\}$$

21.

$$\frac{12}{x} - \frac{12}{x-5} = \frac{2}{x}$$

$$LCD = x(x-5) \quad so\ x \neq 0\ or\ x \neq 5$$

$$x(x-5)\left(\frac{12}{x} - \frac{12}{x-5}\right) = x(x-5)\left(\frac{2}{x}\right)$$

$$x(x-5)\left(\frac{12}{x}\right) - x(x-5)\left(\frac{12}{x-5}\right) = 2x - 10$$

$$12x - 60 - 12x = 2x - 10$$

$$-60 = 2x - 10$$

$$-50 = 2x$$

$$x = -25 \quad \{-25\}$$

23.

$$\frac{3}{a^2} - \frac{4}{a} = -1$$

$$\text{LCD} = a^2 \quad \text{so } a \neq 0$$

$$a^2\left(\frac{3}{a^2} - \frac{4}{a}\right) = a^2(-1)$$

$$a^2\left(\frac{3}{a^2}\right) - a \cdot a\left(\frac{4}{a}\right) = -a^2$$

$$3 - 4a = -a^2$$

$$a^2 - 4a + 3 = 0$$

$$(a-3)(a-1) = 0$$

$$a - 3 = 0 \quad \text{or} \quad a - 1 = 0$$

$$a = 3 \quad \text{or} \quad a = 1$$

$$\{3, 1\}$$

25.

$$\frac{1}{4}a - 4a^{-1} = 0$$

$$\frac{a}{4} - \frac{4}{a} = 0$$

$$\text{LCD} = 4a \quad \text{so } a \neq 0$$

$$4a\left(\frac{a}{4} - \frac{4}{a}\right) = 4a(0)$$

$$4a\left(\frac{a}{4}\right) - 4a\left(\frac{4}{a}\right) = 0$$

$$a^2 - 16 = 0$$

$$(a+4)(a-4) = 0$$

$$a + 4 = 0 \quad \text{or} \quad a - 4 = 0$$

$$a = -4 \quad \text{or} \quad a = 4 \quad \{-4, 4\}$$

27.

$$\frac{y}{y+3} + \frac{2}{y^2 + 3y} = \frac{6}{y}$$

$$\frac{y}{y+3} + \frac{2}{y(y+3)} = \frac{6}{y}$$

$$\text{LCD} = y(y+3) \text{ so } y \neq 0, y \neq -3$$

$$y(y+3)\left(\frac{y}{y+3} + \frac{2}{y(y+3)}\right) = y(y+3)\frac{6}{y}$$

$$y(y+3)\left(\frac{y}{y+3}\right) + y(y+3)\left(\frac{2}{y(y+3)}\right)$$

$$= y(y+3)\frac{6}{y}$$

$$y^2 + 2 = (y+3)6$$

$$y^2 + 2 = 6y + 18$$

$$y^2 - 6y - 16 = 0$$

$$(y-8)(y+2) = 0$$

$$y - 8 = 0 \quad \text{or} \quad y + 2 = 0$$

$$y = 8 \quad \text{or} \quad y = -2 \quad \{8, -2\}$$

29.

$$\frac{4}{t-2} - \frac{8}{t^2 - 2t} = -2$$

$$\frac{4}{t-2} - \frac{8}{t(t-2)} = -2 \qquad \text{LCD} = t(t-2) \qquad \text{so } t \neq 0 \text{ or } t \neq 2$$

$$t(t-2)\left(\frac{4}{t-2} - \frac{8}{t(t-2)}\right) = t(t-2)(-2)$$

$$t(\cancel{t-2})\left(\frac{4}{\cancel{t-2}}\right) - \cancel{t}(\cancel{t-2})\left(\frac{8}{\cancel{t}(\cancel{t-2})}\right) = -2t(t-2)$$

$$4t - 8 = -2t^2 + 4t$$

$$2t^2 - 8 = 0$$

$$2(t^2 - 4) = 0$$

$$2(t+2)(t-2) = 0$$

$$2 \neq 0 \text{ or } t+2 = 0 \text{ or } t-2 = 0$$

$$t = -2 \text{ or } \quad t = 2$$

$\{-2\}$ is the solution. ($t = 2$ does not check because the denominator is zero.)

31.

$$\frac{6}{5y+10} - \frac{1}{y-5} = \frac{4}{y^2 - 3y - 10}$$

$$\frac{6}{5(y+2)} - \frac{1}{y-5} = \frac{4}{(y-5)(y+2)} \qquad \text{LCD} = 5(y-5)(y+2) \qquad \text{so } y \neq 5 \text{ or } y \neq -2$$

$$5(y-5)(y+2)\left(\frac{6}{5(y+2)} - \frac{1}{y-5}\right) = 5(y-5)(y+2)\left(\frac{4}{(y-5)(y+2)}\right)$$

$$\cancel{5}(y-5)(\cancel{y+2})\left(\frac{6}{\cancel{5}(\cancel{y+2})}\right) - 5(\cancel{y-5})(y+2)\left(\frac{1}{\cancel{y-5}}\right) = 5(\cancel{y-5})(\cancel{y+2})\left(\frac{4}{(\cancel{y-5})(\cancel{y+2})}\right)$$

$$6y - 30 - 5y - 10 = 20$$

$$y - 40 = 20$$

$$y = 60$$

$\{60\}$ is the solution.

33.

$$\frac{x}{x-5} + \frac{1}{5} = \frac{5}{x-5} \qquad \text{LCD} = 5(x-5) \quad \text{so } x \neq 5$$

$$5(x-5)\left(\frac{x}{x-5} + \frac{1}{5}\right) = 5(x-5)\left(\frac{5}{x-5}\right)$$

$$5(\cancel{x-5})\left(\frac{x}{\cancel{x-5}}\right) + \cancel{5}(x-5)\left(\frac{1}{\cancel{5}}\right) = 5(\cancel{x-5})\left(\frac{5}{\cancel{x-5}}\right)$$

$$5x + x - 5 = 25$$

$$6x - 5 = 25$$

$$6x = 30$$

$$x = 5$$

$\{\ \}$ No solution. ($x = 5$ does not check because it makes the denominator zero.)

35.

$$\frac{6}{x^2-4x+3} - \frac{1}{x-3} = \frac{1}{4x-4}$$

$$\frac{6}{(x-3)(x-1)} - \frac{1}{x-3} = \frac{1}{4(x-1)} \qquad LCD = 4(x-3)(x-1) \qquad so\ x \neq 3\ or\ x \neq 1$$

$$4(x-3)(x-1)\left(\frac{6}{(x-3)(x-1)} - \frac{1}{x-3}\right) = 4(x-3)(x-1)\left(\frac{1}{4(x-1)}\right)$$

$$4(x-3)(x-1)\left(\frac{6}{(x-3)(x-1)}\right) - 4(x-3)(x-1)\left(\frac{1}{x-3}\right) = x-3$$

$$24 - 4x + 4 = x - 3$$
$$-4x + 28 = x - 3$$
$$-5x = -31$$
$$x = \frac{31}{5} \qquad \left\{\frac{31}{5}\right\}\ \text{is the solution.}$$

37.

$$\frac{1}{k+2} - \frac{4}{k-2} - \frac{k^2}{4-k^2} = 0$$

$$\frac{1}{k+2} - \frac{4}{k-2} + \frac{k^2}{(k+2)(k-2)} = 0 \qquad LCD = (k+2)(k-2) \qquad so\ k \neq -2\ or\ k \neq 2$$

$$(k+2)(k-2)\left(\frac{1}{k+2} - \frac{4}{k-2} + \frac{k^2}{(k+2)(k-2)}\right) = (k+2)(k-2)(0)$$

$$(k+2)(k-2)\left(\frac{1}{k+2}\right) - (k+2)(k-2)\left(\frac{4}{k-2}\right) + (k+2)(k-2)\left(\frac{k^2}{(k+2)(k-2)}\right) = 0$$

$$k - 2 - 4k - 8 + k^2 = 0$$
$$k^2 - 3k - 10 = 0$$
$$(k-5)(k+2) = 0$$
$$k - 5 = 0\ \text{ or }\ k + 2 = 0$$
$$k = 5\ \text{ or }\quad k = -2$$

$\{5\}$ is the solution. ($k = -2$ does not check because the denominator is zero.)

39.

$$\frac{5}{x^2-7x+12}=\frac{2}{x-3}+\frac{5}{x-4}$$

$$\frac{5}{(x-4)(x-3)}=\frac{2}{x-3}+\frac{5}{x-4} \qquad LCD=(x-4)(x-3) \qquad so \ x\neq 4 \ or \ x\neq 3$$

$$(x-4)(x-3)\left(\frac{5}{(x-4)(x-3)}\right)=(x-4)(x-3)\left(\frac{2}{x-3}+\frac{5}{x-4}\right)$$

$$(\cancel{x-4})(\cancel{x-3})\left(\frac{5}{(\cancel{x-4})(\cancel{x-3})}\right)=(x-4)(\cancel{x-3})\left(\frac{2}{\cancel{x-3}}\right)+(\cancel{x-4})(x-3)\left(\frac{5}{\cancel{x-4}}\right)$$

$$5=2x-8+5x-15$$
$$5=7x-23$$
$$28=7x$$
$$x=4$$

$\{ \ \}$ no solution. ($x=4$ does not check because the denominator is zero.)

41.

$$\frac{4}{x^2+7x+12}-\frac{7}{x^2+8x+15}=\frac{1}{x^2+9x+20}$$

$$\frac{4}{(x+4)(x+3)}-\frac{7}{(x+5)(x+3)}=\frac{1}{(x+5)(x+4)} \qquad LCD=(x+4)(x+3)(x+5) \qquad so \ x\neq-4, x\neq-3, x\neq-5$$

$$(x+4)(x+3)(x+5)\left(\frac{4}{(x+4)(x+3)}-\frac{7}{(x+5)(x+3)}\right)=(x+4)(x+3)(x+5)\left(\frac{1}{(x+5)(x+4)}\right)$$

$$(\cancel{x+4})(\cancel{x+3})(x+5)\frac{4}{(\cancel{x+4})(\cancel{x+3})}-(x+4)(\cancel{x+3})(\cancel{x+5})\frac{7}{(\cancel{x+5})(\cancel{x+3})}=(\cancel{x+4})(x+3)(\cancel{x+5})\frac{1}{(\cancel{x+5})(\cancel{x+4})}$$

$$4x+20-7x-28=x+3$$
$$-3x-8=x+3$$
$$-4x=11$$

$$x=-\frac{11}{4} \qquad \left\{-\frac{11}{4}\right\} \text{ is the solution.}$$

43.

$$K=\frac{ma}{F} \qquad \text{for } m$$

$$KF=ma$$

$$\frac{KF}{a}=m$$

45.

$$K=\frac{IR}{E} \qquad \text{for } E$$

$$KE=IR$$

$$E=\frac{IR}{K}$$

47.

$$I=\frac{E}{R+r} \qquad \text{for } R$$

$$I(R+r)=E$$

$$IR+Ir=E$$

$$IR=E-Ir$$

$$R=\frac{E-Ir}{I} \quad \text{or} \quad R=\frac{E}{I}-r$$

49.

$$h=\frac{2A}{B+b} \qquad \text{for } B$$

$$h(B+b)=2A$$

$$hB+hb=2A$$

$$hB=2A-hb$$

$$B=\frac{2A-hb}{h} \quad \text{or} \quad B=\frac{2A}{h}-b$$

51.

$$x = \frac{at + b}{t} \quad \text{for } t$$

$$xt = at + b$$

$$xt - at = b$$

$$t(x - a) = b$$

$$t = \frac{b}{x - a}$$

53.

$$\frac{x - y}{xy} = z \quad \text{for } x$$

$$x - y = xyz$$

$$x - xyz = y$$

$$x(1 - yz) = y$$

$$x = \frac{y}{1 - yz}$$

55.

$$a + b = \frac{2A}{h} \quad \text{for } h$$

$$h(a + b) = 2A$$

$$h = \frac{2A}{a + b}$$

57.

$$\frac{1}{R} = \frac{1}{R_1} + \frac{1}{R_2} \quad \text{for } R$$

$$RR_1R_2 \left(\frac{1}{R}\right) = RR_1R_2 \left(\frac{1}{R_1} + \frac{1}{R_2}\right)$$

$$\cancel{R}R_1R_2 \left(\frac{1}{\cancel{R}}\right) = R\,\cancel{R_1}R_2 \left(\frac{1}{\cancel{R_1}}\right) + RR_1\cancel{R_2}\left(\frac{1}{\cancel{R_2}}\right)$$

$$R_1R_2 = RR_2 + RR_1$$

$$R_1R_2 = R(R_2 + R_1)$$

$$\frac{R_1R_2}{R_2 + R_1} = R$$

59.

$$v = \frac{s_2 - s_1}{t_2 - t_1} \quad \text{for } t_2$$

$$v(t_2 - t_1) = s_2 - s_1$$

$$vt_2 - vt_1 = s_2 - s_1$$

$$vt_2 = s_2 - s_1 + vt_1$$

$$t_2 = \frac{s_2 - s_1 + vt_1}{v}$$

61.

$$\frac{3}{x + 2} + \frac{2}{x} = \frac{-4}{x^2 + 2x}$$

$$\frac{3}{x + 2} + \frac{2}{x} = \frac{-4}{x(x + 2)}$$

$$\text{LCD} = x(x + 2) \text{ so } x \neq 0, \; x \neq -2$$

$$x(x + 2)\left(\frac{3}{x + 2} + \frac{2}{x}\right) = x(x + 2)\frac{-4}{x(x + 2)}$$

$$x\cancel{(x+2)}\frac{3}{\cancel{x+2}} + \cancel{x}(x+2)\frac{2}{\cancel{x}}$$

$$= \cancel{x}\,\cancel{(x+2)}\frac{-4}{\cancel{x}\,\cancel{(x+2)}}$$

$$3x + 2x + 4 = -4$$

$$5x + 4 = -4$$

$$5x = -8$$

$$x = -\frac{8}{5} \qquad \left\{-\frac{8}{5}\right\}$$

63.
$$4c(c+1)=3(c^2+4)$$
$$4c^2+4c=3c^2+12$$
$$c^2+4c-12=0$$
$$(c+6)(c-2)=0$$
$$c+6=0 \quad \text{or} \quad c-2=0$$
$$c=-6 \quad \text{or} \quad c=2 \quad \{-6,\,2\}$$

65.
$$\frac{2}{v-1}-\frac{4}{v+5}=\frac{3}{v^2+4v-5}$$
$$\frac{2}{v-1}-\frac{4}{v+5}=\frac{3}{(v+5)(v-1)}$$

$$\text{LCD}=(v+5)(v-1) \quad \text{so } v\neq-5 \text{ or } v\neq1$$

$$(v+5)(v-1)\left(\frac{2}{v-1}-\frac{4}{v+5}\right)=(v+5)(v-1)\left(\frac{3}{(v+5)(v-1)}\right)$$

$$(v+5)\cancel{(v-1)}\left(\frac{2}{\cancel{v-1}}\right)-\cancel{(v+5)}(v-1)\left(\frac{4}{\cancel{v+5}}\right)=\cancel{(v+5)}\cancel{(v-1)}\left(\frac{3}{\cancel{(v+5)}\cancel{(v-1)}}\right)$$

$$2v+10-4v+4=3$$
$$-2v+14=3$$
$$-2v=-11$$
$$v=\frac{11}{2} \qquad \left\{\frac{11}{2}\right\}$$

67.
$$5(x-9)=3(x+4)-2(4x+1)$$
$$5x-45=3x+12-8x-2$$
$$5x-45=-5x+10$$
$$10x-45=10$$
$$10x=55$$
$$x=\frac{55}{10}=\frac{11}{2} \quad \left\{\frac{11}{2}\right\}$$

69.
$$\frac{3y}{10}-\frac{5}{2y}=\frac{y}{5} \quad \text{LCD}=10y \text{ so } y\neq0$$

$$10y\left(\frac{3y}{10}-\frac{5}{2y}\right)=10y\left(\frac{y}{5}\right)$$

$$\cancel{10}y\left(\frac{3y}{\cancel{10}}\right)-5\cdot\cancel{2y}\left(\frac{5}{\cancel{2y}}\right)=\cancel{5}\cdot2y\left(\frac{y}{\cancel{5}}\right)$$

$$3y^2-25=2y^2$$
$$y^2-25=0$$
$$(y-5)(y+5)=0$$
$$y-5=0 \quad \text{or} \quad y+5=0$$
$$y=5 \quad \text{or} \quad y=-5 \quad \{-5,5\}$$

71. $\frac{1}{2}(4d-1)+\frac{2}{3}(2d+2)=\frac{5}{6}(4d+1)$ LCD = 6

$$6\left[\frac{1}{2}(4d-1)+\frac{2}{3}(2d+2)\right]=6\left[\frac{5}{6}(4d+1)\right]$$

$$3(4d-1)+4(2d+2)=5(4d+1)$$

$$12d-3+8d+8=20d+5$$

$$20d+5=20d+5$$

$$5=5$$

$\{d \mid d \text{ is a real number}\}$

73. $8t^{-1}+2=3t^{-1}$

$$\frac{8}{t}+2=\frac{3}{t} \quad \text{LCD}=t \quad \text{so } t \neq 0$$

$$t\left(\frac{8}{t}+2\right)=t\left(\frac{3}{t}\right)$$

$$\cancel{t}\left(\frac{8}{\cancel{t}}\right)+t(2)=\cancel{t}\left(\frac{3}{\cancel{t}}\right)$$

$$8+2t=3$$

$$2t=-5$$

$$t=-\frac{5}{2} \quad \left\{-\frac{5}{2}\right\}$$

75. $\dfrac{y-1}{11-3}=\dfrac{1}{2}$

$$\frac{y-1}{8}=\frac{1}{2} \quad \text{LCD}=8$$

$$\cancel{8}\left(\frac{y-1}{\cancel{8}}\right)=4\cdot\cancel{2}\left(\frac{1}{\cancel{2}}\right)$$

$$y-1=4$$

$$y=5$$

77. $\dfrac{2-(-2)}{x-4}=4$

$$\frac{4}{x-4}=4 \quad \text{LCD}=x-4$$

$$(\cancel{x-4})\left(\frac{4}{\cancel{x-4}}\right)=(x-4)(4)$$

$$4=4x-16$$

$$20=4x$$

$$x=5$$

Problem Recognition Exercises

1. a. $\dfrac{3}{w-5}+\dfrac{10}{w^2-25}-\dfrac{1}{w+5}=\dfrac{3}{w-5}+\dfrac{10}{(w+5)(w-5)}-\dfrac{1}{w+5}$ LCD $=(w+5)(w-5)$

$$=\frac{3}{w-5}\cdot\frac{w+5}{w+5}+\frac{10}{(w+5)(w-5)}-\frac{1}{w+5}\cdot\frac{w-5}{w-5}=\frac{3w+15+10-w+5}{(w+5)(w-5)}=\frac{2w+30}{(w+5)(w-5)}$$

b. $\dfrac{3}{w-5}+\dfrac{10}{w^2-25}-\dfrac{1}{w+5}=0$

$$\frac{2w+30}{(w+5)(w-5)}=0$$

$$(\cancel{w+5})(\cancel{w-5})\left(\frac{2w+30}{(\cancel{w+5})(\cancel{w-5})}\right)=(w+5)(w-5)(0)$$

$$2w+30=0$$

$$2w=-30$$

$$w=-15 \quad \{-15\}$$

c. The problem in part (a) is an expression, and the problem in part (b) is an equation.

3.

$$\frac{2}{a^2+4a+3}+\frac{1}{a+3}=\frac{2}{(a+3)(a+1)}+\frac{1}{a+3} \qquad \text{LCD}=(a+3)(a+1)$$

$$=\frac{2}{(a+3)(a+1)}+\frac{1}{a+3}\cdot\frac{a+1}{a+1}=\frac{2+a+1}{(a+3)(a+1)}=\frac{\cancel{a+3}}{\cancel{(a+3)}(a+1)}=\frac{1}{a+1}$$

5.

$$\frac{7}{y^2-y-2}+\frac{1}{y+1}-\frac{3}{y-2}=0$$

$$\frac{7}{(y-2)(y+1)}+\frac{1}{y+1}-\frac{3}{y-2}=0 \qquad \text{LCD}=(y-2)(y+1) \qquad \text{so } y\neq2 \text{ or } y\neq-1$$

$$(y-2)(y+1)\left(\frac{7}{(y-2)(y+1)}+\frac{1}{y+1}-\frac{3}{y-2}\right)=(y-2)(y+1)(0)$$

$$\cancel{(y-2)}\cancel{(y+1)}\left(\frac{7}{\cancel{(y-2)}\cancel{(y+1)}}\right)+(y-2)\cancel{(y+1)}\left(\frac{1}{\cancel{y+1}}\right)-\cancel{(y-2)}(y+1)\left(\frac{3}{\cancel{y-2}}\right)=0$$

$$7+y-2-3y-3=0$$
$$-2y+2=0$$
$$-2y=-2$$
$$y=1$$

$\{1\}$ is the solution.

7.

$$\frac{x}{x-1}-\frac{12}{x^2-x}=\frac{x}{x-1}-\frac{12}{x(x-1)}$$

$$\text{LCD}=x(x-1)$$

$$=\frac{x}{x-1}\cdot\frac{x}{x}-\frac{12}{x(x-1)}=\frac{x^2-12}{x(x-1)}$$

9.

$$\frac{3}{w}-5=\frac{7}{w}-1$$

$$\text{LCD}=w \qquad \text{so } w\neq0$$

$$w\left(\frac{3}{w}-5\right)=w\left(\frac{7}{w}-1\right)$$

$$\cancel{w}\left(\frac{3}{\cancel{w}}\right)-w(5)=\cancel{w}\left(\frac{7}{\cancel{w}}\right)-w(1)$$

$$3-5w=7-w$$
$$-4w=4$$
$$w=-1 \qquad \{-1\}$$

11.

$$\frac{4p+1}{8p-12}+\frac{p-3}{2p-3}=\frac{4p+1}{4(2p-3)}+\frac{p-3}{2p-3}$$

$$\text{LCD}=4(2p-3)$$

$$=\frac{4p+1}{4(2p-3)}+\frac{p-3}{2p-3}\cdot\frac{4}{4}$$

$$=\frac{4p+1+4p-12}{4(2p-3)}=\frac{8p-11}{4(2p-3)}$$

13.

$$\frac{1}{2x^2}+\frac{1}{6x} \qquad \text{LCD}=6x^2$$

$$=\frac{1}{2x^2}\cdot\frac{3}{3}+\frac{1}{6x}\cdot\frac{x}{x}=\frac{3+x}{6x^2}$$

15.

$$\frac{3}{2t} + \frac{2}{3t^2} = \frac{-1}{t}$$

$\text{LCD} = 6t^2 \quad$ so $t \neq 0$

$$6t^2 \left(\frac{3}{2t} + \frac{2}{3t^2} \right) = 6t^2 \left(\frac{-1}{t} \right)$$

$$2t \cdot 3t \left(\frac{3}{2t} \right) + 2 \cdot 3t^2 \left(\frac{2}{3t^2} \right) = 6t \cdot t \left(\frac{-1}{t} \right)$$

$$9t + 4 = -6t$$

$$4 = -15t$$

$$-\frac{4}{15} = t \qquad \left\{ -\frac{4}{15} \right\}$$

17.

$$\frac{3}{c^2 + 4c + 3} - \frac{2}{c^2 + 6c + 9}$$

$$= \frac{3}{(c+3)(c+1)} - \frac{2}{(c+3)^2}$$

$\text{LCD} = (c+3)^2 (c+1)$

$$= \frac{3}{(c+3)(c+1)} \cdot \frac{c+3}{c+3} - \frac{2}{(c+3)^2} \cdot \frac{c+1}{c+1}$$

$$= \frac{3c + 9 - 2c - 2}{(c+3)^2 (c+1)} = \frac{c+7}{(c+3)^2 (c+1)}$$

19.

$$\frac{4}{w-4} - \frac{36}{2w^2 - 7w - 4} = \frac{3}{2w+1}$$

$$\frac{4}{w-4} - \frac{36}{(2w+1)(w-4)} = \frac{3}{2w+1} \qquad \text{LCD} = (2w+1)(w-4) \quad \text{so } w \neq -\frac{1}{2} \text{ or } w \neq 4$$

$$(2w+1)(w-4) \left(\frac{4}{w-4} - \frac{36}{(2w+1)(w-4)} \right) = (2w+1)(w-4) \left(\frac{3}{2w+1} \right)$$

$$(2w+1)(w-4) \left(\frac{4}{w-4} \right) - (2w+1)(w-4) \left(\frac{36}{(2w+1)(w-4)} \right) = (2w+1)(w-4) \left(\frac{3}{2w+1} \right)$$

$$8w + 4 - 36 = 3w - 12$$

$$8w - 32 = 3w - 12$$

$$5w - 32 = -12$$

$$5w = 20$$

$$w = 4$$

$\{ \ \}$ no solution. ($w = 4$ does not check because the denominator is zero.)

Section 5.6   Applications of Rational Equations and Proportions

1.   **a.** proportion
        b. proportional

3.

$$2 + \frac{6}{x} = x + 7$$

$\text{LCD} = x \quad$ so $x \neq 0$

$$x \left(2 + \frac{6}{x} \right) = x(x+7)$$

$$x(2) + x \left(\frac{6}{x} \right) = x^2 + 7x$$

$$2x + 6 = x^2 + 7x$$

$$0 = x^2 + 5x - 6$$

$$(x+6)(x-1) = 0$$

$$x + 6 = 0 \text{ or } x - 1 = 0$$

$$x = -6 \text{ or } \quad x = 1 \quad \{-6, 1\}$$

5.

$$\frac{4}{5t-1}+\frac{1}{10t-2}=\frac{4}{5t-1}+\frac{1}{2(5t-1)}$$

$$LCD = 2(5t-1)$$

$$=\frac{4}{5t-1}\cdot\frac{2}{2}+\frac{1}{2(5t-1)}$$

$$=\frac{8+1}{2(5t-1)}=\frac{9}{2(5t-1)}$$

7.

$$\frac{5}{w-2}=7-\frac{10}{w+2}\qquad LCD=(w+2)(w-2)\quad\text{so }w\neq-2\text{ or }w\neq2$$

$$(w+2)(w-2)\left(\frac{5}{w-2}\right)=(w+2)(w-2)\left(7-\frac{10}{w+2}\right)$$

$$(w+2)(\cancel{w-2})\left(\frac{5}{\cancel{w-2}}\right)=(w+2)(w-2)(7)-(\cancel{w+2})(w-2)\left(\frac{10}{\cancel{w+2}}\right)$$

$$5w+10=7(w^2-4)-10w+20$$

$$5w+10=7w^2-28-10w+20$$

$$0=7w^2-15w-18$$

$$(7w+6)(w-3)=0$$

$$7w+6=0\ \text{ or }\ w-3=0$$

$$7w=-6\ \text{ or }\ w=3$$

$$w=-\frac{6}{7}\ \text{ or }\ w=3\quad\left\{-\frac{6}{7},\,3\right\}$$

9.

$$\frac{8p^2-32}{p^2-4p+4}\cdot\frac{3p^2-3p-6}{2p^2+20p+32}=\frac{8(p^2-4)}{p^2-4p+4}\cdot\frac{3(p^2-p-2)}{2(p^2+10p+16)}$$

$$=\frac{4\cdot\cancel{2}\,(p+2)\,\cancel{(p-2)}}{\cancel{(p-2)}\,(p-2)}\cdot\frac{3\,\cancel{(p-2)}\,(p+1)}{\cancel{2}\,(p+8)\,\cancel{(p+2)}}=\frac{12(p+1)}{p+8}$$

11.

$$\frac{y}{6}=\frac{20}{15}\qquad LCD=30$$

$$30\left(\frac{y}{6}\right)=30\left(\frac{20}{15}\right)$$

$$5y=40$$

$$y=8\quad\{8\}$$

13.

$$\frac{9}{75}=\frac{m}{50}\qquad LCD=150$$

$$150\left(\frac{9}{75}\right)=150\left(\frac{m}{50}\right)$$

$$18=3m$$

$$m=6\quad\{6\}$$

15.
$$\frac{p-1}{4} = \frac{p+3}{3} \qquad \text{LCD} = 12$$
$$12\left(\frac{p-1}{4}\right) = 12\left(\frac{p+3}{3}\right)$$
$$3(p-1) = 4(p+3)$$
$$3p-3 = 4p+12$$
$$-15 = p \quad \{-15\}$$

17.
$$\frac{x+1}{5} = \frac{4}{15} \qquad \text{LCD} = 15$$
$$15\left(\frac{x+1}{5}\right) = 15\left(\frac{4}{15}\right)$$
$$3(x+1) = 4$$
$$3x+3 = 4$$
$$3x = 1$$
$$x = \frac{1}{3} \quad \left\{\frac{1}{3}\right\}$$

19.
$$\frac{5-2x}{x} = \frac{1}{4} \qquad \text{LCD} = 4x$$
$$4x\left(\frac{5-2x}{x}\right) = 4x\left(\frac{1}{4}\right)$$
$$4(5-2x) = x$$
$$20-8x = x$$
$$20 = 9x$$
$$x = \frac{20}{9} \quad \left\{\frac{20}{9}\right\}$$

21.
$$\frac{2}{y-1} = \frac{y-3}{4} \qquad \text{LCD} = 4(y-1)$$
$$4(y-1)\left(\frac{2}{y-1}\right) = 4(y-1)\left(\frac{y-3}{4}\right)$$
$$8 = y^2 - 4y + 3$$
$$0 = y^2 - 4y - 5$$
$$(y-5)(y+1) = 0$$
$$y-5 = 0 \quad \text{or} \quad y+1 = 0$$
$$y = 5 \quad \text{or} \quad y = -1 \quad \{5, -1\}$$

23.
$$\frac{1}{49w} = \frac{w}{9} \qquad \text{LCD} = 9 \cdot 49w$$
$$9 \cdot 49w\left(\frac{1}{49w}\right) = 9 \cdot 49w\left(\frac{w}{9}\right)$$
$$9 = 49w^2$$
$$0 = 49w^2 - 9$$
$$(7w+3)(7w-3) = 0$$
$$7w+3 = 0 \quad \text{or} \quad 7w-3 = 0$$
$$7w = -3 \quad \text{or} \quad 7w = 3$$
$$w = -\frac{3}{7} \quad \text{or} \quad w = \frac{3}{7} \quad \left\{-\frac{3}{7}, \frac{3}{7}\right\}$$

25.
$$\frac{x+3}{5x+26} = \frac{2}{x+4}$$
$$\text{LCD} = (5x+26)(x+4)$$
$$(5x+26)(x+4)\left(\frac{x+3}{5x+26}\right)$$
$$= (5x+26)(x+4)\left(\frac{2}{x+4}\right)$$
$$x^2 + 7x + 12 = 10x + 52$$
$$x^2 - 3x - 40 = 0$$
$$(x-8)(x+5) = 0$$
$$x-8 = 0 \quad \text{or} \quad x+5 = 0$$
$$x = 8 \quad \text{or} \quad x = -5 \quad \{8, -5\}$$

27. Let a = the number of adults
$$\frac{3}{1} = \frac{18}{a} \qquad \text{LCD} = a$$
$$a\left(\frac{3}{1}\right) = a\left(\frac{18}{a}\right)$$
$$3a = 18$$
$$a = 6$$
6 adults must be on the staff.

29. Let x = the number of grams of fat
$$\frac{3.5}{21.0} = \frac{14}{x} \qquad \text{LCD} = 21x$$
$$21x\left(\frac{3.5}{21.0}\right) = 21x\left(\frac{14}{x}\right)$$
$$3.5x = 294$$
$$x = 84$$
The 14-oz box of candy contains 84 g of fat.

31. Let x = the number of fish

$$\frac{8}{1840} = \frac{x}{230{,}000}$$

$$LCD = 230{,}000$$

$$230{,}000\left(\frac{8}{1840}\right) = 230{,}000\left(\frac{x}{230{,}000}\right)$$

$$1000 = x$$

1000 swordfish were caught.

33. Let x = the number of gallons of gas

$$\frac{243}{4.5} = \frac{621}{x} \qquad LCD = 4.5x$$

$$4.5x\left(\frac{243}{4.5}\right) = 4.5x\left(\frac{621}{x}\right)$$

$$243x = 2794.5$$

$$x = 11.5$$

Pam needs 11.5 gallons of gas.

35. Let x = the total number of bison

$$\frac{x}{200} = \frac{120}{6} \qquad LCD = 600$$

$$600\left(\frac{x}{200}\right) = 600\left(\frac{120}{6}\right)$$

$$3x = 12000$$

$$x = 4000$$

There are approximately 4000 bison in the park.

37. Let x = the number of men

$186 - x$ = the number of women

$$\frac{1}{5} = \frac{x}{186-x} \qquad LCD = 186-x$$

$$5(186-x)\left(\frac{1}{5}\right) = 5(186-x)\left(\frac{x}{186-x}\right)$$

$$186 - x = 5x$$

$$186 = 6x$$

$$x = 31$$

39. Let x = the number of women

$1095 - x$ = the number of men

$$\frac{119}{100} = \frac{1095-x}{x} \qquad LCD = 100x$$

$$100x\left(\frac{119}{100}\right) = 100x\left(\frac{1095-x}{x}\right)$$

$$119x = 109{,}500 - 100x$$

$$219x = 109{,}500$$

$$x = 500$$

$$1095 - x = 1095 - 500 = 595$$

There are 595 men and 500 women in the group.

41.

$$\frac{11.2}{a} = \frac{14}{10}$$

$$10a\left(\frac{11.2}{a}\right) = 10a\left(\frac{14}{10}\right)$$

$$112 = 14a$$

$$a = 8 \text{ ft}$$

$$\frac{b}{6} = \frac{14}{10}$$

$$30\left(\frac{b}{6}\right) = 30\left(\frac{14}{10}\right)$$

$$5b = 42$$

$$b = 8.4 \text{ ft}$$

43.

$$\frac{1.75}{5} = \frac{4.55}{y}$$

$$5y\left(\frac{1.75}{5}\right) = 5y\left(\frac{4.55}{y}\right)$$

$$1.75y = 22.75$$

$$y = 13 \text{ in}$$

$$(1.75)^2 + z^2 = (4.55)^2$$

$$3.0625 + z^2 = 20.7025$$

$$z^2 = 17.64$$

$$z = 4.2 \text{ in}$$

$$\frac{1.75}{5} = \frac{4.2}{x}$$

$$5x\left(\frac{1.75}{5}\right) = 5x\left(\frac{4.2}{x}\right)$$

$$1.75x = 21$$

$$x = 12 \text{ in}$$

45. Let x = the number

47. Let x = the number

$$\frac{1}{x} + 5 = \frac{16}{3} \quad \text{LCD} = 3x \quad \text{so } x \neq 0$$

$$3x\left(\frac{1}{x} + 5\right) = 3x\left(\frac{16}{3}\right)$$

$$3\cancel{x}\left(\frac{1}{\cancel{x}}\right) + 3x(5) = \cancel{3}x\left(\frac{16}{\cancel{3}}\right)$$

$$3 + 15x = 16x$$

$$3 = x$$

$$7 - \frac{1}{x} = \frac{9}{2}$$

$$\text{LCD} = 2x \quad \text{so } x \neq 0$$

$$2x\left(7 - \frac{1}{x}\right) = 2x\left(\frac{9}{2}\right)$$

$$2x(7) - 2\cancel{x}\left(\frac{1}{\cancel{x}}\right) = \cancel{2}x\left(\frac{9}{\cancel{2}}\right)$$

$$14x - 2 = 9x$$

$$5x = 2$$

$$x = \frac{2}{5}$$

49. **a.** $x + 7$

  b. $\dfrac{48}{x}$

  c. $\dfrac{83}{x + 7}$

51. Let x = the speed in rainstorm
$x + 20$ = the speed in sunny weather

	Distance	Rate	Time
Rain	80	x	$80/x$
Sunny	120	$x + 20$	$120/(x + 20)$

(Time rain) = (Time sunny)

$$\frac{80}{x} = \frac{120}{x + 20} \quad \text{LCD} = x(x + 20)$$

$$x(x + 20)\left(\frac{80}{x}\right) = x(x + 20)\left(\frac{120}{x + 20}\right)$$

$$80x + 1600 = 120x$$

$$1600 = 40x$$

$$x = 40$$

$$x + 20 = 40 + 20 = 60$$

The motorist drives 40 mph in the rainstorm and 60 mph in sunny weather.

53. Let x = the speed of the Broadmoor truck
$x + 6.4$ = the speed of the Wescott truck

	Distance	Rate	Time
Broadmoor	88	x	$88/x$
Wescott	96	$x + 6.4$	$96/(x + 6.4)$

(Time Broadmoor) = (Time Wescott)

$$\frac{88}{x} = \frac{96}{x + 6.4}$$

$$\text{LCD} = x(x + 6.4)$$

$$x(x + 6.4)\left(\frac{88}{x}\right) = x(x + 6.4)\left(\frac{96}{x + 6.4}\right)$$

$$88x + 563.2 = 96x$$

$$563.2 = 8x$$

$$x = 70.4$$

$$x + 6.4 = 70.4 + 6.4 = 76.8$$

The Broadmoor truck travels 70.4 mph and

55. Let x = the speed against the wind
$x + 5$ = the speed with the wind

$$\frac{30}{x} + \frac{30}{x + 5} = 5$$

$$\text{LCD} = x(x + 5)$$

$$x(x + 5)\left(\frac{30}{x} + \frac{30}{x + 5}\right) = x(x + 5)(5)$$

$$30(x + 5) + 30x = 5x(x + 5)$$

$$30x + 150 + 30x = 5x^2 + 25x$$

$$5x^2 - 35x - 150 = 0$$

$$5(x^2 - 7x - 30) = 0$$

$$5(x - 10)(x + 3) = 0$$

$$x - 10 = 0 \quad \text{or} \quad x + 3 = 0$$

$$x = 10 \quad \text{or} \quad x = -3$$

the Wescott truck travels 76.8 mph.

The cyclist rides at a speed of 10 mph against the wind.

57. Let x = Celeste's walking speed
$\quad\quad x + 2$ = speed on moving walkway

	Distance	Rate	Time
Off walkway	100	x	$100/x$
On walkway	140	$x + 2$	$140/(x + 2)$

(Time off walkway)+(Time on walkway) = 40

$$\frac{100}{x} + \frac{140}{x+2} = 40 \quad LCD = x(x+2)$$

$$x(x+2)\left(\frac{100}{x} + \frac{140}{x+2}\right) = x(x+2)(40)$$

$$100(x+2) + 140x = 40x(x+2)$$

$$100x + 200 + 140x = 40x^2 + 80x$$

$$240x + 200 = 40x^2 + 80x$$

$$0 = 40x^2 - 160x - 200$$

$$0 = 40\left(x^2 - 4x - 5\right)$$

$$0 = 40(x-5)(x+1)$$

$$x - 5 = 0 \quad \text{or} \quad x + 1 = 0$$

$$x = 5 \quad \text{or} \quad x \neq -1$$

$$x + 2 = 5 + 2 = 7$$

Celeste walks 5 ft/sec on the ground and travels 7 ft/sec while on the moving walkway.

59. Let x = Joe's speed
$\quad\quad x + 2$ = Beatrice's speed

	Distance	Rate	Time
Joe	12	x	$12/x$
Beatrice	12	$x + 2$	$12/(x + 2)$

(Joe's time) – (Beatrice's time) = 0.5

$$\frac{12}{x} - \frac{12}{x+2} = \frac{1}{2} \quad LCD = 2x(x+2)$$

$$2x(x+2)\left(\frac{12}{x} - \frac{12}{x+2}\right) = 2x(x+2)\left(\frac{1}{2}\right)$$

$$24(x+2) - 24x = x(x+2)$$

$$24x + 48 - 24x = x^2 + 2x$$

$$48 = x^2 + 2x$$

$$0 = x^2 + 2x - 48$$

$$0 = (x-6)(x+8)$$

$$x - 6 = 0 \quad \text{or} \quad x + 8 = 0$$

$$x = 6 \quad \text{or} \quad x \neq -8$$

$$x + 2 = 6 + 2 = 8$$

Joe runs at 6 mph and Beatrice runs at 8 mph.

61.

	Work Rate	Time	Portion of Job Comp
Paint#1	1/6	x	$(1/6)x$
Paint#2	1/8	x	$(1/8)x$

(Paint#1 Part) + (Paint#2 Part) = (1 Job)

$$\frac{1}{6}x + \frac{1}{8}x = 1 \quad LCD = 24$$

$$24\left(\frac{1}{6}x + \frac{1}{8}x\right) = 24(1)$$

$$4x + 3x = 24$$

$$7x = 24$$

$$x = \frac{24}{7} \text{ hr or } 3\frac{3}{7} \text{ hr}$$

Together, the painters can paint the room in $3\frac{3}{7}$ hr.

63.

	Work Rate	Time	Portion of Job Comp
Joel	1/12	x	$(1/12)x$
Michael	1/15	x	$(1/15)x$

(Joel's Part) + (Michael's Part) = (1 Job)

$$\frac{1}{12}x + \frac{1}{15}x = 1 \quad LCD = 60$$

$$60\left(\frac{1}{12}x + \frac{1}{15}x\right) = 60(1)$$

$$5x + 4x = 60$$

$$9x = 60$$

$$x = \frac{60}{9} = \frac{20}{3} \text{ hr or } 6\frac{2}{3} \text{ hr}$$

Together, they can fence the yard in $6\frac{2}{3}$ hr.

65.   a.

	Work Rate	Time	Part of Job Comp
Old	1/30	12	$(1/30)\,12$
New	1/x	12	$(1/x)\,12$

(Old Part) + (New Part) = (1 Job)

	Work Rate	Time	Part of Job Comp
Gus	1/x	4	$(1/x)\,4$
Sid	1/(2x)	4	$(1/(2x))\,4$

(Gus's Part) + (Sid's Part) = (1 Job)

$$\frac{1}{30} \cdot 12 + \frac{1}{x} \cdot 12 = 1 \qquad \text{LCD} = 30x$$

$$30x\left(\frac{1}{30} \cdot 12 + \frac{1}{x} \cdot 12\right) = 30x(1)$$

$$12x + 360 = 30x$$

$$360 = 18x$$

$$x = 20 \text{ hr}$$

b. The new pump will take 20 hr.
The technician should return at noon on Friday.

$$\frac{1}{x} \cdot 4 + \frac{1}{2x} \cdot 4 = 1 \qquad \text{LCD} = 2x$$

$$2x\left(\frac{1}{x} \cdot 4 + \frac{1}{2x} \cdot 4\right) = 2x(1)$$

$$8 + 4 = 2x$$

$$12 = 2x$$

$$x = 6 \text{ hr}$$

$$2x = 2(6) = 12 \text{ hr}$$

Gus would take 6 hr and Sid would take 12 hr to dig the garden.

Section 5.7   Variation

1. **a.** kx

    b. $\dfrac{k}{x}$

    c. kxw

3.
$$\frac{8}{y} = \frac{6}{11}$$

$$11y\left(\frac{8}{y}\right) = 11y\left(\frac{6}{11}\right)$$

$$88 = 6y$$

$$y = \frac{88}{6} = \frac{44}{3}$$

5.
$$\frac{2}{3} = \frac{x-4}{2}$$

$$6\left(\frac{2}{3}\right) = 6\left(\frac{x-4}{2}\right)$$

$$4 = 3x - 12$$

$$16 = 3x$$

$$x = \frac{16}{3}$$

7.
$$\frac{k}{k+1} = \frac{1}{9}$$

$$9(k+1)\left(\frac{k}{k+1}\right) = 9(k+1)\left(\frac{1}{9}\right)$$

$$9k = k+1$$

$$8k = 1$$

$$k = \frac{1}{8}$$

9. Inversely

11. $T = kq$

13. $b = \dfrac{k}{c}$

15. $Q = \dfrac{kx}{y}$

17. $c = kst$

19. $L = kw\sqrt{v}$

21. $x = \dfrac{ky^2}{z}$

23.
$$y = kx$$

$$18 = k(4)$$

$$k = \frac{18}{4} = \frac{9}{2}$$

25.
$$p = \frac{k}{q}$$
$$32 = \frac{k}{16}$$
$$k = 32(16) = 512$$

27.
$$y = kwv$$
$$8.75 = k(50)(0.1)$$
$$8.75 = 5k$$
$$k = \frac{8.75}{5} = 1.75$$

29.
$$x = kp$$
$$50 = k(10)$$
$$k = \frac{50}{10} = 5$$
$$x = 5(14) = 70$$

31.
$$b = \frac{k}{c}$$
$$4 = \frac{k}{3}$$
$$k = 4 \cdot 3 = 12$$
$$b = \frac{12}{2} = 6$$

33.
$$Z = kw^2$$
$$14 = k(4)^2$$
$$14 = 16k$$
$$k = \frac{14}{16} = \frac{7}{8}$$
$$Z = \frac{7}{8}(8)^2 = \frac{7}{8}(64) = 56$$

35.
$$Q = \frac{k}{p^2}$$
$$4 = \frac{k}{3^2}$$
$$k = 4 \cdot 3^2 = 4 \cdot 9 = 36$$
$$Q = \frac{36}{2^2} = \frac{36}{4} = 9$$

37.
$$L = ka\sqrt{b}$$
$$72 = k(8)\sqrt{9}$$
$$72 = k \cdot 8 \cdot 3$$
$$72 = 24k$$
$$k = 3$$
$$L = 3\left(\frac{1}{2}\right)\sqrt{36} = 3\left(\frac{1}{2}\right)(6) = 9$$

39.
$$B = \frac{km}{n}$$
$$20 = \frac{k \cdot 10}{3}$$
$$k = \frac{20 \cdot 3}{10} = \frac{60}{10} = 6$$
$$B = \frac{6 \cdot 15}{12} = \frac{90}{12} = \frac{15}{2}$$

41.
$$m = kw$$
$$3 = k(150)$$
$$k = \frac{3}{150} = \frac{1}{50}$$

a. $m = \frac{1}{50} \cdot 180 = 3.6$ g should be prescribed.

b. $m = \frac{1}{50} \cdot 225 = 4.5$ g should be prescribed.

c. $m = \frac{1}{50} \cdot 120 = 2.4$ g should be prescribed.

43.
$$c = \frac{k}{n}$$
$$0.48 = \frac{k}{5000}$$
$$k = 0.48(5000) = 2400$$

a. $c = \frac{2400}{6000} = 0.40$ Unit cost is $0.40

b. $c = \frac{2400}{8000} = 0.30$ Unit cost is $0.30

c. $c = \frac{2400}{2400} = 1.00$ Unit cost is $1.00

45. Let A = the amount of pollution
P = the number of people
$$A = kP$$
$$56,800 = k(80,000)$$
$$k = \frac{56,800}{80,000} = 0.71$$
$$A = 0.71(500,000) = 355,000 \text{ tons}$$
355,000 tons enter the atmosphere.

47. Let I = the intensity of the light source
d = the distance from the source
$$I = \frac{k}{d^2}$$
$$400 = \frac{k}{5^2}$$
$$k = 400 \cdot 5^2 = 400 \cdot 25 = 10,000$$
$$I = \frac{10,000}{8^2} = \frac{10,000}{64} = 156.25 \text{ lumens/m}^2$$
The intensity is 156.25 lumens/m^2.

49. Let I = the current
V = the voltage
R = the resistance
$$I = \frac{kV}{R}$$
$$9 = \frac{k(90)}{10}$$
$$k = \frac{9 \cdot 10}{90} = \frac{90}{90} = 1$$
$$I = \frac{1 \cdot 185}{10} = \frac{185}{10} = 18.5 \text{ A}$$
The current is 18.5 A.

51. Let I = the interest
P = the principal
T = the time
$$I = kPt$$
$$500 = k(2500)(4)$$
$$500 = 10,000k$$
$$k = \frac{500}{10,000} = 0.05$$
$$I = 0.05(7000)(10) = \$3500$$
\$3500 in interest will be earned.

53. Let d = the stopping distance
s = the speed of the car
$$d = ks^2$$
$$109 = k(40)^2$$
$$k = \frac{109}{40^2} = \frac{109}{1600}$$
$$d = \frac{109}{1600}(25)^2 = \frac{109}{1600}(625) \approx 42.6 \text{ ft}$$
The stopping distance will be 42.6 ft.

55. Let S = the surface area
L = the length of an edge
$$S = kL^2$$
$$24 = k(2)^2$$
$$k = \frac{24}{2^2} = \frac{24}{4} = 6$$
$$S = 6(5)^2 = 6(25) = 150 \text{ ft}^2$$
The surface area is 150 ft^2

57. Let P = the power
I = the current
R = the resistance
$$P = kIR^2$$
$$144 = k(4)(6)^2$$
$$k = \frac{144}{4(6)^2} = \frac{144}{144} = 1$$
$$P = 1(3)(10)^2 = 3 \cdot 100 = 300 \text{ W}$$
The power is 300 W.

59. **a.** $A = kl^2$

b. $A = k(2l)^2 = k(4l^2) = 4kl^2$
The area is 4 times the original.

c. $A = k(3l)^2 = k(9l^2) = 9kl^2$
The area is 9 times the original.

Chapter 5 Group Activity

1. $R = \$150$

$I = \dfrac{0.06}{12} = 0.005$

$n = 30(12) = 360$

$S = 150\left[\dfrac{(1 + 0.005)^{360} - 1}{0.005}\right] = \$150,677.26$

3.

$A = 50,000 \div \left[\dfrac{(1 + 0.09/12)^{180} - 1}{0.09/12}\right] = \$132.13$

Approximately \$133 must be saved each month.

Chapter 5 Review Exercises

Section 5.1

1.

$k(y) = \dfrac{y}{y^2 - 1}$

a.

$k(2) = \dfrac{2}{2^2 - 1} = \dfrac{2}{4 - 1} = \dfrac{2}{3}$

$k(0) = \dfrac{0}{0^2 - 1} = \dfrac{0}{0 - 1} = \dfrac{0}{-1} = 0$

$k(1) = \dfrac{1}{1^2 - 1} = \dfrac{1}{1 - 1} = \dfrac{1}{0}$ is undefined

$k(-1) = \dfrac{-1}{(-1)^2 - 1} = \dfrac{-1}{1 - 1} = \dfrac{-1}{0}$ undefined

$k\left(\dfrac{1}{2}\right) = \dfrac{\dfrac{1}{2}}{\left(\dfrac{1}{2}\right)^2 - 1} = \dfrac{\dfrac{1}{2}}{\dfrac{1}{4} - \dfrac{4}{4}} = \dfrac{\dfrac{1}{2}}{-\dfrac{3}{4}}$

$= \dfrac{1}{2} \cdot \left(-\dfrac{4}{3}\right) = -\dfrac{2}{3}$

b.

$(-\infty,\, -1) \cup (-1,\, 1) \cup (1,\, \infty)$

3. $\dfrac{28a^3b^3}{14a^2b^3} = 2a^{3-2}b^{3-3} = 2a^1b^0 = 2a$

5. $\dfrac{x^2 - 4x + 3}{x - 3} = \dfrac{(x - 3)(x - 1)}{x - 3} = x - 1$

7. $\dfrac{x^3 - 27}{9 - x^2} = \dfrac{x^3 - 3^3}{-(x^2 - 9)} = \dfrac{(x - 3)(x^2 + }{-(x + 3)(x}$

$= -\dfrac{x^2 + 3x + 9}{x + 3}$

9. $\dfrac{2t^2 + 3t - 5}{7 - 6t - t^2} = \dfrac{(2t + 5)(t - 1)}{-(t^2 + 6t - 7)}$

$= \dfrac{(2t + 5)(t - 1)}{-(t + 7)(t - 1)} = -\dfrac{2t + 5}{t + 7}$

11. $f(x) = \dfrac{1}{x - 3}$

$x - 3 = 0$

$x = 3$

$(-\infty,\, 3) \cup (3,\, \infty)$

Graph: c

13.

$$k(x) = \frac{6}{x^2 - 3x} = \frac{6}{x(x-3)}$$

$$x(x-3) = 0$$

$$x = 0 \text{ or } x - 3 = 0$$

$$x = 0 \text{ or } x = 3$$

$$(-\infty, 0) \cup (0, 3) \cup (3, \infty)$$

Graph: b

Section 5.2

15.

$$\frac{3a+9}{a^2} \cdot \frac{a^3}{6a+18} = \frac{3(a+3)}{a^2} \cdot \frac{a^2 \cdot a}{2 \cdot 3(a+3)}$$

$$= \frac{a}{2}$$

17.

$$\frac{x-4y}{x^2+xy} \div \frac{20y-5x}{x^2-y^2} = \frac{x-4y}{x^2+xy} \cdot \frac{x^2-y^2}{20y-5x}$$

$$= \frac{x-4y}{x(x+y)} \cdot \frac{(x+y)(x-y)}{-5(x-4y)} = -\frac{x-y}{5x}$$

19.

$$\frac{7k+28}{2k+4} \cdot \frac{k^2-2k-8}{k^2+2k-8}$$

$$= \frac{7(k+4)}{2(k+2)} \cdot \frac{(k-4)(k+2)}{(k+4)(k-2)} = \frac{7(k-4)}{2(k-2)}$$

21.

$$\frac{x^2+8x-20}{x^2+6x-16} \div \frac{x^2+6x-40}{x^2+3x-40} = \frac{x^2+8x-20}{x^2+6x-16} \cdot \frac{x^2+3x-40}{x^2+6x-40} = \frac{(x+10)(x-2)}{(x+8)(x-2)} \cdot \frac{(x+8)(x-5)}{(x+10)(x-4)} = \frac{x-5}{x-4}$$

23.

$$\frac{2w}{21} \div \frac{3w^2}{7} \cdot \frac{4}{w} = \frac{2w}{21} \cdot \frac{7}{3w^2} \cdot \frac{4}{w} = \frac{2w}{7 \cdot 3} \cdot \frac{7}{3w^2} \cdot \frac{4}{w} = \frac{8}{9w^2}$$

25.

$$\frac{x^2+x-20}{x^2-4x+4} \cdot \frac{x^2+x-6}{12+x-x^2} \div \frac{2x+10}{10-5x} = \frac{x^2+x-20}{x^2-4x+4} \cdot \frac{x^2+x-6}{-(x^2-x-12)} \cdot \frac{10-5x}{2x+10}$$

$$= \frac{(x+5)(x-4)}{(x-2)(x-2)} \cdot \frac{(x+3)(x-2)}{-(x-4)(x+3)} \cdot \frac{-5(x-2)}{2(x+5)} = \frac{-5}{-2} = \frac{5}{2}$$

Section 5.3

27.

$$\frac{1}{x} + \frac{1}{x^2} - \frac{1}{x^3} \qquad \text{LCD} = x^3$$

$$= \frac{1}{x} \cdot \frac{x^2}{x^2} + \frac{1}{x^2} \cdot \frac{x}{x} - \frac{1}{x^3} = \frac{x^2+x-1}{x^3}$$

29.

$$\frac{y}{2y-1} + \frac{3}{1-2y} = \frac{y}{2y-1} + \frac{3}{1-2y} \cdot \frac{(-1)}{(-1)}$$

$$= \frac{y}{2y-1} + \frac{-3}{2y-1} = \frac{y-3}{2y-1}$$

31. $\dfrac{4k}{k^2+2k+1}+\dfrac{3}{k^2-1}=\dfrac{4k}{(k+1)^2}+\dfrac{3}{(k+1)(k-1)}$ $\text{LCD}=(k+1)^2(k-1)$

$\qquad\qquad=\dfrac{4k}{(k+1)^2}\cdot\dfrac{k-1}{k-1}+\dfrac{3}{(k+1)(k-1)}\cdot\dfrac{k+1}{k+1}=\dfrac{4k^2-4k+3k+3}{(k+1)^2(k-1)}=\dfrac{4k^2-k+3}{(k+1)^2(k-1)}$

33. $\dfrac{2}{a+3}+\dfrac{2a^2-2a}{a^2-2a-15}=\dfrac{2}{a+3}+\dfrac{2a^2-2a}{(a-5)(a+3)}=\dfrac{2}{a+3}\cdot\dfrac{a-5}{a-5}+\dfrac{2a^2-2a}{(a-5)(a+3)}=\dfrac{2a-10+2a^2-2a}{(a-5)(a+3)}$

$\qquad\qquad=\dfrac{2a^2-10}{(a-5)(a+3)}=\dfrac{2(a^2-5)}{(a-5)(a+3)}$

35. $\dfrac{2}{3x-5}-8=\dfrac{2}{3x-5}-8\cdot\dfrac{3x-5}{3x-5}=\dfrac{2-24x+40}{3x-5}=\dfrac{-24x+42}{3x-5}=\dfrac{-6(4x-7)}{3x-5}$

37. $\dfrac{6a}{3a^2-7a+2}+\dfrac{2}{1-3a}+\dfrac{3a}{a-2}$

$\qquad=\dfrac{6a}{(3a-1)(a-2)}+\dfrac{-2}{3a-1}+\dfrac{3a}{a-2}$

$\qquad\quad\text{LCD}=(3a-1)(a-2)$

$\qquad=\dfrac{6a}{(3a-1)(a-2)}+\dfrac{-2}{3a-1}\cdot\dfrac{a-2}{a-2}+\dfrac{3a}{a-2}\cdot\dfrac{3a-1}{3a-1}$

$\qquad=\dfrac{6a}{(3a-1)(a-2)}+\dfrac{-2a+4}{(3a-1)(a-2)}+\dfrac{9a^2-3a}{(3a-1)(a-2)}$

$\qquad=\dfrac{6a-2a+4+9a^2-3a}{(3a-1)(a-2)}=\dfrac{9a^2+a+4}{(3a-1)(a-2)}$

Section 5.4

39. $\dfrac{\dfrac{2x}{3x^2-3}}{\dfrac{4x}{6x-6}}=\dfrac{2x}{3(x^2-1)}\cdot\dfrac{6x-6}{4x}$

$\qquad=\dfrac{2x}{3(x+1)(x-1)}\cdot\dfrac{3\cdot2(x-1)}{2\cdot2x}=\dfrac{1}{x+1}$

41. $\dfrac{\dfrac{2}{x}+\dfrac{1}{xy}}{\dfrac{4}{x^2}}$ $\text{LCD}=x^2y$

$\qquad=\dfrac{x^2y\left(\dfrac{2}{x}+\dfrac{1}{xy}\right)}{x^2y\left(\dfrac{4}{x^2}\right)}=\dfrac{x^2y\left(\dfrac{2}{x}\right)+x^2y\left(\dfrac{1}{xy}\right)}{x^2y\left(\dfrac{4}{x^2}\right)}$

$\qquad=\dfrac{2xy+x}{4y}=\dfrac{x(2y+1)}{4y}$

43.

$$\dfrac{\dfrac{1}{a-1}+1}{\dfrac{1}{a+1}-1} \qquad \text{LCD} = (a+1)(a-1)$$

$$= \dfrac{(a+1)(a-1)\left(\dfrac{1}{a-1}+1\right)}{(a+1)(a-1)\left(\dfrac{1}{a+1}-1\right)} = \dfrac{(a+1)(\cancel{a-1})\left(\dfrac{1}{\cancel{a-1}}\right)+(a+1)(a-1)(1)}{(\cancel{a+1})(a-1)\left(\dfrac{1}{\cancel{a+1}}\right)-(a+1)(a-1)(1)}$$

$$= \dfrac{a+1+a^2-1}{a-1-\left(a^2-1\right)} = \dfrac{a^2+a}{a-1-a^2+1} = \dfrac{a^2+a}{-a^2+a} = \dfrac{\cancel{a}(a+1)}{-\cancel{a}(a-1)} = -\dfrac{a+1}{a-1}$$

45.

$$\dfrac{1+xy^{-1}}{x^2y^{-2}-1} = \dfrac{1+\dfrac{x}{y}}{\dfrac{x^2}{y^2}-1} \qquad \text{LCD} = y^2$$

$$= \dfrac{y^2\left(1+\dfrac{x}{y}\right)}{y^2\left(\dfrac{x^2}{y^2}-1\right)} = \dfrac{y^2(1)+y\cdot\cancel{y}\left(\dfrac{x}{\cancel{y}}\right)}{\cancel{y^2}\left(\dfrac{x^2}{\cancel{y^2}}\right)-y^2(1)} = \dfrac{y^2+xy}{x^2-y^2} = \dfrac{y(\cancel{y+x})}{(\cancel{x+y})(x-y)} = \dfrac{y}{x-y}$$

47.

$$m = \dfrac{-\dfrac{5}{3}-\left(-\dfrac{7}{4}\right)}{\dfrac{13}{6}-\dfrac{2}{3}} \qquad \text{LCD} = 12$$

$$= \dfrac{12\left(-\dfrac{5}{3}+\dfrac{7}{4}\right)}{12\left(\dfrac{13}{6}-\dfrac{2}{3}\right)} = \dfrac{-20+21}{26-8} = \dfrac{1}{18}$$

Section 5.5

49.

$$\frac{x+3}{x^2-x}-\frac{8}{x^2-1}=0$$

$$\frac{x+3}{x(x-1)}-\frac{8}{(x+1)(x-1)}=0$$

$LCD=x(x-1)(x+1)$ so $x\neq0$ or $x\neq1$ or $x\neq-1$

$$x(x-1)(x+1)\left(\frac{x+3}{x(x-1)}-\frac{8}{(x+1)(x-1)}\right)=x(x-1)(x+1)(0)$$

$$\cancel{x}(\cancel{x-1})(x+1)\left(\frac{x+3}{\cancel{x}(\cancel{x-1})}\right)-x(\cancel{x-1})(\cancel{x+1})\left(\frac{8}{(\cancel{x+1})(\cancel{x-1})}\right)=0$$

$$x^2+4x+3-8x=0$$
$$x^2-4x+3=0$$
$$(x-3)(x-1)=0$$
$$x-3=0 \text{ or } x-1=0$$
$$x=3 \text{ or } \quad x=1$$

$\{3\}$ is the solution. ($x=1$ does not check because the denominator is zero.)

51.

$$x-9=\frac{72}{x-8} \qquad LCD=(x-8) \qquad \text{so } x\neq8$$

$$(x-8)(x-9)=(\cancel{x-8})\left(\frac{72}{\cancel{x-8}}\right)$$

$$x^2-17x+72=72$$
$$x^2-17x=0$$
$$x(x-17)=0$$
$$x=0 \text{ or } x-17=0$$
$$x=0 \text{ or } \quad x=17 \quad \{0,17\}$$

53.

$$5y^{-2}+1=6y^{-1}$$
$$\frac{5}{y^2}+1=\frac{6}{y}$$

$LCD=y^2$ so $y\neq0$

$$y^2\left(\frac{5}{y^2}+1\right)=y^2\left(\frac{6}{y}\right)$$

$$\cancel{y^2}\left(\frac{5}{\cancel{y^2}}\right)+y^2(1)=y\cdot\cancel{y}\left(\frac{6}{\cancel{y}}\right)$$

$$5+y^2=6y$$
$$y^2-6y+5=0$$
$$(y-5)(y-1)=0$$
$$y-5=0 \text{ or } y-1=0$$
$$y=5 \text{ or } \quad y=1 \quad \{5,1\}$$

55.

$$c=\frac{ax+b}{x} \qquad \text{for } x$$
$$cx=ax+b$$
$$cx-ax=b$$
$$x(c-a)=b$$
$$x=\frac{b}{c-a}$$

Section 5.6

57. $\dfrac{5}{4} = \dfrac{x}{6}$ LCD = 12

$12\left(\dfrac{5}{4}\right) = 12\left(\dfrac{x}{6}\right)$

$15 = 2x$

$x = \dfrac{15}{2}$ $\left\{\dfrac{15}{2}\right\}$

59. $\dfrac{x+2}{3} = \dfrac{5(x+1)}{4}$ LCD = 12

$12\left(\dfrac{x+2}{3}\right) = 12\left(\dfrac{5(x+1)}{4}\right)$

$4(x+2) = 15(x+1)$

$4x + 8 = 15x + 15$

$-11x = 7$

$x = -\dfrac{7}{11}$ $\left\{-\dfrac{7}{11}\right\}$

61. Let y = the number of yards gained

$\dfrac{34}{357} = \dfrac{22}{y}$ LCD = $357y$

$357y\left(\dfrac{34}{357}\right) = 357y\left(\dfrac{22}{y}\right)$

$34y = 7854$

$y = 231$

Manning would gain 231 yd

63. Let x = the speed on first day

$x - 5$ = the speed on second day

	Distance	Rate	Time
First Day	100	x	100/x
Second Day	75	$x - 5$	75/($x - 5$)

(Time first) + (Time second) = 10

$\dfrac{100}{x} + \dfrac{75}{x-5} = 10$ LCD $= x(x-5)$

$x(x-5)\left(\dfrac{100}{x} + \dfrac{75}{x-5}\right) = x(x-5)(10)$

$100x - 500 + 75x = 10x^2 - 50x$

$0 = 10x^2 - 225x + 500$

$0 = 2x^2 - 45x + 100$

$0 = (2x - 5)(x - 20)$

$2x - 5 = 0$ or $x - 20 = 0$

$2x = 5$ or $x = 20$

$x \neq \dfrac{5}{2}$ or $x = 20$

$x - 5 = 20 - 5 = 15$

Tony averaged 20 mph the first day and 15 mph the second day.

65.

	Work Rate	Time	Part of Job Comp
Doug	1/8	x	$(1/8)\,x$
Jean	1/10	x	$(1/10)\,x$

(Doug's Part) + (Jean's Part) = (1 Job)

$$\frac{1}{8}\cdot x+\frac{1}{10}\cdot x=1 \quad LCD=40$$

$$40\left(\frac{1}{8}\cdot x+\frac{1}{10}\cdot x\right)=40(1)$$

$$5x+4x=40$$

$$9x=40$$

$$x=\frac{40}{9}\ hr\approx 4.4\ hr$$

Together it would take about 4.4 hours to complete the job.

Section 5.7

67. a. $F=kd$

b. $6=k(2)$

$k=3$

c. $F=3(4.2)=12.6$ lb

It requires 12.6 lb.

69. $y=kx\sqrt{z}$

$3=k(3)\sqrt{4}$

$3=k(3)(2)$

$k=\frac{3}{3\cdot 2}=\frac{3}{6}=\frac{1}{2}$

$y=\frac{1}{2}(8)\sqrt{9}=4\cdot 3=12$

Chapter 5 Test

1.
$$h(x)=\frac{2x-14}{x^2-49}$$

a. $h(0)=\frac{2(0)-14}{(0)^2-49}=\frac{0-14}{0-49}=\frac{-14}{-49}=\frac{2}{7}$

$h(5)=\frac{2(5)-14}{(5)^2-49}=\frac{10-14}{25-49}=\frac{-4}{-24}=\frac{1}{6}$

$h(7)=\frac{2(7)-14}{(7)^2-49}=\frac{14-14}{49-49}=\frac{0}{0}$
  is undefined

$h(-7)=\frac{2(-7)-14}{(-7)^2-49}=\frac{-14-14}{49-49}=\frac{-28}{0}$
  is undefined

b. $(-\infty,-7)\cup(-7,7)\cup(7,\infty)$

3.
$$f(x)=\frac{2x+6}{x^2-x-12}=\frac{2(x+3)}{(x-4)(x+3)}$$

a. $\{x|x\text{ is a real number and }x\neq 4,\ x\neq -3\}$

b.

$$f(x) = \frac{2x+6}{x^2-x-12} = \frac{2\cancel{(x+3)}}{(x-4)\cancel{(x+3)}}$$

$$= \frac{2}{x-4}$$

5.
$$\frac{9x^2-9}{3x^2+2x-5} = \frac{9\left(x^2-1\right)}{(3x+5)(x-1)}$$

$$= \frac{9(x+1)\cancel{(x-1)}}{(3x+5)\cancel{(x-1)}} = \frac{9(x+1)}{3x+5}$$

7.
$$\frac{2x-5}{25-4x^2} \cdot \left(2x^2-x-15\right)$$

$$= \frac{2x-5}{-\left(4x^2-25\right)} \cdot \frac{(2x+5)(x-3)}{1}$$

$$= \frac{\cancel{2x-5}}{-\cancel{(2x+5)}\cancel{(2x-5)}} \cdot \frac{\cancel{(2x+5)}(x-3)}{1}$$

$$= -(x-3)$$

9.
$$\frac{4x}{x+1} + x + \frac{2}{x+1} = \frac{4x}{x+1} + x \cdot \frac{x+1}{x+1} + \frac{2}{x+1}$$

$$= \frac{4x+x^2+x+2}{x+1} = \frac{x^2+5x+2}{x+1}$$

11.
$$\frac{2u^{-1}+2v^{-1}}{4u^{-3}+4v^{-3}} = \frac{\dfrac{2}{u}+\dfrac{2}{v}}{\dfrac{4}{u^3}+\dfrac{4}{v^3}} \qquad \text{LCD} = u^3v^3$$

$$= \frac{u^3v^3\left(\dfrac{2}{u}+\dfrac{2}{v}\right)}{u^3v^3\left(\dfrac{4}{u^3}+\dfrac{4}{v^3}\right)} = \frac{2u^2v^3+2u^3v^2}{4v^3+4u^3} = \frac{2u^2v^2(v+u)}{4\left(v^3+u^3\right)} = \frac{\cancel{2}u^2v^2\cancel{(v+u)}}{\cancel{2}\cdot 2\cancel{(v+u)}\left(v^2-vu+u^2\right)} = \frac{u^2v^2}{2\left(v^2-vu+u^2\right)}$$

13.
$$\frac{3}{x^2+8x+15} - \frac{1}{x^2+7x+12} - \frac{1}{x^2+9x+20} = \frac{3}{(x+5)(x+3)} - \frac{1}{(x+4)(x+3)} - \frac{1}{(x+5)(x+4)}$$

$$\text{LCD} = (x+5)(x+3)(x+4)$$

$$= \frac{3}{(x+5)(x+3)} \cdot \frac{x+4}{x+4} - \frac{1}{(x+4)(x+3)} \cdot \frac{x+5}{x+5} - \frac{1}{(x+5)(x+4)} \cdot \frac{x+3}{x+3} = \frac{3x+12-x-5-x-3}{(x+5)(x+3)(x+4)}$$

$$= \frac{\cancel{x+4}}{(x+5)(x+3)\cancel{(x+4)}} = \frac{1}{(x+5)(x+3)}$$

15.

$$\frac{3}{y^2-9}+\frac{4}{y+3}=1$$

$$\frac{3}{(y+3)(y-3)}+\frac{4}{y+3}=1$$

$$\text{LCD}=(y+3)(y-3) \quad \text{so } y\neq-3 \text{ or } y\neq3$$

$$(y+3)(y-3)\left(\frac{3}{(y+3)(y-3)}+\frac{4}{y+3}\right)=(y+3)(y-3)(1)$$

$$\cancel{(y+3)}\cancel{(y-3)}\left(\frac{3}{\cancel{(y+3)}\cancel{(y-3)}}\right)+\cancel{(y+3)}(y-3)\left(\frac{4}{\cancel{y+3}}\right)=(y+3)(y-3)$$

$$3+4y-12=y^2-9$$

$$0=y^2-4y$$

$$y(y-4)=0$$

$$y=0 \ \text{ or } \ y-4=0$$

$$y=0 \ \text{ or } \quad y=4 \quad \{0,4\}$$

17.

$$\frac{1+Tv}{T}=p \quad \text{for } T$$

$$1+Tv=pT$$

$$1=pT-Tv$$

$$1=T(p-v)$$

$$\frac{1}{p-v}=T$$

19. Let x = the number

$$\frac{1}{x}+3x=\frac{13}{2}$$

$$\text{LCD}=2x \quad \text{so } x\neq0$$

$$2x\left(\frac{1}{x}+3x\right)=2x\left(\frac{13}{2}\right)$$

$$2\cancel{x}\left(\frac{1}{\cancel{x}}\right)+2x(3x)=\cancel{2}x\left(\frac{13}{\cancel{2}}\right)$$

$$2+6x^2=13x$$

$$6x^2-13x+2=0$$

$$(6x-1)(x-2)=0$$

$$6x-1=0 \ \text{ or } \ x-2=0$$

$$6x=1 \ \text{ or } \quad x=2$$

$$x=\frac{1}{6} \ \text{ or } \quad x=2$$

21. Let x = the actual distance

$$\frac{8.2}{2820}=\frac{5.7}{x} \qquad \text{LCD}=2820x$$

$$2820x\left(\frac{8.2}{2820}\right)=2820x\left(\frac{5.7}{x}\right)$$

$$8.2x=16074$$

$$x\approx1960 \text{ mi}$$

The cities are 1960 mi apart.

23.

	Work Rate	Time	Portion of Job Comp
Barbara	1/4	x	(1/4)x
Jack	1/10	x	(1/10)x

(Barbara's Part) + (Jack's Part) = (1 Job)

$$\frac{1}{4}x+\frac{1}{10}x=1 \qquad \text{LCD}=20$$

$$20\left(\frac{1}{4}x+\frac{1}{10}x\right)=20(1)$$

$$5x+2x=20$$

$$7x=20$$

$$x=\frac{20}{7} \text{ hr or } 2\frac{6}{7} \text{ hr}$$

Together, they can type the chapter in $2\frac{6}{7}$ hr.

25..   Let P = the period of the pendulum
L = the length of the pendulum
$$P = k\sqrt{L}$$
$$2.2 = k\sqrt{4}$$
$$k = \frac{2.2}{\sqrt{4}} = \frac{2.2}{2} = 1.1$$
$$P = 1.1\sqrt{9} = 1.1(3) = 3.3 \text{ sec}$$
The period is 3.3 sec.

Chapters 1 – 5   Cumulative Review Exercises

1.

	-22	π	6	$-\sqrt{2}$
Real numbers	$-$	$-$	$-$	$-$
Irrational numbers		$-$		$-$
Rational numbers	$-$		$-$	
Integers	$-$		$-$	
Whole numbers			$-$	
Natural numbers			$-$	

3. $(2x-3)(x-4)-(x-5)^2$
$$= 2x^2 - 8x - 3x + 12 - \left(x^2 - 10x + 25\right)$$
$$= 2x^2 - 11x + 12 - x^2 + 10x - 25$$
$$= x^2 - x - 13$$

5.   Let w = the width of the pool
$2w - 10$ = the length of the pool
$$2w + 2(2w-10) = 160$$
$$2w + 4w - 20 = 160$$
$$6w = 180$$
$$w = 30$$
$$2w - 10 = 2(30) - 10 = 60 - 10 = 50$$
The length is 50 m and the width is 30 m.

7.   The slope of the line is 3 and the slope of the perpendicular line is $-\frac{1}{3}$.
$$y - 5 = -\frac{1}{3}(x - (-3))$$
$$y - 5 = -\frac{1}{3}x - 1$$
$$y = -\frac{1}{3}x + 4$$

9.
$$s = \frac{k}{t}$$
$$60 = \frac{k}{10}$$
$$600 = k$$

11. $64y^3 - 8z^6 = 8\left(8y^3 - z^6\right)$
$$= 8\left[(2y)^3 - \left(z^2\right)^3\right]$$
$$= 8\left(2y - z^2\right)\left(4y^2 + 2yz^2 + z^4\right)$$

$$s = \frac{600}{8} = 75$$

The speed of the car is 75 mph.

13. $\dfrac{2x^2 + 11x - 21}{4x^2 - 10x + 6} \div \dfrac{2x^2 - 98}{x^2 - x + xa - a} = \dfrac{2x^2 + 11x - 21}{2\left(2x^2 - 5x + 3\right)} \cdot \dfrac{x^2 - x + xa - a}{2x^2 - 98}$

$$= \frac{(2x-3)(x+7)}{2(2x-3)(x-1)} \cdot \frac{x(x-1) + a(x-1)}{2(x^2-49)} = \frac{\cancel{(2x-3)}\cancel{(x+7)}}{2\cancel{(2x-3)}\cancel{(x-1)}} \cdot \frac{\cancel{(x-1)}(x+a)}{2\cancel{(x+7)}(x-7)} = \frac{x+a}{4(x-7)}$$

15. $\dfrac{1 - \dfrac{49}{c^2}}{\dfrac{7}{c} + 1} \qquad LCD = c^2$

$$= \frac{c^2\left(1 - \dfrac{49}{c^2}\right)}{c^2\left(\dfrac{7}{c} + 1\right)} = \frac{c^2(1) - \cancel{c^2}\left(\dfrac{49}{\cancel{c^2}}\right)}{c \cdot \cancel{c}\left(\dfrac{7}{\cancel{c}}\right) + c^2(1)} = \frac{c^2 - 49}{7c + c^2} = \frac{(c+7)(c-7)}{c(c+7)} = \frac{c-7}{c}$$

17. $\dfrac{4y}{y+2} - \dfrac{y}{y-1} = \dfrac{9}{y^2 + y - 2}$

$\dfrac{4y}{y+2} - \dfrac{y}{y-1} = \dfrac{9}{(y+2)(y-1)} \qquad LCD = (y+2)(y-1) \quad$ so $y \neq -2 \ $ or $\ y \neq 1$

$$(y+2)(y-1)\left(\frac{4y}{y+2} - \frac{y}{y-1}\right) = (y+2)(y-1)\left(\frac{9}{(y+2)(y-1)}\right)$$

$$\cancel{(y+2)}(y-1)\left(\frac{4y}{\cancel{y+2}}\right) - (y+2)\cancel{(y-1)}\left(\frac{y}{\cancel{y-1}}\right) = \cancel{(y+2)}\cancel{(y-1)}\left(\frac{9}{\cancel{(y+2)}\cancel{(y-1)}}\right)$$

$$4y^2 - 4y - y^2 - 2y = 9$$
$$3y^2 - 6y - 9 = 0$$
$$3\left(y^2 - 2y - 3\right) = 0$$
$$3(y-3)(y+1) = 0$$
$$y - 3 = 0 \ \text{ or } \ y + 1 = 0$$
$$y = 3 \ \text{ or } \quad y = -1 \quad \{3, -1\}$$

19. a. $x = -5$ is a vertical line. The slope is undefined.

    b. $2y = 8$ is a horizontal line. The slope of the line is 0.

Chapter 6   Radicals and Complex Numbers

Are You Prepared?

1. $(4x-3)(x+5)= 4x^2 +20x -3x -15$
$= 4x^2 +17x -15$

2. $4x^2(3x^2 +5x +1)= 12x^4 +20x^3 +4x^2$

3. $(4x-3)(4x+3)=(4x)^2 -3^2 =16x^2 -9$

4. $(4x-3)^2 =(4x)^2 -2(4x)(3)+3^2$
$=16x^2 -24x +9$

5. $\sqrt{4}=2$

6. $4^2 =16$

7. $\sqrt{-4}$ is not a real number

8. $\sqrt{10^2 -6^2}=\sqrt{100-36}=\sqrt{64}=8$

9. $\left(-3x^4 +9x^2 -15x\right)-\left(-3x^4 -7x^2 +9x +9\right)$
$=-3x^4 +9x^2 -15x +3x^4 +7x^2 -9x -9$
$=16x^2 -24x -9$

10. $(2x)(6x^3)=12x^4$

In this chapter we introduce a new type of number called an **I M A G I N A R Y   N U M B E R** .
4 8 7   1 4 3 7 6 9   3 5   8 10 2 6

Section 6.1   Definition of an *n*th Root

1.
- **a.** b; a
- **b.** principal
- **c.** b^n; a
- **d.** index; radicand
- **e.** cube
- **f.** is not; is
- **g.** even; odd
- **h.** Pythagorean; c^2
- **i.** $[0,\infty)$; $(-\infty,\infty)$
- **j.** −5 and −4

3.
- **a.** 8 is a square root of 64 because $8^2 =64$. −8 is a square root of 64 because
$$(-8)^2 =64.$$
- **b.** $\sqrt{64}=8$
- **c.** There are two square roots for every positive number. $\sqrt{64}$ identifies the positive square root.

5.
- **a.** $\sqrt{81}=9$
- **b.** $-\sqrt{81}=-9$

7. There is no real number b such that $b^2 =-36$.

9. $\sqrt{49}=7$

11. $-\sqrt{49}=-7$

13. $\sqrt{-49}$ is not a real number.

15. $\sqrt{\dfrac{64}{9}}=\dfrac{8}{3}$

17. $\sqrt{0.81}=0.9$

19. $-\sqrt{0.16}=-0.4$

254

21. **a.** $\sqrt{64} = 8$

b. $\sqrt[3]{64} = 4$

c. $-\sqrt{64} = -8$

d. $-\sqrt[3]{64} = -4$

e. $\sqrt{-64}$ is not a real number.

f. $\sqrt[3]{-64} = -4$

23. $\sqrt[3]{-27} = -3$

25. $\sqrt[3]{\dfrac{1}{8}} = \dfrac{1}{2}$

27. $\sqrt[5]{32} = 2$

29. $\sqrt[3]{-\dfrac{125}{64}} = -\dfrac{5}{4}$

31. $\sqrt[4]{-1}$ is not a real number.

33. $\sqrt[6]{1,000,000} = 10$

35. $-\sqrt[3]{0.008} = -0.2$

37. $\sqrt[4]{0.0625} = 0.5$

39. $\sqrt{a^2} = |a|$

41. $\sqrt[3]{a^3} = a$

43. $\sqrt[6]{a^6} = |a|$

45. $\sqrt{(x+1)^2} = |x+1|$

47. $\sqrt{x^2 y^4} = \sqrt{x^2 \left(y^2\right)^2} = |x| y^2$

49. $-\sqrt[3]{\dfrac{x^3}{y^3}} = -\dfrac{x}{y},\ y \neq 0$

51. $\dfrac{2}{\sqrt[4]{x^4}} = \dfrac{2}{|x|},\ x \neq 0$

53. $\sqrt[3]{(-92)^3} = -92$

55. $\sqrt[10]{(-2)^{10}} = |-2| = 2$

57. $\sqrt[7]{(-923)^7} = -923$

59. $\sqrt{y^8} = \sqrt{\left(y^4\right)^2} = y^4$

61. $\sqrt{\dfrac{a^6}{b^2}} = \sqrt{\dfrac{\left(a^3\right)^2}{b^2}} = \dfrac{a^3}{b}$

63. $-\sqrt{\dfrac{25}{q^2}} = -\dfrac{5}{q}$

65. $\sqrt{9x^2 y^4 z^2} = \sqrt{9x^2 \left(y^2\right)^2 z^2} = 3xy^2 z$

67. $\sqrt{\dfrac{h^2 k^4}{16}} = \sqrt{\dfrac{h^2 \left(k^2\right)^2}{16}} = \dfrac{hk^2}{4}$

69. $-\sqrt[3]{\dfrac{t^3}{27}} = -\dfrac{t}{3}$

71. $\sqrt[5]{32 y^{10}} = \sqrt[5]{32 \left(y^2\right)^5} = 2y^2$

73. $\sqrt[6]{64p^{12}q^{18}} = \sqrt[6]{64\left(p^{2}\right)^{6}\left(q^{3}\right)^{6}} = 2p^{2}q^{3}$

75.
$$a^{2} + b^{2} = c^{2}$$
$$12^{2} + b^{2} = 15^{2}$$
$$144 + b^{2} = 225$$
$$b^{2} = 81$$
$$b = 9 \text{ cm}$$

77.
$$a^{2} + b^{2} = c^{2}$$
$$12^{2} + 5^{2} = c^{2}$$
$$144 + 25 = c^{2}$$
$$169 = c^{2}$$
$$c = 13 \text{ ft}$$

79.
$$a^{2} + b^{2} = c^{2}$$
$$4^{2} + 3^{2} = c^{2}$$
$$16 + 9 = c^{2}$$
$$25 = c^{2}$$
$$c = 5 \text{ mi}$$
They were 5 mi apart.

81.
$$a^{2} + b^{2} = c^{2}$$
$$20^{2} + 15^{2} = c^{2}$$
$$400 + 225 = c^{2}$$
$$625 = c^{2}$$
$$c = 25 \text{ mi}$$
They are 25 mi apart.

83. $h(x) = \sqrt{x-2}$

  a. $h(0) = \sqrt{0-2} = \sqrt{-2}$ not a real number

  b. $h(1) = \sqrt{1-2} = \sqrt{-1}$ not a real number

  c. $h(2) = \sqrt{2-2} = \sqrt{0} = 0$

  d. $h(3) = \sqrt{3-2} = \sqrt{1} = 1$

  e. $h(6) = \sqrt{6-2} = \sqrt{4} = 2$

$$x - 2 \geq 0$$
$$x \geq 2$$
Domain: $[2, \infty)$

85. $g(x) = \sqrt[3]{x-2}$

  a. $g(-6) = \sqrt[3]{-6-2} = \sqrt[3]{-8} = -2$

  b. $g(1) = \sqrt[3]{1-2} = \sqrt[3]{-1} = -1$

  c. $g(2) = \sqrt[3]{2-2} = \sqrt[3]{0} = 0$

  d. $g(3) = \sqrt[3]{3-2} = \sqrt[3]{1} = 1$

  Domain: $(-\infty, \infty)$

87. $f(x) = \sqrt{5-2x}$

$$5 - 2x \geq 0$$
$$-2x \geq -5$$
$$x \leq \frac{5}{2}$$
Domain: $\left(-\infty, \dfrac{5}{2}\right]$

89. $k(x) = \sqrt[3]{4x-7}$

Domain: $(-\infty, \infty)$

91. $M(x) = \sqrt{x-5} + 3$

$$x - 5 \geq 0$$
$$x \geq 5$$
Domain: $[5, \infty)$

93. $F(x) = \sqrt[3]{x+7} - 2$

Domain: $(-\infty, \infty)$

95. **a.** $f(x) = \sqrt{1-x}$

$$1 - x \geq 0$$
$$-x \geq -1$$
$$x \leq 1$$
$(-\infty, 1]$

b.   Create a table of ordered pairs where x values are taken to be less than or equal to 1.

x	$f(x)$
1	0
0	1
−3	2
−8	3
−15	4

$f(1) = \sqrt{1-1} = \sqrt{0} = 0$

$f(0) = \sqrt{1-0} = \sqrt{1} = 1$

$f(-3) = \sqrt{1-(-3)} = \sqrt{4} = 2$

$f(-8) = \sqrt{1-(-8)} = \sqrt{9} = 3$

$f(-15) = \sqrt{1-(-15)} = \sqrt{16} = 4$

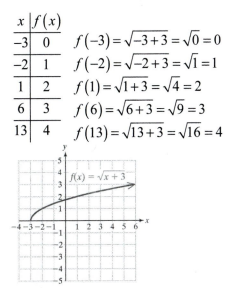

97. a. $f(x) = \sqrt{x+3}$

$x + 3 \geq 0$

$x \geq -3$

$[-3, \infty)$

b.   Create a table of ordered pairs where x values are taken to be greater than or equal to -3.

x	$f(x)$
−3	0
−2	1
1	2
6	3
13	4

$f(-3) = \sqrt{-3+3} = \sqrt{0} = 0$

$f(-2) = \sqrt{-2+3} = \sqrt{1} = 1$

$f(1) = \sqrt{1+3} = \sqrt{4} = 2$

$f(6) = \sqrt{6+3} = \sqrt{9} = 3$

$f(13) = \sqrt{13+3} = \sqrt{16} = 4$

99. a. $f(x) = \sqrt{x} + 2$

$x \geq 0$

$[0, \infty)$

b.   Create a table of ordered pairs where x values are taken to be greater than or equal to 0.

x	$f(x)$
0	2
1	3
4	4
9	5
16	6

$f(0) = \sqrt{0} + 2 = 0 + 2 = 2$

$f(1) = \sqrt{1} + 2 = 1 + 2 = 3$

$f(4) = \sqrt{4} + 2 = 2 + 2 = 4$

$f(9) = \sqrt{9} + 2 = 3 + 2 = 5$

$f(16) = \sqrt{16} + 2 = 4 + 2 = 6$

101. **a.** $f(x) = \sqrt[3]{x-1}$

The index is odd; therefore the domain is all real numbers. $(-\infty, \infty)$

b. Create a table of ordered pairs where x values are taken to be all real numbers.

x	$f(x)$
-7	-2
0	-1
1	0
2	1
9	2

$f(-7) = \sqrt[3]{-7-1} = \sqrt[3]{-8} = -2$

$f(0) = \sqrt[3]{0-1} = \sqrt[3]{-1} = -1$

$f(1) = \sqrt[3]{1-1} = \sqrt[3]{0} = 0$

$f(2) = \sqrt[3]{2-1} = \sqrt[3]{1} = 1$

$f(9) = \sqrt[3]{9-1} = \sqrt[3]{8} = 2$

103. $q + p^2$

105. $\dfrac{6}{\sqrt[3]{x}}$

107. $s^2 = 64$

$s = \sqrt{64} = 8$ in.

109. $\sqrt{69} \approx 8.3066$

111. $2 + \sqrt[3]{5} \approx 2 + 1.7100 = 3.7100$

113. $7\sqrt[4]{25} \approx 15.6525$

115. $\dfrac{3 - \sqrt{19}}{11} \approx -0.1235$

Section 6.2   Rational Exponents

1. **a.** $\sqrt[n]{a}$

b. $\left(\sqrt[n]{a}\right)^m$ or $\sqrt[n]{a^m}$

c. $\dfrac{1}{\sqrt{x}}$

d. $2; \dfrac{1}{2}$

3. $\sqrt{25} = 5$

5. $\sqrt[4]{81} = 3$

7. $144^{1/2} = \sqrt{144} = 12$

9. $-144^{1/2} = -\sqrt{144} = -12$

11. $(-144)^{1/2} = \sqrt{-144}$ is not a real number.

13. $\left(-64\right)^{1/3} = \sqrt[3]{-64} = -4$

15. $25^{-1/2} = \dfrac{1}{25^{1/2}} = \dfrac{1}{\sqrt{25}} = \dfrac{1}{5}$

17. $-49^{-1/2} = -\dfrac{1}{49^{1/2}} = -\dfrac{1}{\sqrt{49}} = -\dfrac{1}{7}$

19. $a^{m/n} = \sqrt[n]{a^m}$; The numerator of the exponent represents the power of the base. The denominator of the exponent represents the index of the radical.

21.
  a. $16^{3/4} = \left(\sqrt[4]{16}\right)^3 = 2^3 = 8$
  b. $-16^{3/4} = -\left(\sqrt[4]{16}\right)^3 = -\left(2^3\right) = -8$
  c. $\left(-16\right)^{3/4} = \left(\sqrt[4]{-16}\right)^3$ is not a real number.
  d. $16^{-3/4} = \dfrac{1}{16^{3/4}} = \dfrac{1}{\left(\sqrt[4]{16}\right)^3} = \dfrac{1}{2^3} = \dfrac{1}{8}$
  e. $-16^{-3/4} = \dfrac{1}{-16^{3/4}} = \dfrac{1}{-\left(\sqrt[4]{16}\right)^3} = \dfrac{1}{-\left(2^3\right)}$
                         $= -\dfrac{1}{8}$
  f. $\left(-16\right)^{-3/4} = \dfrac{1}{\left(-16\right)^{3/4}} = \dfrac{1}{\left(\sqrt[4]{-16}\right)^3}$
                        is not a real number.

23.
  a. $25^{3/2} = \left(\sqrt{25}\right)^3 = 5^3 = 125$
  b. $-25^{3/2} = -\left(\sqrt{25}\right)^3 = -\left(5^3\right) = -125$
  c. $\left(-25\right)^{3/2} = \left(\sqrt{-25}\right)^3$ is not a real number.
  d. $25^{-3/2} = \dfrac{1}{25^{3/2}} = \dfrac{1}{\left(\sqrt{25}\right)^3} = \dfrac{1}{5^3} = \dfrac{1}{125}$
  e. $-25^{-3/2} = \dfrac{1}{-25^{3/2}} = \dfrac{1}{-\left(\sqrt{25}\right)^3} = \dfrac{1}{-\left(5^3\right)}$
                         $= -\dfrac{1}{125}$
  f. $\left(-25\right)^{-3/2} = \dfrac{1}{\left(-25\right)^{3/2}} = \dfrac{1}{\left(\sqrt{-25}\right)^3}$
                        is not a real number.

25. $64^{-3/2} = \dfrac{1}{64^{3/2}} = \dfrac{1}{\left(\sqrt{64}\right)^3} = \dfrac{1}{8^3} = \dfrac{1}{512}$

27. $243^{3/5} = \left(\sqrt[5]{243}\right)^3 = 3^3 = 27$

29. $-27^{-4/3} = \dfrac{1}{-27^{4/3}} = \dfrac{1}{-\left(\sqrt[3]{27}\right)^4} = \dfrac{1}{-(3)^4} = -\dfrac{1}{81}$

31. $\left(\dfrac{100}{9}\right)^{-3/2} = \dfrac{1}{\left(\dfrac{100}{9}\right)^{3/2}} = \dfrac{1}{\left(\sqrt{\dfrac{100}{9}}\right)^3} = \dfrac{1}{\left(\dfrac{10}{3}\right)^3}$
              $= \dfrac{1}{\dfrac{1000}{27}} = \dfrac{27}{1000}$

33. $\left(-4\right)^{-3/2} = \dfrac{1}{\left(-4\right)^{3/2}} = \dfrac{1}{\left(\sqrt{-4}\right)^3}$
            is not a real number

35. $\left(-8\right)^{1/3} = \sqrt[3]{-8} = -2$

37. $-8^{1/3} = -\sqrt[3]{8} = -2$

39. $\dfrac{1}{36^{-1/2}} = 36^{1/2} = \sqrt{36} = 6$

41.

$$\frac{1}{1000^{-1/3}} = 1000^{1/3} = \sqrt[3]{1000} = 10$$

43.

$$\left(\frac{1}{8}\right)^{2/3} + \left(\frac{1}{4}\right)^{1/2} = \left(\sqrt[3]{\frac{1}{8}}\right)^2 + \sqrt{\frac{1}{4}} = \left(\frac{1}{2}\right)^2 + \frac{1}{2}$$

$$= \frac{1}{4} + \frac{1}{2} = \frac{3}{4}$$

45.

$$\left(\frac{1}{16}\right)^{-3/4} - \left(\frac{1}{49}\right)^{-1/2} = 16^{3/4} - 49^{1/2}$$

$$= \left(\sqrt[4]{16}\right)^3 - \sqrt{49} = (2)^3 - 7 = 8 - 7 = 1$$

47.

$$\left(\frac{1}{4}\right)^{1/2} + \left(\frac{1}{64}\right)^{-1/3} = \left(\frac{1}{4}\right)^{1/2} + (64)^{1/3}$$

$$= \sqrt{\frac{1}{4}} + \sqrt[3]{64} = \frac{1}{2} + 4 = \frac{9}{2}$$

49. $q^{2/3} = \sqrt[3]{q^2}$

51. $6y^{3/4} = 6\sqrt[4]{y^3}$

53. $x^{2/3}y^{1/3} = \left(x^2 y\right)^{1/3} = \sqrt[3]{x^2 y}$

55. $6r^{-2/5} = 6 \cdot \dfrac{1}{r^{2/5}} = \dfrac{6}{\sqrt[5]{r^2}}$

57. $\sqrt[3]{x} = x^{1/3}$

59. $10\sqrt{b} = 10b^{1/2}$

61. $\sqrt[3]{y^2} = y^{2/3}$

63. $\sqrt[4]{a^2 b^3} = \left(a^2 b^3\right)^{1/4}$

65. $x^{1/4}x^{-5/4} = x^{1/4 + (-5/4)} = x^{-1} = \dfrac{1}{x}$

67. $\dfrac{p^{5/3}}{p^{2/3}} = p^{(5/3)-(2/3)} = p^1 = p$

69. $\left(y^{1/5}\right)^{10} = y^{(1/5)(10)} = y^2$

71. $6^{-1/5}6^{3/5} = 6^{(-1/5)+(3/5)} = 6^{2/5}$

73. $\dfrac{4t^{-1/3}}{t^{4/3}} = 4t^{(-1/3)-(4/3)} = 4t^{-5/3} = \dfrac{4}{t^{5/3}}$

75. $\left(a^{1/3}a^{1/4}\right)^{12} = \left(a^{1/3}\right)^{12}\left(a^{1/4}\right)^{12} = a^{12/3}a^{12/4}$

$$= a^4 a^3 = a^{3+4} = a^7$$

77. $\left(5a^2 c^{-1/2}d^{1/2}\right)^2 = 5^2\left(a^2\right)^2\left(c^{-1/2}\right)^2\left(d^{1/2}\right)^2$

$$= 5^2 a^4 c^{-2/2}d^{2/2} = 25a^4 c^{-1}d^1 = \frac{25a^4 d}{c}$$

79. $\left(\dfrac{x^{-2/3}}{y^{-3/4}}\right)^{12} = \dfrac{\left(x^{-2/3}\right)^{12}}{\left(y^{-3/4}\right)^{12}} = \dfrac{x^{-24/3}}{y^{-36/4}} = \dfrac{x^{-8}}{y^{-9}} = \dfrac{y^9}{x^8}$

81. $\left(\dfrac{16w^{-2}z}{2wz^{-8}}\right)^{1/3} = \left(8w^{-2-1}z^{1-(-8)}\right)^{1/3} = \left(8w^{-3}z^9\right)^{1/3}$

$$= 8^{1/3}\left(w^{-3}\right)^{1/3}\left(z^9\right)^{1/3} = \sqrt[3]{8}w^{-3/3}z^{9/3}$$

$$= 2w^{-1}z^3 = \frac{2z^3}{w}$$

83. $\left(25x^2 y^4 z^6\right)^{1/2} = 25^{1/2}\left(x^2\right)^{1/2}\left(y^4\right)^{1/2}\left(z^6\right)^{1/2}$

$$= \sqrt{25}\,x^{2/2}y^{4/2}z^{6/2} = 5xy^2 z^3$$

85. $\left(x^2 y^{-1/3}\right)^6 \left(x^{1/2} y z^{2/3}\right)^2$

$= \left(x^2\right)^6 \left(y^{-1/3}\right)^6 \left(x^{1/2}\right)^2 y^2 \left(z^{2/3}\right)^2$

$= x^{12} y^{-2} x^1 y^2 z^{4/3} = x^{12+1} y^{-2+2} z^{4/3}$

$= x^{13} y^0 z^{4/3} = x^{13} z^{4/3}$

87. $\left(\dfrac{x^{3m} y^{2m}}{z^{5m}}\right)^{1/m} = \dfrac{\left(x^{3m}\right)^{1/m} \left(y^{2m}\right)^{1/m}}{\left(z^{5m}\right)^{1/m}}$

$= \dfrac{x^{3m/m} y^{2m/m}}{z^{5m/m}} = \dfrac{x^3 y^2}{z^5}$

89. **a.** $r = \left(\dfrac{A}{P}\right)^{1/t} - 1$

$r = \left(\dfrac{16,802}{10,000}\right)^{1/5} - 1 \approx 0.109 = 10.9\%$

b. $r = \left(\dfrac{18,000}{10,000}\right)^{1/7} - 1 \approx 0.088 = 8.8\%$

c. The account in part (a).

91. $r = \left(\dfrac{3V}{4\pi}\right)^{1/3}$

$r = \left(\dfrac{3(85)}{4\pi}\right)^{1/3} = \sqrt[3]{\dfrac{3(85)}{4\pi}} \approx 2.7$ in

93. $\sqrt[6]{y^3} = y^{3/6} = y^{1/2} = \sqrt{y}$

95. $\sqrt[12]{z^3} = z^{3/12} = z^{1/4} = \sqrt[4]{z}$

97. $\sqrt[9]{x^6} = x^{6/9} = x^{2/3} = \sqrt[3]{x^2}$

99. $\sqrt[6]{x^3 y^6} = \left(x^3 y^6\right)^{1/6} = x^{3/6} y^{6/6} = x^{1/2} y = y\sqrt{x}$

101. $\sqrt{16 x^8 y^6} = 16^{1/2} x^{8/2} y^{6/2} = 4x^4 y^3$

103. $\sqrt[3]{8x^3 y^2 z} = 2x\sqrt[3]{y^2 z}$

103. $\sqrt[3]{8x^3 y^2 z} = 2x\sqrt[3]{y^2 z}$

105. $\sqrt{\sqrt[3]{x}} = \sqrt{x^{1/3}} = \left(x^{1/3}\right)^{1/2} = x^{(1/3)(1/2)} = x^{1/6} = \sqrt[6]{x}$

107. $\sqrt[5]{\sqrt[3]{w}} = \sqrt[5]{w^{1/3}} = \left(w^{1/3}\right)^{1/5} = w^{(1/3)(1/5)} = w^{1/15}$

$= \sqrt[15]{w}$

109. $9^{1/2} = 3$

111. $50^{-1/4} \approx 0.3761$

113. $\sqrt[3]{5^2} \approx 2.9240$

115. $\sqrt{10^3} \approx 31.6228$

Section 6.3   Simplifying Radical Expressions

1. **a.** $\sqrt[n]{a}$; $\sqrt[n]{b}$

b. The exponent within the radicand is greater than the index.

c. is not

d. 3

e. t^{12}

f. No. $\sqrt{2}$ is an irrational number and the decimal form is a nonterminating, nonrepeating decimal.

3. $\left(\dfrac{p^4}{q^{-6}}\right)^{-1/2} \left(p^3 q^{-2}\right) = \dfrac{\left(p^4\right)^{-1/2}}{\left(q^{-6}\right)^{-1/2}} \left(p^3 q^{-2}\right)$

$= \dfrac{p^{-4/2}}{q^{6/2}} \left(p^3 q^{-2}\right) = \dfrac{p^{-2}}{q^3} \cdot p^3 q^{-2}$

$= p^{-2+3} q^{-2-3} = pq^{-5} = \dfrac{p}{q^5}$

5. $x^{4/7} = \sqrt[7]{x^4}$

7 $\sqrt{y^9} = y^{9/2}$

9. $\sqrt{x^5} = \sqrt{x^4 \cdot x} = \sqrt{x^4} \cdot \sqrt{x} = x^2\sqrt{x}$

11. $\sqrt[3]{q^7} = \sqrt[3]{q^6 \cdot q} = \sqrt[3]{q^6} \cdot \sqrt[3]{q} = q^2\sqrt[3]{q}$

13. $\sqrt{a^5 b^4} = \sqrt{a^4 b^4 \cdot a} = \sqrt{a^4 b^4} \cdot \sqrt{a} = a^2 b^2 \sqrt{a}$

15. $-\sqrt[4]{x^8 y^{13}} = -\sqrt[4]{x^8 y^{12} \cdot y} = -\sqrt[4]{x^8 y^{12}} \cdot \sqrt[4]{y}$
$$= -x^2 y^3 \sqrt[4]{y}$$

17. $\sqrt{28} = \sqrt{4 \cdot 7} = \sqrt{4} \cdot \sqrt{7} = 2\sqrt{7}$

19. $\sqrt{20} = \sqrt{4 \cdot 5} = \sqrt{4} \cdot \sqrt{5} = 2\sqrt{5}$

21. $5\sqrt{18} = 5\sqrt{9 \cdot 2} = 5\sqrt{9} \cdot \sqrt{2} = 5 \cdot 3\sqrt{2} = 15\sqrt{2}$

23. $\sqrt[3]{54} = \sqrt[3]{27 \cdot 2} = \sqrt[3]{27} \cdot \sqrt[3]{2} = 3\sqrt[3]{2}$

25. $\sqrt{25ab^3} = \sqrt{25b^2 \cdot ab}$
$$= \sqrt{25b^2} \cdot \sqrt{ab} = 5b\sqrt{ab}$$

27. $\sqrt[3]{40x^7} = \sqrt[3]{8 \cdot 5 \cdot x^6 \cdot x} = \sqrt[3]{8x^6} \cdot \sqrt[3]{5x}$
$$= 2x^2\sqrt[3]{5x}$$

29. $\sqrt[3]{-16x^6 yz^3} = \sqrt[3]{-8x^6 z^3 \cdot 2y}$
$$= \sqrt[3]{-8x^6 z^3} \cdot \sqrt[3]{2y} = -2x^2 z\sqrt[3]{2y}$$

31. $\sqrt[4]{80w^4 z^7} = \sqrt[4]{16w^4 z^4 \cdot 5z^3}$
$$= \sqrt[4]{16w^4 z^4} \cdot \sqrt[4]{5z^3} = 2wz\sqrt[4]{5z^3}$$

33. $\sqrt{\dfrac{x^3}{x}} = \sqrt{x^2} = x$

35. $\sqrt{\dfrac{p^7}{p^3}} = \sqrt{p^4} = p^2$

37. $\sqrt{\dfrac{50}{2}} = \sqrt{25} = 5$

39. $\sqrt[3]{\dfrac{3}{24}} = \sqrt[3]{\dfrac{1}{8}} = \dfrac{1}{2}$

41. $\dfrac{5\sqrt[3]{16}}{6} = \dfrac{5\sqrt[3]{8 \cdot 2}}{6} = \dfrac{5\sqrt[3]{8} \cdot \sqrt[3]{2}}{6}$
$$= \dfrac{5 \cdot 2\sqrt[3]{2}}{6} = \dfrac{5\sqrt[3]{2}}{3}$$

43. $\dfrac{5\sqrt[3]{72}}{12} = \dfrac{5\sqrt[3]{8 \cdot 9}}{12} = \dfrac{5\sqrt[3]{8} \cdot \sqrt[3]{9}}{12}$
$$= \dfrac{5 \cdot 2\sqrt[3]{9}}{12} = \dfrac{5\sqrt[3]{9}}{6}$$

45. $\sqrt{80} = \sqrt{16 \cdot 5} = \sqrt{16} \cdot \sqrt{5} = 4\sqrt{5}$

47. $-6\sqrt{75} = -6\sqrt{25 \cdot 3} = -6\sqrt{25} \cdot \sqrt{3}$
$$= -6 \cdot 5\sqrt{3} = -30\sqrt{3}$$

47. $-6\sqrt{75} = -6\sqrt{25 \cdot 3} = -6\sqrt{25} \cdot \sqrt{3}$
$$= -6 \cdot 5\sqrt{3} = -30\sqrt{3}$$

49. $\sqrt{25x^4 y^3} = \sqrt{25x^4 y^2 \cdot y}$
$$= \sqrt{25x^4 y^2} \cdot \sqrt{y} = 5x^2 y\sqrt{y}$$

51. $\sqrt[3]{27x^2 y^3 z^4} = \sqrt[3]{27y^3 z^3 \cdot x^2 z}$
$$= \sqrt[3]{27y^3 z^3} \cdot \sqrt[3]{x^2 z} = 3yz\sqrt[3]{x^2 z}$$

53. $\sqrt{\dfrac{12w^5}{3w}} = \sqrt{4w^4} = 2w^2$

55. $\sqrt{\dfrac{3y^3}{300y^{15}}} = \sqrt{\dfrac{1}{100y^{12}}} = \dfrac{1}{10y^6}$

57. $\sqrt[3]{\dfrac{16a^2b}{2a^2b^4}} = \sqrt[3]{\dfrac{8}{b^3}} = \dfrac{2}{b}$

57. $\sqrt[3]{\dfrac{16a^2b}{2a^2b^4}} = \sqrt[3]{\dfrac{8}{b^3}} = \dfrac{2}{b}$

59. $\sqrt{2^3 a^{14} b^8 c^{31} d^{22}} = \sqrt{2^2 a^{14} b^8 c^{30} d^{22} \cdot 2c}$
$= \sqrt{2^2 a^{14} b^8 c^{30} d^{22}} \cdot \sqrt{2c}$
$= 2a^7 b^4 c^{15} d^{11} \sqrt{2c}$

61. $\sqrt[3]{54a^6 b^4} = \sqrt[3]{27a^6 b^3 \cdot 2b}$
$= \sqrt[3]{27a^6 b^3} \cdot \sqrt[3]{2b}$
$= 3a^2 b \sqrt[3]{2b}$

63. $-5a\sqrt{12a^3 b^4 c} = -5a\sqrt{4 \cdot 3 \cdot a^2 \cdot a \cdot b^4 \cdot c}$
$= -5a \cdot 2ab^2 \sqrt{3ac}$
$= -10a^2 b^2 \sqrt{3ac}$

65. $\sqrt[4]{7x^5 y} = \sqrt[4]{x^4 \cdot 7xy}$
$= \sqrt[4]{x^4} \cdot \sqrt[4]{7xy}$
$= x\sqrt[4]{7xy}$

67. $\sqrt{54a^4 b^2} = \sqrt{6 \cdot 9a^4 b^2} = 3a^2 b\sqrt{6}$

69. $\dfrac{2\sqrt{27}}{3} = \dfrac{2\sqrt{9 \cdot 3}}{3} = \dfrac{2 \cdot 3\sqrt{3}}{3} = 2\sqrt{3}$

71. $\dfrac{3\sqrt{125}}{20} = \dfrac{3\sqrt{25 \cdot 5}}{20} = \dfrac{3 \cdot 5\sqrt{5}}{20} = \dfrac{3\sqrt{5}}{4}$

73. $\dfrac{1}{\sqrt[3]{w^6}} = \dfrac{1}{w^2}$

75. $\sqrt{k^3} = \sqrt{k^2 \cdot k} = k\sqrt{k}$

77. $a^2 + b^2 = c^2$
$8^2 + 10^2 = c^2$
$64 + 100 = c^2$
$164 = c^2$
$c = \sqrt{164} = \sqrt{4 \cdot 41} = 2\sqrt{41}$ ft

79. $a^2 + b^2 = c^2$
$a^2 + 12^2 = 18^2$
$a^2 + 144 = 324$
$a^2 = 180$
$a = \sqrt{180} = \sqrt{36 \cdot 5} = 6\sqrt{5}$ m

79. $a^2 + b^2 = c^2$
$a^2 + 12^2 = 18^2$
$a^2 + 144 = 324$
$a^2 = 180$
$a = \sqrt{180} = \sqrt{36 \cdot 5} = 6\sqrt{5}$ m

81. $a^2 + b^2 = c^2$
$90^2 + 90^2 = c^2$
$8100 + 8100 = c^2$
$16200 = c^2$
$c = \sqrt{16200} = \sqrt{8100 \cdot 2}$
$= 90\sqrt{2}$ ft ≈ 127.3 ft

The distance is $90\sqrt{2}$ ft or approximately 127.3 ft.

83. Let $b =$ the distance from B to C

$$40^2 + b^2 = 50^2$$
$$1600 + b^2 = 2500$$
$$b^2 = 900$$
$$b = \sqrt{900} = 30 \text{ mi}$$

The distance along the four lane highway is $40 + 30 = 70$ mi. The time from A to C via B is $t = \dfrac{70}{55} = \dfrac{14}{11} \approx 1.27 \text{ hr}$.

The time from A to C along the direct route is $t = \dfrac{50}{35} = \dfrac{10}{7} \approx 1.43 \text{ hr}$.

The route form A to C via B is the faster.

Section 6.4   Addition and Subtraction of Radicals

1.
    a. index; radicand
    b. $2\sqrt{3x}$
    c. cannot; can
    d. $4\sqrt{2}$

3. $-\sqrt[4]{x^7 y^4} = -\sqrt[4]{x^4 y^4 \cdot x^3}$
$$= -\sqrt[4]{x^4 y^4} \cdot \sqrt[4]{x^3} = -xy\sqrt[4]{x^3}$$

5. $\dfrac{\sqrt[3]{7b^8}}{\sqrt[3]{56b^2}} = \sqrt[3]{\dfrac{7b^8}{56b^2}} = \sqrt[3]{\dfrac{b^6}{8}} = \dfrac{b^2}{2}$

7. $\sqrt[4]{x^3 y} = \left(x^3 y\right)^{1/4} = \left(x^3\right)^{1/4} y^{1/4} = x^{3/4} y^{1/4}$

9. $y^{2/3} y^{1/4} = y^{(2/3)+(1/4)} = y^{11/12}$

11.
    a. $\sqrt{2}$ and $\sqrt[3]{2}$ are not like radicals. The indices are different.
    b. $\sqrt{2}$ and $3\sqrt{2}$ are like radicals.
    c. $\sqrt{2}$ and $\sqrt{5}$ are not like radicals. The radicands are different.

13.
    a. $7\sqrt{5} + 4\sqrt{5}$ and $7x + 4x$
    Both expressions can be simplified by using the distributive property.
    b. $-2\sqrt{6} - 9\sqrt{3}$ and $-2x - 9y$
    Neither expression can be simplified because they do not contain like radicals or like terms.

15. $3\sqrt{5} + 6\sqrt{5} = (3+6)\sqrt{5} = 9\sqrt{5}$

17. $3\sqrt[3]{tw} - 2\sqrt[3]{tw} + \sqrt[3]{tw} = (3-2+1)\sqrt[3]{tw} = 2\sqrt[3]{tw}$

19. $6\sqrt{10} - \sqrt{10} = (6-1)\sqrt{10} = 5\sqrt{10}$

21. $\sqrt[4]{3} + 7\sqrt[4]{3} - \sqrt[4]{14} = (1+7)\sqrt[4]{3} - \sqrt[4]{14}$
$$= 8\sqrt[4]{3} - \sqrt[4]{14}$$

23. $8\sqrt{x} + 2\sqrt{y} - 6\sqrt{x} = (8-6)\sqrt{x} + 2\sqrt{y}$
$$= 2\sqrt{x} + 2\sqrt{y}$$

25. $\sqrt[3]{ab} + a\sqrt[3]{b}$ cannot be simplified further.

27. $\sqrt{2t} + \sqrt[3]{2t}$ cannot be simplified further.

29. $\dfrac{5}{6}z\sqrt[3]{6} + \dfrac{7}{9}z\sqrt[3]{6} = \left(\dfrac{5}{6} + \dfrac{7}{9}\right)z\sqrt[3]{6}$

$\qquad = \left(\dfrac{15}{18} + \dfrac{14}{18}\right)z\sqrt[3]{6} = \dfrac{29}{18}z\sqrt[3]{6}$

31. $0.81x\sqrt{y} - 0.11x\sqrt{y} = (0.81 - 0.11)x\sqrt{y}$

$\qquad\qquad\qquad\qquad\quad = 0.70x\sqrt{y}$

33. Simplify each radical. Then add like radicals.

$3\sqrt{2} + 7\sqrt{50} = 3\sqrt{2} + 7\sqrt{25 \cdot 2}$
$\qquad\qquad = 3\sqrt{2} + 7 \cdot 5\sqrt{2} = 3\sqrt{2} + 35\sqrt{2}$
$\qquad\qquad = (3 + 35)\sqrt{2} = 38\sqrt{2}$

35. $\sqrt{36} + \sqrt{81} = 6 + 9 = 15$

37. $2\sqrt{12} + \sqrt{48} = 2\sqrt{4 \cdot 3} + \sqrt{16 \cdot 3}$
$\qquad\qquad = 2 \cdot 2\sqrt{3} + 4\sqrt{3} = 4\sqrt{3} + 4\sqrt{3}$
$\qquad\qquad = (4 + 4)\sqrt{3} = 8\sqrt{3}$

39. $4\sqrt{7} + \sqrt{63} - 2\sqrt{28} = 4\sqrt{7} + \sqrt{9 \cdot 7} - 2\sqrt{4 \cdot 7}$
$\qquad\qquad\qquad = 4\sqrt{7} + 3\sqrt{7} - 2 \cdot 2\sqrt{7}$
$\qquad\qquad\qquad = 4\sqrt{7} + 3\sqrt{7} - 4\sqrt{7}$
$\qquad\qquad\qquad = (4 + 3 - 4)\sqrt{7} = 3\sqrt{7}$

41. $5\sqrt{18} + \sqrt{27} - 4\sqrt{50}$
$\qquad = 5\sqrt{9 \cdot 2} + \sqrt{9 \cdot 3} - 4\sqrt{25 \cdot 2}$
$\qquad = 5 \cdot 3\sqrt{2} + 3\sqrt{3} - 4 \cdot 5\sqrt{2}$
$\qquad = 15\sqrt{2} + 3\sqrt{3} - 20\sqrt{2}$
$\qquad = (15 - 20)\sqrt{2} + 3\sqrt{3}$
$\qquad = -5\sqrt{2} + 3\sqrt{3}$

43. $\sqrt[3]{81} - \sqrt[3]{24} = \sqrt[3]{27 \cdot 3} - \sqrt[3]{8 \cdot 3} = 3\sqrt[3]{3} - 2\sqrt[3]{3}$
$\qquad\qquad = (3 - 2)\sqrt[3]{3} = \sqrt[3]{3}$

45. $3\sqrt{2a} - \sqrt{8a} - \sqrt{72a}$
$\qquad = 3\sqrt{2a} - \sqrt{4 \cdot 2a} - \sqrt{36 \cdot 2a}$
$\qquad = 3\sqrt{2a} - 2\sqrt{2a} - 6\sqrt{2a}$
$\qquad = (3 - 2 - 6)\sqrt{2a} = -5\sqrt{2a}$

47. $2s^2\sqrt[3]{s^2t^6} + 3t^2\sqrt[3]{8s^8}$
$\qquad = 2s^2\sqrt[3]{t^6 \cdot s^2} + 3t^2\sqrt[3]{8s^6 \cdot s^2}$
$\qquad = 2s^2 \cdot t^2\sqrt[3]{s^2} + 3t^2 \cdot 2s^2\sqrt[3]{s^2}$
$\qquad = 2s^2t^2\sqrt[3]{s^2} + 6s^2t^2\sqrt[3]{s^2}$
$\qquad = \left(2s^2t^2 + 6s^2t^2\right)\sqrt[3]{s^2} = 8s^2t^2\sqrt[3]{s^2}$

49. $7\sqrt[3]{x^4} - x\sqrt[3]{x} = 7\sqrt[3]{x^3 \cdot x} - x\sqrt[3]{x}$
$\qquad\qquad = 7x\sqrt[3]{x} - x\sqrt[3]{x}$
$\qquad\qquad = (7x - x)\sqrt[3]{x} = 6x\sqrt[3]{x}$

51. $5p\sqrt{20p^2} + p^2\sqrt{80} = 5p\sqrt{4p^2 \cdot 5} + p^2\sqrt{16 \cdot 5}$
$\qquad\qquad = 5p \cdot 2p\sqrt{5} + p^2 \cdot 4\sqrt{5}$
$\qquad\qquad = 10p^2\sqrt{5} + 4p^2\sqrt{5}$
$\qquad\qquad = \left(10p^2 + 4p^2\right)\sqrt{5} = 14p^2\sqrt{5}$

53. $\sqrt[3]{a^2b} - \sqrt[3]{8a^2b} = \sqrt[3]{a^2b} - 2\sqrt[3]{a^2b}$
$$= (1-2)\sqrt[3]{a^2b} = -\sqrt[3]{a^2b}$$

55. $5x\sqrt{x} + 6\sqrt{x} = (5x+6)\sqrt{x}$

57. $\sqrt{50x^2} - 3\sqrt{8} = \sqrt{25x^2 \cdot 2} - 3\sqrt{4 \cdot 2}$
$$= 5x\sqrt{2} - 3 \cdot 2\sqrt{2}$$
$$= 5x\sqrt{2} - 6\sqrt{2} = (5x-6)\sqrt{2}$$

59. $11\sqrt[3]{54cd^3} - 2\sqrt[3]{2cd^3} + d\sqrt[3]{16c}$
$$= 11\sqrt[3]{27 \cdot 2cd^3} - 2\sqrt[3]{2cd^3} + d\sqrt[3]{8 \cdot 2c}$$
$$= 11 \cdot 3d\sqrt[3]{2c} - 2 \cdot d\sqrt[3]{2c} + d \cdot 2\sqrt[3]{2c}$$
$$= 33d\sqrt[3]{2c} - 2d\sqrt[3]{2c} + 2d\sqrt[3]{2c}$$
$$= (33 - 2 + 2)d\sqrt[3]{2c} = 33d\sqrt[3]{2c}$$

61. $\dfrac{3}{2}ab\sqrt{24a^3} + \dfrac{4}{3}\sqrt{54a^5b^2} - a^2b\sqrt{150a} = \dfrac{3}{2}ab\sqrt{4a^2 \cdot 6a} + \dfrac{4}{3}\sqrt{9a^4b^2 \cdot 6a} - a^2b\sqrt{25 \cdot 6a}$
$$= \dfrac{3}{2}ab \cdot 2a\sqrt{6a} + \dfrac{4}{3} \cdot 3a^2b\sqrt{6a} - a^2b \cdot 5\sqrt{6a} = 3a^2b\sqrt{6a} + 4a^2b\sqrt{6a} - 5a^2b\sqrt{6a}$$
$$= (3 + 4 - 5)a^2b\sqrt{6a} = 2a^2b\sqrt{6a}$$

63. $x\sqrt[3]{16} - 2\sqrt[3]{27x} + \sqrt[3]{54x^3}$
$$= x\sqrt[3]{8 \cdot 2} - 2\sqrt[3]{27x} + \sqrt[3]{27x^3 \cdot 2}$$
$$= x \cdot 2\sqrt[3]{2} - 2 \cdot 3\sqrt[3]{x} + 3x\sqrt[3]{2}$$
$$= 2x\sqrt[3]{2} - 6\sqrt[3]{x} + 3x\sqrt[3]{2}$$
$$= (2+3)x\sqrt[3]{2} - 6\sqrt[3]{x} = 5x\sqrt[3]{2} - 6\sqrt[3]{x}$$

65. $\sqrt{x} + \sqrt{y} = \sqrt{x+y}$ False.
$$\sqrt{9} + \sqrt{16} \neq \sqrt{9+16}$$
$$3 + 4 \neq \sqrt{25}$$
$$7 \neq 5$$

67. $5\sqrt[3]{x} + 2\sqrt[3]{x} = 7\sqrt[3]{x}$ True.

69. $\sqrt{y} + \sqrt{y} = \sqrt{2y}$ False.
$$\sqrt{y} + \sqrt{y} = 2\sqrt{y} \neq \sqrt{2y}$$

71. $2w\sqrt{5} + 4w\sqrt{5} = 6w^2\sqrt{5}$
False: $2w\sqrt{5} + 4w\sqrt{5} = 6w\sqrt{5} \neq 6w^2\sqrt{5}$

73. $\sqrt{48} + \sqrt{12} = \sqrt{16 \cdot 3} + \sqrt{4 \cdot 3} = 4\sqrt{3} + 2\sqrt{3}$
$$= (4+2)\sqrt{3} = 6\sqrt{3}$$

75. $5\sqrt[3]{x^6} - x^2 = 5x^2 - x^2 = 4x^2$

76. $y^3 + \sqrt[4]{y^{12}} = y^3 + y^3 = 2y^3$

77. $\sqrt{18} - 5^2$
The difference of the principal square root of 18 and the square of 5.

79. $\sqrt[4]{x} + y^3$
The sum of the principal fourth root of x and the cube of y..

81.
$$P = 2\sqrt{6} + 2\sqrt{24} + \sqrt{54}$$
$$= 2\sqrt{6} + 2\sqrt{4 \cdot 6} + \sqrt{9 \cdot 6}$$
$$= 2\sqrt{6} + 2 \cdot 2\sqrt{6} + 3\sqrt{6}$$
$$= 2\sqrt{6} + 4\sqrt{6} + 3\sqrt{6}$$
$$= (2 + 4 + 3)\sqrt{6} = 9\sqrt{6} \text{ cm} \approx 22.0 \text{ cm}$$

83.
$$2\sqrt{50} + 2x = 14\sqrt{2}$$
$$2\sqrt{25 \cdot 2} + 2x = 14\sqrt{2}$$
$$2 \cdot 5\sqrt{2} + 2x = 14\sqrt{2}$$
$$10\sqrt{2} + 2x = 14\sqrt{2}$$
$$2x = 4\sqrt{2}$$
$$x = 2\sqrt{2} \text{ ft}$$

85. **a.** Side from (0, 6) to (6, 9):
$$c^2 = 3^2 + 6^2 = 9 + 36 = 45$$
$$c = \sqrt{45} = \sqrt{9 \cdot 5} = 3\sqrt{5}$$
Side from (6, 9) to (7, 7):
$$c^2 = 1^2 + 2^2 = 1 + 4 = 5$$
$$c = \sqrt{5}$$
Side from (7, 7) to (4, 1):
$$c^2 = 3^2 + 6^2 = 9 + 36 = 45$$
$$c = \sqrt{45} = \sqrt{9 \cdot 5} = 3\sqrt{5}$$
Side from (4, 1) to (2, 2):
$$c^2 = 1^2 + 2^2 = 1 + 4 = 5$$
$$c = \sqrt{5}$$

Side from (2, 2) to (0, 6):
$$c^2 = 2^2 + 4^2 = 4 + 16 = 20$$
$$c = \sqrt{20} = \sqrt{4 \cdot 5} = 2\sqrt{5}$$
$$P = 3\sqrt{5} + \sqrt{5} + 3\sqrt{5} + \sqrt{5} + 2\sqrt{5}$$
$$= (3 + 1 + 3 + 1 + 2)\sqrt{5} = 10\sqrt{5} \text{ yd}$$

b. $10\sqrt{5} \approx 22.36$ yd

c. $C = 1.49(22.36)(3) + 0.06(1.49(22.36)(3))$
$$= 99.95 + 6.00 = \$105.95$$

Section 6.5   Multiplication of Radicals

1.
  a. $\sqrt[n]{ab}$
  b. x
  c. a
  d. conjugates
  e. $m - n$
  f. $c + 8\sqrt{c} + 16$

3. $-\sqrt{20a^2b^3c} = -\sqrt{4a^2b^2 \cdot 5bc} = -2ab\sqrt{5bc}$

5. $x^{1/3}y^{1/4}x^{-1/6}y^{1/3} = x^{(1/3)+(-1/6)}y^{(1/4)+(1/3)} = x^{1/6}y^{7/12}$

7. $-2\sqrt[3]{7} + 4\sqrt[3]{7} = (-2 + 4)\sqrt[3]{7} = 2\sqrt[3]{7}$

9. $\sqrt[3]{7} \cdot \sqrt[3]{3} = \sqrt[3]{21}$

11. $\sqrt{2} \cdot \sqrt{10} = \sqrt{20} = \sqrt{4 \cdot 5} = 2\sqrt{5}$

11. $\sqrt{2} \cdot \sqrt{10} = \sqrt{20} = \sqrt{4 \cdot 5} = 2\sqrt{5}$

13. $\sqrt[4]{16} \cdot \sqrt[4]{64} = \sqrt[4]{2^4} \cdot \sqrt[4]{2^6} = 2\sqrt[4]{2^4 \cdot 2^2}$
$$= 2 \cdot 2\sqrt[4]{2^2} = 4\sqrt[4]{4}$$

15. $\left(4\sqrt[3]{4}\right)\left(2\sqrt[3]{5}\right) = (4 \cdot 2)\left(\sqrt[3]{4} \cdot \sqrt[3]{5}\right) = 8\sqrt[3]{20}$

17. $\left(8a\sqrt{b}\right)\left(-3\sqrt{ab}\right) = (8a)(-3)\left(\sqrt{b} \cdot \sqrt{ab}\right)$
$$= -24a\sqrt{ab^2} = -24ab\sqrt{a}$$

19. $\sqrt{30} \cdot \sqrt{12} = \sqrt{360} = \sqrt{36 \cdot 10} = 6\sqrt{10}$

21. $\sqrt{6x}\sqrt{12x} = \sqrt{72x^2} = \sqrt{36x^2 \cdot 2} = 6x\sqrt{2}$

23. $\sqrt{5a^3b^2}\sqrt{20a^3b^3} = \sqrt{100a^6b^5}$
$$= \sqrt{100a^6b^4 \cdot b}$$
$$= 10a^3b^2\sqrt{b}$$

25. $\left(4\sqrt{3xy^3}\right)\left(-2\sqrt{6x^3y^2}\right) = -8\sqrt{18x^4y^5}$
$$= -8\sqrt{9x^4y^4 \cdot 2y} = -8 \cdot 3x^2y^2\sqrt{2y}$$
$$= -24x^2y^2\sqrt{2y}$$

27. $\left(\sqrt[3]{4a^2b}\right)\left(\sqrt[3]{2ab^3}\right)\left(\sqrt[3]{54a^2b}\right)$
$$= \sqrt[3]{8a^3b^3 \cdot b} \cdot \sqrt[3]{27 \cdot 2a^2b}$$
$$= 2ab\sqrt[3]{b} \cdot 3\sqrt[3]{2a^2b} = 6ab\sqrt[3]{2a^2b^2}$$

29. $\sqrt{3}\left(4\sqrt{3} - 6\right) = \sqrt{3} \cdot 4\sqrt{3} - \sqrt{3} \cdot (6)$
$$= 4\sqrt{9} - 6\sqrt{3}$$
$$= 4 \cdot 3 - 6\sqrt{3} = 12 - 6\sqrt{3}$$

31. $\sqrt{2}\left(\sqrt{6} - \sqrt{3}\right) = \sqrt{2} \cdot \sqrt{6} - \sqrt{2} \cdot \sqrt{3}$
$$= \sqrt{12} - \sqrt{6} = \sqrt{4 \cdot 3} - \sqrt{6}$$
$$= 2\sqrt{3} - \sqrt{6}$$

33. $-\frac{1}{3}\sqrt{x}\left(6\sqrt{x} + 7\right) = -\frac{1}{3}\sqrt{x} \cdot 6\sqrt{x} - \frac{1}{3}\sqrt{x} \cdot (7)$
$$= -2\sqrt{x^2} - \frac{7}{3}\sqrt{x} = -2x - \frac{7}{3}\sqrt{x}$$

35. $\left(\sqrt{3} + 2\sqrt{10}\right)\left(4\sqrt{3} - \sqrt{10}\right) = \sqrt{3} \cdot 4\sqrt{3} - \sqrt{3} \cdot \sqrt{10} + 2\sqrt{10} \cdot 4\sqrt{3} - 2\sqrt{10} \cdot \sqrt{10}$
$$= 4\sqrt{9} - \sqrt{30} + 8\sqrt{30} - 2\sqrt{100} = 4 \cdot 3 + 7\sqrt{30} - 2 \cdot 10 = 12 + 7\sqrt{30} - 20 = -8 + 7\sqrt{30}$$

37. $\left(\sqrt{x} + 4\right)\left(\sqrt{x} - 9\right) = \sqrt{x} \cdot \sqrt{x} - \sqrt{x} \cdot 9 + 4 \cdot \sqrt{x} - 4 \cdot 9 = \sqrt{x^2} - 9\sqrt{x} + 4\sqrt{x} - 36 = x - 5\sqrt{x} - 36$

39. $\left(\sqrt[3]{y} + 2\right)\left(\sqrt[3]{y} - 3\right) = \sqrt[3]{y} \cdot \sqrt[3]{y} - \sqrt[3]{y} \cdot 3 + 2 \cdot \sqrt[3]{y} - 2 \cdot 3 = \sqrt[3]{y^2} - 3\sqrt[3]{y} + 2\sqrt[3]{y} - 6 = \sqrt[3]{y^2} - \sqrt[3]{y} - 6$

41. $\left(\sqrt{a} - 3\sqrt{b}\right)\left(9\sqrt{a} - \sqrt{b}\right) = \sqrt{a} \cdot 9\sqrt{a} - \sqrt{a} \cdot \sqrt{b} - 3\sqrt{b} \cdot 9\sqrt{a} + 3\sqrt{b} \cdot \sqrt{b}$
$$= 9\sqrt{a^2} - \sqrt{ab} - 27\sqrt{ab} + 3\sqrt{b^2} = 9a - 28\sqrt{ab} + 3b$$

43. $\left(\sqrt{p} + 2\sqrt{q}\right)\left(8 + 3\sqrt{p} - \sqrt{q}\right) = \sqrt{p} \cdot 8 + \sqrt{p} \cdot 3\sqrt{p} - \sqrt{p} \cdot \sqrt{q} + 2\sqrt{q} \cdot 8 + 2\sqrt{q} \cdot 3\sqrt{p} - 2\sqrt{q} \cdot \sqrt{q}$
$$= 8\sqrt{p} + 3\sqrt{p^2} - \sqrt{pq} + 16\sqrt{q} + 6\sqrt{pq} - 2\sqrt{q^2} = 8\sqrt{p} + 3p + 5\sqrt{pq} + 16\sqrt{q} - 2q$$

45. $\left(\sqrt{15}\right)^2 = 15$

47. $\left(\sqrt{3y}\right)^2 = 3y$

49. $\left(\sqrt[3]{6}\right)^3 = 6$

51. $\sqrt{709} \cdot \sqrt{709} = \left(\sqrt{709}\right)^2 = 709$

53. (a) $(x+y)(x-y) = x^2 - y^2$

(b) $(x+5)(x-5) = x^2 - 5^2 = x^2 - 25$

55. $\left(\sqrt{13}+4\right)^2 = \left(\sqrt{13}\right)^2 + 2\left(\sqrt{13}\right)(4) + 4^2$

$= 13 + 8\sqrt{13} + 16 = 29 + 8\sqrt{13}$

57. $\left(\sqrt{p}-\sqrt{7}\right)^2 = \left(\sqrt{p}\right)^2 - 2\left(\sqrt{p}\right)\left(\sqrt{7}\right) + \left(\sqrt{7}\right)^2$

$= p - 2\sqrt{7p} + 7$

59. $\left(\sqrt{2a}-3\sqrt{b}\right)^2$

$= \left(\sqrt{2a}\right)^2 - 2\left(\sqrt{2a}\right)\left(3\sqrt{b}\right) + \left(3\sqrt{b}\right)^2$

$= 2a - 6\sqrt{2ab} + 9b$

61. $\left(\sqrt{3}+x\right)\left(\sqrt{3}-x\right) = \left(\sqrt{3}\right)^2 - x^2 = 3 - x^2$

63. $\left(\sqrt{6}+\sqrt{2}\right)\left(\sqrt{6}-\sqrt{2}\right) = \left(\sqrt{6}\right)^2 - \left(\sqrt{2}\right)^2$

$= 6 - 2 = 4$

65. $\left(\dfrac{2}{3}\sqrt{x}+\dfrac{1}{2}\sqrt{y}\right)\left(\dfrac{2}{3}\sqrt{x}-\dfrac{1}{2}\sqrt{y}\right)$

$= \left(\dfrac{2}{3}\sqrt{x}\right)^2 - \left(\dfrac{1}{2}\sqrt{y}\right)^2 = \dfrac{4}{9}x - \dfrac{1}{4}y$

67. a. $\left(\sqrt{3}+\sqrt{x}\right)\left(\sqrt{3}-\sqrt{x}\right) = \sqrt{3}^2 - \sqrt{x}^2 = 3 - x$

b. $\left(\sqrt{3}+\sqrt{x}\right)\left(\sqrt{3}+\sqrt{x}\right) = \sqrt{3}^2 + 2\sqrt{3}\sqrt{x} + \sqrt{x}^2$

$= 3 + 2\sqrt{3x} + x$

c. $\left(\sqrt{3}-\sqrt{x}\right)\left(\sqrt{3}-\sqrt{x}\right) = \sqrt{3}^2 - 2\sqrt{3}\sqrt{x} + \sqrt{x}^2$

$= 3 - 2\sqrt{3x} + x$

69. $\sqrt{3}\cdot\sqrt{2} = \sqrt{6}$ True.

71. $\left(x-\sqrt{5}\right)^2 = x - 5$ False.

$\left(x-\sqrt{5}\right)^2 = x^2 - 2x\sqrt{5} + \left(\sqrt{5}\right)^2$

$= x^2 - 2x\sqrt{5} + 5$

73. $5\left(3\sqrt{4x}\right) = 15\sqrt{20x}$ False.

5 is multiplied by 3 only.

75. $\dfrac{3\sqrt{x}}{3} = \sqrt{x}$ True.

77. $\left(-\sqrt{6x}\right)^2 = 6x$

79. $\left(\sqrt{3x+1}\right)^2 = 3x+1$

81. $\left(\sqrt{x+3}-4\right)^2$

$= \left(\sqrt{x+3}\right)^2 - 2\left(\sqrt{x+3}\right)(4) + (4)^2$

$= x + 3 - 8\sqrt{x+3} + 16$

$= x + 19 - 8\sqrt{x+3}$

83. $\left(\sqrt{2t-3}+5\right)^2$

$= \left(\sqrt{2t-3}\right)^2 + 2\left(\sqrt{2t-3}\right)(5) + (5)^2$

$= 2t - 3 + 10\sqrt{2t-3} + 25$

$= 2t + 22 + 10\sqrt{2t-3}$

85. $A = \sqrt{40}\cdot 3\sqrt{2} = 3\sqrt{80} = 3\sqrt{16\cdot 5}$

$= 3\cdot 4\sqrt{5} = 12\sqrt{5}$ ft^2

87. $A = \dfrac{1}{2}\cdot 3\sqrt{5}\cdot 6\sqrt{12} = 9\sqrt{60} = 9\sqrt{4\cdot 15}$

$= 9\cdot 2\sqrt{15} = 18\sqrt{15}$ in.^{2}

89. $\sqrt{x} \cdot \sqrt[4]{x} = x^{1/2} \cdot x^{1/4} = x^{(1/2)+(1/4)} = x^{3/4} = \sqrt[4]{x^3}$

91. $\sqrt[5]{2z} \cdot \sqrt[3]{2z} = (2z)^{1/5} \cdot (2z)^{1/3} = (2z)^{(1/5)+(1/3)}$
$$= (2z)^{8/15} = \sqrt[15]{(2z)^8}$$

93. $\sqrt[3]{p^2} \cdot \sqrt{p^3} = p^{2/3} \cdot p^{3/2} = p^{(2/3)+(3/2)} = p^{13/6}$
$$= \sqrt[6]{p^{13}} = \sqrt[6]{p^{12} \cdot p} = p^2 \sqrt[6]{p}$$

95. $\sqrt[4]{q^3} \cdot \sqrt[3]{q^2} = q^{3/4} \cdot q^{2/3} = q^{(3/4)+(2/3)} = q^{17/12}$
$$= \sqrt[12]{q^{17}} = \sqrt[12]{q^{12} \cdot q^5} = q \sqrt[12]{q^5}$$

97. $\sqrt[3]{x} \cdot \sqrt[6]{y} = x^{1/3} \cdot y^{1/6} = x^{2/6} \cdot y^{1/6}$
$$= \left(x^2 y\right)^{1/6} = \sqrt[6]{x^2 y}$$

99. $\sqrt[4]{8} \cdot \sqrt{3} = \sqrt[4]{2^3} \cdot \sqrt{3} = 2^{3/4} \cdot 3^{1/2} = 2^{3/4} \cdot 3^{2/4}$
$$= \left(2^3 \cdot 3^2\right)^{1/4} = \sqrt[4]{2^3 \cdot 3^2} = \sqrt[4]{8 \cdot 9} = \sqrt[4]{72}$$

101. $\sqrt[3]{2xy} \cdot \sqrt[4]{5xy} = (2xy)^{1/3} (5xy)^{1/4} = 2^{1/3} x^{1/3} y^{1/3} 5^{1/4} x^{1/4} y^{1/4} = 2^{4/12} x^{4/12} y^{4/12} 5^{3/12} x^{3/12} y^{3/12}$
$$= 2^{4/12} \cdot 5^{3/12} x^{7/12} y^{7/12} = \left(2^4 \cdot 5^3 x^7 y^7\right)^{1/12} = \sqrt[12]{2^4 5^3 x^7 y^7}$$

103. $\sqrt[3]{4m^2 n} \cdot \sqrt{6mn} = \left(4m^2 n\right)^{1/3} (6mn)^{1/2} = 4^{1/3} m^{2/3} n^{1/3} 6^{1/2} m^{1/2} n^{1/2} = 4^{2/6} m^{4/6} n^{2/6} 6^{3/6} m^{3/6} n^{3/6}$
$$= 4^{2/6} \cdot 6^{3/6} m^{7/6} n^{5/6} = \left(4^2 \cdot 6^3 m^7 n^5\right)^{1/6} = \sqrt[6]{2^4 2^3 3^3 m^7 n^5} = \sqrt[6]{2^6 \cdot 2 \cdot 3^3 m^6 mn^5}$$
$$= 2m\sqrt[6]{2 \cdot 3^3 mn^5}$$

105. $\left(\sqrt[3]{a} + \sqrt[3]{b}\right)\left(\sqrt[3]{a^2} - \sqrt[3]{ab} + \sqrt[3]{b^2}\right) = \sqrt[3]{a} \cdot \sqrt[3]{a^2} - \sqrt[3]{a} \cdot \sqrt[3]{ab} + \sqrt[3]{a} \cdot \sqrt[3]{b^2} + \sqrt[3]{b} \cdot \sqrt[3]{a^2} - \sqrt[3]{b} \cdot \sqrt[3]{ab} + \sqrt[3]{b} \cdot \sqrt[3]{b^2}$
$$= \sqrt[3]{a^3} - \sqrt[3]{a^2 b} + \sqrt[3]{ab^2} + \sqrt[3]{a^2 b} - \sqrt[3]{ab^2} + \sqrt[3]{b^3} = a + b$$

Problem Recognition Exercises

1. **a.** $\sqrt{24} = \sqrt{4 \cdot 6} = 2\sqrt{6}$

　　b. $\sqrt[3]{24} = \sqrt[3]{8 \cdot 3} = 2\sqrt[3]{3}$

3. **a.** $\sqrt{200y^6} = \sqrt{100 \cdot 2y^6} = 10y^3 \sqrt{2}$

　　b. $\sqrt[3]{200y^6} = \sqrt[3]{8 \cdot 25y^6} = 2y^2 \sqrt[3]{25}$

5. **a.** $\sqrt{80} = \sqrt{16 \cdot 5} = 4\sqrt{5}$

　　b. $\sqrt[3]{80} = \sqrt[3]{8 \cdot 10} = 2\sqrt[3]{10}$

　　c. $\sqrt[4]{80} = \sqrt[4]{16 \cdot 5} = 2\sqrt[4]{5}$

7. **a.** $\sqrt{x^5 y^6} = \sqrt{x^4 \cdot x \cdot y^6} = x^2 y^3 \sqrt{x}$

　　b. $\sqrt[3]{x^5 y^6} = \sqrt[3]{x^3 \cdot x^2 \cdot y^6} = xy^2 \sqrt[3]{x^2}$

　　c. $\sqrt[4]{x^5 y^6} = \sqrt[4]{x^4 \cdot x \cdot y^4 \cdot y^2} = xy \sqrt[4]{xy^2}$

9. **a.** $\sqrt[3]{32s^5 t^6} = \sqrt[3]{8 \cdot 4 \cdot s^3 \cdot s^2 \cdot t^6} = 2st^2 \sqrt[3]{4s^2}$

　　b. $\sqrt[4]{32s^5 t^6} = \sqrt[4]{16 \cdot 2 \cdot s^4 \cdot s \cdot t^4 \cdot t^2} = 2st \sqrt[4]{2st^2}$

　　c. $\sqrt[5]{32s^5 t^6} = \sqrt[5]{32 \cdot s^5 \cdot t^5 \cdot t} = 2st \sqrt[5]{t}$

11. **a.** $\sqrt{5} + \sqrt{5} = 2\sqrt{5}$

　　b. $\sqrt{5} \cdot \sqrt{5} = \sqrt{25} = 5$

13. a. $2\sqrt{6} - 5\sqrt{6} = -3\sqrt{6}$

   b. $2\sqrt{6} \cdot 5\sqrt{6} = 10\sqrt{36} = 10 \cdot 6 = 60$

15. a. $\sqrt{8} + \sqrt{2} = \sqrt{4 \cdot 2} + \sqrt{2} = 2\sqrt{2} + \sqrt{2} = 3\sqrt{2}$

   b. $\sqrt{8} \cdot \sqrt{2} = \sqrt{16} = 4$

17. a. $5\sqrt{18} - 4\sqrt{8} = 5\sqrt{9 \cdot 2} - 4\sqrt{4 \cdot 2}$
$$= 5 \cdot 3\sqrt{2} - 4 \cdot 2\sqrt{2}$$
$$= 15\sqrt{2} - 8\sqrt{2} = 7\sqrt{2}$$

   b. $5\sqrt{18} \cdot 4\sqrt{8} = 20\sqrt{144} = 20 \cdot 12 = 240$

19. a. $4\sqrt[3]{24} + 6\sqrt[3]{3} = 4\sqrt[3]{8 \cdot 3} + 6\sqrt[3]{3}$
$$= 4 \cdot 2\sqrt[3]{3} + 6\sqrt[3]{3}$$
$$= 8\sqrt[3]{3} + 6\sqrt[3]{3} = 14\sqrt[3]{3}$$

   b. $4\sqrt[3]{24} \cdot 6\sqrt[3]{3} = 24\sqrt[3]{72} = 24\sqrt[3]{8 \cdot 9}$
$$= 24 \cdot 2\sqrt[3]{9} = 48\sqrt[3]{9}$$

19. a. $4\sqrt[3]{24} + 6\sqrt[3]{3} = 4\sqrt[3]{8 \cdot 3} + 6\sqrt[3]{3}$
$$= 4 \cdot 2\sqrt[3]{3} + 6\sqrt[3]{3}$$
$$= 8\sqrt[3]{3} + 6\sqrt[3]{3} = 14\sqrt[3]{3}$$

   b. $4\sqrt[3]{24} \cdot 6\sqrt[3]{3} = 24\sqrt[3]{72} = 24\sqrt[3]{8 \cdot 9}$
$$= 24 \cdot 2\sqrt[3]{9} = 48\sqrt[3]{9}$$

Section 6.6   Division of Radicals and Rationalization

1.   **a.** radical

    b. $\dfrac{\sqrt[n]{a}}{\sqrt[b]{b}}$

    c. $\dfrac{4}{x^2}$

    d. rationalizing

    e. is; is not

    f. denominator

3. $2y\sqrt{45} + 3\sqrt{20y^2} = 2y\sqrt{9 \cdot 5} + 3\sqrt{4y^2 \cdot 5}$
$$= 2y \cdot 3\sqrt{5} + 3 \cdot 2y\sqrt{5}$$
$$= 6y\sqrt{5} + 6y\sqrt{5} = 12y\sqrt{5}$$

5. $\left(-6\sqrt{y} + 3\right)\left(3\sqrt{y} + 1\right)$
$$= -6\sqrt{y} \cdot 3\sqrt{y} - 6\sqrt{y} \cdot (1) + 3 \cdot 3\sqrt{y} + 3 \cdot 1$$
$$= -18\sqrt{y^2} - 6\sqrt{y} + 9\sqrt{y} + 3$$
$$= -18y + 3\sqrt{y} + 3$$

7. $\left(8 - \sqrt{t}\right)^2 = 8^2 - 2 \cdot 8 \cdot \sqrt{t} + \left(\sqrt{t}\right)^2$
$$= 64 - 16\sqrt{t} + t$$

9. $\left(\sqrt{2} + \sqrt{7}\right)\left(\sqrt{2} - \sqrt{7}\right) = \left(\sqrt{2}\right)^2 - \left(\sqrt{7}\right)^2$
$$= 2 - 7 = -5$$

11. $\sqrt{\dfrac{49x^4}{y^6}} = \dfrac{\sqrt{49x^4}}{\sqrt{y^6}} = \dfrac{7x^2}{y^3}$

13. $\sqrt{\dfrac{8a^2}{x^6}} = \dfrac{\sqrt{2 \cdot 4a^2}}{\sqrt{x^6}} = \dfrac{2a\sqrt{2}}{x^3}$

15. $\sqrt[3]{\dfrac{-16j^3}{k^3}} = \dfrac{\sqrt[3]{-8j^3 \cdot 2}}{\sqrt[3]{k^3}} = \dfrac{-2j\sqrt[3]{2}}{k}$

17. $\dfrac{\sqrt{72ab^5}}{\sqrt{8ab}} = \sqrt{\dfrac{72ab^5}{8ab}} = \sqrt{9b^4} = 3b^2$

19. $\dfrac{\sqrt[4]{3b^3}}{\sqrt[4]{48b^{11}}} = \sqrt[4]{\dfrac{3b^3}{48b^{11}}} = \sqrt[4]{\dfrac{1}{16b^8}} = \dfrac{\sqrt[4]{1}}{\sqrt[4]{16b^8}} = \dfrac{1}{2b^2}$

21. $\dfrac{\sqrt{3yz^2}}{\sqrt{w^4}} = \dfrac{z\sqrt{3y}}{w^2}$

23. $\dfrac{x}{\sqrt{5}} = \dfrac{x}{\sqrt{5}} \cdot \dfrac{\sqrt{5}}{\sqrt{5}}$

25. $\dfrac{7}{\sqrt[3]{x}} = \dfrac{7}{\sqrt[3]{x}} \cdot \dfrac{\sqrt[3]{x^2}}{\sqrt[3]{x^2}}$

27. $\dfrac{8}{\sqrt{3z}} = \dfrac{8}{\sqrt{3z}} \cdot \dfrac{\sqrt{3z}}{\sqrt{3z}}$

29. $\dfrac{1}{\sqrt[4]{8a^2}} = \dfrac{1}{\sqrt[4]{8a^2}} \cdot \dfrac{\sqrt[4]{2a^2}}{\sqrt[4]{2a^2}}$

31. $\dfrac{1}{\sqrt{3}} = \dfrac{1}{\sqrt{3}} \cdot \dfrac{\sqrt{3}}{\sqrt{3}} = \dfrac{1\sqrt{3}}{\sqrt{3^2}} = \dfrac{\sqrt{3}}{3}$

33. $\sqrt{\dfrac{1}{x}} = \dfrac{\sqrt{1}}{\sqrt{x}} = \dfrac{1}{\sqrt{x}} \cdot \dfrac{\sqrt{x}}{\sqrt{x}} = \dfrac{1\sqrt{x}}{\sqrt{x^2}} = \dfrac{\sqrt{x}}{x}$

35. $\dfrac{6}{\sqrt{2y}} = \dfrac{6}{\sqrt{2y}} \cdot \dfrac{\sqrt{2y}}{\sqrt{2y}} = \dfrac{6\sqrt{2y}}{\sqrt{(2y)^2}}$

$= \dfrac{6\sqrt{2y}}{2y} = \dfrac{3\sqrt{2y}}{y}$

37. $\sqrt{\dfrac{a^3}{2}} = \dfrac{\sqrt{a^3}}{\sqrt{2}} \cdot \dfrac{\sqrt{2}}{\sqrt{2}} = \dfrac{\sqrt{2a \cdot a^2}}{\sqrt{4}} = \dfrac{a\sqrt{2a}}{2}$

39. $\dfrac{6}{\sqrt{8}} = \dfrac{6}{\sqrt{4 \cdot 2}} = \dfrac{6}{2\sqrt{2}} \cdot \dfrac{\sqrt{2}}{\sqrt{2}} = \dfrac{3\sqrt{2}}{\sqrt{(2)^2}} = \dfrac{3\sqrt{2}}{2}$

41. $\dfrac{3}{\sqrt[3]{2}} = \dfrac{3}{\sqrt[3]{2}} \cdot \dfrac{\sqrt[3]{2^2}}{\sqrt[3]{2^2}} = \dfrac{3\sqrt[3]{2^2}}{\sqrt[3]{2^3}} = \dfrac{3\sqrt[3]{4}}{2}$

43. $\dfrac{-6}{\sqrt[4]{x}} = \dfrac{-6}{\sqrt[4]{x}} \cdot \dfrac{\sqrt[4]{x^3}}{\sqrt[4]{x^3}} = \dfrac{-6\sqrt[4]{x^3}}{\sqrt[4]{x^4}} = \dfrac{-6\sqrt[4]{x^3}}{x}$

45. $\dfrac{7}{\sqrt[3]{4}} = \dfrac{7}{\sqrt[3]{2^2}} = \dfrac{7}{\sqrt[3]{2^2}} \cdot \dfrac{\sqrt[3]{2}}{\sqrt[3]{2}} = \dfrac{7\sqrt[3]{2}}{\sqrt[3]{2^3}} = \dfrac{7\sqrt[3]{2}}{2}$

47. $\sqrt[3]{\dfrac{4}{w^2}} = \dfrac{\sqrt[3]{4}}{\sqrt[3]{w^2}} = \dfrac{\sqrt[3]{4}}{\sqrt[3]{w^2}} \cdot \dfrac{\sqrt[3]{w}}{\sqrt[3]{w}} = \dfrac{\sqrt[3]{4}\sqrt[3]{w}}{\sqrt[3]{w^3}} = \dfrac{\sqrt[3]{4w}}{w}$

49. $\sqrt[4]{\dfrac{16}{3}} = \dfrac{\sqrt[4]{16}}{\sqrt[4]{3}} = \dfrac{2}{\sqrt[4]{3}} \cdot \dfrac{\sqrt[4]{3^3}}{\sqrt[4]{3^3}} = \dfrac{2\sqrt[4]{3^3}}{\sqrt[4]{3^4}} = \dfrac{2\sqrt[4]{27}}{3}$

49. $\sqrt[4]{\dfrac{16}{3}} = \dfrac{\sqrt[4]{16}}{\sqrt[4]{3}} = \dfrac{2}{\sqrt[4]{3}} \cdot \dfrac{\sqrt[4]{3^3}}{\sqrt[4]{3^3}} = \dfrac{2\sqrt[4]{3^3}}{\sqrt[4]{3^4}} = \dfrac{2\sqrt[4]{27}}{3}$

51. $\dfrac{2}{\sqrt[3]{4x^2}} = \dfrac{2}{\sqrt[3]{2^2x^2}} \cdot \dfrac{\sqrt[3]{2x}}{\sqrt[3]{2x}} = \dfrac{2\sqrt[3]{2x}}{\sqrt[3]{2^3x^3}}$

$= \dfrac{2\sqrt[3]{2x}}{2x} = \dfrac{\sqrt[3]{2x}}{x}$

53. $\dfrac{8}{7\sqrt{24}} = \dfrac{8}{7\sqrt{4 \cdot 6}} = \dfrac{8}{7 \cdot 2\sqrt{6}} = \dfrac{4}{7\sqrt{6}} \cdot \dfrac{\sqrt{6}}{\sqrt{6}}$

$= \dfrac{4\sqrt{6}}{7\sqrt{6^2}} = \dfrac{4\sqrt{6}}{7 \cdot 6} = \dfrac{2\sqrt{6}}{21}$

55. $\dfrac{1}{\sqrt{x^7}} = \dfrac{1}{\sqrt{x^6 \cdot x}} = \dfrac{1}{x^3\sqrt{x}} = \dfrac{1}{x^3\sqrt{x}} \cdot \dfrac{\sqrt{x}}{\sqrt{x}}$

$= \dfrac{\sqrt{x}}{x^3\sqrt{x^2}} = \dfrac{\sqrt{x}}{x^3 \cdot x} = \dfrac{\sqrt{x}}{x^4}$

57.
$$\frac{2}{\sqrt{8x^5}} = \frac{2}{\sqrt{4x^4 \cdot 2x}} = \frac{2}{2x^2\sqrt{2x}}$$
$$= \frac{1}{x^2\sqrt{2x}} \cdot \frac{\sqrt{2x}}{\sqrt{2x}} = \frac{\sqrt{2x}}{x^2\sqrt{2^2 x^2}}$$
$$= \frac{\sqrt{2x}}{x^2 \cdot 2x} = \frac{\sqrt{2x}}{2x^3}$$

59. $\sqrt{2} + \sqrt{6}$

61. $\sqrt{x} - 23$

63.
$$\frac{4}{\sqrt{2}+3} = \frac{4}{\sqrt{2}+3} \cdot \frac{\sqrt{2}-3}{\sqrt{2}-3} = \frac{4(\sqrt{2}-3)}{(\sqrt{2})^2 - 3^2}$$
$$= \frac{4\sqrt{2}-12}{2-9} = \frac{4\sqrt{2}-12}{-7}$$

65.
$$\frac{8}{\sqrt{6}-2} = \frac{8}{\sqrt{6}-2} \cdot \frac{\sqrt{6}+2}{\sqrt{6}+2} = \frac{8(\sqrt{6}+2)}{(\sqrt{6})^2 - 2^2}$$
$$= \frac{8(\sqrt{6}+2)}{6-4} = \frac{8(\sqrt{6}+2)}{2}$$
$$= 4(\sqrt{6}+2) = 4\sqrt{6}+8$$

67.
$$\frac{\sqrt{7}}{\sqrt{3}+2} = \frac{\sqrt{7}}{\sqrt{3}+2} \cdot \frac{\sqrt{3}-2}{\sqrt{3}-2} = \frac{\sqrt{7}(\sqrt{3}-2)}{(\sqrt{3})^2 - 2^2}$$
$$= \frac{\sqrt{7}\cdot\sqrt{3}-\sqrt{7}\cdot(2)}{3-4} = \frac{\sqrt{21}-2\sqrt{7}}{-1}$$
$$= -\sqrt{21}+2\sqrt{7}$$

69.
$$\frac{-1}{\sqrt{p}+\sqrt{q}} = \frac{-1}{\sqrt{p}+\sqrt{q}} \cdot \frac{\sqrt{p}-\sqrt{q}}{\sqrt{p}-\sqrt{q}}$$
$$= \frac{-1(\sqrt{p}-\sqrt{q})}{(\sqrt{p})^2 - (\sqrt{q})^2} = \frac{-\sqrt{p}+\sqrt{q}}{p-q}$$

71.
$$\frac{x-5}{\sqrt{x}+\sqrt{5}} = \frac{x-5}{\sqrt{x}+\sqrt{5}} \cdot \frac{\sqrt{x}-\sqrt{5}}{\sqrt{x}-\sqrt{5}}$$
$$= \frac{(x-5)(\sqrt{x}-\sqrt{5})}{(\sqrt{x})^2 - (\sqrt{5})^2}$$
$$= \frac{(\cancel{x-5})(\sqrt{x}-\sqrt{5})}{\cancel{x-5}} = \sqrt{x}-\sqrt{5}$$

73.
$$\frac{\sqrt{w}+2}{9-\sqrt{w}} = \frac{\sqrt{w}+2}{9-\sqrt{w}} \cdot \frac{9+\sqrt{w}}{9+\sqrt{w}}$$
$$= \frac{(\sqrt{w}+2)(9+\sqrt{w})}{(9)^2 - (\sqrt{w})^2}$$
$$= \frac{w+9\sqrt{w}+2\sqrt{w}+18}{81-w}$$
$$= \frac{w+11\sqrt{w}+18}{81-w}$$

75.
$$\frac{3\sqrt{x}-\sqrt{y}}{\sqrt{y}+\sqrt{x}} = \frac{3\sqrt{x}-\sqrt{y}}{\sqrt{y}+\sqrt{x}} \cdot \frac{\sqrt{y}-\sqrt{x}}{\sqrt{y}-\sqrt{x}}$$
$$= \frac{(3\sqrt{x}-\sqrt{y})(\sqrt{y}-\sqrt{x})}{(\sqrt{y})^2 - (\sqrt{x})^2}$$
$$= \frac{3\sqrt{xy}-3\sqrt{x^2}-\sqrt{y^2}+\sqrt{xy}}{y-x}$$
$$= \frac{4\sqrt{xy}-3x-y}{y-x}$$

77.

$$\frac{3\sqrt{10}}{2+\sqrt{10}} = \frac{3\sqrt{10}}{2+\sqrt{10}} \cdot \frac{2-\sqrt{10}}{2-\sqrt{10}}$$

$$= \frac{3\sqrt{10}\left(2-\sqrt{10}\right)}{(2)^2 - \left(\sqrt{10}\right)^2} = \frac{6\sqrt{10}-3\sqrt{100}}{4-10}$$

$$= \frac{6\sqrt{10}-3\cdot 10}{-6} = \frac{6\sqrt{10}-30}{-6}$$

$$= \frac{\cancel{6}\left(5-\sqrt{10}\right)}{\cancel{6}} = 5-\sqrt{10}$$

79.

$$\frac{2\sqrt{3}+\sqrt{7}}{3\sqrt{3}-\sqrt{7}} = \frac{2\sqrt{3}+\sqrt{7}}{3\sqrt{3}-\sqrt{7}} \cdot \frac{3\sqrt{3}+\sqrt{7}}{3\sqrt{3}+\sqrt{7}}$$

$$= \frac{\left(2\sqrt{3}+\sqrt{7}\right)\left(3\sqrt{3}+\sqrt{7}\right)}{\left(3\sqrt{3}\right)^2 - \left(\sqrt{7}\right)^2}$$

$$= \frac{6\sqrt{9}+2\sqrt{21}+3\sqrt{21}+\sqrt{49}}{9\cdot 3-7}$$

$$= \frac{6\cdot 3+5\sqrt{21}+7}{27-7} = \frac{18+5\sqrt{21}+7}{20}$$

$$= \frac{25+5\sqrt{21}}{20} = \frac{\cancel{5}\left(5+\sqrt{21}\right)}{\cancel{5}\cdot 4}$$

$$= \frac{5+\sqrt{21}}{4}$$

81.

$$\frac{\sqrt{5}+4}{2-\sqrt{5}} = \frac{\sqrt{5}+4}{2-\sqrt{5}} \cdot \frac{2+\sqrt{5}}{2+\sqrt{5}}$$

$$= \frac{\left(\sqrt{5}+4\right)\left(2+\sqrt{5}\right)}{(2)^2 - \left(\sqrt{5}\right)^2}$$

$$= \frac{2\sqrt{5}+\sqrt{25}+8+4\sqrt{5}}{4-5}$$

$$= \frac{6\sqrt{5}+5+8}{-1} = \frac{6\sqrt{5}+13}{-1}$$

$$= -6\sqrt{5}-13$$

83.

$$\frac{16}{\sqrt[3]{4}} = \frac{16}{\sqrt[3]{2^2}} \cdot \frac{\sqrt[3]{2}}{\sqrt[3]{2}} = \frac{16\sqrt[3]{2}}{\sqrt[3]{2^3}} = \frac{16\sqrt[3]{2}}{2} = 8\sqrt[3]{2}$$

85.

$$\frac{4}{x-\sqrt{2}} = \frac{4}{x-\sqrt{2}} \cdot \frac{x+\sqrt{2}}{x+\sqrt{2}}$$

$$= \frac{4\left(x+\sqrt{2}\right)}{x^2 - \left(\sqrt{2}\right)^2} = \frac{4x+4\sqrt{2}}{x^2-2}$$

87.

$$T(x) = 2\pi\sqrt{\frac{x}{32}}$$

$$T(1) = 2\pi\sqrt{\frac{1}{32}} = 2\pi\frac{1}{\sqrt{16\cdot 2}} = \frac{2\pi}{4\sqrt{2}}$$

$$= \frac{2\pi}{4\sqrt{2}} \cdot \frac{\sqrt{2}}{\sqrt{2}} = \frac{2\pi\sqrt{2}}{4\cdot 2} = \frac{\pi\sqrt{2}}{4}\ \text{sec}$$

$$\approx 1.11\ \text{sec}$$

89.

a. $$\frac{1}{\sqrt{2}} = \frac{1}{\sqrt{2}} \cdot \frac{\sqrt{2}}{\sqrt{2}} = \frac{\sqrt{2}}{\sqrt{2^2}} = \frac{\sqrt{2}}{2}$$

b. $$\frac{1}{\sqrt[3]{2}} = \frac{1}{\sqrt[3]{2}} \cdot \frac{\sqrt[3]{2^2}}{\sqrt[3]{2^2}} = \frac{\sqrt[3]{2^2}}{\sqrt[3]{2^3}} = \frac{\sqrt[3]{4}}{2}$$

91.

a. $$\frac{1}{\sqrt{5a}} = \frac{1}{\sqrt{5a}} \cdot \frac{\sqrt{5a}}{\sqrt{5a}} = \frac{\sqrt{5a}}{\sqrt{(5a)^2}} = \frac{\sqrt{5a}}{5a}$$

b. $\dfrac{1}{\sqrt{5}+a} = \dfrac{1}{\sqrt{5}+a} \cdot \dfrac{\sqrt{5}-a}{\sqrt{5}-a} = \dfrac{\sqrt{5}-a}{\left(\sqrt{5}\right)^2 - a^2}$

$\qquad\qquad = \dfrac{\sqrt{5}-a}{5-a^2}$

93. $\dfrac{\sqrt{6}}{2} + \dfrac{1}{\sqrt{6}} = \dfrac{\sqrt{6}}{2} + \dfrac{1}{\sqrt{6}} \cdot \dfrac{\sqrt{6}}{\sqrt{6}} = \dfrac{\sqrt{6}}{2} + \dfrac{\sqrt{6}}{6}$

$\qquad = \dfrac{\sqrt{6}}{2} \cdot \dfrac{3}{3} + \dfrac{\sqrt{6}}{6} = \dfrac{3\sqrt{6}+\sqrt{6}}{6}$

$\qquad = \dfrac{4\sqrt{6}}{6} = \dfrac{2\sqrt{6}}{3}$

95. $\sqrt{15} - \sqrt{\dfrac{3}{5}} + \sqrt{\dfrac{5}{3}} = \sqrt{15} - \dfrac{\sqrt{3}}{\sqrt{5}} \cdot \dfrac{\sqrt{5}}{\sqrt{5}} + \dfrac{\sqrt{5}}{\sqrt{3}} \cdot \dfrac{\sqrt{3}}{\sqrt{3}}$

$\qquad = \dfrac{\sqrt{15}}{1} - \dfrac{\sqrt{15}}{5} + \dfrac{\sqrt{15}}{3}$

$\qquad = \dfrac{\sqrt{15}}{1} \cdot \dfrac{15}{15} - \dfrac{\sqrt{15}}{5} \cdot \dfrac{3}{3} + \dfrac{\sqrt{15}}{3} \cdot \dfrac{5}{5}$

$\qquad = \dfrac{15\sqrt{15} - 3\sqrt{15} + 5\sqrt{15}}{15} = \dfrac{17\sqrt{15}}{15}$

97. $\sqrt[3]{25} + \dfrac{3}{\sqrt[3]{5}} = \sqrt[3]{5^2} + \dfrac{3}{\sqrt[3]{5}} \cdot \dfrac{\sqrt[3]{5^2}}{\sqrt[3]{5^2}} = \dfrac{\sqrt[3]{5^2}}{1} + \dfrac{3\sqrt[3]{5^2}}{\sqrt[3]{5^3}}$

$\qquad = \dfrac{\sqrt[3]{5^2}}{1} + \dfrac{3\sqrt[3]{5^2}}{5} = \dfrac{\sqrt[3]{5^2}}{1} \cdot \dfrac{5}{5} + \dfrac{3\sqrt[3]{5^2}}{5}$

$\qquad = \dfrac{5\sqrt[3]{5^2} + 3\sqrt[3]{5^2}}{5} = \dfrac{8\sqrt[3]{5^2}}{5} = \dfrac{8\sqrt[3]{25}}{5}$

99. $\dfrac{\sqrt{3}+6}{2} = \dfrac{\sqrt{3}+6}{2} \cdot \dfrac{\sqrt{3}-6}{\sqrt{3}-6} = \dfrac{\left(\sqrt{3}\right)^2 - 6^2}{2\left(\sqrt{3}-6\right)}$

$\qquad = \dfrac{3-36}{2\sqrt{3}-12} = \dfrac{-33}{2\sqrt{3}-12}$

101. $\dfrac{\sqrt{a}-\sqrt{b}}{\sqrt{a}+\sqrt{b}} = \dfrac{\sqrt{a}-\sqrt{b}}{\sqrt{a}+\sqrt{b}} \cdot \dfrac{\sqrt{a}+\sqrt{b}}{\sqrt{a}+\sqrt{b}}$

$\qquad = \dfrac{\left(\sqrt{a}\right)^2 - \left(\sqrt{b}\right)^2}{\left(\sqrt{a}+\sqrt{b}\right)^2}$

$\qquad = \dfrac{a-b}{\left(\sqrt{a}\right)^2 + 2\sqrt{a}\cdot\sqrt{b} + \left(\sqrt{b}\right)^2}$

$\qquad = \dfrac{a-b}{a + 2\sqrt{ab} + b}$

103. $\dfrac{\sqrt{5+3h}-\sqrt{5}}{h} = \dfrac{\sqrt{5+3h}-\sqrt{5}}{h} \cdot \dfrac{\sqrt{5+3h}+\sqrt{5}}{\sqrt{5+3h}+\sqrt{5}}$

$\qquad = \dfrac{\left(\sqrt{5+3h}\right)^2 - \left(\sqrt{5}\right)^2}{h\left(\sqrt{5+3h}+\sqrt{5}\right)}$

$\qquad = \dfrac{5+3h-5}{h\left(\sqrt{5+3h}+\sqrt{5}\right)}$

$\qquad = \dfrac{3h}{h\left(\sqrt{5+3h}+\sqrt{5}\right)} = \dfrac{3}{\sqrt{5+3h}+\sqrt{5}}$

105. $\dfrac{\sqrt{4+5h}-2}{h} = \dfrac{\sqrt{4+5h}-2}{h} \cdot \dfrac{\sqrt{4+5h}+2}{\sqrt{4+5h}+2}$

$$= \dfrac{\left(\sqrt{4+5h}\right)^2 - (2)^2}{h\left(\sqrt{4+5h}+2\right)}$$

$$= \dfrac{4+5h-4}{h\left(\sqrt{4+5h}+2\right)}$$

$$= \dfrac{5h}{h\left(\sqrt{4+5h}+2\right)} = \dfrac{5}{\sqrt{4+5h}+2}$$

Section 6.7   Solving Radical Equations

1.
   a. radical
   b. isolate; 7
   c. extraneous
   d. third

3. $\sqrt{\dfrac{9w^3}{16}} = \sqrt{\dfrac{9w^2 \cdot w}{16}} = \dfrac{3w\sqrt{w}}{4}$

5. $\sqrt[3]{54c^4} = \sqrt[3]{27c^3 \cdot 2c} = 3c\sqrt[3]{2c}$

7. $\left(\sqrt{4x-6}\right)^2 = 4x-6$

9. $\left(\sqrt[3]{9p+7}\right)^3 = 9p+7$

11. $\sqrt{x} = 10$

$\left(\sqrt{x}\right)^2 = (10)^2$

$x = 100$

The solution is $\{100\}$.

Check:
$\sqrt{100} = 10$
$10 = 10$

13. $\sqrt{x} + 4 = 6$

$\sqrt{x} = 2$

$\left(\sqrt{x}\right)^2 = 2^2$

$x = 4$

The solution is $\{4\}$.

Check:
$\sqrt{4} + 4 = 6$
$2 + 4 = 6$
$6 = 6$

15. $\sqrt{5y+1} = 4$

$\left(\sqrt{5y+1}\right)^2 = 4^2$

$5y+1 = 16$

$5y = 15$

$y = 3$

The solution is $\{3\}$.

Check:
$\sqrt{5(3)+1} = 4$
$\sqrt{15+1} = 4$
$\sqrt{16} = 4$
$4 = 4$

17. $6 = \sqrt{2z-3} - 3$

$9 = \sqrt{2z-3}$

$9^2 = \left(\sqrt{2z-3}\right)^2$

$81 = 2z-3$

$2z = 84$

$z = 42$

The solution is $\{42\}$.

Check:
$6 = \left(2(42)-3\right)^{1/2} - 3$
$6 = (84-3)^{1/2} - 3$
$6 = 81^{1/2} - 3$
$6 = 9 - 3$
$6 = 6$

19. $\sqrt{x^2+5} = x+1$

$\left(\sqrt{x^2+5}\right)^2 = (x+1)^2$

$x^2+5 = x^2+2x+1$

$-2x = -4$

$x = 2$

The solution is $\{2\}$.

Check:
$\sqrt{(2)^2+5} = 2+1$
$\sqrt{4+5} = 3$
$\sqrt{9} = 43$
$3 = 3$

21.
$$\sqrt[3]{x-2}-1=2$$
$$\sqrt[3]{x-2}=3$$
$$\left(\sqrt[3]{x-2}\right)^3=3^3$$
$$x-2=27$$
$$x=29$$
Check:
$$\sqrt[3]{29-2}-1=2$$
$$\sqrt[3]{27}-1=2$$
$$3-1=2$$
$$2=2$$
The solution is $\{29\}$.

23.
$$(15-w)^{1/3}+7=2$$
$$(15-w)^{1/3}=-5$$
$$\left(\sqrt[3]{15-w}\right)^3=(-5)^3$$
$$15-w=-125$$
$$140=w$$
Check:
$$(15-w)^{1/3}+7=2$$
$$(15-140)^{1/3}+7=2$$
$$(-125)^{1/3}+7=2$$
$$-5+7=2$$
$$2=2$$
The solution is $\{140\}$.

25.
$$3+\sqrt{x-16}=0$$
$$\sqrt{x-16}=-3$$
$$\left(\sqrt{x-16}\right)^2=(-3)^2$$
$$x-16=9$$
$$x=25$$
$\{\ \}$ ($x=25$ does not check).

Check:
$$3+\sqrt{25-16}=0$$
$$3+\sqrt{9}=0$$
$$3+3=0$$
$$6\neq 0$$

27.
$$2\sqrt{6a+7}-2a=0$$
$$2\sqrt{6a+7}=2a$$
$$\sqrt{6a+7}=a$$
$$\left(\sqrt{6a+7}\right)^2=(a)^2$$
$$6a+7=a^2$$
$$a^2-6a-7=0$$
$$(a-7)(a+1)=0$$
$$a-7=0 \text{ or } a+1=0$$
$$a=7 \text{ or } \quad a=-1$$
The solution is $\{7\}$. ($a=-1$ does not check).

Check $a=7$:
$$2\sqrt{6(7)+7}-2(7)=0$$
$$2\sqrt{42+7}-14=0$$
$$2\sqrt{49}-14=0$$
$$2\cdot 7-14=0$$
$$14-14=0$$
$$0=0$$

Check $a=-1$:
$$2\sqrt{6(-1)+7}-2(-1)=0$$
$$2\sqrt{-6+7}+2=0$$
$$2\sqrt{1}+2=0$$
$$2\cdot 1+2=0$$
$$2+2=0$$
$$4\neq 0$$

29.

$(2x-5)^{1/4} = -1$

$\sqrt[4]{2x-5} = -1$

$\left(\sqrt[4]{2x-5}\right)^4 = (-1)^4$

$2x - 5 = 1$

$2x = 6$

$x = 3$

$\{\ \}$ ($x = 3$ does not check).

Check:

$\sqrt[4]{2(3)-5} = -1$

$\sqrt[4]{6-5} = -1$

$\sqrt[4]{1} = -1$

$1 \neq -1$

31.

$r = \sqrt[3]{\dfrac{3V}{4\pi}}$ for V

$r^3 = \left(\sqrt[3]{\dfrac{3V}{4\pi}}\right)^3$

$r^3 = \dfrac{3V}{4\pi}$

$4\pi r^3 = 3V$

$\dfrac{4\pi r^3}{3} = V$

33.

$r = \pi\sqrt{r^2 + h^2}$ for h^2

$\dfrac{r}{\pi} = \sqrt{r^2 + h^2}$

$\left(\dfrac{r}{\pi}\right)^2 = \left(\sqrt{r^2 + h^2}\right)^2$

$\dfrac{r^2}{\pi^2} = r^2 + h^2$

$\dfrac{r^2}{\pi^2} - r^2 = h^2$

35. $(a+5)^2 = a^2 + 2 \cdot a \cdot 5 + 5^2 = a^2 + 10a + 25$

37.

$\left(\sqrt{5a} - 3\right)^2 = \left(\sqrt{5a}\right)^2 - 2 \cdot \sqrt{5a} \cdot 3 + 3^2$

$= 5a - 6\sqrt{5a} + 9$

39.

$\left(\sqrt{r-3} + 5\right)^2 = \left(\sqrt{r-3}\right)^2 + 2 \cdot \sqrt{r-3} \cdot 5 + 5^2$

$= r - 3 + 10\sqrt{r-3} + 25$

$= r + 22 + 10\sqrt{r-3}$

41.

$\sqrt{a^2 + 2a + 1} = a + 5$

$\left(\sqrt{a^2 + 2a + 1}\right)^2 = (a+5)^2$

$a^2 + 2a + 1 = a^2 + 10a + 25$

$-8a = 24$

$a = -3$

The solution is $\{-3\}$.

Check:

$\sqrt{(-3)^2 + 2(-3) + 1} = -3 + 5$

$\sqrt{9 - 6 + 1} = 2$

$\sqrt{4} = 2$

$2 = 2$

43.

$$\sqrt{25w^2 - 2w - 3} = 5w - 4$$

$$\left(\sqrt{25w^2 - 2w - 3}\right)^2 = (5w - 4)^2$$

$$25w^2 - 2w - 3 = 25w^2 - 40w + 16$$

$$38w = 19$$

$$w = \frac{1}{2}$$

Check:

$$\sqrt{25\left(\frac{1}{2}\right)^2 - 2\left(\frac{1}{2}\right) - 3} = 5\left(\frac{1}{2}\right) - 4$$

$$\sqrt{\frac{25}{4} - 1 - 3} = \frac{5}{2} - 4$$

$$\sqrt{\frac{9}{4}} = -\frac{3}{2}$$

$$\frac{3}{2} \neq -\frac{3}{2}$$

$\{\ \}$ ($w = \frac{1}{2}$ does not check.)

45.

$$4\sqrt{p - 2} - 2 = -p$$

$$4\sqrt{p - 2} = -p + 2$$

$$\left(4\sqrt{p - 2}\right)^2 = (-p + 2)^2$$

$$16(p - 2) = p^2 - 4p + 4$$

$$16p - 32 = p^2 - 4p + 4$$

$$0 = p^2 - 20p + 36$$

$$0 = (p - 18)(p - 2)$$

$$p - 18 = 0 \ \text{ or } p - 2 = 0$$

$$p = 18 \ \text{or} \qquad p = 2$$

The solution is $\{2\}$. ($p = 18$ does not check.)

Check $p = 18$:

$$4\sqrt{18 - 2} - 2 = -18$$

$$4\sqrt{16} - 2 = -18$$

$$4(4) - 2 = -18$$

$$16 - 2 = -18$$

$$14 \neq -18$$

Check $p = 2$:

$$4\sqrt{2 - 2} - 2 = -2$$

$$4\sqrt{0} - 2 = -2$$

$$4(0) - 2 = -2$$

$$0 - 2 = -2$$

$$-2 = -2$$

47.

$$\sqrt[4]{h + 4} = \sqrt[4]{2h - 5}$$

$$\left(\sqrt[4]{h + 4}\right)^4 = \left(\sqrt[4]{2h - 5}\right)^4$$

$$h + 4 = 2h - 5$$

$$9 = h$$

The solution is $\{9\}$.

Check:

$$\sqrt[4]{9 + 4} = \sqrt[4]{2(9) - 5}$$

$$\sqrt[4]{13} = \sqrt[4]{18 - 5}$$

$$\sqrt[4]{13} = \sqrt[4]{13}$$

49.

$$\sqrt[3]{5a + 3} - \sqrt[3]{a - 13} = 0$$

$$\sqrt[3]{5a + 3} = \sqrt[3]{a - 13}$$

$$\left(\sqrt[3]{5a + 3}\right)^3 = \left(\sqrt[3]{a - 13}\right)^3$$

$$5a + 3 = a - 13$$

$$4a = -16$$

$$a = -4$$

The solution is $\{-4\}$.

Check:

$$\sqrt[3]{5(-4) + 3} - \sqrt[3]{-4 - 13} = 0$$

$$\sqrt[3]{-20 + 3} - \sqrt[3]{-17} = 0$$

$$\sqrt[3]{-17} - \sqrt[3]{-17} = 0$$

$$0 = 0$$

51.

$$\sqrt{5a-9} = \sqrt{5a} - 3$$

$$\left(\sqrt{5a-9}\right)^2 = \left(\sqrt{5a} - 3\right)^2$$

$$5a - 9 = 5a - 6\sqrt{5a} + 9$$

$$6\sqrt{5a} = 18$$

$$\sqrt{5a} = 3$$

$$\left(\sqrt{5a}\right)^2 = 3^2$$

$$5a = 9$$

$$a = \frac{9}{5}$$

The solution is $\left\{\frac{9}{5}\right\}$.

Check:

$$\sqrt{5\left(\frac{9}{5}\right) - 9} = \sqrt{5\left(\frac{9}{5}\right)} - 3$$

$$\sqrt{9-9} = \sqrt{9} - 3$$

$$\sqrt{0} = 3 - 3$$

$$0 = 0$$

53.

$$\sqrt{2h+5} - \sqrt{2h} = 1$$

$$\sqrt{2h+5} = \sqrt{2h} + 1$$

$$\left(\sqrt{2h+5}\right)^2 = \left(\sqrt{2h} + 1\right)^2$$

$$2h + 5 = 2h + 2\sqrt{2h} + 1$$

$$4 = 2\sqrt{2h}$$

$$\sqrt{2h} = 2$$

$$\left(\sqrt{2h}\right)^2 = 2^2$$

$$2h = 4$$

$$h = 2$$

The solution is $\{2\}$.

Check:

$$\sqrt{2(2)+5} - \sqrt{2(2)} = 1$$

$$\sqrt{4+5} - \sqrt{4} = 1$$

$$\sqrt{9} - \sqrt{4} = 1$$

$$3 - 2 = 1$$

$$1 = 1$$

55.

$$(t-9)^{1/2} - t^{1/2} = 3$$

$$\sqrt{t-9} - \sqrt{t} = 3$$

$$\sqrt{t-9} = \sqrt{t} + 3$$

$$\left(\sqrt{t-9}\right)^2 = \left(\sqrt{t} + 3\right)^2$$

$$t - 9 = t + 6\sqrt{t} + 9$$

$$-18 = 6\sqrt{t}$$

$$\sqrt{t} = -3$$

$$\left(\sqrt{t}\right)^2 = (-3)^2$$

$$t = 9$$

$\{\ \}$ (The value 9 does not check.)

Check:

$$(9-9)^{1/2} - 9^{1/2} = 3$$

$$\sqrt{9-9} - \sqrt{9} = 3$$

$$\sqrt{0} - \sqrt{9} = 3$$

$$0 - 3 = 3$$

$$-3 \neq 3$$

57.

$$6 = \sqrt{x^2 + 3} - x$$
$$x + 6 = \sqrt{x^2 + 3}$$
$$(x + 6)^2 = \left(\sqrt{x^2 + 3}\right)^2$$
$$x^2 + 12x + 36 = x^2 + 3$$
$$12x = -33$$
$$x = -\frac{33}{12} = -\frac{11}{4}$$

Check:

$$6 = \sqrt{\left(-\frac{11}{4}\right)^2 + 3} - \left(-\frac{11}{4}\right)$$
$$6 = \sqrt{\frac{121}{16} + 3} + \frac{11}{4}$$
$$6 = \sqrt{\frac{169}{16}} + \frac{11}{4}$$
$$6 = \frac{13}{4} + \frac{11}{4}$$
$$6 = \frac{24}{4}$$
$$6 = 6$$

The solution is $\left\{-\frac{11}{4}\right\}$.

59.

$$\sqrt{3t - 7} = 2 - \sqrt{3t + 1}$$
$$\left(\sqrt{3t - 7}\right)^2 = \left(2 - \sqrt{3t + 1}\right)^2$$
$$3t - 7 = 4 - 4\sqrt{3t + 1} + 3t + 1$$
$$-12 = -4\sqrt{3t + 1}$$
$$\sqrt{3t + 1} = 3$$
$$\left(\sqrt{3t + 1}\right)^2 = (3)^2$$
$$3t + 1 = 9$$
$$3t = 8$$
$$t = \frac{8}{3}$$

$\{\ \}$ ($t = \frac{8}{3}$ does not check.)

Check:

$$\sqrt{3\left(\frac{8}{3}\right) - 7} = 2 - \sqrt{3\left(\frac{8}{3}\right) + 1}$$
$$\sqrt{8 - 7} = 2 - \sqrt{8 + 1}$$
$$\sqrt{1} = 2 - \sqrt{9}$$
$$1 = 2 - 3$$
$$1 \neq -1$$

61. $\sqrt{z+1} + \sqrt{2z+3} = 1$

$\sqrt{2z+3} = 1 - \sqrt{z+1}$

$\left(\sqrt{2z+3}\right)^2 = \left(1-\sqrt{z+1}\right)^2$

$2z+3 = 1 - 2\sqrt{z+1} + z + 1$

$z+1 = -2\sqrt{z+1}$

$\left(z+1\right)^2 = \left(-2\sqrt{z+1}\right)^2$

$z^2 + 2z + 1 = 4(z+1)$

$z^2 + 2z + 1 = 4z + 4$

$z^2 - 2z - 3 = 0$

$(z-3)(z+1) = 0$

$z - 3 = 0 \text{ or } z + 1 = 0$

$z = 3 \text{ or } \quad z = -1$

The solution is $\{-1\}$. ($z = 3$ does not check.)

Check $z = 3$:

$\sqrt{3+1} + \sqrt{2(3)+3} = 1$

$\sqrt{4} + \sqrt{6+3} = 1$

$\sqrt{4} + \sqrt{9} = 1$

$2 + 3 = 1$

$5 \neq 1$

Check $z = -1$:

$\sqrt{-1+1} + \sqrt{2(-1)+3} = 1$

$\sqrt{0} + \sqrt{-2+3} = 1$

$\sqrt{0} + \sqrt{1} = 1$

$0 + 1 = 1$

$1 = 1$

63. $\sqrt{6m+7} - \sqrt{3m+3} = 1$

$\sqrt{6m+7} = 1 + \sqrt{3m+3}$

$\left(\sqrt{6m+7}\right)^2 = \left(1+\sqrt{3m+3}\right)^2$

$6m+7 = 1 + 2\sqrt{3m+3} + 3m + 3$

$3m+3 = 2\sqrt{3m+3}$

$\left(3m+3\right)^2 = \left(2\sqrt{3m+3}\right)^2$

$9m^2 + 18m + 9 = 4(3m+3)$

$9m^2 + 18m + 9 = 12m + 12$

$9m^2 + 6m - 3 = 0$

$3\left(3m^2 + 2m - 1\right) = 0$

$3(3m-1)(m+1) = 0$

$3m - 1 = 0 \text{ or } m + 1 = 0$

$3m = 1 \text{ or } \quad m = -1$

$m = \dfrac{1}{3} \text{ or } \quad m = -1$

The solution is $\left\{\frac{1}{3}, -1\right\}$.

Check $m = \dfrac{1}{3}$:

$\sqrt{6\left(\frac{1}{3}\right)+7} - \sqrt{3\left(\frac{1}{3}\right)+3} = 1$

$\sqrt{2+7} - \sqrt{1+3} = 1$

$\sqrt{9} - \sqrt{4} = 1$

$3 - 2 = 1$

$1 = 1$

Check $m = -1$:

$\sqrt{6(-1)+7} - \sqrt{3(-1)+3} = 1$

$\sqrt{-6+7} - \sqrt{-3+3} = 1$

$\sqrt{1} - \sqrt{0} = 1$

$1 - 0 = 1$

$1 = 1$

65.

$$2 + 2\sqrt{2t+3} + 2\sqrt{3t-5} = 0$$

$$2 + 2\sqrt{2t+3} = -2\sqrt{3t-5}$$

$$1 + \sqrt{2t+3} = -\sqrt{3t-5}$$

$$\left(1 + \sqrt{2t+3}\right)^2 = \left(-\sqrt{3t-5}\right)^2$$

$$1 + 2\sqrt{2t+3} + 2t + 3 = 3t - 5$$

$$2\sqrt{2t+3} = t - 9$$

$$\left(2\sqrt{2t+3}\right)^2 = (t-9)^2$$

$$4(2t+3) = t^2 - 18t + 81$$

$$8t + 12 = t^2 - 18t + 81$$

$$t^2 - 26t + 69 = 0$$

$$(t-3)(t-23) = 0$$

$$t - 3 = 0 \text{ or } t - 23 = 0$$

$$t = 3 \text{ or } \quad t = 23$$

$\{\,\}$ ($t = 3$ and $t = 23$ do not check.)

Check $t = 3$:

$$2 + 2\sqrt{2(3)+3} + 2\sqrt{3(3)-5} = 0$$

$$2 + 2\sqrt{6+3} + 2\sqrt{9-5} = 0$$

$$2 + 2\sqrt{9} + 2\sqrt{4} = 0$$

$$2 + 2\cdot 3 + 2\cdot 2 = 0$$

$$2 + 6 + 4 = 0$$

$$12 \neq 0$$

Check $t = 23$:

$$2 + 2\sqrt{2(23)+3} + 2\sqrt{3(23)-5} = 0$$

$$2 + 2\sqrt{46+3} + 2\sqrt{69-5} = 0$$

$$2 + 2\sqrt{49} + 2\sqrt{64} = 0$$

$$2 + 2\cdot 7 + 2\cdot 8 = 0$$

$$2 + 14 + 16 = 0$$

$$32 \neq 0$$

67.

$$3\sqrt{y-3} = \sqrt{4y+3}$$

$$\left(3\sqrt{y-3}\right)^2 = \left(\sqrt{4y+3}\right)^2$$

$$9(y-3) = 4y+3$$

$$9y - 27 = 4y + 3$$

$$5y = 30$$

$$y = 6$$

The solution is $\{6\}$.

Check:

$$3\sqrt{6-3} = \sqrt{4(6)+3}$$

$$3\sqrt{3} = \sqrt{24+3}$$

$$3\sqrt{3} = \sqrt{27}$$

$$3\sqrt{3} = 3\sqrt{3}$$

69.

$$\sqrt{p+7} = \sqrt{2p}+1$$

$$\left(\sqrt{p+7}\right)^2 = \left(\sqrt{2p}+1\right)^2$$

$$p + 7 = 2p + 2\sqrt{2p} + 1$$

$$-p + 6 = 2\sqrt{2p}$$

$$(-p+6)^2 = \left(2\sqrt{2p}\right)^2$$

$$p^2 - 12p + 36 = 4(2p)$$

$$p^2 - 12p + 36 = 8p$$

$$p^2 - 20p + 36 = 0$$

$$(p-18)(p-2) = 0$$

$$p - 18 = 0 \text{ or } p - 2 = 0$$

$$p = 18 \text{ or } \quad p = 2$$

The solution is $\{2\}$. ($p = 18$ does not check.)

Check $p = 18$:

$$\sqrt{18+7} = \sqrt{2\cdot 18}+1$$

$$\sqrt{25} = \sqrt{36}+1$$

$$5 = 6 + 1$$

$$5 \neq 7$$

Check $p = 2$:

$$\sqrt{2+7} = \sqrt{2\cdot 2}+1$$

$$\sqrt{9} = \sqrt{4}+1$$

$$3 = 2 + 1$$

$$3 = 3$$

71. $v = \sqrt{2gh}$

   a. $44 = \sqrt{2(32)h}$

       $44 = \sqrt{64h}$

       $44 = 8\sqrt{h}$

       $11 = 2\sqrt{h}$

       $11^2 = \left(2\sqrt{h}\right)^2$

       $121 = 4h$

       $h = \dfrac{121}{4} = 30.25$ ft

   b. $26 = \sqrt{2(9.8)h}$

       $26 = \sqrt{19.6h}$

       $26^2 = \left(\sqrt{19.6h}\right)^2$

       $676 = 19.6h$

       $h = \dfrac{676}{19.6} \approx 34.5$ m

73. $C(x) = \sqrt{0.3x + 1}$

   a. $C(x) = \sqrt{0.3(10) + 1} = \sqrt{3 + 1} = \sqrt{4}$

       $= \$2$ million

   b. $P(x) = R(x) - C(x)$

       $P(x) = 320(10,000) - 2,000,000$

           $= 3,200,000 - 2,000,000$

           $= \$1.2$ million

   c. $4 = \sqrt{0.3x + 1}$

       $4^2 = \left(\sqrt{0.3x + 1}\right)^2$

       $16 = 0.3x + 1$

       $15 = 0.3x$

       $x = \dfrac{15}{0.3} = 50$ (50,000 passengers)

75. $t(x) = 0.90\sqrt[5]{x^3}$

   a. $4 = 0.90\sqrt[5]{x^3}$

       $\dfrac{4}{0.90} = \sqrt[5]{x^3}$

       $\left(\dfrac{4}{0.90}\right)^5 = \left(\sqrt[5]{x^3}\right)^5$

       $1734.15 = x^3$

         $x = \sqrt[3]{1734.15} \approx 12$ lb

   b. $t(18) = 0.90\sqrt[5]{18^3} \approx 5.1$ hr

       An 18-lb turkey will take about 5.1 hr to cook.

77. $a^2 + b^2 = c^2$

      $h^2 + b^2 = 5^2$

        $b^2 = 5^2 - h^2$

         $b = \sqrt{25 - h^2}$

79. $a^2 + b^2 = c^2$

      $a^2 + 14^2 = k^2$

        $a^2 = k^2 - 14^2$

         $a = \sqrt{k^2 - 196}$

Section 6.8   Complex Numbers

1.
  a. imaginary
  b. $\sqrt{-1}$; -1
  c. $i\sqrt{b}$
  d. $a+bi$; $\sqrt{-1}$
  e. real; b
  f. $a+bi$
  g. True
  h. True

3. $\left(3-\sqrt{x}\right)\left(3+\sqrt{x}\right)=3^2-\left(\sqrt{x}\right)^2=9-x$

5. $\sqrt[3]{3p+7}-\sqrt[3]{2p-1}=0$

$$\sqrt[3]{3p+7}=\sqrt[3]{2p-1}$$
$$\left(\sqrt[3]{3p+7}\right)^3=\left(\sqrt[3]{2p-1}\right)^3$$
$$3p+7=2p-1$$
$$p=-8$$

The solution is $\{-8\}$.

Check:
$$\sqrt[3]{3(-8)+7}-\sqrt[3]{2(-8)-1}=0$$
$$\sqrt[3]{-24+7}-\sqrt[3]{-16-1}=0$$
$$\sqrt[3]{-17}-\sqrt[3]{-17}=0$$
$$0=0$$

7. $\sqrt{4a+29}=2\sqrt{a}+5$

$$\left(\sqrt{4a+29}\right)^2=\left(2\sqrt{a}+5\right)^2$$
$$4a+29=4a+20\sqrt{a}+25$$
$$4=20\sqrt{a}$$
$$5\sqrt{a}=1$$
$$\left(5\sqrt{a}\right)^2=1^2$$
$$25a=1$$
$$a=\frac{1}{25}$$

The solution is $\left\{\frac{1}{25}\right\}$.

Check:
$$\sqrt{4\left(\frac{1}{25}\right)+29}=2\sqrt{\frac{1}{25}}+5$$
$$\sqrt{\frac{4}{25}+29}=2\cdot\frac{1}{5}+5$$
$$\sqrt{\frac{729}{25}}=\frac{2}{5}+5$$
$$\frac{27}{5}=\frac{27}{5}$$

9. $\sqrt{-1}=i$ and $-\sqrt{1}=-1$

11. $\sqrt{-144}=i\sqrt{144}=12i$

13. $\sqrt{-3}=i\sqrt{3}$

15. $-\sqrt{-20}=-i\sqrt{20}=-i\sqrt{4\cdot5}=-2i\sqrt{5}$

17. $2\sqrt{-25}\cdot3\sqrt{-4}=2i\sqrt{25}\cdot3i\sqrt{4}=5\cdot2i\cdot2\cdot3i$
$$=10i\cdot6i=60i^2=60(-1)=-60$$

19. $7\sqrt{-63}-4\sqrt{-28}=7\sqrt{-9\cdot7}-4\sqrt{-4\cdot7}$
$$=7\cdot3i\sqrt{7}-4\cdot2i\sqrt{7}$$
$$=21i\sqrt{7}-8i\sqrt{7}=13i\sqrt{7}$$

21. $\sqrt{-7}\cdot\sqrt{-7}=i\sqrt{7}\cdot i\sqrt{7}=i^2\sqrt{49}=-1\cdot7=-7$

23. $\sqrt{-11}\cdot\sqrt{-11}=i\sqrt{11}\cdot i\sqrt{11}=i^2\sqrt{121}$
$$=-1\cdot11=-11$$

25. $\sqrt{-15} \cdot \sqrt{-6} = i\sqrt{15} \cdot i\sqrt{6} = i^2\sqrt{90}$

$\qquad = -1\sqrt{9 \cdot 10} = -3\sqrt{10}$

27. $\dfrac{\sqrt{-50}}{\sqrt{25}} = \dfrac{\sqrt{-25 \cdot 2}}{5} = \dfrac{5i\sqrt{2}}{5} = i\sqrt{2}$

29. $\dfrac{\sqrt{-90}}{\sqrt{-10}} = \dfrac{\sqrt{-9 \cdot 10}}{\sqrt{-1 \cdot 10}} = \dfrac{3i\sqrt{10}}{i\sqrt{10}} = 3$

31. $i^7 = i^4 \cdot i^3 = 1(-i) = -i$

33. $i^{64} = \left(i^4\right)^{16} = 1^{64} = 1$

35. $i^{41} = i^{40} \cdot i = \left(i^4\right)^{10} \cdot i = 1^{10} \cdot i = 1 \cdot i = i$

37. $i^{52} = \left(i^4\right)^{13} = 1^{13} = 1$

39. $i^{23} = i^{20} \cdot i^3 = \left(i^4\right)^5 \cdot i^3 = 1^5(-i) = 1(-i) = -i$

41. $i^6 = i^4 \cdot i^2 = 1(-1) = -1$

43. $a - bi$

45. $-5 + 12i$
Real part: -5;   Imaginary part: 12

47. $-6i = 0 - 6i$
Real part: 0;   Imaginary part: -6

49. $35 = 35 + 0i$
Real part: 35;   Imaginary part: 0

51. $\dfrac{3}{5} + i$

Real part: $\dfrac{3}{5}$;   Imaginary part: 1

53. $(2 - i) + (5 + 7i) = (2 + 5) + (-1 + 7)i$

$\qquad = 7 + 6i$

55. $\left(\dfrac{1}{2} + \dfrac{2}{3}i\right) - \left(\dfrac{1}{5} - \dfrac{5}{6}i\right) = \dfrac{1}{2} + \dfrac{2}{3}i - \dfrac{1}{5} + \dfrac{5}{6}i$

$\qquad = \left(\dfrac{1}{2} - \dfrac{1}{5}\right) + \left(\dfrac{2}{3} + \dfrac{5}{6}\right)i$

$\qquad = \left(\dfrac{5}{10} - \dfrac{2}{10}\right) + \left(\dfrac{4}{6} + \dfrac{5}{6}\right)i$

$\qquad = \dfrac{3}{10} + \dfrac{9}{6}i = \dfrac{3}{10} + \dfrac{3}{2}i$

57. $\sqrt{-98} - \sqrt{-8} = i\sqrt{98} - i\sqrt{8}$

$\qquad = i\sqrt{7^2 \cdot 2} - i\sqrt{2^2 \cdot 2}$

$\qquad = 7i\sqrt{2} - 2i\sqrt{2}$

$\qquad = 5i\sqrt{2}$

59. $(2 + 3i) - (1 - 4i) + (-2 + 3i)$

$\qquad = 2 + 3i - 1 + 4i - 2 + 3i$

$\qquad = (2 - 1 - 2) + (3 + 4 + 3)i$

$\qquad = -1 + 10i$

61. $(8i)(3i) = 24i^2 = 24(-1) = -24 + 0i$

63. $6i(1 - 3i) = 6i(1) - 6i(3i) = 6i - 18i^2$

$\qquad = 6i - 18(-1) = 18 + 6i$

65. $(2-10i)(3+2i)$

$$= 2(3)+2(2i)-10i(3)-10i(2i)$$
$$= 6+4i-30i-20i^2$$
$$= 6-26i-20(-1)$$
$$= 6-26i+20 = 26-26i$$

67. $(-5+2i)(5+2i)$

$$= -5(5)-5(2i)+2i(5)+(2i)(2i)$$
$$= -25-10i+10i+4i^2$$
$$= -25+4(-1)=-25-4$$
$$= -29+0i$$

69. $(4+5i)^2 = 4^2+2\cdot4\cdot5i+(5i)^2$

$$= 16+40i+25i^2 = 16+40i+25(-1)$$
$$= 16+40i-25 = -9+40i$$

71. $(2+i)(3-2i)(4+3i)$

$$= \left[2\cdot3-2\cdot2i+i\cdot3-i(2i)\right](4+3i)$$
$$= \left(6-4i+3i-2i^2\right)(4+3i)$$
$$= \left(6-i-2(-1)\right)(4+3i)$$
$$= (6-i+2)(4+3i)=(8-i)(4+3i)$$
$$= 8\cdot4+8\cdot3i-i\cdot4-i(3i)$$
$$= 32+24i-4i-3i^2$$
$$= 32+20i-3(-1)$$
$$= 32+20i+3 = 35+20i$$

73. $(-4-6i)^2 = (-4)^2+2\cdot4\cdot6i+(6i)^2$

$$= 16+48i+36i^2$$
$$= 16+48i+36(-1)$$
$$= 16+48i-36 = -20+48i$$

75. $\left(-\dfrac{1}{2}-\dfrac{3}{4}i\right)\left(-\dfrac{1}{2}+\dfrac{3}{4}i\right)=\left(-\dfrac{1}{2}\right)^2-\left(\dfrac{3}{4}i\right)^2$

$$= \dfrac{1}{4}-\dfrac{9}{16}i^2 = \dfrac{1}{4}-\dfrac{9}{16}(-1)$$
$$= \dfrac{1}{4}+\dfrac{9}{16} = \dfrac{4}{16}+\dfrac{9}{16} = \dfrac{13}{16}+0i$$

77. $\dfrac{2}{1+3i} = \dfrac{2}{1+3i}\cdot\dfrac{1-3i}{1-3i} = \dfrac{2(1-3i)}{1^2-(3i)^2} = \dfrac{2(1-3i)}{1-9i^2}$

$$= \dfrac{2(1-3i)}{1-9(-1)} = \dfrac{2(1-3i)}{1+9} = \dfrac{2(1-3i)}{10}$$
$$= \dfrac{1-3i}{5} = \dfrac{1}{5}-\dfrac{3}{5}i$$

79. $\dfrac{-i}{4-3i} = \dfrac{-i}{4-3i}\cdot\dfrac{4+3i}{4+3i} = \dfrac{-i(4+3i)}{4^2-(3i)^2}$

$$= \dfrac{-i\cdot4-i(3i)}{16-9i^2} = \dfrac{-4i-3i^2}{16-9(-1)}$$
$$= \dfrac{-4i-3(-1)}{16+9} = \dfrac{3-4i}{25} = \dfrac{3}{25}-\dfrac{4}{25}i$$

81.
$$\frac{5+2i}{5-2i} = \frac{5+2i}{5-2i} \cdot \frac{5+2i}{5+2i}$$
$$= \frac{5\cdot 5 + 5\cdot 2i + 2i\cdot 5 + 2i\cdot 2i}{5^2 - (2i)^2}$$
$$= \frac{25 + 10i + 10i + 4i^2}{25 - 4i^2}$$
$$= \frac{25 + 20i + 4(-1)}{25 - 4(-1)} = \frac{25 + 20i - 4}{25 + 4}$$
$$= \frac{21 + 20i}{29} = \frac{21}{29} + \frac{20}{29}i$$

83.
$$\frac{3+7i}{-2-4i} = \frac{3+7i}{-2-4i} \cdot \frac{-2+4i}{-2+4i}$$
$$= \frac{3(-2) + 3\cdot 4i - 7i\cdot 2 + 7i\cdot 4i}{(-2)^2 - (4i)^2}$$
$$= \frac{-6 + 12i - 14i + 28i^2}{4 - 16i^2}$$
$$= \frac{-6 - 2i + 28(-1)}{4 - 16(-1)} = \frac{-6 - 2i - 28}{4 + 16}$$
$$= \frac{-34 - 2i}{20} = \frac{2(-17 - i)}{20}$$
$$= \frac{-17 - i}{10} = -\frac{17}{10} - \frac{1}{10}i$$

85.
$$\frac{13i}{-5-i} = \frac{13i}{-5-i} \cdot \frac{-5+i}{-5+i} = \frac{13i(-5) + 13i\cdot i}{(-5)^2 - i^2}$$
$$= \frac{-65i + 13i^2}{25 - (-1)} = \frac{-65i + 13(-1)}{25 + 1}$$
$$= \frac{-13 - 65i}{26} = \frac{13(-1 - 5i)}{26}$$
$$= \frac{-1 - 5i}{2} = -\frac{1}{2} - \frac{5}{2}i$$

87.
$$\frac{2+3i}{6i} = \frac{2+3i}{6i} \cdot \frac{-6i}{-6i} = \frac{-12i - 18i^2}{-36i^2}$$
$$= \frac{-12i - 18(-1)}{-36(-1)} = \frac{18 - 12i}{36}$$
$$= \frac{6(3 - 2i)}{36} = \frac{3 - 2i}{6} = \frac{1}{2} - \frac{1}{3}i$$

89.
$$\frac{-10+i}{i} = \frac{-10+i}{i} \cdot \frac{-i}{-i} = \frac{10i - i^2}{-i^2}$$
$$= \frac{10i - (-1)}{-(-1)} = \frac{1 + 10i}{1} = 1 + 10i$$

91.
$$\frac{2 + \sqrt{-16}}{8} = \frac{2 + 4i}{8} = \frac{2(1 + 2i)}{8}$$
$$= \frac{1 + 2i}{4} = \frac{1}{4} + \frac{1}{2}i$$

93.
$$\frac{-6 + \sqrt{-72}}{6} = \frac{-6 + \sqrt{-36\cdot 2}}{6} = \frac{-6 + 6i\sqrt{2}}{6}$$
$$= \frac{6(-1 + i\sqrt{2})}{6} = -1 + i\sqrt{2}$$

95.
$$\frac{-8 - \sqrt{-48}}{4} = \frac{-8 - \sqrt{-16\cdot 3}}{4} = \frac{-8 - 4i\sqrt{3}}{4}$$
$$= \frac{4(-2 - i\sqrt{3})}{4} = -2 - i\sqrt{3}$$

97.
$$\frac{-5 + \sqrt{-50}}{10} = \frac{-5 + \sqrt{-25\cdot 2}}{10} = \frac{-5 + 5i\sqrt{2}}{10}$$
$$= \frac{5(-1 + i\sqrt{2})}{5\cdot 2} = -\frac{1}{2} + \frac{\sqrt{2}}{2}i$$

99.
$$x^2 - 4x + 5 = 0 \qquad x = 2 + i$$
$$(2+i)^2 - 4(2+i) + 5 = 0$$
$$4 + 4i + i^2 - 8 - 4i + 5 = 0$$
$$4 + 4i - 1 - 8 - 4i + 5 = 0$$
$$0 = 0$$

$2 + i$ is a solution.

101.
$$x^2 + 12 = 0 \qquad x = -2i\sqrt{3}$$

$$\left(-2i\sqrt{3}\right)^2 + 12 = 0$$

$$4i^2 \cdot 3 + 12 = 0$$

$$-12 + 12 = 0$$

$$0 = 0$$

$-2i\sqrt{3}$ is a solution.

Chapter 6 Group Activity

1.–6.   Answers will vary.

Chapter 6 Review Exercises

Section 6.1

1. **a.** False; $\sqrt{0} = 0$ is not positive.
    b. False; $\sqrt[3]{-8} = -2$

3. **a.** False
    b. True

5. $\sqrt[4]{625} = \sqrt[4]{5^4} = 5$

7.
$$f(x) = \sqrt{x-1}$$
  a. $f(10) = \sqrt{10-1} = \sqrt{9} = 3$
  b. $f(1) = \sqrt{1-1} = \sqrt{0} = 0$
  c. $f(8) = \sqrt{8-1} = \sqrt{7}$
  d. $x - 1 \geq 0$
$$x \geq 1 \qquad [1, \infty)$$

9. $\dfrac{\sqrt[3]{2x}}{\sqrt[4]{2x}} + 4$

11.
  a. $\sqrt{4y^2} = 2|y|$
  b. $\sqrt[3]{27y^3} = 3y$
  c. $\sqrt[100]{y^{100}} = |y|$
  d. $\sqrt[101]{y^{101}} = y$

Section 6.2

13. Yes, provided the expressions are well defined. For example: $x^5 \cdot x^3 = x^8$ and $x^{1/5} \cdot x^{2/5} = x^{3/5}$

15. Take the reciprocal of the base and change the exponent to positive.

17. $16^{-1/4} = \left(\dfrac{1}{16}\right)^{1/4} = \sqrt[4]{\dfrac{1}{16}} = \dfrac{1}{2}$

19. $\left(b^{1/2} \cdot b^{1/3}\right)^{12} = \left(b^{1/2}\right)^{12} \left(b^{1/3}\right)^{12} = b^6 \cdot b^4$
$$= b^{6+4} = b^{10}$$

21.
$$\left(\frac{a^{12}b^{-4}c^{7}}{a^{3}b^{2}c^{4}}\right)^{1/3} = \left(a^{12-3}b^{-4-2}c^{7-4}\right)^{1/3}$$
$$= \left(a^{9}b^{-6}c^{3}\right)^{1/3} = \left(a^{9}\right)^{1/3}\left(b^{-6}\right)^{1/3}\left(c^{3}\right)^{1/3}$$
$$= a^{3}b^{-2}c^{1} = \frac{a^{3}c}{b^{2}}$$

23. $\sqrt[3]{2y^{2}} = \left(2y^{2}\right)^{1/3}$

25. $17.8^{2/3} \approx 6.8173$

Section 6.3

27. For a radical expression to be simplified the following conditions must be met:

　　1. Factors of the radicand must have powers less than the index.

　　2. There may be no fractions in the radicand.

　　3. There may be no radical in the denominator of a fraction.

29. $\sqrt[4]{x^{5}yz^{4}} = \sqrt[4]{x^{4}z^{4} \cdot xy} = xz\sqrt[4]{xy}$

31.
$$\sqrt[3]{\frac{-16a^{4}}{2ab^{3}}} = \sqrt[3]{\frac{-8a^{3}}{b^{3}}} = -\frac{2a}{b}$$

33. Let h = the height of the bulge

$\frac{1}{2}$ length of the bridge $= \frac{1}{8}$ mi $= \frac{1}{8}(5280)$

　　　$= 660$ ft

$h^{2} + 660^{2} = 660.75^{2}$

　　$h^{2} = 660.75^{2} - 660^{2}$

　　$h^{2} = 436,590.5625 - 435,600$

　　$h^{2} = 990.5625$

　　$h = \sqrt{990.5625} \approx 31$ ft

Section 6.4

35. $\sqrt[3]{2x} - 2\sqrt{2x}$ cannot be combined; the indices are different.

37. $\sqrt[4]{3xy} + 2\sqrt[4]{3xy} = 3\sqrt[4]{3xy}$ can be combined.

39. $4\sqrt{7} - 2\sqrt{7} + 3\sqrt{7} = (4 - 2 + 3)\sqrt{7} = 5\sqrt{7}$.

41. $\sqrt{50} + 7\sqrt{2} - \sqrt{8} = \sqrt{25 \cdot 2} + 7\sqrt{2} - \sqrt{4 \cdot 2}$
$$= 5\sqrt{2} + 7\sqrt{2} - 2\sqrt{2} = 10\sqrt{2}$$

43. False; 5 and $3\sqrt{x}$ are not like radicals.

Section 6.5

45. $\sqrt{3} \cdot \sqrt{12} = \sqrt{36} = 6$

47. $-2\sqrt{3}\left(\sqrt{7} - 3\sqrt{11}\right) = -2\sqrt{3} \cdot \sqrt{7} + 2\sqrt{3} \cdot 3\sqrt{11}$
$$= -2\sqrt{21} + 6\sqrt{33}$$

49. $\left(2\sqrt{x} - 3\right)\left(2\sqrt{x} + 3\right) = \left(2\sqrt{x}\right)^2 - 3^2 = 4x - 9$

51. $\left(\sqrt{7y} - \sqrt{3x}\right)^2$
$$= \left(\sqrt{7y}\right)^2 - 2 \cdot \sqrt{7y} \cdot \sqrt{3x} + \left(\sqrt{3x}\right)^2$$
$$= 7y - 2\sqrt{21xy} + 3x$$

53. $\left(-\sqrt{z} - \sqrt{6}\right)\left(2\sqrt{z} + 7\sqrt{6}\right) = \left(-\sqrt{z}\right)\left(2\sqrt{z}\right) - \sqrt{z}\left(7\sqrt{6}\right) - \sqrt{6}\left(2\sqrt{z}\right) - \sqrt{6}\left(7\sqrt{6}\right)$
$$= -2\sqrt{z^2} - 7\sqrt{6z} - 2\sqrt{6z} - 7\sqrt{36} = -2z - 9\sqrt{6z} - 7 \cdot 6 = -2z - 9\sqrt{6z} - 42$$

55. $\sqrt[3]{u} \cdot \sqrt{u^5} = u^{1/3} \cdot u^{5/2} = u^{(1/3)+(5/2)} = u^{17/6}$
$$= \sqrt[6]{u^{17}} = \sqrt[6]{u^{12} \cdot u^5} = u^2 \sqrt[6]{u^5}$$

Section 6.6

57. $\sqrt{\dfrac{3y^5}{25x^6}} = \dfrac{\sqrt{3y \cdot y^4}}{\sqrt{25x^6}} = \dfrac{y^2 \sqrt{3y}}{5x^3}$

59. $\dfrac{\sqrt{324w^7}}{\sqrt{4w^3}} = \sqrt{\dfrac{324w^7}{4w^3}} = \sqrt{81w^4} = 9w^2$

61. $\sqrt{\dfrac{7}{2y}} = \dfrac{\sqrt{7}}{\sqrt{2y}} = \dfrac{\sqrt{7}}{\sqrt{2y}} \cdot \dfrac{\sqrt{2y}}{\sqrt{2y}} = \dfrac{\sqrt{14y}}{\sqrt{4y^2}} = \dfrac{\sqrt{14y}}{2y}$

63. $\dfrac{4}{\sqrt[3]{9p^2}} = \dfrac{4}{\sqrt[3]{3^2 \cdot p^2}} \cdot \dfrac{\sqrt[3]{3p}}{\sqrt[3]{3p}} = \dfrac{4\sqrt[3]{3p}}{\sqrt[3]{3^3 p^3}} = \dfrac{4\sqrt[3]{3p}}{3p}$

65. $\dfrac{-5}{\sqrt{15} + \sqrt{10}} = \dfrac{-5}{\sqrt{15} + \sqrt{10}} \cdot \dfrac{\sqrt{15} - \sqrt{10}}{\sqrt{15} - \sqrt{10}}$
$$= \dfrac{-5\left(\sqrt{15} - \sqrt{10}\right)}{\left(\sqrt{15}\right)^2 - \left(\sqrt{10}\right)^2} = \dfrac{-5\left(\sqrt{15} - \sqrt{10}\right)}{15 - 10}$$
$$= \dfrac{-5\left(\sqrt{15} - \sqrt{10}\right)}{5} = -\sqrt{15} + \sqrt{10}$$

67. $\dfrac{t-3}{\sqrt{t} - \sqrt{3}} = \dfrac{t-3}{\sqrt{t} - \sqrt{3}} \cdot \dfrac{\sqrt{t} + \sqrt{3}}{\sqrt{t} + \sqrt{3}}$
$$= \dfrac{(t-3)\left(\sqrt{t} + \sqrt{3}\right)}{\left(\sqrt{t}\right)^2 - \left(\sqrt{3}\right)^2} = \dfrac{\left(\cancel{t-3}\right)\left(\sqrt{t} + \sqrt{3}\right)}{\cancel{t-3}}$$
$$= \sqrt{t} + \sqrt{3}$$

69. The quotient of the principal square root of 2 and the square of x.

Section 6.7

71.
$$\sqrt{a-6}-5=0$$
$$\sqrt{a-6}=5$$
$$\left(\sqrt{a-6}\right)^2=5^2$$
$$a-6=25$$
$$a=31$$
The solution is $\{31\}$.

Check:
$$\sqrt{31-6}-5=0$$
$$\sqrt{25}-5=0$$
$$5-5=0$$
$$0=0$$

73.
$$\sqrt[4]{p+12}-\sqrt[4]{5p-16}=0$$
$$\sqrt[4]{p+12}=\sqrt[4]{5p-16}$$
$$\left(\sqrt[4]{p+12}\right)^4=\left(\sqrt[4]{5p-16}\right)^4$$
$$p+12=5p-16$$
$$-4p=-28$$
$$p=7$$

Check:
$$\sqrt[4]{7+12}-\sqrt[4]{5(7)-16}=0$$
$$\sqrt[4]{19}-\sqrt[4]{35-16}=0$$
$$\sqrt[4]{19}-\sqrt[4]{19}=0$$
$$0=0$$
The solution is $\{7\}$.

75.
$$\sqrt{8x+1}=-\sqrt{x-13}$$
$$\left(\sqrt{8x+1}\right)^2=\left(-\sqrt{x-13}\right)^2$$
$$8x+1=x-13$$
$$7x=-14$$
$$x=-2$$
$\{\ \}$ ($x=-2$ does not check.)

Check:
$$\sqrt{8(-2)+1}=-\sqrt{-2-13}$$
$$\sqrt{-16+1}=-\sqrt{-15}$$
$$\sqrt{-15}\neq-\sqrt{-15}$$

77.
$$\sqrt{x+2}=1-\sqrt{2x+5}$$
$$\left(\sqrt{x+2}\right)^2=\left(1-\sqrt{2x+5}\right)^2$$
$$x+2=1-2\sqrt{2x+5}+2x+5$$
$$-x-4=-2\sqrt{2x+5}$$
$$\left(-x-4\right)^2=\left(-2\sqrt{2x+5}\right)^2$$
$$x^2+8x+16=4(2x+5)$$
$$x^2+8x+16=8x+20$$
$$x^2-4=0$$
$$(x+2)(x-2)=0$$
$$x+2=0 \text{ or } x-2=0$$
$$x=-2 \text{ or } x=2$$
The solution is $\{-2\}$. ($x=2$ does not check.)

Check $x=-2$:
$$\sqrt{-2+2}=1-\sqrt{2(-2)+5}$$
$$\sqrt{0}=1-\sqrt{-4+5}$$
$$0=1-\sqrt{1}$$
$$0=1-1$$
$$0=0$$

Check $x=2$:
$$\sqrt{2+2}=1-\sqrt{2(2)+5}$$
$$\sqrt{4}=1-\sqrt{4+5}$$
$$2=1-\sqrt{9}$$
$$2=1-3$$
$$2\neq-2$$

79.
$$v(d) = \sqrt{32d}$$

(a) $v(20) = \sqrt{32(20)} = \sqrt{640} = \sqrt{64 \cdot 10}$

$$= 8\sqrt{10} \text{ ft/sec} \approx 25.3 \text{ ft/sec}$$

When the water is 20 ft deep, a wave travels about 25.3 ft/sec.

(b) $16 = \sqrt{32d}$

$$16^2 = \left(\sqrt{32d}\right)^2$$

$$256 = 32d$$

$$d = 8 \text{ ft}$$

Section 6.8

81. $a + bi$, where $b \neq 0$.

82.
To simplify the expression $\dfrac{3}{4 + 6i}$, first multiply the numerator and denominator by $4 - 6i$, which is the complex conjugate of the denominator.

83. $\sqrt{-16} = 4i$

85. $\sqrt{-75} \cdot \sqrt{-3} = \sqrt{-25 \cdot 3} \cdot \sqrt{-3} = 5i\sqrt{3} \cdot i\sqrt{3}$
$$= 5i^2\sqrt{9} = 5(-1)(3) = -15$$

87. $i^{38} = i^{36} \cdot i^2 = \left(i^4\right)^9 \cdot i^2 = 1^9(-1) = 1(-1) = -1$

89. $i^{19} = i^{16} \cdot i^3 = \left(i^4\right)^4 \cdot i^3 = 1^4(-i) = 1(-i) = -i$

91. $(-3 + i) - (2 - 4i) = -3 + i - 2 + 4i$
$$= (-3 - 2) + (1 + 4)i = -5 + 5i$$

93. $(4 - 3i)(4 + 3i) = 4^2 - (3i)^2 = 16 - 9i^2$
$$= 16 - 9(-1) = 16 + 9 = 25 + 0i$$

95. $\dfrac{17 - 4i}{-4} = -\dfrac{17}{4} + \dfrac{4}{4}i = -\dfrac{17}{4} + i$

Real part: $-\dfrac{17}{4}$

Imaginary part: 1

97. $\dfrac{2 - i}{3 + 2i} = \dfrac{2 - i}{3 + 2i} \cdot \dfrac{3 - 2i}{3 - 2i}$

$$= \dfrac{2 \cdot 3 - 2 \cdot 2i - i \cdot 3 + i \cdot 2i}{3^2 - (2i)^2}$$

$$= \dfrac{6 - 4i - 3i + 2i^2}{9 - 4i^2} = \dfrac{6 - 7i + 2(-1)}{9 - 4(-1)}$$

$$= \dfrac{6 - 7i - 2}{9 + 4} = \dfrac{4 - 7i}{13} = \dfrac{4}{13} - \dfrac{7}{13}i$$

99. $\dfrac{5+3i}{-2i} = \dfrac{5+3i}{-2i} \cdot \dfrac{2i}{2i} = \dfrac{5 \cdot 2i + 3i \cdot 2i}{-(2i)^2} = \dfrac{10i + 6i^2}{-4i^2}$

$= \dfrac{10i + 6(-1)}{-4(-1)} = \dfrac{-6 + 10i}{4} = -\dfrac{6}{4} + \dfrac{10}{4}i$

$= -\dfrac{3}{2} + \dfrac{5}{2}i$

101. $\dfrac{-8 + \sqrt{-40}}{12} = \dfrac{-8 + \sqrt{-4 \cdot 10}}{12} = \dfrac{-8 + 2i\sqrt{10}}{12}$

$= \dfrac{2\left(-4 + i\sqrt{10}\right)}{12} = \dfrac{-4 + i\sqrt{10}}{6}$

$= -\dfrac{4}{6} + \dfrac{\sqrt{10}}{6}i = -\dfrac{2}{3} + \dfrac{\sqrt{10}}{6}i$

Chapter 6 Test

1. a. $\sqrt{36} = 6$

 b. $-\sqrt{36} = -6$

3. a. $\sqrt[3]{y^3} = y$

 b. $\sqrt[4]{y^4} = |y|$

5. $\sqrt{\dfrac{16}{9}} = \dfrac{4}{3}$

7. $\sqrt{a^4 b^3 c^5} = \sqrt{a^4 b^2 c^4 \cdot bc} = a^2 bc^2 \sqrt{bc}$

9. $\sqrt{\dfrac{32w^6}{2w}} = \sqrt{16w^5} = \sqrt{16w^4 \cdot w} = 4w^2\sqrt{w}$

11. $\dfrac{2\sqrt{72}}{8} = \dfrac{2\sqrt{36 \cdot 2}}{8} = \dfrac{2 \cdot 6\sqrt{2}}{8} = \dfrac{12\sqrt{2}}{8} = \dfrac{3\sqrt{2}}{2}$

13. $\dfrac{-3 - \sqrt{5}}{17} \approx -0.3080$

15. $8^{2/3} \cdot \left(\dfrac{25x^4 y^6}{z^2}\right)^{1/2} = \left(\sqrt[3]{8}\right)^2 \cdot \dfrac{25^{1/2}\left(x^4\right)^{1/2}\left(y^6\right)^{1/2}}{\left(z^2\right)^{1/2}}$

$= 2^2 \cdot \dfrac{5x^2 y^3}{z} = \dfrac{20x^2 y^3}{z}$

17. $\dfrac{\sqrt[3]{10}}{\sqrt[4]{10}} = \dfrac{10^{1/3}}{10^{1/4}} = 10^{(1/3)-(1/4)} = 10^{1/12} = \sqrt[12]{10}$

19. $3\sqrt{x}\left(\sqrt{2} - \sqrt{5}\right) = 3\sqrt{x} \cdot \sqrt{2} - 3\sqrt{x} \cdot \sqrt{5}$

$= 3\sqrt{2x} - 3\sqrt{5x}$

21. $\dfrac{-2}{\sqrt[3]{x}} = \dfrac{-2}{\sqrt[3]{x}} \cdot \dfrac{\sqrt[3]{x^2}}{\sqrt[3]{x^2}} = \dfrac{-2\sqrt[3]{x^2}}{\sqrt[3]{x^3}} = \dfrac{-2\sqrt[3]{x^2}}{x}$

23. a. $\sqrt{-8} = \sqrt{-4 \cdot 2} = 2i\sqrt{2}$

 b. $2\sqrt{-16} = 2 \cdot 4i = 8i$

 c. $\dfrac{2 + \sqrt{-8}}{4} = \dfrac{2 + \sqrt{-4 \cdot 2}}{4} = \dfrac{2 + 2i\sqrt{2}}{4}$

$= \dfrac{2\left(1 + i\sqrt{2}\right)}{4} = \dfrac{1 + i\sqrt{2}}{2} = \dfrac{1}{2} + \dfrac{\sqrt{2}}{2}i$

25. $(4 + i)(8 + 2i) = 4 \cdot 8 + 4 \cdot 2i + i \cdot 8 + i \cdot 2i$

$= 32 + 8i + 8i + 2i^2$

$= 32 + 16i - 2 = 30 + 16i$

27. $(4 - 7i)^2 = 4^2 - 2 \cdot 4 \cdot 7i + (7i)^2$

$= 16 - 56i + 49i^2$

$= 16 - 56i + 49(-1)$

$= 16 - 56i - 49 = -33 - 56i$

29. $\dfrac{3-2i}{3-4i} = \dfrac{3-2i}{3-4i} \cdot \dfrac{3+4i}{3+4i}$

$= \dfrac{3\cdot 3 + 3\cdot 4i - 2i\cdot 3 - 2i\cdot 4i}{3^2 - (4i)^2}$

$= \dfrac{9+12i-6i-8i^2}{9-16i^2} = \dfrac{9+6i-8(-1)}{9-16(-1)}$

$= \dfrac{9+6i+8}{9+16} = \dfrac{17+6i}{25} = \dfrac{17}{25} + \dfrac{6}{25}i$

31.
$r(V) = \sqrt[3]{\dfrac{3V}{4\pi}}$

$r(10) = \sqrt[3]{\dfrac{3(10)}{4\pi}} = \sqrt[3]{\dfrac{30}{4\pi}} \approx 1.34$

The radius of a sphere of volume 10 cubic units is approximately 1.34 units.

33. $\sqrt[3]{2x+5} = -3$ Check:

$\left(\sqrt[3]{2x+5}\right)^3 = (-3)^3$ $\sqrt[3]{2(-16)+5} = -3$

$2x+5 = -27$ $\sqrt[3]{-32+5} = -3$

$2x = -32$ $\sqrt[3]{-27} = -3$

$x = -16$ $-3 = -3$

The solution is $\{-16\}$.

34. $\sqrt{5x+8} = \sqrt{5x-1}+1$

$\left(\sqrt{5x+8}\right)^2 = \left(\sqrt{5x-1}+1\right)^2$

$5x+8 = 5x-1+2\sqrt{5x-1}+1$

$8 = 2\sqrt{5x-1}$

$4 = \sqrt{5x-1}$

$(4)^2 = \left(\sqrt{5x-1}\right)^2$

$16 = 5x-1$

$17 = 5x$

$x = \dfrac{17}{5}$

The solution is $\left\{\dfrac{17}{5}\right\}$.

Check:

$\sqrt{5\left(\dfrac{17}{5}\right)+8} = \sqrt{5\left(\dfrac{17}{5}\right)-1}+1$

$\sqrt{17+8} = \sqrt{17-1}+1$

$\sqrt{25} = \sqrt{16}+1$

$5 = 4+1$

$5 = 5$

35. $\sqrt{t+7} - \sqrt{2t-3} = 2$

$$\sqrt{t+7} = 2 + \sqrt{2t-3}$$

$$\left(\sqrt{t+7}\right)^2 = \left(2 + \sqrt{2t-3}\right)^2$$

$$t + 7 = 4 + 4\sqrt{2t-3} + 2t - 3$$

$$-t + 6 = 4\sqrt{2t-3}$$

$$\left(-t+6\right)^2 = \left(4\sqrt{2t-3}\right)^2$$

$$t^2 - 12t + 36 = 16\left(2t-3\right)$$

$$t^2 - 12t + 36 = 32t - 48$$

$$t^2 - 44t + 84 = 0$$

$$\left(t-42\right)\left(t-2\right) = 0$$

$$t - 42 = 0 \text{ or } t - 2 = 0$$

$$t = 42 \quad \text{or} \quad t = 2$$

The solution is $\{2\}$. ($t = 42$ does not check.)

Check $t = 42$:

$$\sqrt{42+7} - \sqrt{2(42)-3} = 2$$

$$\sqrt{49} - \sqrt{84-3} = 2$$

$$7 - \sqrt{81} = 2$$

$$7 - 9 = 2$$

$$-2 \neq 2$$

Check $t = 2$:

$$\sqrt{2+7} - \sqrt{2(2)-3} = 2$$

$$\sqrt{9} - \sqrt{4-3} = 2$$

$$3 - \sqrt{1} = 2$$

$$3 - 1 = 2$$

$$2 = 2$$

Chapters 1 – 6   Cumulative Review Exercises

1. $6^2 - 2\left[5 - 8(3-1) + 4 \div 2\right]$

$$= 6^2 - 2\left[5 - 8(2) + 4 \div 2\right]$$

$$= 6^2 - 2\left[5 - 16 + 2\right]$$

$$= 6^2 - 2\left[-9\right] = 36 - 2\left[-9\right]$$

$$= 36 + 18 = 54$$

3. $9(2y+8) = 20 - (y+5)$

$$18y + 72 = 20 - y - 5$$

$$18y + 72 = -y + 15$$

$$19y = -57$$

$$y = -3 \quad \{-3\}$$

5. $11 \geq -x + 2 > 5$

$$9 \geq -x > 3$$

$$-9 \leq x < -3 \quad [-9, -3)$$

7. $2x + y = 9$

$$y = -2x + 9$$

The slope is –2 so the slope of the parallel line is also –2.

$$y - (-1) = -2(x-3)$$

$$y + 1 = -2x + 6$$

$$y = -2x + 5$$

9.
$$\begin{array}{r} 2x - 3y = 0 \\ -4x + 3y = -1 \\ \hline -2x = -1 \end{array}$$

$$x = \frac{1}{2}$$

$$2\left(\frac{1}{2}\right) - 3y = 0$$

$$1 - 3y = 0$$

$$-3y = -1$$

$$y = \frac{1}{3}$$

The solution is $\left\{\left(\dfrac{1}{2}, \dfrac{1}{3}\right)\right\}$.

11.

	Work Rate	Time	Portion of Job Comp
Bennette	1/3	x	$(1/3)x$
Pepe	1/5	x	$(1/5)x$

(Bennette Part) + (Pepe Part) = (1 Job)

$$\frac{1}{3}x + \frac{1}{5}x = 1 \qquad \text{LCD} = 15$$

$$15\left(\frac{1}{3}x + \frac{1}{5}x\right) = 15(1)$$

$$5x + 3x = 15$$

$$8x = 15$$

$$x = \frac{15}{8} \text{ hr or } 1\frac{7}{8} \text{ hr}$$

Together, it will take them $1\frac{7}{8}$ hr.

13.
$$\left(\frac{a^{3/2}b^{-1/4}c^{1/3}}{ab^{-5/4}c^0}\right)^{12} = \left(a^{(3/2)-1}b^{(-1/4)-(-5/4)}c^{1/3}\right)^{12}$$

$$= \left(a^{1/2}b^1c^{1/3}\right)^{12}$$

$$= \left(a^{1/2}\right)^{12}\left(b^{12}\right)\left(c^{1/3}\right)^{12}$$

$$= a^6 b^{12} c^4$$

15.
$$(2x+5)(x-3) = 2x \cdot x - 2x \cdot 3 + 5 \cdot x - 5 \cdot 3$$

$$= 2x^2 - 6x + 5x - 15$$

$$= 2x^2 - x - 15$$

The product is a second degree polynomial.

17.
$$\frac{x^2 - x - 12}{x+3} = \frac{(x-4)(\cancel{x+3})}{\cancel{x+3}} = x - 4$$

19.
$$\sqrt[3]{\frac{54c^4}{cd^3}} = \sqrt[3]{\frac{27c^3 \cdot 2\cancel{c}}{\cancel{c}d^3}} = \frac{3c\sqrt[3]{2}}{d}$$

21.
$$\frac{13i}{3+2i} = \frac{13i}{3+2i} \cdot \frac{3-2i}{3-2i} = \frac{13i(3-2i)}{3^2 - (2i)^2}$$

$$= \frac{13i(3-2i)}{9 - 4i^2} = \frac{13(3i - 2i^2)}{9 - 4(-1)}$$

$$= \frac{13(3i - 2(-1))}{9+4} = \frac{\cancel{13}(3i+2)}{\cancel{13}}$$

$$= 2 + 3i$$

23.
$$\frac{3}{x^2 + 5x} + \frac{-2}{x^2 - 25}$$

$$= \frac{3}{x(x+5)} + \frac{-2}{(x+5)(x-5)}$$

$$\text{LCD} = x(x+5)(x-5)$$

$$= \frac{3}{x(x+5)} \cdot \frac{x-5}{x-5} + \frac{-2}{(x+5)(x-5)} \cdot \frac{x}{x}$$

$$= \frac{3x - 15 - 2x}{x(x+5)(x-5)} = \frac{x-15}{x(x+5)(x-5)}$$

25.
$$\left(-5x^2 - 4x + 8\right) - \left(3x - 5\right)^2$$

$$= \left(-5x^2 - 4x + 8\right) - \left(9x^2 - 30x + 25\right)$$

$$= -5x^2 - 4x + 8 - 9x^2 + 30x - 25$$

$$= -14x^2 + 26x - 17$$

27.
$$\frac{4}{3-5i} = \frac{4}{3-5i} \cdot \frac{3+5i}{3+5i} = \frac{12+20i}{9-25i^2}$$

$$= \frac{12+20i}{9-25(-1)} = \frac{12+20i}{9+25} = \frac{12+20i}{34}$$

$$= \frac{12}{34} + \frac{20}{34}i = \frac{6}{17} + \frac{10}{17}i$$

29.
$$x^2 + 6x + 9 - y^2 = (x+3)^2 - y^2$$

$$= (x+3-y)(x+3+y)$$

Chapter 7　Quadratic Equations, Functions, and Inequalities

Are You Prepared?

1.
$$x^2 - 9 = 0$$
$$(x-3)(x+3) = 0$$
$$x - 3 = 0 \text{ or } x + 3 = 0$$
$$x = 3 \text{ or } \quad x = -3 \quad \{3, -3\}$$

2.
$$x^2 - 6x + 8 = 0$$
$$(x-4)(x-2) = 0$$
$$x - 4 = 0 \text{ or } x - 2 = 0$$
$$x = 4 \text{ or } \quad x = 2 \quad \{2, 4\}$$

3.
$$3x(x+1) = 60$$
$$3x^2 + 3x = 60$$
$$3x^2 + 3x - 60 = 0$$
$$3(x^2 + x - 20) = 0$$
$$3(x-4)(x+5) = 0$$
$$x - 4 = 0 \text{ or } x + 5 = 0$$
$$x = 4 \text{ or } \quad x = -5 \quad \{4, -5\}$$

4.
$$\frac{-5 \pm \sqrt{50}}{10} = \frac{-5 \pm 5\sqrt{2}}{10}$$
$$= \frac{5(-1 \pm \sqrt{2})}{10}$$
$$= \frac{-1 \pm \sqrt{2}}{2}$$

5.
$$\frac{7 \pm 21i}{7} = \frac{7(1 \pm 3i)}{7} = 1 \pm 3i$$

6.
$$\frac{2 \pm \sqrt{-12}}{4} = \frac{2 \pm 2i\sqrt{3}}{4} = \frac{2(1 \pm i\sqrt{3})}{4}$$
$$= \frac{1 \pm i\sqrt{3}}{2} = \frac{1}{2} \pm \frac{\sqrt{3}}{2}i$$

7.
$$f(x) = x^2 - 6x + 5 = 0$$
$$(x-5)(x-1) = 0$$
$$x - 5 = 0 \text{ or } x - 1 = 0$$
$$x = 5 \text{ or } \quad x = 1$$
$$(5, 0) \text{ and } (1, 0)$$

8.
$$f(x) = x^2 - 6x + 5$$
$$f(0) = 0^2 - 6(0) + 5 = 0 - 0 + 5 = 5$$
$$(0, 5)$$

A function defined by $f(x) = ax^2 + bx + c$ $(a \neq 0)$ is called a $\underline{Q\ U\ A\ D\ R\ A\ T\ I\ C}$ function.
$$2\ 1\ 6\ 5\ 8\ 6\ 4\ 7\ 3$$

Section 7.1   Square Root Property and Completing the Square

1. (a) 0; 0
    (b) 0
    (c) $\sqrt{k}$; $-\sqrt{k}$
    (d) 2; $\{3, -3\}$
    (e) Completing
    (f) 100
    (g) 4; 1
    (h) 8

3. $y^2 = 4$

$y = \pm\sqrt{4}$

$y = \pm 2 \quad \{\pm 2\}$

5. $k^2 - 7 = 0$

$k^2 = 7$

$k = \pm\sqrt{7} \quad \{\pm\sqrt{7}\}$

7. $36u^2 = 121$

$u^2 = \dfrac{121}{36}$

$u = \pm\sqrt{\dfrac{121}{36}}$

$u = \pm\dfrac{11}{6} \quad \left\{\dfrac{11}{6}, -\dfrac{11}{6}\right\}$

9. $-2m^2 = 50$

$m^2 = -25$

$m = \pm\sqrt{-25}$

$m = \pm 5i \quad \{\pm 5i\}$

11. $(q+3)^2 = 4$

$q + 3 = \pm\sqrt{4}$

$q + 3 = \pm 2$

$q = -3 \pm 2$

$q = -1 \text{ or } q = -5 \quad \{-1, -5\}$

13. $(2y+3)^2 - 7 = 0$

$(2y+3)^2 = 7$

$2y + 3 = \pm\sqrt{7}$

$2y = -3 \pm \sqrt{7}$

$y = \dfrac{-3 \pm \sqrt{7}}{2} \quad \left\{\dfrac{-3 \pm \sqrt{7}}{2}\right\}$

15. $(t+5)^2 = -18$

$t + 5 = \pm\sqrt{-18}$

$t + 5 = \pm 3i\sqrt{2}$

$t = -5 \pm 3i\sqrt{2} \quad \left\{-5 \pm 3i\sqrt{2}\right\}$

17. $15 = 4 + 3w^2$

$3w^2 = 11$

$w^2 = \dfrac{11}{3}$

$w = \pm\sqrt{\dfrac{11}{3}} \cdot \sqrt{\dfrac{3}{3}}$

$w = \pm\dfrac{\sqrt{33}}{3} \quad \left\{\pm\dfrac{\sqrt{33}}{3}\right\}$

19.

$$\left(m+\frac{4}{5}\right)^2+\frac{3}{25}=0$$

$$\left(m+\frac{4}{5}\right)^2=-\frac{3}{25}$$

$$m+\frac{4}{5}=\pm\sqrt{-\frac{3}{25}}$$

$$m+\frac{4}{5}=\pm\frac{i\sqrt{3}}{5}$$

$$x=-\frac{4}{5}\pm\frac{\sqrt{3}}{5}i \quad \left\{-\frac{4}{5}\pm\frac{\sqrt{3}}{5}i\right\}$$

21.

$$-y^2-2=14$$

$$-y^2=14+2$$

$$-y^2=16$$

$$y^2=-16$$

$$y=\sqrt{-16}$$

$$y=\pm4i \quad \{4i,-4i\}$$

23. 1. Factoring and applying the zero product rule.

$$x^2-36=0$$

$$(x+6)(x-6)=0$$

$$x+6=0 \text{ or } x-6=0$$

$$x=-6 \text{ or } x=6 \quad \{\pm6\}$$

2. Applying the square root property.

$$x^2-36=0$$

$$x^2=36$$

$$x=\pm\sqrt{36}$$

$$x=\pm6 \quad \{\pm6\}$$

25.   **a.**

$$\sqrt{x}=4$$

$$\left(\sqrt{x}\right)^2=(4)^2$$

$$x=16 \quad \{16\}$$

b. $\quad x^2=4$

$$x=\pm\sqrt{4}$$

$$x=\pm2 \quad \{2,-2\}$$

27. x^2-6x+n

$$n=\left(\frac{1}{2}b\right)^2=\left(\frac{1}{2}\cdot(-6)\right)^2=(-3)^2=9$$

$$x^2-6x+9=(x-3)^2$$

29. t^2+8t+n

$$n=\left(\frac{1}{2}b\right)^2=\left(\frac{1}{2}\cdot8\right)^2=(4)^2=16$$

$$t^2+8t+16=(t+4)^2$$

31. c^2-c+n

$$n=\left(\frac{1}{2}b\right)^2=\left(\frac{1}{2}\cdot(-1)\right)^2=\left(-\frac{1}{2}\right)^2=\frac{1}{4}$$

$$c^2-c+\frac{1}{4}=\left(c-\frac{1}{2}\right)^2$$

33. y^2+5y+n

$$n=\left(\frac{1}{2}b\right)^2=\left(\frac{1}{2}\cdot5\right)^2=\left(\frac{5}{2}\right)^2=\frac{25}{4}$$

$$y^2+5y+\frac{25}{4}=\left(y+\frac{5}{2}\right)^2$$

35. $b^2+\frac{2}{5}b+n$

$$n=\left(\frac{1}{2}b\right)^2=\left(\frac{1}{2}\cdot\frac{2}{5}\right)^2=\left(\frac{1}{5}\right)^2=\frac{1}{25}$$

$$b^2+\frac{2}{5}b+\frac{1}{25}=\left(b+\frac{1}{5}\right)^2$$

37. $p^2-\frac{2}{3}p+n$

$$n=\left(\frac{1}{2}b\right)^2=\left(\frac{1}{2}\cdot\left(-\frac{2}{3}\right)\right)^2=\left(-\frac{1}{3}\right)^2=\frac{1}{9}$$

$$p^2-\frac{2}{3}p+\frac{1}{9}=\left(p-\frac{1}{3}\right)^2$$

39. 1. Write the equation in the form
$ax^2 + bx + c = 0$.
2. Divide each term by a.
3. Isolate the variable terms.
4. Complete the square and factor.
5. Apply the square root property.

41.
$$t^2 + 8t + 15 = 0$$
$$t^2 + 8t = -15$$
$$t^2 + 8t + 16 = -15 + 16$$
$$(t+4)^2 = 1$$
$$t + 4 = \pm\sqrt{1}$$
$$t + 4 = \pm 1$$
$$t = -4 \pm 1$$
$$t = -3 \text{ or } t = -5 \quad \{-3, -5\}$$

43.
$$x^2 + 6x = -16$$
$$x^2 + 6x + 9 = -16 + 9$$
$$(x+3)^2 = -7$$
$$x + 3 = \pm\sqrt{-7}$$
$$x + 3 = \pm i\sqrt{7}$$
$$x = -3 \pm i\sqrt{7} \quad \{-3 \pm i\sqrt{7}\}$$

45.
$$p^2 + 4p + 6 = 0$$
$$p^2 + 4p = -6$$
$$p^2 + 4p + 4 = -6 + 4$$
$$(p+2)^2 = -2$$
$$p + 2 = \pm\sqrt{-2}$$
$$p + 2 = \pm i\sqrt{2}$$
$$p = -2 \pm i\sqrt{2} \quad \{-2 \pm i\sqrt{2}\}$$

47.
$$-3y - 10 = -y^2$$
$$y^2 - 3y - 10 = 0$$
$$y^2 - 3y = 10$$
$$y^2 - 3y + \frac{9}{4} = 10 + \frac{9}{4}$$
$$\left(y - \frac{3}{2}\right)^2 = \frac{49}{4}$$
$$y - \frac{3}{2} = \pm\sqrt{\frac{49}{4}}$$
$$y - \frac{3}{2} = \pm\frac{7}{2}$$
$$y = \frac{3}{2} \pm \frac{7}{2}$$
$$y = 5 \text{ or } y = -2 \quad \{5, -2\}$$

49.
$$2a^2 + 4a + 5 = 0$$
$$\frac{2a^2}{2} + \frac{4a}{2} + \frac{5}{2} = \frac{0}{2}$$
$$a^2 + 2a + \frac{5}{2} = 0$$
$$a^2 + 2a = -\frac{5}{2}$$
$$a^2 + 2a + 1 = -\frac{5}{2} + 1$$
$$(a+1)^2 = -\frac{3}{2}$$
$$a + 1 = \pm\sqrt{-\frac{3}{2}} \cdot \sqrt{\frac{2}{2}}$$
$$a + 1 = \pm\frac{i\sqrt{6}}{2}$$
$$a = -1 \pm \frac{\sqrt{6}}{2} i \quad \left\{-1 \pm \frac{\sqrt{6}}{2} i\right\}$$

301

51. $9x^2 - 36x + 40 = 0$

$$\frac{9x^2}{9} - \frac{36x}{9} + \frac{40}{9} = \frac{0}{9}$$

$$x^2 - 4x + \frac{40}{9} = 0$$

$$x^2 - 4x = -\frac{40}{9}$$

$$x^2 - 4x + 4 = -\frac{40}{9} + 4$$

$$(x-2)^2 = -\frac{4}{9}$$

$$x - 2 = \pm\sqrt{-\frac{4}{9}}$$

$$x - 2 = \pm\frac{2i}{3}$$

$$x = 2 \pm \frac{2}{3}i \qquad \left\{ 2 \pm \frac{2}{3}i \right\}$$

53. $25p^2 - 10p = 2$

$$p^2 - \frac{2}{5}p = \frac{2}{25}$$

$$p^2 - \frac{2}{5}p + \frac{1}{25} = \frac{2}{25} + \frac{1}{25}$$

$$\left(p - \frac{1}{5} \right)^2 = \frac{3}{25}$$

$$p - \frac{1}{5} = \pm\sqrt{\frac{3}{25}}$$

$$p - \frac{1}{5} = \pm\frac{\sqrt{3}}{5}$$

$$p = \frac{1}{5} \pm \frac{\sqrt{3}}{5} \qquad \left\{ \frac{1}{5} \pm \frac{\sqrt{3}}{5} \right\}$$

55. $(2w+5)(w-1) = 2$

$$2w^2 + 3w - 5 = 2$$

$$2w^2 + 3w - 7 = 0$$

$$\frac{2w^2}{2} + \frac{3w}{2} - \frac{7}{2} = \frac{0}{2}$$

$$w^2 + \frac{3}{2}w - \frac{7}{2} = 0$$

$$w^2 + \frac{3}{2}w = \frac{7}{2}$$

$$w^2 + \frac{3}{2}w + \frac{9}{16} = \frac{7}{2} + \frac{9}{16}$$

$$\left(w + \frac{3}{4} \right)^2 = \frac{65}{16}$$

$$w + \frac{3}{4} = \pm\sqrt{\frac{65}{16}}$$

$$w + \frac{3}{4} = \pm\frac{\sqrt{65}}{4}$$

$$w = -\frac{3}{4} \pm \frac{\sqrt{65}}{4} \qquad \left\{ -\frac{3}{4} \pm \frac{\sqrt{65}}{4} \right\}$$

57. $n(n-4) = 7$

$$n^2 - 4n = 7$$

$$n^2 - 4n + 4 = 7 + 4$$

$$(n-2)^2 = 11$$

$$n - 2 = \pm\sqrt{11}$$

$$n = 2 \pm \sqrt{11} \qquad \left\{ 2 \pm \sqrt{11} \right\}$$

59.
$$2x(x+6)=14$$
$$x(x+6)=7$$
$$x^2+6x=7$$
$$x^2+6x+9=7+9$$
$$(x+3)^2=16$$
$$x+3=\pm\sqrt{16}$$
$$x=-3\pm4$$
$$x=1 \text{ or } x=-7 \quad \{1,-7\}$$

61. a.
$$d=16t^2$$
$$t^2=\frac{d}{16}$$
$$t=\frac{\sqrt{d}}{4}$$

b.
$$t=\frac{\sqrt{1024}}{4}=\frac{32}{4}=8 \text{ sec}$$

63. $A=\pi r^2$ for r
$$r^2=\frac{A}{\pi}$$
$$r=\sqrt{\frac{A}{\pi}} \text{ or } r=\frac{\sqrt{A\pi}}{\pi}$$

65. $a^2+b^2+c^2=d^2$ for a
$$a^2=d^2-b^2-c^2$$
$$a=\sqrt{d^2-b^2-c^2}$$

67. $V=\frac{1}{3}\pi r^2 h$ for r
$$3V=\pi r^2 h$$
$$r^2=\frac{3V}{\pi h}$$
$$r=\sqrt{\frac{3V}{\pi h}} \text{ or } r=\frac{\sqrt{3V\pi h}}{\pi h}$$

69.
$$x^2+x^2=6^2$$
$$2x^2=36$$
$$x^2=18$$
$$x=\sqrt{18}=3\sqrt{2} \text{ ft} \approx 4.2 \text{ ft}$$
The shelf extends 4.2 ft.

71. $x^2=50$
$$x=\sqrt{50}=5\sqrt{2}\approx 7.1$$
The sides are 7.1 in.

73. a.
$$P(x)=-\frac{1}{8}x^2+5x$$
$$-\frac{1}{8}x^2+5x=20$$
$$-8\left(-\frac{1}{8}x^2+5x\right)=-8(20)$$
$$x^2-40x=-160$$
$$x^2-40x+400=-160+400$$
$$(x-20)^2=240$$
$$x-20=\pm\sqrt{240}$$
$$x=20\pm\sqrt{240}$$
$$x\approx 20\pm15.5$$
$$x\approx 4.5 \text{ or } x\approx 35.5$$
4.5 thousand or 35.5 thousand textbooks are sold for a profit of $20,000.

b. Profit increases to a point as more books are produced. Beyond that point, the market is "flooded", and profit decreases. Hence there are two points at which the profit is $20,000. Producing 4.5 thousand books makes the same profit using fewer resources as producing 35.5 thousand books.

Section 7.2   Quadratic Formula

1. **a.** quadratic; $\dfrac{-b \pm \sqrt{b^2 - 4ac}}{2a}$

b. $ax^2 + bx + c = 0$

c. 8; -42; -27

d. 7; 97

e. $b^2 - 4ac$; discriminant

f. Imaginary

g. Real

h. less than

3. $\dfrac{16 - \sqrt{320}}{4} = \dfrac{16 - \sqrt{64 \cdot 5}}{4} = \dfrac{16 - 8\sqrt{5}}{4}$

$= \dfrac{\cancel{4}\left(4 - 2\sqrt{5}\right)}{\cancel{4}} = 4 - 2\sqrt{5}$

5. $\dfrac{14 - \sqrt{-147}}{7} = \dfrac{14 - \sqrt{-49 \cdot 3}}{7} = \dfrac{14 - 7i\sqrt{3}}{7}$

$= \dfrac{\cancel{7}\left(2 - i\sqrt{3}\right)}{\cancel{7}} = 2 - i\sqrt{3}$

7.
$$2(x - 5) + x^2 = 3x(x - 4) - 2x^2$$
$$2x - 10 + x^2 = 3x^2 - 12x - 2x^2$$
$$x^2 - 3x^2 + 2x^2 + 2x + 12x - 10 = 0$$
$$14x - 10 = 0$$

The equation has degree one, so it is linear.

9. $x^2 + 11x - 12 = 0$

$a = 1,\ b = 11,\ c = -12$

$x = \dfrac{-(11) \pm \sqrt{(11)^2 - 4(1)(-12)}}{2(1)}$

$= \dfrac{-11 \pm \sqrt{121 + 48}}{2} = \dfrac{-11 \pm \sqrt{169}}{2}$

$= \dfrac{-11 \pm 13}{2}$

$x = \dfrac{2}{2} = 1$ or $x = -\dfrac{24}{2} = -12$ $\quad \{1, -12\}$

11. $9y^2 - 2y + 5 = 0$

$a = 9,\ b = -2,\ c = 5$

$y = \dfrac{-(-2) \pm \sqrt{(-2)^2 - 4(9)(5)}}{2(9)}$

$= \dfrac{2 \pm \sqrt{4 - 180}}{18} = \dfrac{2 \pm \sqrt{-176}}{18} = \dfrac{2 \pm 4i\sqrt{11}}{18}$

$= \dfrac{\cancel{2}\left(1 \pm 2i\sqrt{11}\right)}{\cancel{2} \cdot 9} = \dfrac{1 \pm 2i\sqrt{11}}{9} = \dfrac{1}{9} \pm \dfrac{2\sqrt{11}}{9}i$

$$\left\{ \dfrac{1}{9} \pm \dfrac{2\sqrt{11}}{9}i \right\}$$

13. $12p^2 - 4p + 5 = 0$

$a = 12,\ b = -4,\ c = 5$

$$p = \frac{-(-4) \pm \sqrt{(-4)^2 - 4(12)(5)}}{2(12)}$$

$$= \frac{4 \pm \sqrt{16 - 240}}{24} = \frac{4 \pm \sqrt{-224}}{24}$$

$$= \frac{4 \pm 4i\sqrt{14}}{24} = \frac{\cancel{4}\left(1 \pm i\sqrt{14}\right)}{\cancel{4} \cdot 6} = \frac{1 \pm i\sqrt{14}}{6}$$

$$= \frac{1}{6} \pm \frac{\sqrt{14}}{6}i \qquad \left\{\frac{1}{6} \pm \frac{\sqrt{14}}{6}i\right\}$$

15. $-z^2 = -2z - 35$

$z^2 - 2z - 35 = 0$

$a = 1,\ b = -2,\ c = -35$

$$z = \frac{-(-2) \pm \sqrt{(-2)^2 - 4(1)(-35)}}{2(1)}$$

$$= \frac{2 \pm \sqrt{4 + 140}}{2} = \frac{2 \pm \sqrt{144}}{2} = \frac{2 \pm 12}{2}$$

$$z = \frac{14}{2} = 7 \text{ or } z = -\frac{10}{2} = -5 \quad \{7, -5\}$$

17. $y^2 + 3y = 8$

$y^2 + 3y - 8 = 0$

$a = 1,\ b = 3,\ c = -8$

$$y = \frac{-(3) \pm \sqrt{(3)^2 - 4(1)(-8)}}{2(1)} = \frac{-3 \pm \sqrt{9 + 32}}{2}$$

$$= \frac{-3 \pm \sqrt{41}}{2} \qquad \left\{\frac{-3 \pm \sqrt{41}}{2}\right\}$$

19. $25x^2 - 20x + 4 = 0$

$a = 25,\ b = -20,\ c = 4$

$$x = \frac{-(-20) \pm \sqrt{(-20)^2 - 4(25)(4)}}{2(25)}$$

$$= \frac{20 \pm \sqrt{400 - 400}}{50} = \frac{20 \pm \sqrt{0}}{50} = \frac{20}{50} = \frac{2}{5}$$

$$\left\{\frac{2}{5}\right\}$$

21. $w(w - 6) = -14$

$w^2 - 6w + 14 = 0$

$a = 1,\ b = -6,\ c = 14$

$$w = \frac{-(-6) \pm \sqrt{(-6)^2 - 4(1)(14)}}{2(1)}$$

$$= \frac{6 \pm \sqrt{36 - 56}}{2} = \frac{6 \pm \sqrt{-20}}{2} = \frac{6 \pm 2i\sqrt{5}}{2}$$

$$= \frac{\cancel{2}\left(3 \pm i\sqrt{5}\right)}{\cancel{2}} = 3 \pm i\sqrt{5} \quad \left\{3 \pm i\sqrt{5}\right\}$$

23. $(x + 2)(x - 3) = 1$

$x^2 - x - 6 = 1$

$x^2 - x - 7 = 0$

$a = 1,\ b = -1,\ c = -7$

$$x = \frac{-(-1) \pm \sqrt{(-1)^2 - 4(1)(-7)}}{2(1)}$$

$$= \frac{1 \pm \sqrt{1 + 28}}{2} = \frac{1 \pm \sqrt{29}}{2} \quad \left\{\frac{1 \pm \sqrt{29}}{2}\right\}$$

25. $-4a^2 - 2a + 3 = 0$

$a = -4, b = -2, c = 3$

$a = \dfrac{-(-2) \pm \sqrt{(-2)^2 - 4(-4)(3)}}{2(-4)}$

$= \dfrac{2 \pm \sqrt{4 + 48}}{-8} = \dfrac{2 \pm \sqrt{52}}{-8} = \dfrac{2 \pm 2\sqrt{13}}{-8}$

$= \dfrac{\cancel{2}\left(1 \pm \sqrt{13}\right)}{\cancel{2}(-4)} = \dfrac{1 \pm \sqrt{13}}{-4} = \dfrac{-1 \pm \sqrt{13}}{4}$

$\left\{ \dfrac{-1 \pm \sqrt{13}}{4} \right\}$

27. $\dfrac{1}{2}y^2 + \dfrac{2}{3} = -\dfrac{2}{3}y$

$6\left(\dfrac{1}{2}y^2 + \dfrac{2}{3}\right) = 6\left(-\dfrac{2}{3}y\right)$

$3y^2 + 4 = -4y$

$3y^2 + 4y + 4 = 0$

$a = 3, b = 4, c = 4$

$y = \dfrac{-(4) \pm \sqrt{(4)^2 - 4(3)(4)}}{2(3)}$

$= \dfrac{-4 \pm \sqrt{16 - 48}}{6} = \dfrac{-4 \pm \sqrt{-32}}{6}$

$= \dfrac{-4 \pm 4i\sqrt{2}}{6} = -\dfrac{4}{6} \pm \dfrac{4\sqrt{2}}{6}i = -\dfrac{2}{3} \pm \dfrac{2\sqrt{2}}{3}i$

$\left\{ -\dfrac{2}{3} \pm \dfrac{2\sqrt{2}}{3}i \right\}$

29. $\dfrac{1}{5}h^2 + h + \dfrac{3}{5} = 0$

$5\left(\dfrac{1}{5}h^2 + h + \dfrac{3}{5}\right) = 5(0)$

$h^2 + 5h + 3 = 0$

$a = 1, b = 5, c = 3$

$h = \dfrac{-(5) \pm \sqrt{(5)^2 - 4(1)(3)}}{2(1)}$

$= \dfrac{-5 \pm \sqrt{25 - 12}}{2} = \dfrac{-5 \pm \sqrt{13}}{2}$ $\left\{ \dfrac{-5 \pm \sqrt{13}}{2} \right\}$

31. $0.01x^2 + 0.06x + 0.08 = 0$

$100\left(0.01x^2 + 0.06x + 0.08\right) = 100(0)$

$x^2 + 6x + 8 = 0$

$a = 1, b = 6, c = 8$

$x = \dfrac{-(6) \pm \sqrt{(6)^2 - 4(1)(8)}}{2(1)}$

$= \dfrac{-6 \pm \sqrt{36 - 32}}{2} = \dfrac{-6 \pm \sqrt{4}}{2} = \dfrac{-6 \pm 2}{2}$

$x = \dfrac{-4}{2} = -2$ or $x = \dfrac{-8}{2} = -4$ $\{-2, -4\}$

33. $0.3t^2 + 0.7t - 0.5 = 0$

$10\left(0.3t^2 + 0.7t - 0.5\right) = 10(0)$

$3t^2 + 7t - 5 = 0$

$a = 3, b = 7, c = -5$

$t = \dfrac{-(7) \pm \sqrt{(7)^2 - 4(3)(-5)}}{2(3)}$

$= \dfrac{-7 \pm \sqrt{49 + 60}}{6} = \dfrac{-7 \pm \sqrt{109}}{6}$

$\left\{ \dfrac{-7 \pm \sqrt{109}}{6} \right\}$

35. a. $x^3 - 27 = x^3 - 3^3 = (x - 3)(x^2 + 3x + 9)$

b. $x^3 - 27 = 0$

$(x - 3)(x^2 + 3x + 9) = 0$

$x - 3 = 0$ or $x^2 + 3x + 9 = 0$

$x = 3$ or $a = 1, b = 3, c = 9$

$x = \dfrac{-(3) \pm \sqrt{(3)^2 - 4(1)(9)}}{2(1)}$

$= \dfrac{-3 \pm \sqrt{9 - 36}}{2} = \dfrac{-3 \pm \sqrt{-27}}{2}$

$$x = 3 \text{ or } x = \frac{-3 \pm 3i\sqrt{3}}{2}$$

$$\left\{ 3, -\frac{3}{2} \pm \frac{3\sqrt{3}}{2} i \right\}$$

37. a. $3x^3 - 6x^2 + 6x = 3x\left(x^2 - 2x + 2\right)$

b. $3x^3 - 6x^2 + 6x = 0$

$\left(3x\right)\left(x^2 - 2x + 2\right) = 0$

$3x = 0 \text{ or } x^2 - 2x + 2 = 0$

$x = 0 \text{ or } a = 1, \ b = -2, \ c = 2$

$$x = \frac{-(-2) \pm \sqrt{(-2)^2 - 4(1)(2)}}{2(1)}$$

$$= \frac{2 \pm \sqrt{4 - 8}}{2} = \frac{2 \pm \sqrt{-4}}{2}$$

$x = 0 \text{ or } x = \dfrac{2 \pm 2i}{2} = 1 \pm i \quad \{0, 1 \pm i\}$

39. $s^3 = 27$

$s = \sqrt[3]{27} = 3$

The length of a side of the cube is 3 ft.

41. Let $x =$ one leg of the triangle

$x + 2 =$ the other leg of the triangle

$$x^2 + \left(x + 2\right)^2 = 4^2$$

$$x^2 + x^2 + 4x + 4 = 16$$

$$2x^2 + 4x - 12 = 0$$

$a = 2, b = 4, c = -12$

$$x = \frac{-(4) \pm \sqrt{(4)^2 - 4(2)(-12)}}{2(2)}$$

$$x = \frac{-4 \pm \sqrt{16 + 96}}{4}$$

$$= \frac{-4 \pm \sqrt{112}}{4}$$

$$= \frac{-4 \pm 10.58}{4}$$

$x \approx 1.6 \text{ or } x \approx 3.6$

The legs are 1.6 in and 3.6 in.

43. Let $x =$ one leg of the triangle

$x - 2.1 =$ the other leg of the triangle

$$x^2 + \left(x - 2.1\right)^2 = 10.2^2$$

$$x^2 + x^2 - 4.2x + 4.41 = 104.04$$

$$2x^2 - 4.2x - 99.63 = 0$$

$$100\left(2x^2 - 4.2x - 99.63\right) = 100(0)$$

$$200x^2 - 420x - 9963 = 0$$

$a = 200, \ b = -420, \ c = -9963$

$$x = \frac{-(-420) \pm \sqrt{(-420)^2 - 4(200)(-9963)}}{2(200)}$$

$$x = \frac{420 \pm \sqrt{176,400 + 7,970,400}}{400}$$

$$= \frac{420 \pm \sqrt{8,146,800}}{400}$$

$$= \frac{420 \pm 2854.259974}{400}$$

$x \approx 8.2 \text{ or } x \approx -6.1$

$x - 2.1 = 8.2 - 2.1 \approx 6.1$

The legs are 8.2 m and 6.1 m.

45. a. $F(x) = 0.0036x^2 - 0.35x + 9.2$

$F(x) = 0.0036(16)^2 - 0.35(16) + 9.2 = 4.5216 \approx 4.5$

The fatality rate is approximately 4.5 fatalities per 100 million miles driven.

b. $F(x) = 0.0036(40)^2 - 0.35(40) + 9.2 = 0.96 \approx 1$

The fatality rate is approximately 1 fatality per 100 million miles driven.

c. $F(x) = 0.0036(80)^2 - 0.35(80) + 9.2 = 4.24 \approx 4.2$

The fatality rate is approximately 4.2 fatalities per 100 million miles driven.

d. $F(x) = 0.0036(80)^2 - 0.35(80) + 9.2 = 4.24 \approx 4.2$ fatalities per 100 million miles driven

$0.0036x^2 - 0.35x + 9.2 = 2.5$

$0.0036x^2 - 0.35x + 6.7 = 0$

$a = 0.0036, \ b = -0.35, \ c = 6.7$

$$t = \frac{-(-0.35) \pm \sqrt{(-0.35)^2 - 4(0.0036)(6.7)}}{2(0.0036)} = \frac{0.35 \pm \sqrt{0.02602}}{0.0072} = \frac{0.35 \pm 0.1613}{0.0072}$$

$t \approx 71 \ $ or $\ t \approx 26$

The fatality rate is 2.5 for drivers 26 years old or 71 years old.

47. $h(t) = -16t^2 + 48t + 48$

$64 = -16t^2 + 48t + 48$

$16t^2 - 48t + 16 = 0$

$t^2 - 3t + 1 = 0$

$a = 1, \ b = -3, \ c = 1$

$$t = \frac{-(-3) \pm \sqrt{(-3)^2 - 4(1)(1)}}{2(1)} = \frac{3 \pm \sqrt{5}}{2}$$

$t = \dfrac{3 + \sqrt{5}}{2} \approx 2.62$ sec

or $t = \dfrac{3 - \sqrt{5}}{2} \approx 0.38$ sec

49. a. $x^2 + 2x = -1$

$x^2 + 2x + 1 = 0$

b. $b^2 - 4ac = 2^2 - 4(1)(1) = 4 - 4 = 0$

c. 1 rational solution

51. a. $19m^2 = 8m$

$19m^2 - 8m + 0 = 0$

b. $b^2 - 4ac = (-8)^2 - 4(19)(0) = 64 - 0 = 64$

c. 2 rational solutions

53. a. $5p^2 - 21 = 0$

$5p^2 + 0p - 21 = 0$

b. $b^2 - 4ac = (0)^2 - 4(5)(-21)$

$= 0 + 420 = 420$

c. 2 irrational solutions

55. a. $4n(n-2) - 5n(n-1) = 4$

$4n^2 - 8n - 5n^2 + 5n = 4$

$-n^2 - 3n - 4 = 0$

$n^2 + 3n + 4 = 0$

b. $b^2 - 4ac = (3)^2 - 4(1)(4) = 9 - 16 = -7$

c. 2 imaginary solutions

57. $f(x) = x^2 - 6x + 5$

$b^2 - 4ac = (-6)^2 - 4(1)(5) = 36 - 20 = 16$

two x-intercepts

59. $h(x) = 4x^2 + 12x + 9$

$b^2 - 4ac = (12)^2 - 4(4)(9) = 144 - 144 = 0$

one x-intercept

61. $p(x) = 2x^2 + 3x + 6$

$b^2 - 4ac = (3)^2 - 4(2)(6) = 9 - 48 = -39$

no x-intercepts

63. $f(x) = x^2 - 5x + 3 = 0$

$a = 1, b = -5, c = 3$

$$x = \frac{-(-5) \pm \sqrt{(-5)^2 - 4(1)(3)}}{2(1)}$$

$$= \frac{5 \pm \sqrt{25 - 12}}{2} = \frac{5 \pm \sqrt{13}}{2}$$

x - intercepts : $\left(\frac{5 + \sqrt{13}}{2}, 0 \right), \left(\frac{5 - \sqrt{13}}{2}, 0 \right)$

$f(0) = 0^2 - 5(0) + 3 = 3$

y - intercept : $(0, 3)$

65. $g(x) = -x^2 + x - 1 = 0$

$a = -1, b = 1, c = -1$

$$x = \frac{-(1) \pm \sqrt{(1)^2 - 4(-1)(-1)}}{2(-1)}$$

$$= \frac{-1 \pm \sqrt{1 - 4}}{-2} = \frac{-1 \pm \sqrt{-3}}{-2} = \frac{-1 \pm i\sqrt{3}}{-2}$$

x-intercepts: none - solutions imaginary

$g(0) = -(0)^2 + 0 - 1 = -1$

y-intercept: $(0, -1)$

67. $p(x) = 2x^2 + 5x - 2 = 0$

$a = 2, b = 5, c = -2$

$$x = \frac{-(5) \pm \sqrt{(5)^2 - 4(2)(-2)}}{2(2)}$$

$$= \frac{-5 \pm \sqrt{25 + 16}}{4} = \frac{-5 \pm \sqrt{41}}{4}$$

x - intercepts : $\left(\frac{-5 + \sqrt{41}}{4}, 0 \right), \left(\frac{-5 - \sqrt{41}}{4}, 0 \right)$

$p(0) = 2(0)^2 + 5(0) - 2 = -2$

y - intercept : $(0, -2)$

69. $a^2 + 3a + 4 = 0$

$a = 1, b = 3, c = 4$

$$a = \frac{-(3) \pm \sqrt{(3)^2 - 4(1)(4)}}{2(1)}$$

$$= \frac{-3 \pm \sqrt{9 - 16}}{2} = \frac{-3 \pm \sqrt{-7}}{2} = \frac{-3 \pm i\sqrt{7}}{2}$$

$$\left\{ -\frac{3}{2} \pm \frac{\sqrt{7}}{2} i \right\}$$

71.

$$(x - 2)^2 + 2x^2 - 13x = 10$$

$$x^2 - 4x + 4 + 2x^2 - 13x - 10 = 0$$

$$3x^2 - 17x - 6 = 0$$

$$(3x + 1)(x - 6) = 0$$

$$3x + 1 = 0 \ \text{ or } \ x - 6 = 0$$

$$3x = -1 \ \text{ or } \ x = 6$$

$$x = -\frac{1}{3} \ \text{ or } \ x = 6$$

$$\left\{ -\frac{1}{3}, 6 \right\}$$

73. $4y^2 - 20y + 43 = 0$

$a = 4, b = -20, c = 43$

$$y = \frac{-(-20) \pm \sqrt{(-20)^2 - 4(4)(43)}}{2(4)}$$

$$= \frac{20 \pm \sqrt{400 - 688}}{8}$$

$$= \frac{20 \pm \sqrt{-288}}{8}$$

$$= \frac{20 \pm \sqrt{-144(2)}}{8}$$

$$= \frac{20 \pm 12i\sqrt{2}}{8} = \frac{5}{2} \pm \frac{3\sqrt{2}}{2}i \qquad \left\{\frac{5}{2} \pm \frac{3\sqrt{2}}{2}i\right\}$$

75.

$$\left(x + \frac{1}{2}\right)^2 + 4 = 0$$

$$\left(x + \frac{1}{2}\right)^2 = -4$$

$$x + \frac{1}{2} = \pm\sqrt{-4}$$

$$x + \frac{1}{2} = \pm 2i$$

$$x = -\frac{1}{2} \pm 2i \qquad \left\{-\frac{1}{2} \pm 2i\right\}$$

77.

$$2y(y-3) = -1$$

$$2y^2 - 6y = -1$$

$$2y^2 - 6y + 1 = 0$$

$$a = 2, \ b = -6, \ c = 1$$

$$y = \frac{-(-6) \pm \sqrt{(-6)^2 - 4(2)(1)}}{2(2)} = \frac{6 \pm \sqrt{36-8}}{4}$$

$$= \frac{6 \pm \sqrt{28}}{4} = \frac{6 \pm 2\sqrt{7}}{4} = \frac{3 \pm \sqrt{7}}{2} \qquad \left\{\frac{3 \pm \sqrt{7}}{2}\right\}$$

79.

$$(2t+5)(t-1) = (t-3)(t+8)$$

$$2t^2 + 3t - 5 = t^2 + 5t - 24$$

$$t^2 - 2t + 19 = 0$$

$$a = 1, \ b = -2, \ c = 19$$

$$t = \frac{-(-2) \pm \sqrt{(-2)^2 - 4(1)(19)}}{2(1)}$$

$$= \frac{2 \pm \sqrt{4-76}}{2} = \frac{2 \pm \sqrt{-72}}{2} = \frac{2 \pm 6i\sqrt{2}}{2}$$

$$= 1 \pm 3i\sqrt{2} \qquad \left\{1 \pm 3i\sqrt{2}\right\}$$

81.

$$\frac{1}{8}x^2 - \frac{1}{2}x + \frac{1}{4} = 0$$

$$8\left(\frac{1}{8}x^2 - \frac{1}{2}x + \frac{1}{4}\right) = 8(0)$$

$$x^2 - 4x + 2 = 0$$

$$x^2 - 4x = -2$$

$$x^2 - 4x + 4 = -2 + 4$$

$$(x-2)^2 = 2$$

$$x - 2 = \pm\sqrt{2}$$

$$x = 2 \pm \sqrt{2} \qquad \left\{2 \pm \sqrt{2}\right\}$$

83.

$$32z^2 - 20z - 3 = 0$$

$$(8z+1)(4z-3) = 0$$

$$8z + 1 = 0 \ \text{ or } \ 4z - 3 = 0$$

$$8z = -1 \ \text{ or } \ \ 4z = 3$$

$$z = -\frac{1}{8} \ \text{ or } \ z = \frac{3}{4} \quad \left\{-\frac{1}{8}, \frac{3}{4}\right\}$$

85.

$$4p^2 - 21 = 0$$

$$4p^2 = 21$$

$$p^2 = \frac{21}{4}$$

$$p = \pm\sqrt{\frac{21}{4}}$$

$$p = \pm\frac{\sqrt{21}}{2} \qquad \left\{\pm\frac{\sqrt{21}}{2}\right\}$$

310

87. a.

$$x^2 + 6x = 5$$
$$x^2 + 6x + 9 = 5 + 9$$
$$(x+3)^2 = 14$$
$$x + 3 = \pm\sqrt{14}$$
$$x = -3 \pm \sqrt{14} \quad \left\{-3 \pm \sqrt{14}\right\}$$

b.

$$x^2 + 6x = 5$$
$$x^2 + 6x - 5 = 0$$
$$a = 1,\ b = 6,\ c = -5$$
$$x = \frac{-6 \pm \sqrt{6^2 - 4(1)(-5)}}{2(1)} = \frac{-6 \pm \sqrt{36 + 20}}{2}$$
$$= \frac{-6 \pm \sqrt{56}}{2} = \frac{-6 \pm 2\sqrt{14}}{2}$$
$$= \frac{2\left(-3 \pm \sqrt{14}\right)}{2} = -3 \pm \sqrt{14} \quad \left\{-3 \pm \sqrt{14}\right\}$$

c.   Answers will vary.

Section 7.3   Equations in Quadratic Form

1. **(a)** Quadratic
(b) $3x - 1$
(c) $p^{1/3}$

3.

$$\left(x - \frac{3}{2}\right)^2 = \frac{7}{4}$$
$$x - \frac{3}{2} = \pm\sqrt{\frac{7}{4}}$$
$$x = \frac{3}{2} \pm \frac{\sqrt{7}}{2} \quad \left\{\frac{3}{2} \pm \frac{\sqrt{7}}{2}\right\}$$

5.

$$x(x + 8) = -16$$
$$x^2 + 8x = -16$$
$$x^2 + 8x + 16 = 0$$
$$(x + 4)^2 = 0$$
$$x + 4 = 0$$
$$x = -4 \quad \left\{-4\right\}$$

7.

$$2x^2 - 8x - 44 = 0$$
$$a = 2,\ b = -8,\ c = -44$$
$$k = \frac{-(-8) \pm \sqrt{(-8)^2 - 4(2)(-44)}}{2(2)}$$
$$= \frac{8 \pm \sqrt{64 + 352}}{4} = \frac{8 \pm \sqrt{416}}{4}$$
$$= \frac{8 \pm 4\sqrt{26}}{4} = 2 \pm \sqrt{26} \quad \left\{2 \pm \sqrt{26}\right\}$$

9. a. $u^2 + 10u + 24 = 0$

$(u+6)(u+4) = 0$

$u + 6 = 0$ or $u + 4 = 0$

$u = -6$ or $u = -4$ $\{-6, -4\}$

b. $\left(y^2 + 5y\right)^2 + 10\left(y^2 + 5y\right) + 24 = 0$

Let $u = y^2 + 5y$

$u^2 + 10u + 24 = 0$

$(u+6)(u+4) = 0$

$u + 6 = 0$ or $u + 4 = 0$

$u = -6$ or $u = -4$

$y^2 + 5y = -6$ or $y^2 + 5y = -4$

$y^2 + 5y + 6 = 0$ or $y^2 + 5y + 4 = 0$

$(y+3)(y+2) = 0$ or $(y+4)(y+1) = 0$

$y + 3 = 0$ or $y + 2 = 0$ or $y + 4 = 0$ or $y + 1 = 0$

$y = -3$ or $y = -2$ or $y = -4$ or $y = -1$

$\{-3, -2, -4, -1\}$

11. $\left(x^2 - 2x\right)^2 + 2\left(x^2 - 2x\right) = 3$

Let $u = x^2 - 2x$

$u^2 + 2u = 3$

$u^2 + 2u - 3 = 0$

$(u+3)(u-1) = 0$

$u + 3 = 0$ or $u - 1 = 0$

$x^2 - 2x + 3 = 0$ or $x^2 - 2x - 1 = 0$

$x = \dfrac{2 \pm \sqrt{(-2)^2 - 4(1)(3)}}{2(1)}$

or $x = \dfrac{2 \pm \sqrt{(-2)^2 - 4(1)(-1)}}{2(1)}$

$x = \dfrac{2 \pm \sqrt{-8}}{2}$ or $x = \dfrac{2 \pm \sqrt{8}}{2}$

$x = \dfrac{2 \pm 2i\sqrt{2}}{2}$ or $x = \dfrac{2 \pm 2\sqrt{2}}{2}$

$x = 1 \pm i\sqrt{2}$ or $x = 1 \pm \sqrt{2}$ $\left\{1 \pm \sqrt{2}, 1 \pm i\sqrt{2}\right\}$

13. $\left(y^2 - 4y\right)^2 - \left(y^2 - 4y\right) = 20$

Let $u = y^2 - 4y$

$u^2 - u = 20$

$u^2 - u - 20 = 0$

$(u+4)(u-5) = 0$

$u + 4 = 0$ or $u - 5 = 0$

$y^2 - 4y + 4 = 0$ or $y^2 - 4y - 5 = 0$

$(y-2)^2 = 0$ or $(y-5)(y+1) = 0$

$y - 2 = 0$ or $y - 5 = 0$ or $y + 1 = 0$

$y = 2$ or $y = 5$ or $y = -1$

$\{2, 5, -1\}$

15. $m^{2/3} - m^{1/3} - 6 = 0$

$\left(m^{1/3}\right)^2 - m^{1/3} - 6 = 0$

Let $u = m^{1/3}$

$u^2 - u - 6 = 0$

$(u-3)(u+2) = 0$

$u - 3 = 0$ or $u + 2 = 0$

$u = 3$ or $u = -2$

$m^{1/3} = 3$ or $m^{1/3} = -2$

$\left(m^{1/3}\right)^3 = 3^3$ or $\left(m^{1/3}\right)^3 = (-2)^3$

$m = 27$ or $m = -8$ $\{27, -8\}$

17.

$$2t^{2/5} + 7t^{1/5} + 3 = 0$$

$$2\left(t^{1/5}\right)^2 + 7t^{1/5} + 3 = 0$$

Let $u = t^{1/5}$

$$2u^2 + 7u + 3 = 0$$

$$(2u + 1)(u + 3) = 0$$

$2u + 1 = 0$ or $\quad u + 3 = 0$

$\quad 2u = -1$ or $\qquad u = -3$

$\quad u = -\dfrac{1}{2}$ or $\qquad u = -3$

$\quad t^{1/5} = -\dfrac{1}{2}$ or $\qquad t^{1/5} = -3$

$$\left(t^{1/5}\right)^5 = \left(-\dfrac{1}{2}\right)^5 \text{ or } \left(t^{1/5}\right)^5 = (-3)^5$$

$\quad t = -\dfrac{1}{32}$ or $\qquad t = -243$

$$\left\{-\dfrac{1}{32}, -243\right\}$$

19.

$$y + 6\sqrt{y} = 16$$

$$\left(\sqrt{y}\right)^2 + 6\sqrt{y} = 16$$

Let $u = \sqrt{y}$

$$u^2 + 6u = 16$$

$$u^2 + 6u - 16 = 0$$

$$(u + 8)(u - 2) = 0$$

$u + 8 = 0$ or $u - 2 = 0$

$u = -8$ or $\quad u = 2$

$\sqrt{y} = -8$ or $\sqrt{y} = 2$

$y = 64$ or $\quad y = 4$

Check:

$y = 64$:	$y = 4$:
$64 + 6\sqrt{64} = 16$	$4 + 6\sqrt{4} = 16$
$64 + 6(8) = 16$	$4 + 6(2) = 16$
$64 + 48 = 16$	$4 + 12 = 16$
$112 \neq 16$	$16 = 16$

Solution: $\{4\}$ ($y = 64$ does not check.)

21.

$$2x + 3\sqrt{x} - 2 = 0$$

$$2\left(\sqrt{x}\right)^2 + 3\sqrt{x} - 2 = 0$$

Let $u = \sqrt{x}$

$$2u^2 + 3u - 2 = 0$$

$$(2u - 1)(u + 2) = 0$$

$2u - 1 = 0$ or $u + 2 = 0$

$\quad 2u = 1$ or $\quad u = -2$

$\quad u = \dfrac{1}{2}$ or $\quad u = -2$

$\quad \sqrt{x} = \dfrac{1}{2}$ or $\sqrt{x} = -2$

$\quad x = \dfrac{1}{4}$ or $\quad x = 4$

Check:

$x = \dfrac{1}{4}$:

$$2\left(\dfrac{1}{4}\right) + 3\sqrt{\dfrac{1}{4}} - 2 = 0$$

$$\dfrac{1}{2} + 3\left(\dfrac{1}{2}\right) - 2 = 0$$

$$\dfrac{1}{2} + \dfrac{3}{2} - \dfrac{4}{2} = 0$$

$$0 = 0$$

$x = 4$:

$$2(4) + 3\sqrt{4} - 2 = 0$$

$$8 + 3(2) - 2 = 0$$

$$8 + 6 - 2 = 0$$

$$12 \neq 0$$

Solution: $\left\{\frac{1}{4}\right\}$ ($x = 4$ does not check.)

23.

$$16\left(\frac{x+6}{4}\right)^2 + 8\left(\frac{x+6}{4}\right) + 1 = 0$$

Let $u = \dfrac{x+6}{4}$

$$16u^2 + 8u + 1 = 0$$

$$(4u+1)(4u+1) = 0$$

$$4u+1 = 0 \quad \text{or} \quad 4u+1 = 0$$

$$4u = -1 \quad \text{or} \qquad 4u = -1$$

$$4\left(\frac{x+6}{4}\right) = -1 \quad \text{or} \quad 4\left(\frac{x+6}{4}\right) = -1$$

$$x+6 = -1 \quad \text{or} \qquad x+6 = -1$$

$$x = -7 \quad \text{or} \qquad x = -7 \quad \{-7\}$$

25.

$$x - \sqrt{x} - 12 = 0$$

$$x - 12 = \sqrt{x}$$

$$(x-12)^2 = \left(\sqrt{x}\right)^2$$

$$x^2 - 24x + 144 = x$$

$$x^2 - 25x + 144 = 0$$

$$(x-9)(x-16) = 0$$

$$x - 9 = 0 \quad \text{or} \quad x - 16 = 0$$

$$x = 9 \quad \text{or} \qquad x = 16$$

Check:

$$9 - \sqrt{9} - 12 = 0 \qquad 16 - \sqrt{16} - 12 = 0$$

$$9 - 3 - 12 = 0 \qquad 16 - 4 - 12 = 0$$

$$-6 \neq 0 \qquad\qquad 0 = 0$$

The solution is $\{16\}$. ($x = 9$ does not check.)
Yes, we obtain the same answers.

27.

$$w^4 + 4w^2 - 45 = 0$$

$$(w^2+9)(w^2-5) = 0$$

$$w^2 + 9 = 0 \quad \text{or} \quad w^2 - 5 = 0$$

$$w^2 = -9 \quad \text{or} \quad w^2 = 5$$

$$w = \pm\sqrt{-9} \quad \text{or} \quad w = \pm\sqrt{5}$$

$$w = \pm 3i \quad \text{or} \quad w = \pm\sqrt{5}$$

$$\left\{\pm 3i, \pm\sqrt{5}\right\}$$

29.

$$y^2\left(4y^2 + 17\right) = 15$$

$$4y^4 + 17y^2 - 15 = 0$$

$$(4y^2 - 3)(y^2 + 5) = 0$$

$$4y^2 - 3 = 0 \quad \text{or} \quad y^2 + 5 = 0$$

$$4y^2 = 3 \quad \text{or} \qquad y^2 = -5$$

$$2y = \pm\sqrt{3} \quad \text{or} \quad y = \pm\sqrt{-5}$$

$$y = \pm\frac{\sqrt{3}}{2} \quad \text{or} \quad y = \pm i\sqrt{5}$$

$$\left\{\pm\frac{\sqrt{3}}{2},\ \pm i\sqrt{5}\right\}$$

31.
$$1 + \frac{5}{x} = -\frac{3}{x^2}$$
$$x^2\left(1 + \frac{5}{x}\right) = x^2\left(-\frac{3}{x^2}\right)$$
$$x^2 + 5x = -3$$
$$x^2 + 5x + 3 = 0$$
$$a = 1, \ b = 5, \ c = 3$$
$$x = \frac{-(5) \pm \sqrt{(5)^2 - 4(1)(3)}}{2(1)}$$
$$x = \frac{-5 \pm \sqrt{13}}{2} \qquad \left\{\frac{-5 \pm \sqrt{13}}{2}\right\}$$

33.
$$\frac{3x}{x+1} - \frac{2}{x-3} = 1$$
$$(x+1)(x-3)\left(\frac{3x}{x+1} - \frac{2}{x-3}\right) = (x+1)(x-3)\cdot 1$$
$$3x(x-3) - 2(x+1) = x^2 - 2x - 3$$
$$3x^2 - 9x - 2x - 2 = x^2 - 2x - 3$$
$$2x^2 - 9x + 1 = 0$$
$$a = 2, \ b = -9, \ c = 1$$
$$x = \frac{-(-9) \pm \sqrt{(-9)^2 - 4(2)(1)}}{2(2)}$$
$$x = \frac{9 \pm \sqrt{73}}{4} \qquad \left\{\frac{9 \pm \sqrt{73}}{4}\right\}$$

35.
$$\frac{x}{2x-1} = \frac{1}{x-2}$$
$$(2x-1)(x-2)\left(\frac{x}{2x-1}\right) = (2x-1)(x-2)\left(\frac{1}{x-2}\right)$$
$$x(x-2) = 1(2x-1)$$
$$x^2 - 2x = 2x - 1$$
$$x^2 - 4x = -1$$
$$x^2 - 4x + 4 = -1 + 4$$
$$(x-2)^2 = 3$$
$$x - 2 = \pm\sqrt{3}$$
$$x = 2 \pm \sqrt{3} \qquad \left\{2 \pm \sqrt{3}\right\}$$

37.
$$x^4 - 16 = 0$$
$$(x^2 - 4)(x^2 + 4) = 0$$
$$x^2 - 4 = 0 \ \text{ or } \ x^2 + 4 = 0$$
$$x^2 = 4 \ \text{ or } \quad x^2 = -4$$
$$x = \pm\sqrt{4} \ \text{ or } \ x = \pm\sqrt{-4}$$
$$x = \pm 2 \ \text{ or } \quad x = \pm 2i$$
Solutions: $\{2, -2, 2i, -2i\}$

39. $(4x+5)^2 + 3(4x+5) + 2 = 0$

Let $u = 4x + 5$

$u^2 + 3u + 2 = 0$

$(u+2)(u+1) = 0$

$u + 2 = 0$ or $u + 1 = 0$

$u = -2$ or $u = -1$

$4x + 5 = -2$ or $4x + 5 = -1$

$4x = -7$ or $4x = -6$

$x = -\dfrac{7}{4}$ or $x = -\dfrac{6}{4} = -\dfrac{3}{2}$

$$\left\{ -\dfrac{7}{4}, -\dfrac{3}{2} \right\}$$

41. $4m^4 - 9m^2 + 2 = 0$

$4\left(m^2\right)^2 - 9m^2 + 2 = 0$

Let $u = m^2$

$4u^2 - 9u + 2 = 0$

$(4u-1)(u-2) = 0$

$4u - 1 = 0$ or $u - 2 = 0$

$4m^2 - 1 = 0$ or $m^2 - 2 = 0$

$4m^2 = 1$ or $m^2 = 2$

$m^2 = \dfrac{1}{4}$ or $m = \pm\sqrt{2}$

$m = \pm\dfrac{1}{2}$ or $m = \pm\sqrt{2}$

$$\left\{ \dfrac{1}{2}, -\dfrac{1}{2}, \sqrt{2}, -\sqrt{2} \right\}$$

43. $x^6 - 9x^3 + 8 = 0$

$\left(x^3\right)^2 - 9x^3 + 8 = 0$

Let $u = x^3$

$u^2 - 9u + 8 = 0$

$(u-8)(u-1) = 0$

$u - 8 = 0$ or $u - 1 = 0$

$x^3 - 8 = 0$ or $x^3 - 1 = 0$

$(x-2)(x^2 + 2x + 4) = 0$ or $(x-1)(x^2 + x + 1) = 0$

$x - 2 = 0$ or $x^2 + 2x + 4 = 0$ or $x - 1 = 0$ or $x^2 + x + 1 = 0$

$x = 2$ or $x = \dfrac{-2 \pm \sqrt{2^2 - 4(1)(4)}}{2(1)}$ or $x = 1$ or $x = \dfrac{-1 \pm \sqrt{1^2 - 4(1)(1)}}{2(1)}$

$x = 2$ or $x = \dfrac{-2 \pm \sqrt{-12}}{2}$ or $x = 1$ or $x = \dfrac{-1 \pm \sqrt{-3}}{2}$

$x = 2$ or $x = \dfrac{-2 \pm 2i\sqrt{3}}{2}$ or $x = 1$ or $x = \dfrac{-1 \pm i\sqrt{3}}{2}$

$x = 2$ or $x = -1 \pm i\sqrt{3}$ or $x = 1$ or $x = \dfrac{-1 \pm i\sqrt{3}}{2}$ $\left\{ 2, 1, -1 \pm i\sqrt{3}, -\dfrac{1}{2} \pm \dfrac{\sqrt{3}}{2}i \right\}$

45.

$$\sqrt{x^2 + 20} = 3\sqrt{x}$$

$$\left(\sqrt{x^2 + 20}\right)^2 = \left(3\sqrt{x}\right)^2$$

$$x^2 + 20 = 9x$$

$$x^2 - 9x + 20 = 0$$

$$(x - 5)(x - 4) = 0$$

$$x - 5 = 0 \text{ or } x - 4 = 0$$

$$x = 5 \text{ or } \quad x = 4$$

Check:

$x = 5:$ $x = 4:$

$$\sqrt{5^2 + 20} = 3\sqrt{5} \qquad \sqrt{4^2 + 20} = 3\sqrt{4}$$

$$\sqrt{45} = 3\sqrt{5} \qquad\qquad \sqrt{36} = 3(2)$$

$$3\sqrt{5} = 3\sqrt{5} \qquad\qquad\quad 6 = 6$$

Solutions: $\{5,\ 4\}$

47.

$$\sqrt{4t + 1} = t + 1$$

$$\left(\sqrt{4t + 1}\right)^2 = (t + 1)^2$$

$$4t + 1 = t^2 + 2t + 1$$

$$t^2 - 2t = 0$$

$$t(t - 2) = 0$$

$$t = 0 \text{ or } t - 2 = 0$$

$$t = 0 \text{ or } \quad t = 2$$

Check:

$t = 0:$ $t = 2:$

$$\sqrt{4(0) + 1} = 0 + 1 \qquad \sqrt{4(2) + 1} = 2 + 1$$

$$\sqrt{1} = 1 \qquad\qquad\qquad \sqrt{9} = 3$$

$$1 = 1 \qquad\qquad\qquad\quad 3 = 3$$

Solutions: $\{0, 2\}$

49.

$$2\left(\frac{t - 4}{3}\right)^2 - \left(\frac{t - 4}{3}\right) - 3 = 0$$

Let $u = \dfrac{t - 4}{3}$

$$2u^2 - u - 3 = 0$$

$$(2u - 3)(u + 1) = 0$$

$$2u - 3 = 0 \text{ or } \quad u + 1 = 0$$

$$2u = 3 \text{ or } \qquad u = -1$$

$$u = \frac{3}{2} \text{ or } \qquad u = -1$$

$$\frac{t - 4}{3} = \frac{3}{2} \text{ or } \quad \frac{t - 4}{3} = -1$$

$$t - 4 = \frac{9}{2} \text{ or } \quad t - 4 = -3$$

$$t = \frac{17}{2} \text{ or } \qquad t = 1 \quad \left\{\frac{17}{2}, 1\right\}$$

51.

$$x^{2/3} + x^{1/3} = 20$$

$$\left(x^{1/3}\right)^2 + x^{1/3} - 20 = 0$$

Let $u = x^{1/3}$

$$u^2 + u - 20 = 0$$

$$(u + 5)(u - 4) = 0$$

$$u + 5 = 0 \text{ or } \qquad u - 4 = 0$$

$$u = -5 \text{ or } \qquad u = 4$$

$$x^{1/3} = -5 \text{ or } \qquad x^{1/3} = 4$$

$$\left(x^{1/3}\right)^3 = (-5)^3 \text{ or } \left(x^{1/3}\right)^3 = (4)^3$$

$$x = -125 \text{ or } \qquad x = 64$$

$$\{-125, 64\}$$

53.

$$m^4 + 2m^2 - 8 = 0$$

$$(m^2 + 4)(m^2 - 2) = 0$$

$$m^2 + 4 = 0 \text{ or } m^2 - 2 = 0$$

$$m^2 = -4 \text{ or } \quad m^2 = 2$$

$$m = \pm\sqrt{-4} \text{ or } m = \pm\sqrt{2}$$

$$m = \pm 2i \text{ or } \quad m = \pm\sqrt{2}$$

$$\left\{\pm 2i, \pm\sqrt{2}\right\}$$

55.

$$a^3 + 16a - a^2 - 16 = 0$$

$$a(a^2 + 16) - 1(a^2 + 16) = 0$$

$$(a^2 + 16)(a - 1) = 0$$

$$a^2 + 16 = 0 \text{ or } a - 1 = 0$$

$$a^2 = -16 \text{ or } a = 1$$

$$a = \pm 4i \text{ or } a = 1$$

$$\{\pm 4i, 1\}$$

57. $x^3 + 5x - 4x^2 - 20 = 0$

$x(x^2 + 5) - 4(x^2 + 5) = 0$

$(x^2 + 5)(x - 4) = 0$

$x^2 + 5 = 0$ or $x - 4 = 0$

$x^2 = -5$ or $x = 4$

$x = \pm i\sqrt{5}$ or $x = 4$

$\left\{ \pm i\sqrt{5},\, 4 \right\}$

59. $\left(\dfrac{2}{x-3} \right)^2 + 8\left(\dfrac{2}{x-3} \right) + 12 = 0$

Let $u = \dfrac{2}{x-3}$

$u^2 + 8u + 12 = 0$

$(u + 6)(u + 2) = 0$

$u + 6 = 0$ or $u + 2 = 0$

$u = -6$ or $u = -2$

$\dfrac{2}{x-3} = -6$ or $\dfrac{2}{x-3} = -2$

$2 = -6(x - 3)$ or $2 = -2(x - 3)$

$2 = -6x + 18$ or $2 = -2x + 6$

$-16 = -6x$ or $-4 = -2x$

$x = \dfrac{8}{3}$ or $x = 2$

$\left\{ \dfrac{8}{3},\, 2 \right\}$

Problem Recognition Exercises

1. **a.** $x^2 + 10x + 3 = 0$

$x^2 + 10x = -3$

$x^2 + 10x + 25 = -3 + 25$

$(x + 5)^2 = 22$

$x + 5 = \pm\sqrt{22}$

$x = -5 \pm \sqrt{22}$ $\left\{ -5 \pm \sqrt{22} \right\}$

b. $x^2 + 10x + 3 = 0$

$a = 1,\ b = 10,\ c = 3$

$x = \dfrac{-10 \pm \sqrt{10^2 - 4(1)(3)}}{2(1)}$

$= \dfrac{-10 \pm \sqrt{100 - 12}}{2} = \dfrac{-10 \pm \sqrt{88}}{2}$

$= \dfrac{-10 \pm 2\sqrt{22}}{2} = -5 \pm \sqrt{22}$ $\left\{ -5 \pm \sqrt{22} \right\}$

3. **a.**
$$3t^2 + t + 4 = 0$$
$$3t^2 + t = -4$$
$$t^2 + \frac{1}{3}t = -\frac{4}{3}$$
$$t^2 + \frac{1}{3}t + \frac{1}{36} = -\frac{4}{3} + \frac{1}{36}$$
$$\left(t + \frac{1}{6}\right)^2 = -\frac{47}{36}$$
$$t + \frac{1}{6} = \pm\frac{i\sqrt{47}}{6}$$
$$t = \frac{-1 \pm i\sqrt{47}}{6} \quad \left\{-\frac{1}{6} \pm \frac{\sqrt{47}}{6}i\right\}$$

b.
$$3t^2 + t + 4 = 0$$
$$a = 3, \, b = 1, \, c = 4$$
$$t = \frac{-(1) \pm \sqrt{(1)^2 - 4(3)(4)}}{2(3)} = \frac{-1 \pm \sqrt{1 - 48}}{6}$$
$$= \frac{-1 \pm \sqrt{-47}}{6} = \frac{-1 \pm i\sqrt{47}}{6}$$
$$\left\{-\frac{1}{6} \pm \frac{\sqrt{47}}{6}i\right\}$$

5. **a.** Quadratic equation
b.
$$t^2 + 5t - 14 = 0$$
$$(t + 7)(t - 2) = 0$$
$$t + 7 = 0 \quad \text{or} \quad t - 2 = 0$$
$$t = -7 \quad \text{or} \quad t = 2 \quad \{-7, 2\}$$

7. **a.** Quadratic in form
b.
$$a^4 - 10a^2 + 9 = 0$$
$$\text{Let } t = a^2$$
$$t^2 - 10t + 9 = 0$$
$$(t - 9)(t - 1) = 0$$
$$t - 9 = 0 \quad \text{or} \quad t - 1 = 0$$
$$t = 9 \quad \text{or} \quad t = 1$$
$$a^2 = 9 \quad \text{or} \quad a^2 = 1$$
$$a = \pm 3 \quad \text{or} \quad a = \pm 1$$
$$\{-3, 3, -1, 1\}$$

9. **a.** Quadratic in form
b.
$$x - 3x^{1/2} - 4 = 0$$
$$\text{Let } a = x^{1/2}$$
$$a^2 - 3a - 4 = 0$$
$$(a - 4)(a + 1) = 0$$
$$a - 4 = 0 \quad \text{or} \quad a + 1 = 0$$
$$a = 4 \quad \text{or} \quad a = -1$$
$$x^{1/2} = 4 \quad \text{or} \quad x^{1/2} = -1$$
$$x = 16 \quad \text{or} \quad x = 1$$
Check:
$$x = 16:$$
$$16 - 3(16)^{1/2} - 4 = 0$$
$$16 - 3(4) - 4 = 0$$
$$16 - 12 - 4 = 0$$
$$0 = 0$$

$$x = 1:$$
$$1 - 3(1)^{1/2} - 4 = 0$$
$$1 - 3(1) - 4 = 0$$
$$1 - 3 - 4 = 0$$
$$-6 \neq 0$$
Solutions: $\{16\}$ ($x = 1$ does not check.)

11. a. Linear equation

b. $8b(b+1)+2(3b-4)=4b(2b+3)$

$$8b^2+8b+6b-8=8b^2+12b$$
$$14b-8=12b$$
$$2b=8$$
$$b=4 \quad \{4\}$$

13. a. Quadratic equation

b. $5a(a+6)=10(3a-1)$

$$5a^2+30a=30a-10$$
$$5a^2=-10$$
$$a^2=-2$$
$$a=\pm\sqrt{-2}=\pm i\sqrt{2} \quad \left\{\pm i\sqrt{2}\right\}$$

15. a. Rational equation

b.

$$\frac{t}{t+5}+\frac{3}{t-4}=\frac{17}{t^2+t-20}$$
$$\frac{t}{t+5}+\frac{3}{t-4}=\frac{17}{(t+5)(t-4)}$$
$$(t+5)(t-4)\left(\frac{t}{t+5}+\frac{3}{t-4}\right)$$
$$=(t+5)(t-4)\left(\frac{17}{(t+5)(t-4)}\right)$$
$$t(t-4)+3(t+5)=17$$
$$t^2-4t+3t+15=17$$
$$t^2-t-2=0$$
$$(t-2)(t+1)=0$$
$$t-2=0 \text{ or } t+1=0$$
$$t=2 \text{ or } \quad t=-1 \quad \{2,-1\}$$

17. a. Quadratic equation

b. $c^2-20c-1=0$

$$a=1, \ b=-20, \ c=-1$$
$$c=\frac{-(-20)\pm\sqrt{(-20)^2-4(1)(-1)}}{2(1)}$$
$$=\frac{20\pm\sqrt{400+4}}{2}=\frac{20\pm\sqrt{404}}{2}$$
$$=\frac{20\pm2\sqrt{101}}{2}=10\pm\sqrt{101} \quad \left\{10\pm\sqrt{101}\right\}$$

19. a. Quadratic equation

b. $2u(u-3)=4(2-u)$

$$2u^2-6u=8-4u$$
$$2u^2-2u-8=0$$
$$u^2-u-4=0$$
$$a=1, \ b=-1, \ c=-4$$
$$u=\frac{-(-1)\pm\sqrt{(-1)^2-4(1)(-4)}}{2(1)}$$
$$=\frac{1\pm\sqrt{1+16}}{2}=\frac{1\pm\sqrt{17}}{2} \quad \left\{\frac{1\pm\sqrt{17}}{2}\right\}$$

21. a. Radical equation

b.
$$\sqrt{2b+3}=b$$
$$\left(\sqrt{2b+3}\right)^2=(b)^2$$
$$2b+3=b^2$$
$$b^2-2b-3=0$$
$$(b-3)(b+1)=0$$
$$b-3=0 \text{ or } b+1=0$$
$$b=3 \text{ or } b=-1$$

Check: Check:

$$\sqrt{2(3)+3}=3 \qquad \sqrt{2(-1)+3}=-1$$
$$\sqrt{6+3}=3 \qquad \sqrt{-2+3}=-1$$
$$\sqrt{9}=3 \qquad \sqrt{1}=-1$$
$$3=3 \qquad 1\ne-1$$

$\{3\}$ ($b=-1$ does not check.)

23. a. Quadratic in form (or radical)

b. $x^{2/3} + 2x^{1/3} - 15 = 0$

$$\text{Let } t = x^{1/3}$$

$$t^2 + 2t - 15 = 0$$

$$(t-3)(t+5) = 0$$

$$t - 3 = 0 \quad \text{or} \quad t + 5 = 0$$

$$t = 3 \quad \text{or} \quad t = -5$$

$$x^{1/3} = 3 \quad \text{or} \quad x^{1/3} = -5$$

$$x = 3^3 \quad \text{or} \quad x = (-5)^3$$

$$x = 27 \quad \text{or} \quad x = -125$$

$$\{27, -125\}$$

Section 7.4   Graphs of Quadratic Functions

1. (a) Parabola

(b) $>; <$

(c) lowest; highest

(d) (h, k); upward; downward

(e) vertex; $x = h$

3. $(y-3)^2 = -4$

$$y - 3 = \pm\sqrt{-4}$$

$$y - 3 = \pm 2i$$

$$y = 3 \pm 2i \quad \{3 \pm 2i\}$$

5. $5t(t-2) = -3$

$$5t^2 - 10t = -3$$

$$5t^2 - 10t + 3 = 0$$

$$a = 5, \, b = -10, \, c = 3$$

$$t = \frac{-(-10) \pm \sqrt{(-10)^2 - 4(5)(3)}}{2(5)}$$

$$= \frac{10 \pm \sqrt{100 - 60}}{10} = \frac{10 \pm \sqrt{40}}{10}$$

$$= \frac{10 \pm 2\sqrt{10}}{10} = \frac{\cancel{2}\left(5 \pm \sqrt{10}\right)}{\cancel{2} \cdot 5} = \frac{5 \pm \sqrt{10}}{5}$$

$$\left\{\frac{5 \pm \sqrt{10}}{5}\right\}$$

7. $x^{2/3} + 5x^{1/3} + 6 = 0$

$$\left(x^{1/3}\right)^2 + 5x^{1/3} + 6 = 0$$

$$\text{Let } u = x^{1/3}$$

$$u^2 + 5u + 6 = 0$$

$$(u+3)(u+2) = 0$$

$$u + 3 = 0 \quad \text{or} \quad u + 2 = 0$$

$$u = -3 \quad \text{or} \quad u = -2$$

$$x^{1/3} = -3 \quad \text{or} \quad x^{1/3} = -2$$

$$\left(x^{1/3}\right)^3 = (-3)^3 \quad \text{or} \quad \left(x^{1/3}\right)^3 = (-2)^3$$

$$x = -27 \quad \text{or} \quad x = -8 \quad \{-27, -8\}$$

9. The value of k shifts the graph of $y = x^2$ vertically.

11. $f(x) = x^2 + 2$

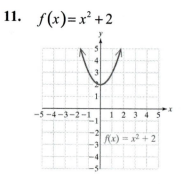

13. $q(x) = x^2 - 4$

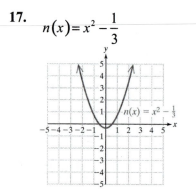

15. $S(x) = x^2 + \dfrac{3}{2}$

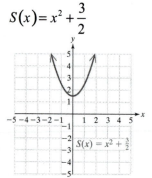

17. $n(x) = x^2 - \dfrac{1}{3}$

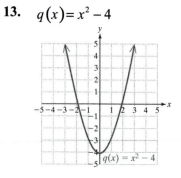

19. $r(x) = (x+1)^2$

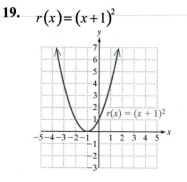

21. $k(x) = (x-3)^2$

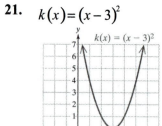

23. $A(x) = \left(x + \dfrac{3}{4}\right)^2$

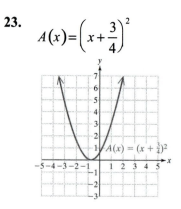

25. $W(x) = (x - 1.25)^2$

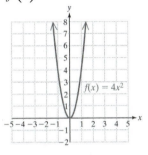

27. The value of a vertically stretches or shrinks the graph of $y = x^2$.

29. $f(x) = 4x^2$

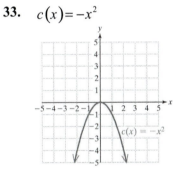

31. $h(x) = \frac{1}{4}x^2$

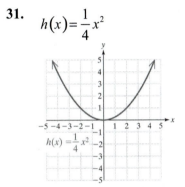

33. $c(x) = -x^2$

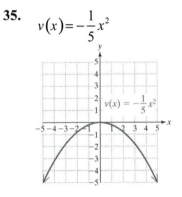

35. $v(x) = -\frac{1}{5}x^2$

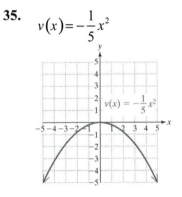

37. d

39. g

41. a

43. e

45. $f(x) = (x - 3)^2 + 2$
Domain: $(-\infty, \infty)$; range: $[2, \infty)$

47. $f(x) = (x + 1)^2 - 3$
Domain: $(-\infty, \infty)$; range: $[-3, \infty)$

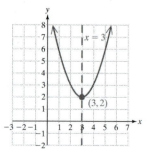

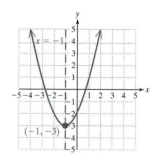

49. $f(x) = -(x-4)^2 - 2$

Domain: $(-\infty, \infty)$; range: $(-\infty, -2]$

51. $f(x) = -(x+3)^2 + 3$

Domain: $(-\infty, \infty)$; range: $(-\infty, 3]$

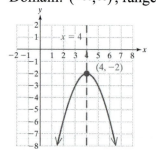

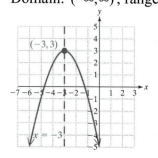

53. $f(x) = (x+1)^2 + 1$

Domain: $(-\infty, \infty)$; range: $[1, \infty)$

55. $f(x) = 3(x-1)^2$

Domain: $(-\infty, \infty)$; range: $[0, \infty)$

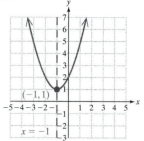

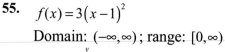

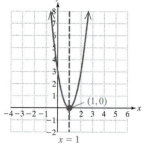

57. $f(x) = -4x^2 + 3$

Domain: $(-\infty, \infty)$; range: $(-\infty, 3]$

59. $f(x) = 2(x+3)^2 - 1$

Domain: $(-\infty, \infty)$; range: $[-1, \infty)$

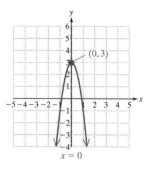

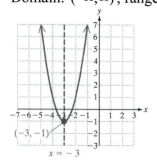

61. $f(x) = -\dfrac{1}{4}(x-1)^2 + 2$

Domain: $(-\infty, \infty)$; range: $(-\infty, 2]$

63. $f(x) = \dfrac{1}{3}(x-2)^2 + 1$

Domain: $(-\infty, \infty)$; range: $[1, \infty)$

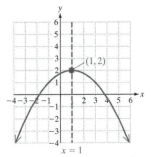

$x = 1$

65. a. $y = x^2 + 3$ is $y = x^2$ shifted up 3 units.

b. $y = (x + 3)^2$ is $y = x^2$ shifted left 3 units.

c. $y = 3x^2$ is $y = x^2$ with a vertical stretch.

69.

$$p(x) = -\frac{2}{5}(x - 2)^2 + 5$$

Vertex: (2, 5) is a maximum point with maximum value of 5.

73.

$$n(x) = -6x^2 + \frac{21}{4}$$

Vertex: $\left(0, \dfrac{21}{4}\right)$ is a maximum point with

maximum value of $\dfrac{21}{4}$.

77. $F(x) = 7x^2$

Vertex: $(0, 0)$ is a minimum point with minimum value of 0.

81. False, since the minimum value corresponds to the y-value of 8.

85.

$$h(t) = -16t^2 + 96t + 6$$

a. $h(3) = -16(3)^2 + 96(3) + 6$

$= -16(9) + 288 + 6$

$= -144 + 288 + 6 = 150$

The fireworks will explode at a height of 150 ft.

67. $f(x) = 4(x - 6)^2 - 9$

Vertex: (6, –9) is a minimum point with minimum value of –9.

71.

$$k(x) = \frac{1}{2}(x + 8)^2$$

Vertex: (–8, 0) is a minimum point with minimum value of 0.

75.

$$A(x) = 2(x - 7)^2 - \frac{3}{2}$$

Vertex: $\left(7, -\dfrac{3}{2}\right)$ is a minimum point with

minimum value of $-\dfrac{3}{2}$.

79. True, since the parabola opens down.

83.

$$H(x) = \frac{1}{90}(x - 60)^2 + 30$$

a. Vertex: $(60, 30)$

b. The minimum height is 30 ft.

c. $H(0) = \dfrac{1}{90}(0 - 60)^2 + 30$

$= \dfrac{1}{90}(3600) + 30 = 40 + 30 = 70$ ft

The towers are 70 ft high.

b. $h(t) = -16t^2 + 96t + 6$

$= -16(t^2 - 6t) + 6$

$= -16(t^2 - 6t + 9) + 6 + 16(9)$

$= -16(t - 3)^2 + 150$

Yes, because the ordered pair (3, 150) is the vertex.

Section 7.5   Vertex of a Parabola: Applications and Modeling

1. **(a)** $\dfrac{-b}{2a}; \dfrac{-b}{2a}$

(b) True

(c) True

(d) True

(e) False

3. The graph of p is the graph of $y = x^2$ shrunk vertically by a factor of $\dfrac{1}{4}$.

5. The graph of r is the graph of $y = x^2$ shifted up 7 units.

7. The graph of t is the graph of $y = x^2$ shifted to the left 10 units.

9. $x^2 - 8x + n$

$$n = \left(\frac{1}{2}b\right)^2 = \left(\frac{1}{2}\cdot(-8)\right)^2 = (-4)^2 = 16$$

$$x^2 - 8x + 16 = (x-4)^2$$

11. $y^2 + 7y + n$

$$n = \left(\frac{1}{2}b\right)^2 = \left(\frac{1}{2}\cdot(7)\right)^2 = \left(\frac{7}{2}\right)^2 = \frac{49}{4}$$

$$y^2 + 7y + \frac{49}{4} = \left(y + \frac{7}{2}\right)^2$$

13. $b^2 + \dfrac{2}{9}b + n$

$$n = \left(\frac{1}{2}b\right)^2 = \left(\frac{1}{2}\cdot\left(\frac{2}{9}\right)\right)^2 = \left(\frac{1}{9}\right)^2 = \frac{1}{81}$$

$$b^2 + \frac{2}{9}b + \frac{1}{81} = \left(b + \frac{1}{9}\right)^2$$

15. $t^2 - \dfrac{1}{3}t + n$

$$n = \left(\frac{1}{2}b\right)^2 = \left(\frac{1}{2}\cdot\left(-\frac{1}{3}\right)\right)^2 = \left(-\frac{1}{6}\right)^2 = \frac{1}{36}$$

$$t^2 - \frac{1}{3}t + \frac{1}{36} = \left(t - \frac{1}{6}\right)^2$$

17. $g(x) = x^2 - 8x + 5$

$$= 1\left(x^2 - 8x\right) + 5$$
$$= 1\left(x^2 - 8x + 16 - 16\right) + 5$$
$$= 1\left(x^2 - 8x + 16\right) - 16 + 5$$
$$g(x) = (x-4)^2 - 11$$

Vertex: $(4, -11)$

19. $n(x) = 2x^2 + 12x + 13$

$$= 2\left(x^2 + 6x\right) + 13$$
$$= 2\left(x^2 + 6x + 9 - 9\right) + 13$$
$$= 2\left(x^2 + 6x + 9\right) - 18 + 13$$
$$n(x) = 2(x+3)^2 - 5$$

Vertex: $(-3, -5)$

21.
$$p(x) = -3x^2 + 6x - 5$$
$$= -3(x^2 - 2x) - 5$$
$$= -3(x^2 - 2x + 1 - 1) - 5$$
$$= -3(x^2 - 2x + 1) + 3 - 5$$
$$p(x) = -3(x-1)^2 - 2$$
Vertex: $(1, -2)$

23.
$$k(x) = x^2 + 7x - 10$$
$$= 1(x^2 + 7x) - 10$$
$$= 1\left(x^2 + 7x + \frac{49}{4} - \frac{49}{4}\right) - 10$$
$$= 1\left(x^2 + 7x + \frac{49}{4}\right) - \frac{49}{4} - \frac{40}{4}$$
$$k(x) = \left(x + \frac{7}{2}\right)^2 - \frac{89}{4}$$
Vertex: $\left(-\frac{7}{2}, -\frac{89}{4}\right)$

25.
$$F(x) = 5x^2 + 10x + 1$$
$$= 5(x^2 + 2x) + 1$$
$$= 5(x^2 + 2x + 1 - 1) + 1$$
$$= 5(x^2 + 2x + 1) - 5 + 1$$
$$F(x) = 5(x+1)^2 - 4$$
Vertex: $(-1, -4)$

27.
$$P(x) = -2x^2 + x$$
$$= -2\left(x^2 - \frac{1}{2}x\right)$$
$$= -2\left(x^2 - \frac{1}{2}x + \frac{1}{16} - \frac{1}{16}\right)$$
$$= -2\left(x^2 - \frac{1}{2}x + \frac{1}{16}\right) + \frac{1}{8}$$
$$P(x) = -2\left(x - \frac{1}{4}\right)^2 + \frac{1}{8}$$
Vertex: $\left(\frac{1}{4}, \frac{1}{8}\right)$

29.
$$Q(x) = x^2 - 4x + 7$$
$$a = 1,\ b = -4,\ c = 7$$
$$\frac{-b}{2a} = \frac{-(-4)}{2(1)} = \frac{4}{2} = 2$$
$$Q(2) = 2^2 - 4(2) + 7 = 4 - 8 + 7 = 3$$
Vertex: $(2, 3)$

31.
$$r(x) = -3x^2 - 6x - 5$$
$$a = -3,\ b = -6,\ c = -5$$
$$\frac{-b}{2a} = \frac{-(-6)}{2(-3)} = \frac{6}{-6} = -1$$
$$r(-1) = -3(-1)^2 - 6(-1) - 5$$
$$= -3 + 6 - 5 = -2$$
Vertex: $(-1, -2)$

33.
$$N(x) = x^2 + 8x + 1$$
$$a = 1,\ b = 8,\ c = 1$$
$$\frac{-b}{2a} = \frac{-(8)}{2(1)} = \frac{-8}{2} = -4$$
$$N(-4) = (-4)^2 + 8(-4) + 1$$
$$= 16 - 32 + 1 = -15$$
Vertex: $(-4, -15)$

35.
$$m(x) = \frac{1}{2}x^2 + x + \frac{5}{2}$$
$$a = \frac{1}{2},\ b = 1,\ c = \frac{5}{2}$$
$$\frac{-b}{2a} = \frac{-(1)}{2\left(\frac{1}{2}\right)} = \frac{-1}{1} = -1$$
$$m(-1) = \frac{1}{2}(-1)^2 + (-1) + \frac{5}{2} = \frac{1}{2} - 1 + \frac{5}{2} = 2$$
Vertex: $(-1, 2)$

37. $k(x) = -x^2 + 2x + 2$

$a = -1,\ b = 2,\ c = 2$

$\dfrac{-b}{2a} = \dfrac{-(2)}{2(-1)} = \dfrac{-2}{-2} = 1$

$k(1) = -(1)^2 + 2(1) + 2 = -1 + 2 + 2 = 3$

Vertex: $(1,\ 3)$

39.

$A(x) = -\dfrac{1}{3}x^2 + x$

$a = -\dfrac{1}{3},\ b = 1,\ c = 0$

$\dfrac{-b}{2a} = \dfrac{-(1)}{2\left(-\dfrac{1}{3}\right)} = \dfrac{-1}{-\dfrac{2}{3}} = \dfrac{3}{2}$

$A\left(\dfrac{3}{2}\right) = -\dfrac{1}{3}\left(\dfrac{3}{2}\right)^2 + \left(\dfrac{3}{2}\right) = -\dfrac{3}{4} + \dfrac{3}{2} = \dfrac{3}{4}$

Vertex: $\left(\dfrac{3}{2},\ \dfrac{3}{4}\right)$

41. a. $p(x) = x^2 + 8x + 1$

$p(x) = \left(x^2 + 8x + 16\right) + 1 - 16$

$p(x) = (x + 4)^2 - 15$

Vertex: $(-4,\ -15)$

b. $p(x) = x^2 + 8x + 1$

$a = 1,\ b = 8,\ c = 1$

$\dfrac{-b}{2a} = \dfrac{-(8)}{2(1)} = \dfrac{-8}{2} = -4$

$p(-4) = (-4)^2 + 8(-4) + 1 = 16 - 32 + 1$

$\qquad = -15$

Vertex: $(-4,\ -15)$

43. a. $f(x) = 2x^2 + 4x + 6$

$f(x) = 2\left(x^2 + 2x + 1\right) + 6 - 2$

$f(x) = 2(x + 1)^2 + 4$

Vertex: $(-1,\ 4)$

b. $f(x) = 2x^2 + 4x + 6$

$a = 2,\ b = 4,\ c = 6$

$\dfrac{-b}{2a} = \dfrac{-(4)}{2(2)} = \dfrac{-4}{4} = -1$

$f(-1) = 2(-1)^2 + 4(-1) + 6 = 2 - 4 + 6 = 4$

Vertex: $(-1,\ 4)$

45. a. $f(x) = x^2 + 2x - 3$

$a = 1, b = 2, c = -3$

$\dfrac{-b}{2a} = \dfrac{-(2)}{2(1)} = \dfrac{-2}{2} = -1$

$f(-1) = (-1)^2 + 2(-1) - 3 = 1 - 2 - 3 = -4$

Vertex: $(-1, -4)$

b.

$y = (0)^2 + 2(0) - 3 = 0 + 0 - 3 = -3$

y-intercept: $(0,\ -3)$

c. $x^2 + 2x - 3 = 0$

$(x + 3)(x - 1) = 0$

$x + 3 = 0\ $ or $\ x - 1 = 0$

$x = -3\ $ or $\quad x = 1$

x-intercepts: $(-3,\ 0), (1,\ 0)$

d.

47. a. $f(x) = 2x^2 - 2x + 4$

$a = 2, b = -2, c = 4$

$\dfrac{-b}{2a} = \dfrac{-(-2)}{2(2)} = \dfrac{2}{4} = \dfrac{1}{2}$

$f\left(\dfrac{1}{2}\right) = 2\left(\dfrac{1}{2}\right)^2 - 2\left(\dfrac{1}{2}\right) + 4$

$\qquad = \dfrac{1}{2} - 1 + 4 = \dfrac{7}{2}$

Vertex: $\left(\dfrac{1}{2}, \dfrac{7}{2}\right)$

b. $y = 2(0)^2 - 2(0) + 4 = 0 - 0 + 4 = 4$

y-intercept: $(0, 4)$

c. $2x^2 - 2x + 4 = 0$

$x = \dfrac{-(-2) \pm \sqrt{(-2)^2 - 4(2)(4)}}{2(2)}$

$\quad = \dfrac{2 \pm \sqrt{4 - 32}}{4} = \dfrac{2 \pm \sqrt{-28}}{4}$

No x-intercepts (Complex solutions)

d.

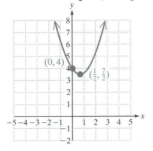

49. a. $f(x) = -x^2 + 3x - \dfrac{9}{4}$

$a = -1, b = 3, c = -\dfrac{9}{4}$

$\dfrac{-b}{2a} = \dfrac{-(3)}{2(-1)} = \dfrac{-3}{-2} = \dfrac{3}{2}$

$f\left(\dfrac{3}{2}\right) = -\left(\dfrac{3}{2}\right)^2 + 3\left(\dfrac{3}{2}\right) - \dfrac{9}{4}$

$\qquad = -\dfrac{9}{4} + \dfrac{9}{2} - \dfrac{9}{4} = 0$

Vertex: $\left(\dfrac{3}{2}, 0\right)$

b. $y = -(0)^2 + 3(0) - \dfrac{9}{4} = -0 + 0 - \dfrac{9}{4} = -\dfrac{9}{4}$

y-intercept: $\left(0, -\dfrac{9}{4}\right)$

c. $-x^2 + 3x - \dfrac{9}{4} = 0$

$4x^2 - 12x + 9 = 0$

$(2x - 3)^2 = 0$

$2x - 3 = 0$

$2x = 3$

$x = \dfrac{3}{2}$

x-intercept: $\left(\dfrac{3}{2}, 0\right)$

d.

329

51. a.
$$f(x) = -x^2 - 2x + 3$$
$$a = -1, b = -2, c = 3$$
$$\frac{-b}{2a} = \frac{-(-2)}{2(-1)} = \frac{2}{-2} = -1$$
$$f(-1) = -(-1)^2 - 2(-1) + 3 = -1 + 2 + 3$$
$$= 4$$
Vertex: $(-1, 4)$

b.
$$y = -(0)^2 - 2(0) + 3 = -0 + 0 + 3 = 3$$
y-intercept: $(0, 3)$

c.
$$-x^2 - 2x + 3 = 0$$
$$x^2 + 2x - 3 = 0$$
$$(x + 3)(x - 1) = 0$$
$$x + 3 = 0 \ \text{ or } \ x - 1 = 0$$
$$x = -3 \ \text{ or } \ x = 1$$
x-intercept: $(-3, 0), (1, 0)$

d.

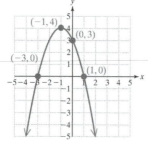

53.
$$C(x) = 2x^2 - 40x + 2200$$
$$a = 2, \ b = -40, \ c = 2200$$
$$\frac{-b}{2a} = \frac{-(-40)}{2(2)} = \frac{40}{4} = 10$$
$$C(10) = 2(10)^2 - 40(10) + 2200$$
$$= 200 - 400 + 2200 = 2000$$
Vertex: $(10, 2000)$

Mia must package 10 MP3 players to minimize her average cost at $2000.

55. a.
$$h(t) = -16t^2 + 100t + 8$$
$$a = -16, \ b = 100, \ c = 8$$
$$\frac{-b}{2a} = \frac{-(100)}{2(-16)} = \frac{-100}{-32} = 3.125$$
$$h(3.125) = -16(3.125)^2 + 100(3.125) + 8$$
$$= -156.25 + 312.5 + 8 = 164.25$$

b. The maximum height it 164.25 ft.
The fuses should be set for 3.125 sec.

57. a.
$$m(x) = -0.04x^2 + 3.6x - 49$$
$$a = -0.04, \ b = 3.6, \ c = -49$$
$$\frac{-b}{2a} = \frac{-(3.6)}{2(-0.04)} = \frac{-3.6}{-0.08} = 45$$
The maximum gas mileage will occur at a speed of 45 mph.

b.
$$m(45) = -0.04(45)^2 + 3.6(45) - 49$$
$$= -81 + 162 - 49 = 32$$
The maximum gas mileage is 32 mpg.

59. a.
$$b(t) = -\frac{1}{1152}t^2 + \frac{1}{12}t$$
$$a = -\frac{1}{1152}, \ b = \frac{1}{12}, \ c = 0$$
$$\frac{-b}{2a} = \frac{-\left(\frac{1}{12}\right)}{2\left(-\frac{1}{1152}\right)} = \frac{-\frac{1}{12}}{-\frac{1}{576}} = \frac{576}{12} = 48$$

61. Substitute each ordered pair for x and y into the standard form of a parabola to get three equations in a, b, and c:
$$4 = a(0)^2 + b(0) + c$$
$$4 = c$$
$$0 = a(1)^2 + b(1) + c$$
$$0 = a + b + c$$

b. The maximum yield occurs at 48 hours.

$$b(48) = -\frac{1}{1152}(48)^2 + \frac{1}{12}(48)$$
$$= -2 + 4 = 2$$

The maximum yield is 2 grams.

$$-10 = a(-1)^2 + b(-1) + c$$
$$-10 = a - b + c$$

Substitute $c = 4$ and solve for a:

$$a + b + 4 = 0 \quad \rightarrow a + b = -4$$
$$\underline{a - b + 4 = -10 \rightarrow a - b = -14}$$
$$2a = -18$$
$$a = -9$$

Solve for b:

$$-9 + b = -4$$
$$b = 5$$

The equation is: $y = -9x^2 + 5x + 4$.

63. Substitute each ordered pair for x and y into the standard form of a parabola to get three equations in a, b, and c:

$$1 = a(2)^2 + b(2) + c$$
$$1 = 4a + 2b + c \qquad \text{(A)}$$
$$5 = a(-2)^2 + b(-2) + c$$
$$5 = 4a - 2b + c \qquad \text{(B)}$$
$$-4 = a(1)^2 + b(1) + c$$
$$-4 = a + b + c \qquad \text{(C)}$$

Subtract (B) from (A) and solve for b:

$$4a + 2b + c = 1$$
$$\underline{-(4a - 2b + c = 5)}$$
$$4b \quad = -4$$
$$b = -1$$

(continued on the right)

Substitute $b = -1$ into (B) and (C), subtract and solve for a:

$$5 = 4a - 2(-1) + c \rightarrow 4a + c = 3$$
$$-4 = \ a + (-1) + c \rightarrow \underline{a + c = -3}$$
$$3a = 6$$
$$a = 2$$

Solve for c:

$$2 + (-1) + c = -4$$
$$1 + c = -4$$
$$c = -5$$

The equation is: $y = 2x^2 - x - 5$.

65. Substitute each ordered pair for x and y into the standard form of a parabola to get three equations in a, b, and c:

$$-4 = a(2)^2 + b(2) + c$$
$$-4 = 4a + 2b + c \qquad \text{(A)}$$
$$1 = a(1)^2 + b(1) + c$$
$$1 = a + b + c \qquad \text{(B)}$$
$$-7 = a(-1)^2 + b(-1) + c$$
$$-7 = a - b + c \qquad \text{(C)}$$

Subtract (C) from (B) and solve for b:

67. a. The sum of the three sides must equal the total amount of fencing.

b. $A = x(200 - 2x)$

c. $A = 200x - 2x^2$
$$A = -2x^2 + 200x$$
$$a = -2, \ b = 200, \ c = 0$$
$$x = \frac{-b}{2a} = \frac{-(200)}{2(-2)} = \frac{-200}{-4} = 50$$
$$y = 200 - 2(50) = 200 - 100 = 100$$

The dimensions of the corral are 50 ft by 100 ft.

$$a+b+c= 1$$
$$\underline{-(a-b+c=-7)}$$
$$2b \quad\;=8$$
$$b=4$$

Substitute $b=4$ into (A) and (B), subtract and solve for a:

$$4a+2(4)+c=-4 \rightarrow 4a+c=-12$$
$$a+(4)+c= 1\rightarrow \underline{\;a+c=-3\;}$$
$$3a=-9$$
$$a=-3$$

Solve for c:

$$-3+4+c=1$$
$$1+c=1$$
$$c=0$$

The equation is: $y=-3x^2+4x$.

Section 7.6   Polynomial and Rational Inequalities

1. **(a)** quadratic

    (b) solutions; undefined

    (c) test; point

    (d) true

    (e) rational

    (f) denominator

    (g) $\{\ \}$; $(-\infty,\infty)$

    (h) excludes; includes

3. $2t^2-3t+4=0$

$a=2,\ b=-3,\ c=4$

$$t=\frac{-(-3)\pm\sqrt{(-3)^2-4(2)(4)}}{2(2)}=\frac{3\pm\sqrt{9-32}}{4}$$

$$=\frac{3\pm\sqrt{-23}}{4}=\frac{3\pm i\sqrt{23}}{4}\qquad \left\{\frac{3}{4}\pm\frac{\sqrt{23}}{4}i\right\}$$

5. $2m(m-4)(m+6)=0$

$2m=0$ or $m-4=0$ or $m+6=0$

$m=0$ or $\quad m=4$ or $\quad m=-6$

$\{0,\ 4,\ -6\}$

7. $\dfrac{14}{x-3}=7 \qquad x\neq3$

$14=7(x-3)$

$14=7x-21$

$35=7x$

$5=x \quad \{5\}$

9.   **a.** $(-\infty,-2)\cup(3,\infty)$

    b. $(-2,3)$

    c. $[-2,3]$

    d. $(-\infty,-2]\cup[3,\infty)$

11.   **a.** $(-2,0)\cup(3,\infty)$

    b. $(-\infty,-2)\cup(0,3)$

    c. $(-\infty,-2]\cup[0,3]$

    d. $[-2,0]\cup[3,\infty)$

13. a. $3(4-x)(2x+1)=0$

$$4-x=0 \quad \text{or} \quad 2x+1=0$$
$$-x=-4 \quad \text{or} \quad 2x=-1$$
$$x=4 \quad \text{or} \quad x=-\frac{1}{2} \quad \left\{4,-\frac{1}{2}\right\}$$

b. $3(4-x)(2x+1)<0$ The boundary points are 4 and $-\frac{1}{2}$.

Use test points $x=-1$, $x=0$, and $x=5$.

Test $x=-1$: $3(4-(-1))(2(-1)+1)=3(5)(-1)=-15<0$ True
Test $x=0$: $3(4-0)(2(0)+1)=3(4)(1)=12<0$ False
Test $x=5$: $3(4-5)(2(5)+1)=3(-1)(11)=-33<0$ True

The boundary points are not included.

The solution is $\left\{x\middle|x<-\frac{1}{2} \text{ or } x>4\right\}$ or $\left(-\infty,-\frac{1}{2}\right)\cup(4,\infty)$.

c. $3(4-x)(2x+1)>0$ The boundary points are 4 and $-\frac{1}{2}$.

Use test points $x=-1$, $x=0$, and $x=5$.

Test $x=-1$: $3(4-(-1))(2(-1)+1)=3(5)(-1)=-15>0$ False
Test $x=0$: $3(4-0)(2(0)+1)=3(4)(1)=12>0$ True
Test $x=5$: $3(4-5)(2(5)+1)=3(-1)(11)=-33>0$ False

The boundary points are not included. The solution is $\left\{x\middle|-\frac{1}{2}<x<4\right\}$ or $\left(-\frac{1}{2},4\right)$.

15. a.

$$x^2+7x=30$$
$$x^2+7x-30=0$$
$$(x+10)(x-3)=0$$
$$x+10=0 \quad \text{or} \quad x-3=0$$
$$x=-10 \quad \text{or} \quad x=3 \quad \{-10,3\}$$

b. $x^2+7x<30$ The boundary points are -10 and 3.
Use test points $x=-11$, $x=0$, and $x=4$.

Test $x=-11$: $(-11)^2+7(-11)=121-77=44<30$ False

Test $x=0$: $(0)^2+7(0)=0+0=0<30$ True

Test $x=4$: $(4)^2+7(4)=16+28=44<30$ False

The boundary points are not included. The solution is $\{x|-10<x<3\}$ or $(-10,3)$.

c. $x^2+7x>30$ The boundary points are -10 and 3.
Use test points $x=-11$, $x=0$, and $x=4$.

Test $x=-11$: $(-11)^2+7(-11)=121-77=44>30$ True

Test $x=0$: $(0)^2+7(0)=0+0=0>30$ False

Test $x=4$: $(4)^2+7(4)=16+28=44>30$ True

The boundary points are not included.
The solution is $\{x|x<-10 \text{ or } x>3\}$ or $(-\infty,-10)\cup(3,\infty)$.

17. a.
$$2p(p-2) = p + 3$$
$$2p^2 - 4p = p + 3$$
$$2p^2 - 5p - 3 = 0$$
$$(2p+1)(p-3) = 0$$
$$2p + 1 = 0 \quad \text{or} \quad p - 3 = 0$$
$$2p = -1 \quad \text{or} \quad p = 3$$
$$p = -\frac{1}{2} \quad \text{or} \quad p = 3 \quad \left\{-\frac{1}{2}, 3\right\}$$

b. $2p(p-2) \le p + 3$ The boundary points are $-\frac{1}{2}$ and 3.

Use test points $p = -1$, $p = 0$, and $p = 4$.

Test $p = -1$: $\quad 2(-1)(-1-2) = -2(-3) = 6 \le -1 + 3 = 2$ $\qquad$ False

Test $p = 0$: $\quad 2(0)(0-2) = 0(-2) = 0 \le 0 + 3 = 3$ $\qquad$ True

Test $p = 4$: $\quad 2(4)(4-2) = 8(2) = 16 \le 4 + 3 = 7$ $\qquad$ False

The boundary points are included.

The solution is $\left\{p \,\middle|\, -\dfrac{1}{2} \le p \le 3\right\}$ or $\left[-\dfrac{1}{2}, 3\right]$.

c. $2p(p-2) \ge p + 3$ The boundary points are $-\frac{1}{2}$ and 3.

Use test points $p = -1$, $p = 0$, and $p = 4$.

Test $p = -1$: $\quad 2(-1)(-1-2) = -2(-3) = 6 \ge -1 + 3 = 2$ $\qquad$ True

Test $p = 0$: $\quad 2(0)(0-2) = 0(-2) = 0 \ge 0 + 3 = 3$ $\qquad$ False

Test $p = 4$: $\quad 2(4)(4-2) = 8(2) = 16 \ge 4 + 3 = 7$ $\qquad$ True

The boundary points are included. The solution is $\left\{p \,\middle|\, p \le -\dfrac{1}{2} \text{ or } p \ge 3\right\}$ or $\left(-\infty, -\dfrac{1}{2}\right] \cup [3, \infty)$.

19. $(t-7)(t-1) < 0$
$$(t-7)(t-1) = 0$$
$$t - 7 = 0 \quad \text{or} \quad t - 1 = 0$$
$$t = 7 \quad \text{or} \quad t = 1$$
The boundary points are 1 and 7.
Use test points $t = -2$, $t = 2$, and $t = 8$.

Test $t = -2$: $\quad (-2-7)(-2+1) = -9(-1) = 9 < 0$ $\qquad$ False

Test $t = 2$: $\quad (2-7)(2+1) = -5(3) = -15 < 0$ $\qquad$ True

Test $t = 8$: $\quad (8-7)(8+1) = 1(9) = 9 < 0$ $\qquad$ False

The boundary points are not included.
The solution is $\left\{t \,\middle|\, 1 < t < 7\right\}$ or $(1, 7)$.

21. $-6(4+2x)(5-x) > 0$
$$-6(4+2x)(5-x) = 0$$
$$4 + 2x = 0 \quad \text{or} \quad 5 - x = 0$$
$$2x = -4 \quad \text{or} \quad -x = -5$$
$$x = -2 \quad \text{or} \quad x = 5$$
The boundary points are -2 and 5.

Use test points $x = -3$, $x = 0$, and $x = 6$.

Test $x = -3$: $-6(4 + 2(-3))(5 - (-3)) = -6(-2)(8) = 96 > 0$ True

Test $x = 0$: $-6(4 + 2(0))(5 - 0) = -6(4)(5) = -120 > 0$ False

Test $x = 6$: $-6(4 + 2(6))(5 - 6) = -6(16)(-1) = 96 > 0$ True

The boundary points are not included.

The solution is $\{x | x < -2 \text{ or } x > 5\}$ or $(-\infty, -2) \cup (5, \infty)$.

23. $m(m + 1)^2 (m + 5) \le 0$

$m(m + 1)^2 (m + 5) = 0$

$m = 0$ or $m + 1 = 0$ or $m + 5 = 0$

$m = 0$ or $m = -1$ or $m = -5$

The boundary points are -5, -1, and 0.

Use test points $m = -6$, $m = -2$, $m = -0.5$, and $m = 1$.

Test $m = -6$: $-6(-6 + 1)^2 (-6 + 5) = -6(-5)^2 (-1) = 150 \le 0$ False

Test $m = -2$: $-2(-2 + 1)^2 (-2 + 5) = -2(-1)^2 (3) = -6 \le 0$ True

Test $m = -0.5$: $-0.5(-0.5 + 1)^2 (-0.5 + 5) = -0.5(0.5)^2 (4.5) = -0.5625 \le 0$ True

Test $m = 1$: $1(1 + 1)^2 (1 + 5) = 1(2)^2 (6) = 24 \le 0$ False

The boundary points are included.

The solution is $\{m | -5 \le m \le 0\}$ or $[-5, 0]$.

25. $a^2 - 12a \le -32$

$a^2 - 12a = -32$

$a^2 - 12a + 32 = 0$

$(a - 8)(a - 4) = 0$

$a - 8 = 0$ or $a - 4 = 0$

$a = 8$ or $a = 4$

The boundary points are 4 and 8.

Use test points $a = 0$, $a = 5$, and $a = 9$.

Test $a = 0$: $0^2 - 12(0) = 0 - 0 = 0 \le -32$ False

Test $a = 5$: $5^2 - 12(5) = 25 - 60 = -35 \le -32$ True

Test $a = 9$: $9^2 - 12(9) = 81 - 108 = -27 \le -32$ False

The boundary points are included. The solution is $\{a | 4 \le a \le 8\}$ or $[4, 8]$.

27. $5x^2 - 2x - 1 > 0$

$a = 5,\ b = -2,\ c = -1$

$$x = \frac{-(-2) \pm \sqrt{(-2)^2 - 4(5)(-1)}}{2(5)}$$

$$= \frac{2 \pm \sqrt{4 + 20}}{10} = \frac{2 \pm \sqrt{24}}{10} = \frac{2 \pm 2\sqrt{6}}{10}$$

$$= \frac{2\left(1 \pm \sqrt{6}\right)}{10} = \frac{1 \pm \sqrt{6}}{5}$$

The boundary points are $\dfrac{1 - \sqrt{6}}{5}$ and $\dfrac{1 + \sqrt{6}}{5}$.

Use test points $x = -1,\ x = 0,$ and $x = 1.$

Test $x = -1$: $5(-1)^2 - 2(-1) - 1 = 5 + 2 - 1 = 6 > 0$ True

Test $x = 0$: $5(0)^2 - 2(0) - 1 = 0 + 0 - 1 = -1 > 0$ False

Test $x = 1$: $5(1)^2 - 2(1) - 1 = 5 - 2 - 1 = 2 > 0$ True

The boundary points are not included.

The solution is $\left\{ x \,\middle|\, x < \dfrac{1 - \sqrt{6}}{5} \text{ or } x > \dfrac{1 + \sqrt{6}}{5} \right\}$ or $\left(-\infty,\ \dfrac{1 - \sqrt{6}}{5} \right) \cup \left(\dfrac{1 + \sqrt{6}}{5},\ \infty \right).$

---

29. $x^2 + 3x \le 6$

$x^2 + 3x - 6 \le 0$

$a = 1,\ b = 3,\ c = -6$

$$x = \frac{-(3) \pm \sqrt{(3)^2 - 4(1)(-6)}}{2(1)}$$

$$= \frac{-3 \pm \sqrt{9 + 24}}{2} = \frac{-3 \pm \sqrt{33}}{2}$$

The boundary points are $\dfrac{-3 - \sqrt{33}}{2}$ and $\dfrac{-3 + \sqrt{33}}{2}$.

Use test points $x = -5,\ x = 0,$ and $x = 2.$

Test $x = -5$: $(-5)^2 + 3(-5) = 25 - 15 = 10 \le 6$ False

Test $x = 0$: $(0)^2 + 3(0) = 0 - 0 = 0 \le 6$ True

Test $x = 2$: $(2)^2 + 3(2) = 4 + 6 = 10 \le 6$ False

The boundary points are included.

The solution is $\left\{ x \,\middle|\, \dfrac{-3 - \sqrt{33}}{2} \le x \le \dfrac{-3 + \sqrt{33}}{2} \right\}$ or $\left[\dfrac{-3 - \sqrt{33}}{2},\ \dfrac{-3 + \sqrt{33}}{2} \right].$

31.

$$b^2 - 121 < 0$$
$$b^2 - 121 = 0$$
$$(b+11)(b-11) = 0$$
$$b+11 = 0 \quad \text{or} \quad b-11 = 0$$
$$b = -11 \quad \text{or} \quad b = 11$$

The boundary points are –11 and 11.
Use test points $b = -12$, $b = 0$, and $b = 12$.

Test $b = -12$: $\quad (-12)^2 - 121 = 144 - 121 = 23 < 0 \qquad$ False

Test $b = 0$: $\qquad (0)^2 - 121 = 0 - 121 = -121 < 0 \qquad$ True

Test $b = 12$: $\quad (12)^2 - 121 = 144 - 121 = 23 < 0 \qquad$ False

The boundary points are not included. The solution is $\{b \mid -11 < b < 11\}$ or $(-11, 11)$.

33.

$$3p(p-2) - 3 \geq 2p$$
$$3p(p-2) - 3 = 2p$$
$$3p^2 - 6p - 3 = 2p$$
$$3p^2 - 8p - 3 = 0$$
$$(3p+1)(p-3) = 0$$
$$3p+1 = 0 \quad \text{or} \quad p-3 = 0$$
$$3p = -1 \quad \text{or} \quad p = 3$$
$$p = -\frac{1}{3} \quad \text{or} \quad p = 3$$

The boundary points are $-\frac{1}{3}$ and 3.
Use test points $p = -1$, $p = 0$, and $p = 4$.

Test $p = -1$: $\quad 3(-1)(-1-2) - 3 = -3(-3) - 3 = 6 \geq 2(-1) = -2 \qquad$ True

Test $p = 0$: $\quad 3(0)(0-2) - 3 = 0(-2) - 3 = -3 \geq 2(0) = 0 \qquad$ False

Test $p = 4$: $\quad 3(4)(4-2) - 3 = 12(2) - 3 = 21 \geq 2(4) = 8 \qquad$ True

The boundary points are included.

The solution is $\left\{ p \mid p \leq -\dfrac{1}{3} \text{ or } p \geq 3 \right\}$ or $\left(-\infty, -\dfrac{1}{3} \right] \cup [3, \infty)$.

35.

$$x^3 - x^2 \leq 12x$$
$$x^3 - x^2 = 12x$$
$$x^3 - x^2 - 12x = 0$$
$$x(x^2 - x - 12) = 0$$
$$x(x+3)(x-4) = 0$$
$$x = 0 \quad \text{or} \quad x+3 = 0 \quad \text{or} \quad x-4 = 0$$
$$x = 0 \quad \text{or} \quad x = -3 \quad \text{or} \quad x = 4$$

The boundary points are –3, 0, and 4.
Use test points $x = -4$, $x = -1$, $x = 1$, and $x = 5$.

Test $x = -4$: $(-4)^3 - (-4)^2 = -64 - 16 = -80 \le 12(-4) = -48$ True

Test $x = -1$: $(-1)^3 - (-1)^2 = -1 - 1 = -2 \le 12(-1) = -12$ False

Test $x = 1$: $(1)^3 - (1)^2 = 1 - 1 = 0 \le 12(1) = 12$ True

Test $x = 5$: $(5)^3 - (5)^2 = 125 - 25 = 100 \le 12(5) = 60$ False

The boundary points are included.

The solution is $\{x \mid x \le -3 \text{ or } 0 \le x \le 4\}$ or $(-\infty, -3] \cup [0, 4]$.

37.
$$w^3 + w^2 > 4w + 4$$
$$w^3 + w^2 = 4w + 4$$
$$w^3 + w^2 - 4w - 4 = 0$$
$$w^2(w+1) - 4(w+1) = 0$$
$$(w+1)(w^2 - 4) = 0$$
$$(w+1)(w+2)(w-2) = 0$$
$$w + 1 = 0 \quad \text{or} \quad w + 2 = 0 \quad \text{or} \quad w - 2 = 0$$
$$w = -1 \quad \text{or} \quad w = -2 \quad \text{or} \quad w = 2$$

The boundary points are -2, -1, and 2.

Use test points $w = -3$, $w = -1.5$, $w = 0$, and $w = 3$.

Test $w = -3$: $(-3)^3 + (-3)^2 = -27 + 9 = -18 > 4(-3) + 4 = -12 + 4 = -8$ False

Test $w = -1.5$: $(-1.5)^3 + (-1.5)^2 = -3.375 + 2.25 = -1.125 > 4(-1.5) + 4 = -6 + 4 = -2$ True

Test $w = 0$: $(0)^3 + (0)^2 = 0 + 0 = 0 > 4(0) + 4 = 0 + 4 = 4$ False

Test $w = 3$: $(3)^3 + (3)^2 = 27 + 9 = 36 > 4(3) + 4 = 12 + 4 = 16$ True

The boundary points are not included.

The solution is $\{w \mid -2 < w < -1 \text{ or } w > 2\}$ or $(-2, -1) \cup (2, \infty)$.

39. a.
$$\frac{10}{x-5} = 5$$
$$10 = 5(x-5)$$
$$10 = 5x - 25$$
$$35 = 5x$$
$$x = 7 \quad \{7\}$$

b.
$$\frac{10}{x-5} < 5$$ The boundary points are 7 and 5 (undefined).

Use test points $x = 0$, $x = 6$, and $x = 8$.

Test $x = 0$: $\dfrac{10}{0-5} = \dfrac{10}{-5} = -2 < 5$ True

Test $x = 6$: $\dfrac{10}{6-5} = \dfrac{10}{1} = 10 < 5$ False

Test $x = 8$: $\dfrac{10}{8-5} = \dfrac{10}{3} < 5$ True

The boundary points are not included. The solution is $\{x | x < 5 \text{ or } x > 7\}$ or $(-\infty, 5) \cup (7, \infty)$.

c. $\dfrac{10}{x-5} > 5$ The boundary points are 7 and 5 (undefined).

Use test points $x = 0$, $x = 6$, and $x = 8$.

Test $x = 0$: $\dfrac{10}{0-5} = \dfrac{10}{-5} = -2 > 5$ False

Test $x = 6$: $\dfrac{10}{6-5} = \dfrac{10}{1} = 10 > 5$ True

Test $x = 8$: $\dfrac{10}{8-5} = \dfrac{10}{3} > 5$ False

The boundary points are not included. The solution is $\{x | 5 < x < 7\}$ or $(5, 7)$.

41. a. $\dfrac{z+2}{z-6} = -3$

$z + 2 = -3(z - 6)$

$z + 2 = -3z + 18$

$4z = 16$

$z = 4$ $\{4\}$

b. $\dfrac{z+2}{z-6} \le -3$ The boundary points are 4 and 6 (undefined).

Use test points $z = 0$, $z = 5$, and $z = 7$.

Test $z = 0$: $\dfrac{0+2}{0-6} = \dfrac{2}{-6} = -\dfrac{1}{3} \le -3$ False

Test $z = 5$: $\dfrac{5+2}{5-6} = \dfrac{7}{-1} = -7 \le -3$ True

Test $z = 7$: $\dfrac{7+2}{7-6} = \dfrac{9}{1} = 9 \le -3$ False

The boundary point at $z = 4$ is included. The solution is $\{z | 4 \le z < 6\}$ or $[4, 6)$.

c. $\dfrac{z+2}{z-6} \ge -3$ The boundary points are 4 and 6 (undefined).

Use test points $z = 0$, $z = 5$, and $z = 7$.

Test $z = 0$: $\dfrac{0+2}{0-6} = \dfrac{2}{-6} = -\dfrac{1}{3} \ge -3$ True

Test $z = 5$: $\dfrac{5+2}{5-6} = \dfrac{7}{-1} = -7 \ge -3$ False

Test $z = 7$: $\dfrac{7+2}{7-6} = \dfrac{9}{1} = 9 \ge -3$ True

The boundary point at $z = 4$ is included.

The solution is $\{z | z \le 4 \text{ or } z > 6\}$ or $(-\infty, 4] \cup (6, \infty)$.

43.

$$\frac{2}{x-1} \geq 0$$

$$\frac{2}{x-1} = 0$$

$$2 = 0(x-1)$$

$$2 = 0 \quad \text{No Solution}$$

The boundary point is 1 (undefined).
Use test points $x = 0$ and $x = 2$.

Test $x = 0$: $\quad \dfrac{2}{0-1} = \dfrac{2}{-1} = -2 \geq 0 \quad$ False

Test $x = 2$: $\quad \dfrac{2}{2-1} = \dfrac{2}{1} = 2 \geq 0 \quad$ True

The boundary point is not included.
The solution is $\{x \mid x > 1\}$ or $(1, \infty)$.

45.

$$\frac{b+4}{b-4} > 0$$

$$\frac{b+4}{b-4} = 0$$

$$b+4 = 0$$

$$b = -4$$

Boundary points are −4 and 4 (undefined).
Use test points $b = -5$, $b = 0$, and $b = 5$.

Test $b = -5$: $\quad \dfrac{-5+4}{-5-4} = \dfrac{-1}{-9} = \dfrac{1}{9} > 0 \quad$ True

Test $b = 0$: $\quad \dfrac{0+4}{0-4} = \dfrac{4}{-4} = -1 > 0 \quad$ False

Test $b = 5$: $\quad \dfrac{5+4}{5-4} = \dfrac{9}{1} = 9 > 0 \quad$ True

The boundary points are not included.
The solution is
$$\{b \mid b < -4 \text{ or } b > 4\} \text{ or } (-\infty, -4) \cup (4, \infty).$$

47.

$$\frac{3}{2x-7} < -1$$

$$\frac{3}{2x-7} = -1$$

$$3 = -1(2x-7)$$

$$3 = -2x+7$$

$$2x = 4$$

$$x = 2$$

Boundary points are 2 and $\frac{7}{2}$ (undefined).
Use test points $x = 0$, $x = 3$, and $x = 4$.

Test $x = 0$: $\quad \dfrac{3}{2(0)-7} = \dfrac{3}{-7} = -\dfrac{3}{7} < -1 \quad$ False

Test $x = 3$: $\quad \dfrac{3}{2(3)-7} = \dfrac{3}{-1} = -3 < -1 \quad$ True

Test $x = 4$: $\quad \dfrac{3}{2(4)-7} = \dfrac{3}{1} = 3 < -1 \quad$ False

The boundary points are not included.

The solution is $\left\{x \mid 2 < x < \dfrac{7}{2}\right\}$ or $\left(2, \dfrac{7}{2}\right)$.

49.

$$\frac{x+1}{x-5} \geq 4$$

$$\frac{x+1}{x-5} = 4$$

$$x+1 = 4(x-5)$$

$$x+1 = 4x-20$$

$$-3x = -21$$

$$x = 7$$

Boundary points are 7 and 5 (undefined).
Use test points $x = 0$, $x = 6$, and $x = 8$.

Test $x = 0$: $\quad \dfrac{0+1}{0-5} = \dfrac{1}{-5} = -\dfrac{1}{5} \geq 4 \quad$ False

Test $x = 6$: $\quad \dfrac{6+1}{6-5} = \dfrac{7}{1} = 7 \geq 4 \quad$ True

Test $x = 8$: $\quad \dfrac{8+1}{8-5} = \dfrac{9}{3} = 3 \geq 4 \quad$ False

The boundary point $x = 7$ is included.
The solution is $\{x \mid 5 < x \leq 7\}$ or $(5, 7]$.

51. $\dfrac{1}{x} \le 2$

$\dfrac{1}{x} = 2$

$1 = 2x$

$x = \dfrac{1}{2}$

Boundary points are $\frac{1}{2}$ and 0 (undefined).
Use test points $x = -1$, $x = 0.1$, and $x = 1$.

Test $x = -1$: $\dfrac{1}{-1} = -1 \le 2$ True

Test $x = 0.1$: $\dfrac{1}{0.1} = 10 \le 2$ False

Test $x = 1$: $\dfrac{1}{1} = 1 \le 2$ True

The boundary point $x = \frac{1}{2}$ is included.
The solution is

$\left\{ x \middle| x < 0 \text{ or } x \ge \dfrac{1}{2} \right\}$ or $(-\infty, 0) \cup \left[\dfrac{1}{2}, \infty \right)$.

53. $\dfrac{(x+2)^2}{x} > 0$

$\dfrac{(x+2)^2}{x} = 0$

$(x+2)^2 = 0$

$x + 2 = 0$

$x = -2$

Boundary points are –2 and 0 (undefined).
Use test points $x = -3$, $x = -1$, and $x = 1$.

Test $x = -3$: $\dfrac{(-3+2)^2}{-3} = \dfrac{1}{-3} > 0$ False

Test $x = -1$: $\dfrac{(-1+2)^2}{-1} = \dfrac{1}{-1} = -1 > 0$ False

Test $x = 1$: $\dfrac{(1+2)^2}{1} = \dfrac{9}{1} = 9 > 0$ True

The boundary points are not included.
The solution is $\{ x | x > 0 \}$ or $(0, \infty)$.

55. $x^2 + 10x + 25 \ge 0$

$(x+5)^2 \ge 0$

The quantity $(x+5)^2$ is greater than or
equal to zero for all real numbers. The
solution is all real numbers, $(-\infty, \infty)$.

57. $x^2 + 2x + 1 < 0$

$(x+1)^2 < 0$

The quantity $(x+1)^2$ is greater than or equal to
zero for all real numbers. There is no solution.
$\{\ \}$

59. $x^4 + 3x^2 \le 0$

The expression $x^4 + 3x^2$ is greater than zero
for all real numbers, x, except 0 which
makes the expression equal to 0. The
solution is $\{0\}$ since the inequality is "$\le$".

61. $x^2 + 12x + 36 < 0$

$(x+6)^2 < 0$

The quantity $(x+6)^2$ is greater than or equal
to zero for all real numbers. There is no
solution. $\{\ \}$

63. $x^2 + 3x + 5 < 0$

$a = 1, b = 3, c = 5$

$x = \dfrac{-3 \pm \sqrt{3^2 - 4(1)(5)}}{2(1)} = \dfrac{-3 \pm \sqrt{9 - 20}}{2}$

$= \dfrac{-3 \pm \sqrt{-11}}{2} = \dfrac{-3 \pm i\sqrt{11}}{2}$

Since the solutions for the related equation
are complex numbers, there are either no
real number solutions or all real numbers as
solutions. Test to check.

65. $-5x^2 + x < 1$

$-5x^2 + x - 1 < 0$

$a = -5, b = 1, c = -1$

$x = \dfrac{-1 \pm \sqrt{1^2 - 4(-5)(-1)}}{2(-5)} = \dfrac{-1 \pm \sqrt{1 - 20}}{-10}$

$= \dfrac{-1 \pm \sqrt{-19}}{-10} = \dfrac{-1 \pm i\sqrt{19}}{-10}$

Since the solutions for the related equation are
complex numbers, there are either no real
number solutions or all real numbers as

Test $x = 0$:

$0^2 + 3(0) + 5 = 0 + 0 + 5 = 5 < 0$ False

There are no real number solutions. $\{\ \}$

solutions. Test to check.

Test $x = 0$:

$-5(0)^2 + (0) = 0 + 0 = 0 < 1$ True

The solution is all real numbers. $(-\infty, \infty)$

67. $x^2 + 22x + 121 > 0$

$(x + 11)^2 > 0$

The expression $(x + 11)^2$ is greater than zero for all real numbers, x, except -11 which makes the expression equal to 0. The solution is $(-\infty, -11) \cup (-11, \infty)$.

69. $4t^2 - 12t \le -9$

$4t^2 - 12t + 9 \le 0$

$(2t - 3)^2 \le 0$

The expression $(2t - 3)^2$ is greater than zero for all real numbers, t, except $\frac{3}{2}$ which makes the expression equal to 0. The solution is $\left\{ \frac{3}{2} \right\}$.

71. $2y^2 - 8 \le 24$ Quadratic

$2y^2 - 8 = 24$

$2y^2 - 32 = 0$

$2(y^2 - 16) = 0$

$2(y + 4)(y - 4) = 0$

$y + 4 = 0$ or $y - 4 = 0$

$y = -4$ or $y = 4$

The boundary points are -4 and 4.

Use test points $y = -5$, $y = 0$, and $y = 5$.

Test $y = -5$:

$2(-5)^2 - 8 = 50 - 8 = 42 \le 24$ False

Test $y = 0$:

$2(0)^2 - 8 = 0 - 8 = -8 \le 24$ True

Test $y = 5$:

$2(5)^2 - 8 = 50 - 8 = 42 \le 24$ False

The boundary points are included.

The solution is $\{y | -4 \le y \le 4\}$ or $[-4, 4]$.

73. $(5x + 2)^2 > -4$ Quadratic

The quantity $(5x + 2)^2$ is greater than or equal to zero which is greater than -4 for all real numbers. The solution is all real numbers, $(-\infty, \infty)$.

75. $4(x - 2) < 6x - 3$ Linear

$4x - 8 < 6x - 3$

$-2x < 5$

$x > -\dfrac{5}{2}$ $\left(-\dfrac{5}{2}, \infty \right)$

77. $\dfrac{2x + 3}{x + 1} \le 2$ Rational

$\dfrac{2x + 3}{x + 1} = 2$

$2x + 3 = 2(x + 1)$

$2x + 3 = 2x + 2$

$3 = 2$ No Solution

Boundary point is -1 (undefined).

Use test points $x = -2$ and $x = 0$.

Test $x = -2$: $\dfrac{2(-2)+3}{-2+1} = \dfrac{-1}{-1} = 1 \le 2$ True

Test $x = 0$: $\dfrac{2(0)+3}{0+1} = \dfrac{3}{1} = 3 \le 2$ False

The boundary point is not included.

The solution is $\{x \mid x < -1\}$ or $(-\infty, -1)$.

79. $4x^3 - 40x^2 + 100x > 0$ Polynomial (degree > 2)

$4x^3 - 40x^2 + 100x = 0$

$4x(x^2 - 10x + 25) = 0$

$4x(x-5)^2 = 0$

$4x = 0$ or $x - 5 = 0$

$x = 0$ or $x = 5$

The boundary points are 0 and 5.

Use test points $x = -1$, $x = 1$, and $x = 6$.

Test $x = -1$: $4(-1)^3 - 40(-1)^2 + 100(-1) = -4 - 40 - 100 = -144 > 0$ False

Test $x = 1$: $4(1)^3 - 40(1)^2 + 100(1) = 4 - 40 + 100 = 64 > 0$ True

Test $x = 6$: $4(6)^3 - 40(6)^2 + 100(6) = 864 - 1440 + 600 = 24 > 0$ True

The boundary points are not included.

The solution is $\{x \mid 0 < x < 5 \text{ or } x > 5\}$ or $(0, 5) \cup (5, \infty)$.

81. $2p^3 > 4p^2$ Polynomial (degree > 2)

$2p^3 = 4p^2$

$2p^3 - 4p^2 = 0$

$2p^2(p-2) = 0$

$p^2 = 0$ or $p - 2 = 0$

$p = 0$ or $p = 2$

The boundary points are 0 and 2.

Use test points $p = -1$, $p = 1$, and $p = 3$.

Test $p = -1$: $2(-1)^3 = -2 > 4(-1)^2 = 4$ False

Test $p = 1$: $2(1)^3 = 2 > 4(1)^2 = 4$ False

Test $p = 3$: $2(3)^3 = 54 > 4(3)^2 = 36$ True

The boundary points are not included.

The solution is $\{p \mid p > 2\}$ or $(2, \infty)$.

83. $-3(x+4)^2(x-5) \ge 0$ Polynomial (degree > 2)

$x + 4 = 0$ or $x - 5 = 0$

$x = -4$ or $x = 5$

The boundary points are –4 and 5.
Use test points $x = -5$, $x = 0$, and $x = 6$.

Test $x = -5$: $-3(-5+4)^2(-5-5) = -3(-1)^2(-10) = -3(1)(-10) = 30 \geq 0$ True

Test $x = 0$: $-3(0+4)^2(0-5) = -3(4)^2(-5) = -3(16)(-5) = 240 \geq 0$ True

Test $x = 6$: $-3(6+4)^2(6-5) = -3(10)^2(1) = -3(100)(1) = -300 \geq 0$ False

The boundary points are included.
The solution is $\{x \mid x \leq 5\}$ or $(-\infty, 5]$.

85. $x^2 - 2 < 0$ Quadratic

$x^2 - 2 = 0$

$\qquad x^2 = 2$

$\qquad x = \pm\sqrt{2}$

The boundary points are $-\sqrt{2}$ and $\sqrt{2}$.
Use test points $x = -2$, $x = 0$, and $x = 2$.

Test $x = -2$: $(-2)^2 - 2 = 4 - 2 = 2 < 0$ False

Test $x = 0$: $(0)^2 - 2 = 0 - 2 = -2 < 0$ True

Test $x = 2$: $(2)^2 - 2 = 4 - 2 = 2 < 0$ False

The boundary points are not included.
The solution is $\{x \mid -\sqrt{2} < x < \sqrt{2}\}$ or $(-\sqrt{2}, \sqrt{2})$.

87. $x^2 + 5x - 2 \geq 0$ Quadratic

$x^2 + 5x - 2 = 0$

$$x = \frac{-5 \pm \sqrt{5^2 - 4(1)(-2)}}{2(1)} = \frac{-5 \pm \sqrt{25 + 8}}{2} = \frac{-5 \pm \sqrt{33}}{2}$$

The boundary points are $\dfrac{-5 - \sqrt{33}}{2}$ and $\dfrac{-5 + \sqrt{33}}{2}$.

Use test points $x = -6$, $x = 0$, and $x = 1$.

Test $x = -6$: $(-6)^2 + 5(-6) - 2 = 36 - 30 - 2 = 4 \geq 0$ True

Test $x = 0$: $(0)^2 + 5(0) - 2 = 0 + 0 - 2 = -2 \geq 0$ False

Test $x = 1$: $(1)^2 + 5(1) - 2 = 1 + 5 - 2 = 4 \geq 0$ True

The boundary points are included.

The solution is $\left\{ x \mid x \leq \dfrac{-5 - \sqrt{33}}{2} \text{ or } x \geq \dfrac{-5 + \sqrt{33}}{2} \right\}$ or $\left(-\infty, \dfrac{-5 - \sqrt{33}}{2}\right] \cup \left[\dfrac{-5 + \sqrt{33}}{2}, \infty\right)$.

89. $\dfrac{a+2}{a-5} \geq 0$ Rational

$\dfrac{a+2}{a-5} = 0$

$a+2 = 0$

$a = -2$

Boundary points are –2 and 5 (undefined).
Use test points $a = -3$, $a = 0$, and $a = 6$.

Test $a = -3$: $\dfrac{-3+2}{-3-5} = \dfrac{-1}{-8} = \dfrac{1}{8} \geq 0$ True

Test $a = 0$: $\dfrac{0+2}{0-5} = \dfrac{2}{-5} = -\dfrac{2}{5} \geq 0$ False

Test $a = 6$: $\dfrac{6+2}{6-5} = \dfrac{8}{1} = 8 \geq 0$ True

The boundary points $a = -2$ is included.
The solution is

$\left\{ a \,\middle|\, a \leq -2 \text{ or } a > 5 \right\}$ or $(-\infty, -2] \cup (5, \infty)$.

93. $4x^2 + 2 \geq -x$ Quadratic

$4x^2 + x + 2 \geq 0$

$a = 4$, $b = 1$, $c = 2$

$x = \dfrac{-1 \pm \sqrt{1^2 - 4(4)(2)}}{2(4)} = \dfrac{-1 \pm \sqrt{1-32}}{8}$

$= \dfrac{-1 \pm \sqrt{-31}}{8} = \dfrac{-1 \pm i\sqrt{31}}{8}$

Since the solutions for the related equation are complex numbers, there are either no real number solutions or all real numbers as solutions. Test to check.
Test $x = 0$:

$4(0)^2 + 2 = 0 + 2 = 2 \geq 0$ True

The solution is all real numbers. $(-\infty, \infty)$

91. $2 \geq t - 3$ Linear

$5 \geq t$

$t \leq 5$ $(-\infty, 5]$

95. $h(t) = -16t^2 + 108t + 8 \geq 120$ Quadratic

$-16t^2 + 108 - 112 \geq 0$

$-4\left(4t^2 - 27t + 28\right) \geq 0$

$t = \dfrac{-(-27) \pm \sqrt{(-27)^2 - 4(4)(28)}}{2(4)} = \dfrac{27 \pm \sqrt{729 - 448}}{8} = \dfrac{27 \pm \sqrt{281}}{8} \approx 1.3 \text{ or } 5.4$

The boundary points are $\dfrac{27 - \sqrt{281}}{8}$ and $\dfrac{27 + \sqrt{281}}{8}$.

Use test points $t = 0$, $t = 2$, and $t = 6$.

Test $t = 0$: $-16(0)^2 + 108(0) - 112 = 0 + 0 - 112 = -112 \ge 0$ False

Test $t = 2$: $-16(2)^2 + 108(2) - 112 = -64 + 216 - 112 = 40 \ge 0$ True

Test $t = 6$: $-16(6)^2 + 108(6) - 112 = -576 + 648 - 112 = -40 \ge 0$ False

The boundary points are included.

The solution is $\left\{ t \left| \dfrac{27 - \sqrt{281}}{8} \le t \le \dfrac{27 + \sqrt{281}}{8} \right. \right\}$ or $\left[\dfrac{27 - \sqrt{281}}{8}, \dfrac{27 + \sqrt{281}}{8} \right] \approx [1.3,\, 5.5]$.

The fuses should be set for between 1.3 sec and 5.5 sec after launch.

Problem Recognition Exercises

1. **a.** Equation quadratic in form and polynomial equation

    b. $\left(z^2 - 4\right)^2 - \left(z^2 - 4\right) - 12 = 0$

        Let $u = z^2 - 4$

        $u^2 - u - 12 = 0$

        $(u - 4)(u + 3) = 0$

        $u - 4 = 0$ or $u + 3 = 0$

        $u = 4$ or $u = -3$

        $z^2 - 4 = 4$ or $z^2 - 4 = -3$

        $z^2 = 8$ or $z^2 = 1$

        $z = \pm\sqrt{8}$ or $z = \pm 1$

        $z = \pm 2\sqrt{2}$ or $z = \pm 1$

        $\left\{ \pm 2\sqrt{2},\, \pm 1 \right\}$

3. **a.** Polynomial inequality

    b. $2y(y - 4) \le 5 + y$

        $2y^2 - 8y = 5 + y$

        $2y^2 - 9y - 5 = 0$

        $(2y + 1)(y - 5) = 0$

        $2y + 1 = 0$ or $y - 5 = 0$

        $y = -\dfrac{1}{2}$ or $y = 5$

    The boundary points are $-\dfrac{1}{2}$ and 5.

    Use test points $y = -1$, $y = 0$, and $y = 6$.

    Test $y = -1$: $2(-1)^2 - 9(-1) - 5 = 2 + 9 - 5 = 6 \le 0$ False

    Test $y = 0$: $2(0)^2 - 9(0) - 5 = 0 - 0 - 5 = -5 \le 0$ True

    Test $y = 6$: $2(6)^2 - 9(6) - 5 = 72 - 54 - 5 = 13 \le 0$ False

The boundary points are included.

The solution is $\left\{y \mid -\dfrac{1}{2} \le y \le 5\right\}$ or $\left[-\dfrac{1}{2}, 5\right]$.

5. a. Absolute value equation

b. $$-5 = -|w-4|$$
$$5 = |w-4|$$
$$w-4=5 \text{ or } w-4=-5$$
$$w=9 \text{ or } \qquad w=-1 \quad \{9, -1\}$$

7. a. Polynomial inequality

b. $$m^3 + 5m^2 - 4m - 20 \ge 0$$
$$m^2(m+5) - 4(m+5) = 0$$
$$(m+5)(m^2-4) = 0$$
$$(m+5)(m-2)(m+2) = 0$$
$$m+5=0 \quad \text{or } m-2=0 \text{ or } m+2=0$$
$$m=-5 \text{ or } \qquad m=2 \text{ or } \qquad m=-2$$

The boundary points are -5, -2, and 2.

Use test points $m=-6$, $m=-3$, $m=0$, and $m=3$.

Test $m=-6$: $\quad (-6)^3 + 5(-6)^2 - 4(-6) - 20 = -216 + 180 + 24 - 20 = -32 \ge 0$ False

Test $m=-3$: $\quad (-3)^3 + 5(-3)^2 - 4(-3) - 20 = -27 + 45 + 12 - 20 = 10 \ge 0$ True

Test $m=0$: $\quad (0)^3 + 5(0)^2 - 4(0) - 20 = 0 + 0 - 0 - 20 = -20 \ge 0$ False

Test $m=3$: $\quad (3)^3 + 5(3)^2 - 4(3) - 20 = 27 + 45 - 12 - 20 = 40 \ge 0$ True

The boundary points are included.

The solution is $\{m \mid -5 \le y \le -2 \text{ or } y \ge 2\}$ or $[-5, -2] \cup [2, \infty)$.

9. a. Linear inequality

b. $$5 - 2[3 - (x-4)] \le 3x + 14$$
$$5 - 2[3 - x + 4] \le 3x + 14$$
$$5 - 2[-x + 7] \le 3x + 14$$
$$5 + 2x - 14 \le 3x + 14$$
$$2x - 9 \le 3x + 14$$
$$-x \le 23$$
$$x \ge -23 \quad [-23, \infty)$$

11. a. Rational inequality

b. $$\frac{3}{x-2} \le 1$$
$$\frac{3}{x-2} = 1$$
$$3 = x - 2$$
$$5 = x$$

Boundary points are 5 and 2 (undefined).

Use test points $a=-3$, $a=0$, and $a=6$.

Test $x=0$: $\quad \dfrac{3}{0-2} = \dfrac{3}{-2} = -\dfrac{3}{2} \le 1$ True

Test $x=3$: $\quad \dfrac{3}{3-2} = \dfrac{3}{1} = 3 \le 1$ False

Test $x=6$: $\quad \dfrac{3}{6-2} = \dfrac{3}{4} \le 1$ True

The boundary points $x = 5$ is included.

The solution is

$$\{x\,|\,x<2 \text{ or } x\ge5\} \text{ or } (-\infty,\,2)\cup[5,\,\infty).$$

13. a. Radical equation

b.
$$\sqrt{t+8}-6=t$$
$$\sqrt{t+8}=t+6$$
$$\left(\sqrt{t+8}\right)^2=(t+6)^2$$
$$t+8=t^2+12t+36$$
$$0=t^2+11t+28$$
$$0=(t+4)(t+7)$$
$$t+4=0 \quad \text{or } t+7=0$$
$$t=-4 \text{ or } \qquad t=-7$$

Check:
$$t=-4:$$
$$\sqrt{-4+8}-6=-4$$
$$\sqrt{4}-6=-4$$
$$2-6=-4$$
$$-4=-4$$

$$t=-7:$$
$$\sqrt{-7+8}-6=-7$$
$$\sqrt{1}-6=-7$$
$$1-6=-7$$
$$-5\ne-7$$
$$\{-4\} \quad (t=-7 \text{ does not check.})$$

17. a. Polynomial inequality

b.
$$x^2-10x\le-25$$
$$x^2-10x+25\le0$$
$$(x-5)^2\le0$$

The expression $(x-5)^2$ is greater than zero for all real numbers, t, except 5 which makes the expression equal to 0. The solution is $\{5\}$.

15. a. Compound inequality

b.
$$-4-x>2 \quad \text{or } 8<2x$$
$$-x>6 \quad \text{or } 4<x$$
$$x<-6 \text{ or } x>4$$
$$(-\infty,\,-6)\cup(4,\,\infty)$$

19. a. Radical equation and equation quadratic in form

b.
$$x-13\sqrt{x}+36=0$$
Let $u=\sqrt{x}$
$$u^2-13u+36=0$$
$$(u-9)(u-4)=0$$
$$u-9=0 \quad \text{or } u-4=0$$
$$u=9 \quad \text{or} \qquad u=4$$
$$\sqrt{x}=9 \quad \text{or} \quad \sqrt{x}=4$$
$$x=81 \text{ or} \qquad x=16$$
Check:

$$x = 81:$$

$$81 - 13\sqrt{81} + 36 = 0$$

$$81 - 13(9) + 36 = 0$$

$$81 - 117 + 36 = 0$$

$$0 = 0$$

$$x = 16:$$

$$16 - 13\sqrt{16} + 36 = 0$$

$$16 - 13(4) + 36 = 0$$

$$16 - 52 + 36 = 0$$

$$0 = 0$$

$$\{81, 16\}$$

Chapter 7 Group Activity

1.

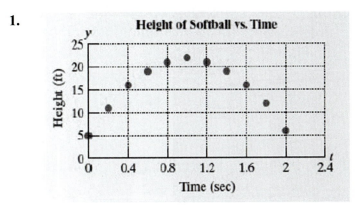

Height of Softball vs. Time

3. $y = a(t - 1)^2 + 22$

5. Answers will vary, but should be close to
$$y = -16(t - 1)^2 + 22.$$

7. Answers will vary, but should be close to the observed value of 12 ft.

Chapter 7 Review Exercises

Section 7.1

1. $x^2 = 5$

$$x = \pm\sqrt{5} \quad \left\{\pm\sqrt{5}\right\}$$

3. $a^2 = 81$

$$a = \pm 9 \quad \{\pm 9\}$$

5. $(x - 2)^2 = 72$

$$x - 2 = \pm\sqrt{72}$$

$$x - 2 = \pm 6\sqrt{2}$$

$$x = 2 \pm 6\sqrt{2} \quad \left\{2 \pm 6\sqrt{2}\right\}$$

7. $(3y - 1)^2 = 3$

$$3y - 1 = \pm\sqrt{3}$$

$$3y = 1 \pm\sqrt{3}$$

$$y = \frac{1 \pm\sqrt{3}}{3} \quad \left\{\frac{1 \pm\sqrt{3}}{3}\right\}$$

9. Let h = the height of the triangle
$$5^2 + h^2 = 10^2$$

$$25 + h^2 = 100$$

$$h^2 = 75$$

$$h = \pm\sqrt{75} = \pm 5\sqrt{3} \approx 8.7$$

The height of the triangle is about 8.7 in.

11. Let s = the length of a side of the square
$$s^2 = 150$$

$$s = \pm\sqrt{150} = \pm 5\sqrt{6} \approx 12.2$$

The length of a side of the square is about 12.2 in.

13. $x^2 - 9x + n$

$$n = \left(\frac{1}{2}b\right)^2 = \left(\frac{1}{2} \cdot (-9)\right)^2 = \left(-\frac{9}{2}\right)^2 = \frac{81}{4}$$

$$x^2 - 9k + \frac{81}{4} = \left(x - \frac{9}{2}\right)^2$$

15. $z^2 - \frac{2}{5}z + n$

$$n = \left(\frac{1}{2}b\right)^2 = \left(\frac{1}{2} \cdot \left(-\frac{2}{5}\right)\right)^2 = \left(-\frac{1}{5}\right)^2 = \frac{1}{25}$$

$$z^2 - \frac{2}{5}z + \frac{1}{25} = \left(z - \frac{1}{5}\right)^2$$

17. $4y^2 - 8y - 20 = 0$

$$y^2 - 2y - 5 = 0$$
$$y^2 - 2y = 5$$
$$y^2 - 2y + 1 = 5 + 1$$
$$(y - 1)^2 = 6$$
$$y - 1 = \pm\sqrt{6}$$
$$y = 1 \pm \sqrt{6} \quad \left\{1 \pm \sqrt{6}\right\}$$

19. $-t^2 + 8t - 25 = 0$

$$t^2 - 8t + 25 = 0$$
$$t^2 - 8t = -25$$
$$t^2 - 8t + 16 = -25 + 16$$
$$(t - 4)^2 = -9$$
$$t - 4 = \pm\sqrt{-9}$$
$$t - 4 = \pm 3i$$
$$t = 4 \pm 3i \quad \left\{4 \pm 3i\right\}$$

21.

$$b^2 + \frac{7}{2}b = 2$$

$$b^2 + \frac{7}{2}b + \frac{49}{16} = 2 + \frac{49}{16}$$

$$\left(b + \frac{7}{4}\right)^2 = \frac{81}{16}$$

$$b + \frac{7}{4} = \pm\sqrt{\frac{81}{16}}$$

$$b + \frac{7}{4} = \pm\frac{9}{4}$$

$$b = -\frac{7}{4} \pm \frac{9}{4}$$

$$b = \frac{2}{4} = \frac{1}{2} \quad \text{or} \quad b = -\frac{16}{4} = -4$$

$$\left\{\frac{1}{2}, -4\right\}$$

23. $A = 6s^2,$ Solve for s:

$$\frac{A}{6} = s^2$$

$$s = \sqrt{\frac{A}{6}} \quad \text{or} \quad s = \frac{\sqrt{6A}}{6}$$

Section 7.2

25. $x^2 - 5x = -6$

$$x^2 - 5x + 6 = 0$$
$$b^2 - 4ac = (-5)^2 - 4(1)(6) = 25 - 24 = 1$$
2 rational solutions

27. $z^2 + 23 = 17z$

$$z^2 - 17z + 23 = 0$$
$$b^2 - 4ac = (-17)^2 - 4(1)(23)$$
$$= 289 - 92 = 197$$
2 irrational solutions

29.
$$10b+1=-25b^2$$
$$25b^2+10b+1=0$$
$$b^2-4ac=(10)^2-4(25)(1)=100-100=0$$
1 rational solution

31.
$$y^2-4y+1=0$$
$$a=1,\ b=-4,\ c=1$$
$$y=\frac{-(-4)\pm\sqrt{(-4)^2-4(1)(1)}}{2(1)}$$
$$=\frac{4\pm\sqrt{16-4}}{2}=\frac{4\pm\sqrt{12}}{2}=\frac{4\pm2\sqrt{3}}{2}$$
$$=\frac{\cancel{2}\left(2\pm\sqrt{3}\right)}{\cancel{2}}=2\pm\sqrt{3}\quad\left\{2\pm\sqrt{3}\right\}$$

33.
$$6a(a-1)=10+a$$
$$6a^2-6a=10+a$$
$$6a^2-7a-10=0$$
$$a=6,\ b=-7,\ c=-10$$
$$a=\frac{-(-7)\pm\sqrt{(-7)^2-4(6)(-10)}}{2(6)}$$
$$=\frac{7\pm\sqrt{49+240}}{12}=\frac{7\pm\sqrt{289}}{12}=\frac{7\pm17}{12}$$
$$a=\frac{24}{12}=2\ \text{ or }\ a=-\frac{10}{12}=-\frac{5}{6}\quad\left\{2,-\frac{5}{6}\right\}$$

35.
$$b^2-\frac{4}{25}=\frac{3}{5}b$$
$$b^2-\frac{3}{5}b-\frac{4}{25}=0$$
$$25b^2-15b-4=0$$
$$a=25,\ b=-15,\ c=-4$$
$$b=\frac{-(-15)\pm\sqrt{(-15)^2-4(25)(-4)}}{2(25)}$$
$$=\frac{15\pm\sqrt{225+400}}{50}=\frac{15\pm\sqrt{625}}{50}=\frac{15\pm25}{50}$$
$$b=\frac{40}{50}=\frac{4}{5}\ \text{ or }\ b=-\frac{10}{50}=-\frac{1}{5}\quad\left\{\frac{4}{5},-\frac{1}{5}\right\}$$

37.
$$-32+4x-x^2=0$$
$$x^2-4x+32=0$$
$$a=1,\ b=-4,\ c=32$$
$$x=\frac{-(-4)\pm\sqrt{(-4)^2-4(1)(32)}}{2(1)}$$
$$=\frac{4\pm\sqrt{16-128}}{2}=\frac{4\pm\sqrt{-112}}{2}=\frac{4\pm4i\sqrt{7}}{2}$$
$$=\frac{2\left(2\pm2i\sqrt{7}\right)}{2}=2\pm2i\sqrt{7}\quad\left\{2\pm2i\sqrt{7}\right\}$$

39.
$$3x^2-4x=6$$
$$3x^2-4x-6=0$$
$$a=3,\ b=-4,\ c=-6$$
$$x=\frac{-(-4)\pm\sqrt{(-4)^2-4(3)(-6)}}{2(3)}$$
$$=\frac{4\pm\sqrt{16+72}}{6}=\frac{4\pm\sqrt{88}}{6}=\frac{4\pm2\sqrt{22}}{6}$$
$$=\frac{2\left(2\pm\sqrt{22}\right)}{6}=\frac{2\pm\sqrt{22}}{3}\quad\left\{\frac{2\pm\sqrt{22}}{3}\right\}$$

41.
$$y^2 + 14y = -46$$
$$y^2 + 14y + 46 = 0$$
$$a = 1, \, b = 14, \, c = 46$$
$$x = \frac{-(14) \pm \sqrt{(14)^2 - 4(1)(46)}}{2(1)}$$
$$= \frac{-14 \pm \sqrt{196 - 184}}{2} = \frac{-14 \pm \sqrt{12}}{2}$$
$$= \frac{-14 \pm 2\sqrt{3}}{2} = \frac{2(-7 \pm \sqrt{3})}{2} = -7 \pm \sqrt{3}$$
$$\left\{ -7 \pm \sqrt{3} \right\}$$

43. a.
$$D(s) = \frac{1}{10}s^2 - 3s + 22$$
$$D(150) = \frac{1}{10}(150)^2 - 3(150) + 22$$
$$= 2250 - 450 + 22 = 1822$$
The landing distance is 1822 ft.

b.
$$1000 = \frac{1}{10}s^2 - 3s + 22$$
$$10,000 = s^2 - 30s + 220$$
$$0 = s^2 - 30s - 9780$$
$$a = 1, \, b = -30, \, c = -9780$$
$$x = \frac{-(-30) \pm \sqrt{(-30)^2 - 4(1)(-9780)}}{2(1)}$$
$$= \frac{30 \pm \sqrt{900 + 39120}}{2} = \frac{30 \pm \sqrt{40020}}{2}$$
$$x \approx 115 \text{ or } x \approx -85$$
The landing speed is about 115 ft/sec.

45. Let x = the width of the island
$2x + 1$ = the length of the island
$$x(2x + 1) = 22.32$$
$$2x^2 + x - 22.32 = 0$$
$$a = 2, \, b = 1, \, c = -22.32$$
$$x = \frac{-(1) \pm \sqrt{(1)^2 - 4(2)(-22.32)}}{2(2)}$$
$$= \frac{-1 \pm \sqrt{1 + 178.56}}{4} = \frac{-1 \pm \sqrt{179.56}}{4}$$
$$= \frac{-1 \pm 13.4}{4} = 3.1 \text{ or } -3.6$$
$$2x + 1 = 2(3.1) + 1 = 7.2$$
The dimensions are approximately 3.1 ft by 7.2 ft.

Section 7.3

47.
$$x - 4\sqrt{x} - 21 = 0$$
Let $u = \sqrt{x}$
$$u^2 - 4u - 21 = 0$$
$$(u - 7)(u + 3) = 0$$
$$u - 7 = 0 \text{ or } u + 3 = 0$$
$$u = 7 \text{ or } \quad u = -3$$
$$\sqrt{x} = 7 \text{ or } \sqrt{x} = -3$$
$$x = 49 \text{ or } \quad x = 9$$

Check:
$x = 49$:

$$49 - 4\sqrt{49} - 21 = 0$$
$$49 - 28 - 21 = 0$$
$$0 = 0$$

$x = 9$:

$$9 - 4\sqrt{9} - 21 = 0$$
$$9 - 12 - 21 = 0$$
$$-24 \neq 0$$

Solution: $\{49\}$ ($x = 9$ does not check.)

49.
$$y^4 - 11y^2 + 18 = 0$$
$$\left(y^2\right)^2 - 11y^2 + 18 = 0$$
Let $u = y^2$
$$u^2 - 11u + 18 = 0$$
$$(u-9)(u-2) = 0$$
$$u - 9 = 0 \quad \text{or} \quad u - 2 = 0$$
$$y^2 - 9 = 0 \quad \text{or} \quad y^2 - 2 = 0$$
$$y^2 = 9 \quad \text{or} \quad y^2 = 2$$
$$y = \pm 3 \quad \text{or} \quad y = \pm\sqrt{2}$$
$$\left\{\pm 3,\ \pm\sqrt{2}\right\}$$

51.
$$t^{2/5} + t^{1/5} - 6 = 0$$
$$\left(t^{1/5}\right)^2 + t^{1/5} - 6 = 0$$
Let $u = t^{1/5}$
$$u^2 + u - 6 = 0$$
$$(u-2)(u+3) = 0$$
$$u - 2 = 0 \quad \text{or} \quad u + 3 = 0$$
$$u = 2 \text{ or} \quad u = -3$$
$$t^{1/5} = 2 \quad \text{or} \quad t^{1/5} = -3$$
$$\left(t^{1/5}\right)^5 = (2)^5 \quad \text{or} \quad \left(t^{1/5}\right)^5 = (-3)^5$$
$$t = 32 \quad \text{or} \quad t = -243$$
$$\{32,\ -243\}$$

53.
$$\frac{2t}{t+1} + \frac{-3}{t-2} = 1$$
$$(t+1)(t-2)\left(\frac{2t}{t+1} + \frac{-3}{t-2}\right) = (t+1)(t-2)1$$
$$2t(t-2) - 3(t+1) = t^2 - t - 2$$
$$2t^2 - 4t - 3t - 3 = t^2 - t - 2$$
$$t^2 - 6t = 1$$
$$t^2 - 6t + 9 = 1 + 9$$
$$(t-3)^2 = 10$$
$$t - 3 = \pm\sqrt{10}$$
$$t = 3 \pm \sqrt{10} \quad \left\{3 \pm \sqrt{10}\right\}$$

55.
$$\left(x^2 + 5\right)^2 + 2\left(x^2 + 5\right) - 8 = 0$$
Let $u = x^2 + 5$
$$u^2 + 2u - 8 = 0$$
$$(u+4)(u-2) = 0$$
$$u + 4 = 0 \quad \text{or} \quad u - 2 = 0$$
$$x^2 + 5 + 4 = 0 \quad \text{or} \quad x^2 + 5 - 2 = 0$$
$$x^2 + 9 = 0 \quad \text{or} \quad x^2 + 3 = 0$$
$$x^2 = -9 \quad \text{or} \quad x^2 = -3$$
$$x = \pm 3i \quad \text{or} \quad x = \pm i\sqrt{3}$$
$$\left\{\pm 3i,\ \pm i\sqrt{3}\right\}$$

Section 7.4

57. $g(x) = x^2 - 5$

Domain: $(-\infty, \infty)$; range: $[-5, \infty)$

59. $h(x) = (x-5)^2$

Domain: $(-\infty, \infty)$; range: $[0, \infty)$

61. $m(x) = -2x^2$

Domain: $(-\infty, \infty)$; range: $(-\infty, 0]$

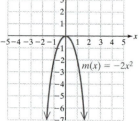

63. $p(x) = -2(x-5)^2 - 5$

Domain: $(-\infty, \infty)$; range: $(-\infty, -5]$

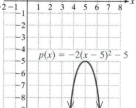

65. $t(x) = \dfrac{1}{3}(x-4)^2 + \dfrac{5}{3}$

Vertex: $\left(4, \dfrac{5}{3}\right)$ is a minimum point with

minimum value of $\dfrac{5}{3}$.

67. $a(x) = -\dfrac{3}{2}\left(x + \dfrac{2}{11}\right)^2 - \dfrac{14}{3}$

Axis of symmetry: $x = -\dfrac{2}{11}$

Section 7.5

69. $z(x) = x^2 - 6x + 7$

$\quad = (x^2 - 6x) + 7$

$\quad = (x^2 - 6x + 9 - 9) + 7$

$\quad = (x^2 - 6x + 9) - 9 + 7$

$z(x) = (x-3)^2 - 2$

Vertex: $(3, -2)$

71. $p(x) = -5x^2 - 10x - 13$

$\quad = -5(x^2 + 2x) - 13$

$\quad = -5(x^2 + 2x + 1 - 1) - 13$

$\quad = -5(x^2 + 2x + 1) + 5 - 13$

$p(x) = -5(x+1)^2 - 8$

Vertex: $(-1, -8)$

73. $f(x) = -2x^2 + 4x - 17$

$a = -2,\ b = 4,\ c = -17$

$\dfrac{-b}{2a} = \dfrac{-(4)}{2(-2)} = \dfrac{-4}{-4} = 1$

$f(1) = -2(1)^2 + 4(1) - 17$

$\quad = -2 + 4 - 17 = -15$

Vertex: $(1, -15)$

75. $m(x) = 3x^2 - 3x + 11$

$a = 3,\ b = -3,\ c = 11$

$\dfrac{-b}{2a} = \dfrac{-(-3)}{2(3)} = \dfrac{3}{6} = \dfrac{1}{2}$

$m\left(\dfrac{1}{2}\right) = 3\left(\dfrac{1}{2}\right)^2 - 3\left(\dfrac{1}{2}\right) + 11$

$\quad = \dfrac{3}{4} - \dfrac{3}{2} + 11 = \dfrac{41}{4}$

Vertex: $\left(\dfrac{1}{2}, \dfrac{41}{4}\right)$

77. a.
$$y = \frac{3}{4}x^2 - 3x$$

$$y = \frac{3}{4}(x^2 - 4x + 4) - 3$$

$$y = \frac{3}{4}(x - 2)^2 - 3$$

Vertex: $(2, -3)$

b.
$$0 = \frac{3}{4}x^2 - 3x$$

$$\frac{3}{4}x(x - 4) = 0$$

$$\frac{3}{4}x = 0 \text{ or } x - 4 = 0$$

$$x = 0 \text{ or } \quad x = 4$$

x-intercepts: $(0, 0), (4, 0)$

$$y = \frac{3}{4}(0)^2 - 3(0) = 0 - 0 = 0$$

y-intercept: $(0, 0)$

c.

81. Substitute each ordered pair for x and y into the standard form of a parabola to get three equations in a, b, and c:

$$-4 = a(-3)^2 + b(-3) + c$$
$$-4 = 9a - 3b + c \qquad \text{(A)}$$
$$-5 = a(-2)^2 + b(-2) + c$$
$$-5 = 4a - 2b + c \qquad \text{(B)}$$
$$4 = a(1)^2 + b(1) + c$$
$$4 = a + b + c \qquad \text{(C)}$$

Subtract (B) from (A) to eliminate c:

$$9a - 3b + c = -4$$
$$\underline{-(4a - 2b + c = -5)}$$
$$5a - b \quad = 1 \qquad \text{(D)}$$

Subtract (C) from (B) to eliminate c:

$$4a - 2b + c = -5$$
$$\underline{-(\ a + b + c = \ 4)}$$
$$3a - 3b \quad = -9 \rightarrow a - b = -3 \qquad \text{(E)}$$

79. a. $h(t) = -16t^2 + 96t$

$a = -16$, $b = 96$, $c = 0$

$$\frac{-b}{2a} = \frac{-(96)}{2(-16)} = \frac{-96}{-32} = 3$$

The projectile reaches its maximum height at 3 sec.

b. $h(3) = -16(3)^2 + 96(3)$
$$= -144 + 288 = 144$$
The maximum height is 144 ft.

Subtract (E) from (D) to eliminate b:

$$5a - b = 1$$
$$\underline{-(\ a - b = -3)}$$
$$4a \quad = 4$$
$$a = 1$$

Substitute $a = 1$ into (E) and solve for b:

$$1 - b = -3$$
$$b = 4$$

Solve for c:

$$1 + 4 + c = 4$$
$$5 + c = 4$$
$$c = -1$$

The equation is: $y = x^2 + 4x - 1$.

Section 7.6

83. a.

$$x^2 - 4 = 0$$
$$(x + 2)(x - 2) = 0$$
$$x + 2 = 0 \quad \text{or } x - 2 = 0$$
$$x = -2 \quad \text{or} \quad x = 2 \quad \{-2, 2\}; \quad (-2, 0) \text{ and } (2, 0) \text{ are the } x\text{-intercepts.}$$

b. $x^2 - 4 < 0$ The boundary points are -2 and 2.
Use test points $x = -3$, $x = 0$, and $x = 3$.

Test $x = -3$: $(-3)^2 - 4 = 9 - 4 = 5 < 0$ False

Test $x = 0$: $(0)^2 - 4 = 0 - 4 = -4 < 0$ True

Test $x = 3$: $(3)^2 - 4 = 9 - 4 = 5 < 0$ False

The boundary points are not included. The solution is $\{x \mid -2 < x < 2\}$ or $(-2, 2)$.

On the interval $(-2, 2)$ the graph is below the x-axis.

c. $x^2 - 4 > 0$ The boundary points are -2 and 2.
Use test points $x = -3$, $x = 0$, and $x = 3$.

Test $x = -3$: $(-3)^2 - 4 = 9 - 4 = 5 > 0$ True

Test $x = 0$: $(0)^2 - 4 = 0 - 4 = -4 > 0$ False

Test $x = 3$: $(3)^2 - 4 = 9 - 4 = 5 > 0$ True

The boundary points are not included. The solution is $\{x \mid x < -2 \text{ or } x > 2\}$ or

$(-\infty, -2) \cup (2, \infty)$. On the intervals $(-\infty, -2)$ and $(2, \infty)$ the graph is above the x-axis.

85. $w^2 - 4w - 12 < 0$

$$w^2 - 4w - 12 = 0$$
$$(w + 2)(w - 6) = 0$$
$$w + 2 = 0 \quad \text{or } w - 6 = 0$$
$$w = -2 \quad \text{or} \quad w = 6$$

The boundary points are -2 and 6.
Use test points $w = -3$, $w = 0$, and $w = 7$.

Test $w = -3$: $(-3)^2 - 4(-3) - 12 = 9 + 12 - 12 = 9 < 0$ False

Test $w = 0$: $(0)^2 - 4(0) - 12 = 0 - 0 - 12 = -12 < 0$ True

Test $w = 7$: $(7)^2 - 4(7) - 12 = 49 - 28 - 12 = 9 < 0$ False

The boundary points are not included.

The solution is $\{w \mid -2 < w < 6\}$ or $(-2, 6)$.

87. $\dfrac{12}{x+2} \le 6$

$\dfrac{12}{x+2} = 6$

$\quad 12 = 6(x+2)$

$\quad 12 = 6x + 12$

$\quad 0 = 6x$

$\quad x = 0$

Boundary points are 0 and -2 (undefined).

Use test points $x = -3$, $x = -1$, and $x = 1$.

Test $x = -3$: $\dfrac{12}{-3+2} = \dfrac{12}{-1} = -12 \le 6$ True

Test $x = -1$: $\dfrac{12}{-1+2} = \dfrac{12}{1} = 12 \le 6$ False

Test $x = 1$: $\dfrac{12}{1+2} = \dfrac{12}{3} = 4 \le 6$ True

The boundary point $x = 0$ is included.

The solution is

$\{x \mid x < -2 \text{ or } x \ge 0\}$ or $(-\infty, -2) \cup [0, \infty)$.

89. $3y(y-5)(y+2) > 0$

$3y(y-5)(y+2) = 0$

$\qquad\qquad 3y = 0$ or $y - 5 = 0$ or $y + 2 = 0$

$\qquad\qquad y = 0$ or $\quad y = 5$ or $\quad y = -2$

The boundary points are -2, 0 and 5.

Use test points $y = -3$, $y = -1$, $y = 1$, and $y = 6$.

Test $y = -3$: $3(-3)(-3-5)(-3+2) = -9(-8)(-1) = -72 > 0$ False

Test $y = -1$: $3(-1)(-1-5)(-1+2) = -3(-6)(1) = 18 > 0$ True

Test $y = 1$: $3(1)(1-5)(1+2) = 3(-4)(3) = -36 > 0$ False

Test $y = 6$: $3(6)(6-5)(6+2) = 18(1)(8) = 144 > 0$ True

The boundary points are not included.

The solution is $\{y \mid -2 < y < 0 \text{ or } y > 5\}$ or $(-2, 0) \cup (5, \infty)$.

91.
$$-x^2 - 4x \le 1$$
$$-x^2 - 4x - 1 = 0$$
$$a = -1, \ b = -4, \ c = -1$$
$$x = \frac{-(-4) \pm \sqrt{(-4)^2 - 4(-1)(-1)}}{2(-1)} = \frac{4 \pm \sqrt{12}}{-2}$$
$$= \frac{4 \pm 2\sqrt{3}}{-2} = \frac{-2\left(-2 \pm \sqrt{3}\right)}{-2} = -2 \pm \sqrt{3}$$

The boundary points are $-2 - \sqrt{3}$ and $-2 + \sqrt{3}$.

Use test points $x = -4$, $x = -1$, and $x = 0$.

Test $x = -4$:
$$-(-4)^2 - 4(-4) = -16 + 16 = 0 \le 1 \quad \text{True}$$

Test $x = -1$:
$$-(-1)^2 - 4(-1) = -1 + 4 = 3 \le 1 \quad \text{False}$$

Test $x = 0$:
$$-(0)^2 - 4(0) = 0 - 0 = 0 \le 1 \quad \text{True}$$

The boundary points are included. The solution is $\left\{ x \mid x \le -2 - \sqrt{3} \text{ or } x \ge -2 + \sqrt{3} \right\}$

or $\left(-\infty, -2 - \sqrt{3}\right] \cup \left[-2 + \sqrt{3}, \infty\right)$.

93.
$$\frac{w+1}{w-3} > 1$$
$$\frac{w+1}{w-3} = 1$$
$$w + 1 = w - 3$$
$$1 = -3 \quad \text{contradiction}$$

Boundary point is 3 (undefined).
Use test points $w = 0$ and $w = 4$.

Test $w = 0$: $\dfrac{0+1}{0-3} = \dfrac{1}{-3} = -\dfrac{1}{3} > 1 \quad$ False

Test $w = 4$: $\dfrac{4+1}{4-3} = \dfrac{5}{1} = 5 > 1 \quad$ True

The boundary point is not included.

The solution is $\left\{ w \mid w > 3 \right\}$ or $(3, \infty)$.

95.
$$t^2 + 10t + 25 \le 0$$
$$(t+5)^2 \le 0$$

The quantity $(t+5)^2$ is greater than zero for all real numbers. except $t = -5$, for which it is zero. The solution is $\{-5\}$.

Chapter 7   Test

1.
$$(x+3)^2 = 25$$
$$x + 3 = \pm 5$$
$$x = -3 \pm 5$$
$$x = 2 \ \text{ or } \ x = -8 \quad \{2, -8\}$$

3.
$$(m+1)^2 = -1$$
$$m + 1 = \pm \sqrt{-1}$$
$$m + 1 = \pm i$$
$$m = -1 \pm i \quad \{-1 \pm i\}$$

5. $2x^2 + 12x - 36 = 0$

$x^2 + 6x - 18 = 0$

$x^2 + 6x = 18$

$x^2 + 6x + 9 = 18 + 9$

$(x+3)^2 = 27$

$x + 3 = \pm\sqrt{27}$

$x + 3 = \pm 3\sqrt{3}$

$x = -3 \pm 3\sqrt{3}$ $\left\{-3 \pm 3\sqrt{3}\right\}$

7. a. $x^2 - 3x = -12$

$x^2 - 3x + 12 = 0$

b. $a = 1,\ b = -3,\ c = 12$

c. $b^2 - 4ac = (-3)^2 - 4(1)(12)$

$= 9 - 48 = -39$

d. Two imaginary solutions

9. $3x^2 - 4x + 1 = 0$

$a = 3,\ b = -4,\ c = 1$

$x = \dfrac{-(-4) \pm \sqrt{(-4)^2 - 4(3)(1)}}{2(3)}$

$= \dfrac{4 \pm \sqrt{16 - 12}}{6} = \dfrac{4 \pm \sqrt{4}}{6} = \dfrac{4 \pm 2}{6}$

$x = \dfrac{6}{6} = 1$ or $x = \dfrac{2}{6} = \dfrac{1}{3}$ $\left\{1, \dfrac{1}{3}\right\}$

11. Let h = the height of the triangle

$2h - 3$ = the base of the triangle

$\dfrac{1}{2}(2h - 3)(h) = 14$

$(2h - 3)(h) = 28$

$2h^2 - 3h = 28$

$2h^2 - 3h - 28 = 0$

$a = 2,\ b = -3,\ c = -28$

$h = \dfrac{-(-3) \pm \sqrt{(-3)^2 - 4(2)(-28)}}{2(2)}$

$= \dfrac{3 \pm \sqrt{9 + 224}}{4} = \dfrac{3 \pm \sqrt{233}}{4}$

$h \approx 4.6$ or $h \approx -3.1$

$2h - 3 = 2(4.6) - 3 = 9.2 - 3 = 6.2$

The height is approximately 4.6 ft and the base is 6.2 ft.

13. $x - \sqrt{x} - 6 = 0$

Let $u = \sqrt{x}$

$u^2 - u - 6 = 0$

$(u - 3)(u + 2) = 0$

$u - 3 = 0$ or $u + 2 = 0$

$u = 3$ or $u = -2$

$\sqrt{x} = 3$ or $\sqrt{x} = -2$

$x = 9$ or $x = 4$

Check:

15. $(3y - 8)^2 - 13(3y - 8) + 30 = 0$

Let $u = 3y - 8$

$u^2 - 13u + 30 = 0$

$(u - 3)(u - 10) = 0$

$u - 3 = 0$ or $u - 10 = 0$

$3y - 8 - 3 = 0$ or $3y - 8 - 10 = 0$

$3y - 11 = 0$ or $3y - 18 = 0$

$3y = 11$ or $3y = 18$

$y = \dfrac{11}{3}$ or $y = 6$ $\left\{\dfrac{11}{3}, 6\right\}$

$x = 9:$ $x = 4:$

$9 - \sqrt{9} - 6 = 0$ $4 - \sqrt{4} - 6 = 0$

$9 - 3 - 6 = 0$ $4 - 2 - 6 = 0$

$0 = 0$ $-4 \neq 0$

Solution: $\{9\}$ ($x = 4$ does not check.)

17.
$$3 = \frac{y}{2} - \frac{1}{y+1}$$

$$3 \cdot 2(y+1) = \left(\frac{y}{2} - \frac{1}{y+1}\right) 2(y+1)$$

$$6y + 6 = y(y+1) - 2$$

$$6y + 6 = y^2 + y - 2$$

$$0 = y^2 - 5y - 8$$

$$a = 1,\ b = -5,\ c = -8$$

$$y = \frac{-(-5) \pm \sqrt{(-5)^2 - 4(1)(-8)}}{2(1)}$$

$$= \frac{5 \pm \sqrt{25 + 32}}{2} = \frac{5 \pm \sqrt{57}}{2} \quad \left\{\frac{5 \pm \sqrt{57}}{2}\right\}$$

19.
$$x^2 - 8x + 1 = 0$$

$$a = 1, b = -8, c = 1$$

$$x = \frac{-(-8) \pm \sqrt{(-8)^2 - 4(1)(1)}}{2(1)}$$

$$= \frac{8 \pm \sqrt{64 - 4}}{2}$$

$$= \frac{8 \pm \sqrt{60}}{2}$$

$$= \frac{8 \pm \sqrt{4(15)}}{2}$$

$$= \frac{8 \pm 2\sqrt{15}}{2} = 4 \pm \sqrt{15} \quad \left\{4 \pm \sqrt{15}\right\}$$

21.
$$x(x - 12) = -13$$

$$x^2 - 12x + 13 = 0$$

$$a = 1, b = -12, c = 13$$

$$x = \frac{-(-12) \pm \sqrt{(-12)^2 - 4(1)(13)}}{2(1)}$$

$$= \frac{12 \pm \sqrt{144 - 52}}{2}$$

$$= \frac{12 \pm \sqrt{92}}{2}$$

$$= \frac{12 \pm \sqrt{4(23)}}{2}$$

$$= \frac{12 \pm 2\sqrt{23}}{2} = 6 \pm \sqrt{23} \quad \left\{6 \pm \sqrt{23}\right\}$$

23. $h(x) = x^2 - 4$

Domain: $(-\infty, \infty)$; Range: $[-4, \infty)$

25.

$$g(x) = \frac{1}{2}(x+2)^2 - 3$$

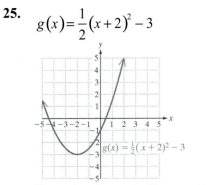

Domain: $(-\infty, \infty)$; Range: $[-3, \infty)$

27. a.

$$P(t) = 0.135t^2 + 12.6t + 600$$

$$P(40) = 0.135(40)^2 + 12.6(40) + 600$$

$$\approx 1320$$

The population in 2014 will be approximately 1,320,000,000.

b.

$$1000 = 0.135t^2 + 12.6t + 600$$

$$0.135t^2 + 12.6t - 400 = 0$$

$$t = \frac{-12.6 \pm \sqrt{(12.6)^2 - 4(0.135)(-400)}}{2(0.135)}$$

$$= \frac{-12.6 \pm \sqrt{158.76 + 216}}{0.27}$$

$$= \frac{-12.6 \pm \sqrt{374.76}}{0.27}$$

$$t \approx 25.03 \approx 25$$

The population reached 1 billion in 1999.

29. The graph of $y = (x+3)^2$ is the graph of $y = x^2$ shifted 3 units to the left.

31. a.

$$f(x) = -(x-4)^2 + 2$$

Vertex: $(4, 2)$

b. The parabola opens downward.

c. The vertex is the maximum point of the function.

d. The maximum value of the function is 2.

e. The axis of symmetry is $x = 4$.

33. a.

$$f(x) = x^2 + 4x - 12$$

$$f(x) = (x^2 + 4x + 4) - 12 - 4$$

$$f(x) = (x+2)^2 - 16$$

b. The vertex is $(-2, -16)$.

c.

$$x^2 + 4x - 12 = 0$$

$$(x-2)(x+6) = 0$$

$$x - 2 = 0 \quad \text{or} \quad x + 6 = 0$$

$$x = 2 \quad \text{or} \quad x = -6$$

The x-intercepts are: $(2, 0), (-6, 0)$

$$f(0) = 0^2 + 4(0) - 12 = 0 + 0 - 12 = -12$$

The y-intercept is: $(0, -12)$

d. The minimum value is -16.

35.

$$\frac{2x-1}{x-6} \leq 0$$

$$\frac{2x-1}{x-6} = 0$$

$$2x - 1 = 0$$

$$2x = 1$$

$$x = \frac{1}{2}$$

Boundary points are $\frac{1}{2}$ and 6 (undefined).
Use test points $x = 0$, $x = 1$, and $x = 7$.

e.   The axis of symmetry is $x = -2$.

Test $x = 0$: $\dfrac{2(0)-1}{0-6} = \dfrac{-1}{-6} = \dfrac{1}{6} \le 0$ False

Test $x = 1$: $\dfrac{2(1)-1}{1-6} = \dfrac{1}{-5} = -\dfrac{1}{5} \le 0$ True

Test $x = 7$: $\dfrac{2(7)-1}{7-6} = \dfrac{13}{1} = 13 \le 0$ False

The boundary point $x = \frac{1}{2}$ is included.

The solution is $\left\{ x \mid \dfrac{1}{2} \le x < 6 \right\}$ or $\left[\dfrac{1}{2}, 6 \right)$.

37. $y^3 + 3y^2 - 4y - 12 < 0$

$y^3 + 3y^2 - 4y - 12 = 0$

$y^2(y+3) - 4(y+3) = 0$

$(y+3)(y^2 - 4) = 0$

$(y+3)(y+2)(y-2) = 0$

$y+3 = 0$ or $y+2 = 0$ or $y-2 = 0$

$y = -3$ or $y = -2$ or $y = 2$

The boundary points are -3, -2, and 2. Use test points $y = -4$, $y = -2.5$, $y = 0$, and $y = 3$.

Test $y = -4$: $(-4)^3 + 3(-4)^2 - 4(-4) - 12 = -64 + 48 + 16 - 12 = -12 < 0$ True

Test $y = -2.5$: $(-2.5)^3 + 3(-2.5)^2 - 4(-2.5) - 12 = -15.625 + 18.75 + 10 - 12 = 1.125 < 0$ False

Test $y = 0$: $(0)^3 + 3(0)^2 - 4(0) - 12 = 0 + 0 - 0 - 12 = -12 < 0$ True

Test $y = 3$: $(3)^3 + 3(3)^2 - 4(3) - 12 = 27 + 27 - 12 - 12 = 30 < 0$ False

The boundary points are not included.

The solution is $\left\{ u \mid y < -3 \text{ or } -2 < y < 2 \right\}$ or $(-\infty, -3) \cup (-2, 2)$.

39. $5x^2 - 2x + 2 < 0$

$a = 5,\ b = -2,\ c = 2$

$x = \dfrac{-(-2) \pm \sqrt{(-2)^2 - 4(5)(2)}}{2(5)}$

$= \dfrac{2 \pm \sqrt{4 - 40}}{10} = \dfrac{2 \pm \sqrt{-36}}{10} = \dfrac{2 \pm 6i}{10}$

Since the solutions for the related equation are complex numbers, there are either no real number solutions or all real numbers as solutions. Test to check.

Test $x = 0$:

$5(0)^2 - 2(0) + 2 = 0 - 0 + 2 = 2 < 0$ False

There are no real number solutions. $\{\ \}$

Chapters 1 – 7   Cumulative Review Exercises

1. **a.** $A \cup B = \{2,4,6,8,10,12,16\}$

    b. $A \cap B = \{2,8\}$

3.

$$4^0 - \left(\frac{1}{2}\right)^{-3} - 81^{1/2} = 1 - 2^3 - \sqrt{81}$$
$$= 1 - 8 - 9 = -16$$

5. **a.** $x^3 + 2x^2 - 9x - 18 = x^2(x+2) - 9(x+2)$
$$= (x+2)(x^2-9)$$
$$= (x+2)(x+3)(x-3)$$

    b.

$$\begin{array}{r}
x^2 + 5x + 6 \\
x-3 \overline{) \ x^3 + 2x^2 - \ 9x - 18} \\
\underline{-\left(x^3 - 3x^2\right)} \\
5x^2 - \ 9x \\
\underline{-\left(5x^2 - 15x\right)} \\
6x - 18 \\
\underline{-(6x-18)} \\
0
\end{array}$$

Quotient: $x^2 + 5x + 6$ Remainder: 0

7.

$$\frac{4}{\sqrt{2x}} = \frac{4}{\sqrt{2x}} \cdot \frac{\sqrt{2x}}{\sqrt{2x}} = \frac{4\sqrt{2x}}{2x} = \frac{2\sqrt{2x}}{x}$$

9.   Multiply each equation by the LCD:

$$\frac{1}{9}x - \frac{1}{3}y = -\frac{13}{9} \ \rightarrow \ x - 3y = -13$$
$$x - \frac{1}{2}y = \frac{9}{2} \ \rightarrow \ 2x - y = 9$$

Multiply the first equation by –2, add to the second equation, and solve for y:

$$x - 3y = -13 \xrightarrow{\times -2} -2x + 6y = 26$$
$$2x - y = 9 \longrightarrow \underline{\ \ 2x - \ y = 9 \ \ }$$
$$5y = 35$$
$$y = 7$$

Substitute into the first equation and solve:

$$x - 3(7) = -13$$
$$x - 21 = -13$$
$$x = 8$$

The solution is $(8, 7)$.

11. $(x-3)^2 + 16 = 0$
$$(x-3)^2 = -16$$
$$x - 3 = \pm\sqrt{-16}$$
$$x - 3 = \pm 4i$$
$$x = 3 \pm 4i \quad \{3 \pm 4i\}$$

13. $x^2 + 10x + n$

$$n = \left(\frac{1}{2}b\right)^2 = \left(\frac{1}{2} \cdot (10)\right)^2 = (5)^2 = 25$$

$$x^2 + 10x + 25 = (x+5)^2$$

15. $3x - 5y = 10$

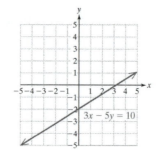

17. The domain element 3 has more than one corresponding range element.

19.

$$y = k\frac{x}{z}$$

$$15 = k\left(\frac{50}{10}\right)$$

$$15 = 5k$$

$$k = 3$$

$$y = 3\left(\frac{65}{5}\right) = 3(13) = 39$$

21. a. $m(x) = \sqrt{x+4}$

Domain: $[-4, \infty)$

b. $n(x) = x^2 + 2$

Domain: $(-\infty, \infty)$

23.

$$\sqrt{8x+5} = \sqrt{2x} + 2$$

$$\left(\sqrt{8x+5}\right)^2 = \left(\sqrt{2x}+2\right)^2$$

$$8x + 5 = 2x + 4\sqrt{2x} + 4$$

$$6x + 1 = 4\sqrt{2x}$$

$$\left(6x+1\right)^2 = \left(4\sqrt{2x}\right)^2$$

$$36x^2 + 12x + 1 = 16(2x)$$

$$36x^2 + 12x + 1 = 32x$$

$$36x^2 - 20x + 1 = 0$$

$$(18x-1)(2x-1) = 0$$

$$18x - 1 = 0 \quad \text{or} \quad 2x - 1 = 0$$

$$18x = 1 \quad \text{or} \quad 2x = 1$$

$$x = \frac{1}{18} \quad \text{or} \quad x = \frac{1}{2}$$

Check:

$$x = \frac{1}{18}: \quad \sqrt{8\left(\frac{1}{18}\right)+5} = \sqrt{2\left(\frac{1}{18}\right)} + 2$$

$$\sqrt{\frac{49}{9}} = \sqrt{\frac{1}{9}} + 2$$

$$\frac{7}{3} = \frac{1}{3} + 2$$

$$\frac{7}{3} = \frac{7}{3}$$

$$x = \frac{1}{2}: \quad \sqrt{8\left(\frac{1}{2}\right)+5} = \sqrt{2\left(\frac{1}{2}\right)} + 2$$

$$\sqrt{9} = \sqrt{1} + 2$$

$$3 = 1 + 2$$

$$3 = 3$$

Solutions: $\left\{\dfrac{1}{18}, \dfrac{1}{2}\right\}$

25.
$$\frac{15}{t^2 - 2t - 8} = \frac{1}{t-4} + \frac{2}{t+2}$$

LCD: $(t-4)(t+2)$

$$(t-4)(t+2)\left(\frac{15}{(t-4)(t+2)}\right)$$
$$= (t-4)(t+2)\left(\frac{1}{t-4} + \frac{2}{t+2}\right)$$
$$15 = 1(t+2) + 2(t-4)$$
$$15 = t + 2 + 2t - 8$$
$$15 = 3t - 6$$
$$3t = 21$$
$$t = 7 \quad \{7\}$$

27. a. $f(x) = 2(x-3)^2 + 1$

Vertex: $(3, 1)$

b.   The graph opens upward.

c. $f(0) = 2(0-3)^2 + 1 = 2(-3)^2 + 1$
$$= 18 + 1 = 19$$

y-intercept: $(0, 19)$

d. $2(x-3)^2 + 1 = 0$
$$2(x-3)^2 = -1$$
$$(x-3)^2 = -\frac{1}{2}$$
$$x - 3 = \pm\sqrt{-\frac{1}{2}}$$

There are no x-intercepts.

e.

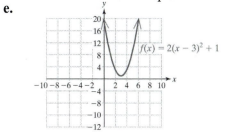

$$f(x) = 2(x-3)^2 + 1$$

29. a.
$$2x^2 + x - 10 \geq 0$$
$$2x^2 + x - 10 = 0$$
$$(2x+5)(x-2) = 0$$
$$2x + 5 = 0 \quad \text{or} \quad x - 2 = 0$$
$$2x = -5 \quad \text{or} \qquad x = 2$$
$$x = -\frac{5}{2} \quad \text{or} \qquad x = 2$$

The boundary points are $-\frac{5}{2}$ and 2.
Use test points $x = -3$, $x = 0$, and $x = 3$.
Test $x = -3$:

$$2(-3)^2 + (-3) - 10 = 18 - 3 - 10 = 5 \geq 0 \text{ True}$$

Test $x = 0$:

$$2(0)^2 + (0) - 10 = 0 + 0 - 10 = -10 \geq 0 \text{ False}$$

Test $x = 3$:

$$2(3)^2 + (3) - 10 = 18 + 3 - 10 = 11 \geq 0 \quad \text{True}$$

The boundary points are included.
The solution is

b. $\left\{ x \middle| x \le -\dfrac{5}{2} \text{ or } x \ge 2 \right\}$ or $\left(-\infty, -\dfrac{5}{2} \right] \cup [2, \infty)$.

On these intervals, the graph is on or above the x-axis.

Chapter 8   Exponential and Logarithmic Functions and Applications

Are You Prepared?

A. $5^x = 25$
$5^x = 5^2$
$x = 2$

B. $3^x = 81$
$3^x = 3^4$
$x = 4$

C. $8^{1/x} = 2$
$\left(2^3\right)^{1/x} = 2$
$2^{3/x} = 2^1$
$\dfrac{3}{x} = 1$
$3 = x$

D. $x^4 = 16$
$x^4 = 2^4$
$x = 2$

E. $2^x = \dfrac{1}{16}$
$2^x = 2^{-4}$
$x = -4$
$|x| = |-4| = 4$

F. $2^x = 64$
$2^x = 2^6$
$x = 6$

G. $5^0 = x$
$1 = x$

H. $e^0 = x$
$1 = x$

5	1	^{B}4	^{C}3	6	2
3	^{F}6	2	1	4	5
^{G}1	5	3	6	^{D}2	4
^{A}2	4	6	5	1	3
6	2	5	^{E}4	3	1
4	3	^{H}1	2	5	6

Section 8.1   Algebra of Functions and Composition

1. **a.** $f(x)$; $g(x)$
b. $g(x)$
c. $f(g(x))$

3. $(f+g)(x) = f(x) + g(x)$
$= (x+4) + (2x^2 + 4x)$
$= x + 4 + 2x^2 + 4x$
$= 2x^2 + 5x + 4$

5. $(g-f)(x)=g(x)-f(x)$

$\qquad =\left(2x^2+4x\right)-\left(x+4\right)$

$\qquad =2x^2+4x-x-4$

$\qquad =2x^2+3x-4$

7. $(f\cdot h)(x)=f(x)\cdot h(x)$

$\qquad =\left(x+4\right)\left(x^2+1\right)$

$\qquad =x^3+4x^2+x+4$

9. $(g\cdot f)(x)=g(x)\cdot f(x)$

$\qquad =\left(2x^2+4x\right)\left(x+4\right)$

$\qquad =2x^3+8x^2+4x^2+16x$

$\qquad =2x^3+12x^2+16x$

11. $\left(\dfrac{h}{f}\right)(x)=\dfrac{h(x)}{f(x)}=\dfrac{x^2+1}{x+4},x\neq -4$

13. $\left(\dfrac{f}{g}\right)(x)=\dfrac{f(x)}{g(x)}=\dfrac{x+4}{2x^2+4x},x\neq 0,x\neq -2$

15. $(f\circ g)(x)=f\left(g(x)\right)$

$\qquad =f\left(2x^2+4x\right)=2x^2+4x+4$

17. $(g\circ f)(x)=g\left(f(x)\right)=g\left(x+4\right)$

$\qquad =2\left(x+4\right)^2+4\left(x+4\right)$

$\qquad =2\left(x^2+8x+16\right)+4x+16$

$\qquad =2x^2+16x+32+4x+16$

$\qquad =2x^2+20x+48$

19. $(k\circ h)(x)=k\left(h(x)\right)=k\left(x^2+1\right)=\dfrac{1}{x^2+1}$

21. $(k\circ g)(x)=k\left(g(x)\right)=k\left(2x^2+4x\right)$

$\qquad =\dfrac{1}{2x^2+4x},\quad x\neq 0,x\neq -2$

23. No

25. $f(x)=x^2-3x+1\quad g(x)=5x$

$(f\circ g)(x)=f\left(g(x)\right)=f(5x)$

$\qquad =(5x)^2-3(5x)+1$

$\qquad =25x^2-15x+1$

$(g\circ f)(x)=g\left(f(x)\right)=g\left(x^2-3x+1\right)$

$\qquad =5\left(x^2-3x+1\right)$

$\qquad =5x^2-15x+5$

27. $h(x)=5x-4$

$(h\circ h)(x)=h\left(h(x)\right)=h\left(5x-4\right)$

$\qquad =5\left(5x-4\right)-4=25x-20-4$

$\qquad =25x-24$

29. $f(x)=|x|\quad g(x)=x^3-1$

$(f\circ g)(x)=f\left(g(x)\right)=f\left(x^3-1\right)$

$\qquad =\left|x^3-1\right|$

$(g\circ f)(x)=g\left(f(x)\right)=g\left(|x|\right)$

$\qquad =|x|^3-1$

31. $(m\cdot r)(0)=m(0)\cdot r(0)$

$\qquad =0^3\cdot\sqrt{0+4}=0\sqrt{4}=0$

33. $(m+r)(-4) = m(-4) + r(-4)$
$$= (-4)^3 + \sqrt{-4+4} = -64 + \sqrt{0}$$
$$= -64 + 0 = -64$$

35. $(r \circ n)(3) = r(n(3)) = r(3-3) = r(0)$
$$= \sqrt{0+4} = \sqrt{4} = 2$$

37. $(p \circ m)(-1) = p(m(-1)) = p((-1)^3) = p(-1)$
$$= \frac{1}{-1+2} = \frac{1}{1} = 1$$

39. $(m \circ p)(2) = m(p(2)) = m\left(\frac{1}{2+2}\right) = m\left(\frac{1}{4}\right)$
$$= \left(\frac{1}{4}\right)^3 = \frac{1}{64}$$

41. $(r+p)(-3) = r(-3) + p(-3)$
$$= \sqrt{-3+4} + \frac{1}{-3+2}$$
$$= \sqrt{1} + \frac{1}{-1} = 1 - 1 = 0$$

43. $(m \circ p)(-2) = m(p(-2)) = m\left(\frac{1}{-2+2}\right)$
$$= m\left(\frac{1}{0}\right) = \text{Undefined}$$

45. $\left(\frac{r}{n}\right)(12) = \frac{r(12)}{n(12)} = \frac{\sqrt{12+4}}{12-3} = \frac{\sqrt{16}}{9} = \frac{4}{9}$

47. $f(-4) = -2$

49. $g(-2) = 2$

51. $(f+g)(2) = f(2) + g(2) = 2 + (-2) = 0$

53. $(f \cdot g)(-1) = f(-1) \cdot g(-1) = 1 \cdot 1 = 1$

55. $\left(\frac{g}{f}\right)(0) = \frac{g(0)}{f(0)} = \frac{0}{2} = 0$

57. $\left(\frac{f}{g}\right)(0) = \frac{f(0)}{g(0)} = \frac{2}{0} = \text{Undefined}$

59. $(g \circ f)(-1) = g(f(-1)) = g(1) = -1$

61. $(f \circ g)(-4) = f(g(-4)) = f(2) = 2$

63. $(g \circ g)(2) = g(g(2)) = g(-2) = 2$

65. $a(-3) = -1$

67. $b(-1) = -2$

69. $(a-b)(-1) = a(-1) - b(-1) = 1 - (-2) = 3$

71. $(b \cdot a)(1) = b(1) \cdot a(1) = -2 \cdot 3 = -6$

73. $(b \circ a)(0) = b(a(0)) = b(2) = -1$

75. $(a \circ b)(-4) = a(b(-4)) = a(4) = 4$

77. $\left(\frac{b}{a}\right)(3) = \frac{b(3)}{a(3)} = \frac{0}{5} = 0$

79. $(a \circ a)(-2) = a(a(-2)) = a(0) = 2$

81. a. $P(x) = R(x) - C(x)$

$$= 5.98x - (2.2x + 1)$$
$$= 5.98x - 2.2x - 1$$
$$= 3.78x - 1$$

   b. $P(50) = 3.78(50) - 1 = 189 - 1 = \$188$

83. a. $F(t) = D(t) - R(t)$

$$= (0.925t + 26.958) - (0.725t + 20.558)$$
$$= 0.925t + 26.958 - 0.725t - 20.558$$
$$= 0.2t + 6.4$$

   F represents the amount of child support (in billion dollars) not paid.

   b. $F(4) = 0.2(4) + 6.4 = 7.2$ means that in 2004, \$7.2 billion of child support was not paid.

85. a. $(D \circ r)(t) = D(r(t)) = D(80t)$

$$= 7(80t) = 560t$$

   This function represents the total distance Joe travels as a function of time.

   b. $(D \circ r)(10) = 560(10) = 5600 \, \text{ft}$

Section 8.2   Inverse Functions

1. a. $\{(2, 1), (3, 2), (4, 3)\}$

   b. one-to-one; y

   c. is not

   d. is

   e. $y = x$

   f. x; x

   g. f^{-1}

   h. (b, a)

3. The relation is a function.

5. The relation is not a function.

7. The relation is a function.

9. $g = \{(3,5),(8,1),(-3,9),(0,2)\}$
$g^{-1} = \{(5,3),(1,8),(9,-3),(2,0)\}$

11. $r = \{(a,3),(b,6),(c,9)\}$
$r^{-1} = \{(3,a),(6,b),(9,c)\}$

13. The function is not one-to-one.

15. The function is one-to-one.

17. The function is not one-to-one.

19. The function is one-to-one.

21. $f(x) = 6x + 1$; $g(x) = \dfrac{x-1}{6}$

23. $f(x) = \dfrac{\sqrt[3]{x}}{2}$; $g(x) = 8x^3$

a.
$$(f \circ g)(x) = f(g(x)) = f\left(\frac{x-1}{6}\right)$$
$$= 6\left(\frac{x-1}{6}\right) + 1 = x - 1 + 1 = x$$

b.
$$(g \circ f)(x) = g(f(x)) = g(6x+1)$$
$$= \frac{(6x+1)-1}{6} = \frac{6x}{6} = x$$

a.
$$(f \circ g)(x) = f(g(x)) = f(8x^3)$$
$$= \frac{\sqrt[3]{8x^3}}{2} = \frac{2x}{2} = x$$

b.
$$(g \circ f)(x) = g(f(x)) = g\left(\frac{\sqrt[3]{x}}{2}\right)$$
$$= 8\left(\frac{\sqrt[3]{x}}{2}\right)^3 = \frac{8x}{8} = x$$

25. $f(x) = x^2 + 1, x \geq 0 \; ; g(x) = \sqrt{x-1}, x \geq 1$

a. $(f \circ g)(x) = f(g(x)) = f\left(\sqrt{x-1}\right)$
$$= \left(\sqrt{x-1}\right)^2 + 1 = x - 1 + 1 = x$$

b. $(g \circ f)(x) = g(f(x)) = g(x^2 + 1)$
$$= \sqrt{(x^2+1)-1} = \sqrt{x^2} = x$$

27.
$$h(x) = x + 4$$
$$y = x + 4$$
$$x = y + 4$$
$$x - 4 = y$$
$$h^{-1}(x) = x - 4$$

29.
$$m(x) = \frac{1}{3}x - 2$$
$$y = \frac{1}{3}x - 2$$
$$x = \frac{1}{3}y - 2$$
$$x + 2 = \frac{1}{3}y$$
$$3(x+2) = y$$
$$m^{-1}(x) = 3(x+2)$$

31.
$$p(x) = -x + 10$$
$$y = -x + 10$$
$$x = -y + 10$$
$$x - 10 = -y$$
$$-x + 10 = y$$
$$p^{-1}(x) = -x + 10$$

33.
$$n(x) = \frac{3x+2}{5}$$
$$y = \frac{3x+2}{5}$$
$$x = \frac{3y+2}{5}$$
$$5x = 3y + 2$$
$$5x - 2 = 3y$$
$$\frac{5x-2}{3} = y$$
$$n^{-1}(x) = \frac{5x-2}{3}$$

35.
$$h(x) = \frac{4x-1}{3}$$
$$y = \frac{4x-1}{3}$$
$$x = \frac{4y-1}{3}$$
$$3x = 4y - 1$$
$$3x + 1 = 4y$$
$$\frac{3x+1}{4} = y$$
$$h^{-1}(x) = \frac{3x+1}{4}$$

37. $f(x) = x^3 + 1$

   $y = x^3 + 1$

   $x = y^3 + 1$

   $x - 1 = y^3$

   $\sqrt[3]{x-1} = y$

   $f^{-1}(x) = \sqrt[3]{x-1}$

39. $g(x) = \sqrt[3]{2x-1}$

   $y = \sqrt[3]{2x-1}$

   $x = \sqrt[3]{2y-1}$

   $x^3 = 2y - 1$

   $x^3 + 1 = 2y$

   $\dfrac{x^3 + 1}{2} = y$

   $g^{-1}(x) = \dfrac{x^3 + 1}{2}$

41. $g(x) = x^2 + 9, x \geq 0$

   $y = x^2 + 9$

   $x = y^2 + 9$

   $x - 9 = y^2$

   $\sqrt{x-9} = y$

   $g^{-1}(x) = \sqrt{x-9}$

43.   **a.** $f(x) = 0.3048x$

   $f(4) = 0.3048(4) = 1.2192$ m

   $f(50) = 0.3048(50) = 15.24$ m

   b. $f(x) = 0.3048x$

   $y = 0.3048x$

   $x = 0.3048y$

   $\dfrac{x}{0.3048} = y$

   $f^{-1}(x) = \dfrac{x}{0.3048}$

   c. $f^{-1}(1500) = \dfrac{1500}{0.3048} = 4921.3$ ft

45.   False, $x = 2$ is not a function.

47.   True, any function of the form
   $f(x) = mx + b \, (m \neq 0)$ has an inverse.

49.   False, $k(1) = 1$ and $k(-1) = 1$.

51.   True

53. $(b, 0)$

55.   **a.** $f(x) = \sqrt{x-1}$
   Domain: $[1, \infty)$; Range: $[0, \infty)$

   b. $f^{-1}(x) = x^2 + 1, x \geq 0$
   Domain: $[0, \infty)$; Range: $[1, \infty)$

57.
  a.   Domain f: $[-4,0]$

  b.   Range f: $[0,2]$

  c.   Domain f^{-1}: $[0,2]$

  d.   Range f^{-1}: $[-4,0]$

  e.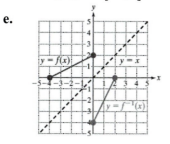

59.
  a.   Domain f: $[0,2]$

  b.   Range f: $[0,4]$

  c.   Domain f^{-1}: $[0,4]$

  d.   Range f^{-1}: $[0,2]$

  e.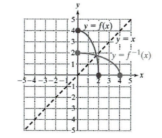

61.
$$q(x)=\sqrt{x+4}$$
$$y=\sqrt{x+4}$$
$$x=\sqrt{y+4}$$
$$x^2=y+4, x\geq 0$$
$$x^2-4=y, x\geq 0$$
$$q^{-1}(x)=x^2-4, x\geq 0$$

63.
$$z(x)=-\sqrt{x+4}$$
$$y=-\sqrt{x+4}$$
$$x=-\sqrt{y+4}$$
$$x^2=y+4, x\leq 0$$
$$x^2-4=y, x\leq 0$$
$$z^{-1}(x)=x^2-4, x\leq 0$$

65.
$$f(x)=\frac{x-1}{x+1}$$
$$y=\frac{x-1}{x+1}$$
$$x=\frac{y-1}{y+1}$$
$$x(y+1)=y-1$$
$$xy+x=y-1$$
$$xy-y=-x-1$$
$$y(x-1)=-x-1$$
$$y=\frac{-x-1}{x-1}=\frac{x+1}{1-x}$$
$$f^{-1}(x)=\frac{x+1}{1-x}$$

67.
$$t(x)=\frac{2}{x-1}$$
$$y=\frac{2}{x-1}$$
$$x=\frac{2}{y-1}$$
$$x(y-1)=2$$
$$xy-x=2$$
$$xy=x+2$$
$$y=\frac{x+2}{x}$$
$$t^{-1}(x)=\frac{x+2}{x}$$

373

69.

$$n(x) = x^2 + 9, x \le 0$$

$$y = x^2 + 9$$

$$x = y^2 + 9$$

$$x - 9 = y^2$$

$$-\sqrt{x - 9} = y$$

$$n^{-1}(x) = -\sqrt{x - 9}$$

Section 8.3   Exponential Functions

1.
  a. b^x
  b. is not; is
  c. increasing
  d. decreasing
  e. $(-\infty, \infty)$; $(0, \infty)$
  f. $(0, 1)$
  g. $y = 0$
  h. is not

3.
$$(g - f)(x) = g(x) - f(x)$$
$$= (3x - 1) - (2x^2 + x + 2)$$
$$= 3x - 1 - 2x^2 - x - 2$$
$$= -2x^2 + 2x - 3$$

5.
$$\left(\frac{g}{f}\right)(x) = \frac{g(x)}{f(x)} = \frac{3x - 1}{2x^2 + x + 2}$$

7.
$$(g \circ f)(x) = g(f(x)) = g(2x^2 + x + 2)$$
$$= 3(2x^2 + x + 2) - 1$$
$$= 6x^2 + 3x + 6 - 1$$
$$= 6x^2 + 3x + 5$$

9. $5^2 = 5 \cdot 5 = 25$

11.
$$10^{-3} = \frac{1}{10^3} = \frac{1}{10 \cdot 10 \cdot 10} = \frac{1}{1000}$$

13. $36^{1/2} = \sqrt{36} = 6$

15.
$$16^{3/4} = \left(16^{1/4}\right)^3 = \left(\sqrt[4]{16}\right)^3 = 2^3 = 8$$

17. $5^{1.1} \approx 5.8731$

19. $10^\pi \approx 1385.4557$

21. $36^{-\sqrt{2}} \approx 0.0063$

23. $16^{-0.04} \approx 0.8950$

25.
  a. $3^x = 9$
$$3^x = 3^2$$
$$x = 2$$
  b. $3^x = 27$
$$3^x = 3^3$$
$$x = 3$$
  c. Between 2 and 3

27.
  a. $2^x = 16$
$$2^x = 2^4$$
$$x = 4$$
  b. $2^x = 32$
$$2^x = 2^5$$
$$x = 5$$
  c. Between 4 and 5

29.
$$f(x) = \left(\tfrac{1}{5}\right)^x$$
$$f(0) = \left(\tfrac{1}{5}\right)^0 = 1$$
$$f(1) = \left(\tfrac{1}{5}\right)^1 = \frac{1}{5}$$
$$f(2) = \left(\tfrac{1}{5}\right)^2 = \frac{1}{25}$$
$$f(-1) = \left(\tfrac{1}{5}\right)^{-1} = 5$$
$$f(-2) = \left(\tfrac{1}{5}\right)^{-2} = 5^2 = 25$$

31.
$$h(x) = 3^x$$
$$h(0) = 3^0 = 1$$
$$h(1) = 3^1 = 3$$
$$h(-1) = 3^{-1} = \frac{1}{3} \approx 0.33$$
$$h\left(\sqrt{2}\right) = 3^{\sqrt{2}} \approx 4.73$$
$$h(\pi) = 3^\pi \approx 31.54$$

33. If $b > 1$, the graph is increasing. If $0 < b < 1$, the graph is decreasing.

35. $f(x) = 4^x$

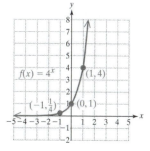

37.
$$m(x) = \left(\frac{1}{8}\right)^x$$

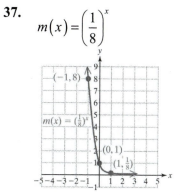

39. $h(x) = 2^{x+1}$

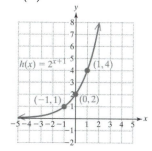

41. $g(x) = 5^{-x}$

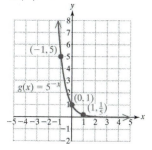

43.
$$A(t) = (0.5)^{t/3.8}$$

a. $A(7.6) = (0.5)^{7.6/3.8} = (0.5)^2 = 0.25\,\text{g}$

b. $A(10) = (0.5)^{10/3.8} \approx 0.16\,\text{g}$

45. $A(t) = 1,000,000(2)^{-t/5}$

    a. $A(2) = 1,000,000(2)^{-2/5} \approx 758,000$

    b. $A(7) = 1,000,000(2)^{-7/5} \approx 379,000$

    c. $A(14) = 1,000,000(2)^{-14/5} \approx 144,000$

47.   **a.** $P(t) = P_0(1+r)^t$

        $P(t) = 153,000,000(1+0.0125)^t$

        $P(t) = 153,000,000(1.0125)^t$

    b. $P(41) = 153,000,000(1.0125)^{41}$

        $\approx 255,000,000$

49. $A(t) = 1000(2)^{t/7}$

    a. $A(5) = 1000(2)^{5/7} = \$1640.67$

    b. $A(10) = 1000(2)^{10/7} = \$2691.80$

    c. $A(0) = 1000(2)^{0/7} = \$1000$

    The initial amount of the investment is $1000.

    $A(7) = 1000(2)^{7/7} = \$2000$

    The amount of the investment doubles in 7 years.

Section 8.4   Logarithmic Functions

1.   **a.**   logarithmic; b

    b.   logarithm; base; argument

    c. $(0, \infty)$; $(-\infty, \infty)$

    d.   exponential

    e.   common

    f.   increasing; decreasing

    g.   2, 3, and 4

    h. $y = x$

3.   Graph i is increasing.

5.   **a.** $s(x) = \left(\frac{2}{5}\right)^x$

        $s(-2) = \left(\frac{2}{5}\right)^{-2} = \left(\frac{5}{2}\right)^2 = \frac{25}{4}$

        $s(-1) = \left(\frac{2}{5}\right)^{-1} = \frac{5}{2}$

        $s(0) = \left(\frac{2}{5}\right)^0 = 1$

        $s(1) = \left(\frac{2}{5}\right)^1 = \frac{2}{5}$

        $s(2) = \left(\frac{2}{5}\right)^2 = \frac{4}{25}$

7. $g(x) = 3^x$

    $g(-2) = 3^{-2} = \frac{1}{3^2} = \frac{1}{9}$

    $g(-1) = 3^{-1} = \frac{1}{3}$

    $g(0) = 3^0 = 1$

    $g(1) = 3^1 = 3$

    $g(2) = 3^2 = 9$

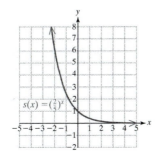

9. $y = \log_b x \leftrightarrow b^y = x$

11. $\log_5 625 = 4 \leftrightarrow 5^4 = 625$

13. $\log_{10} 0.0001 = -4 \leftrightarrow 10^{-4} = 0.0001$

15. $\log_6 36 = 2 \leftrightarrow 6^2 = 36$

17. $\log_b 15 = x \leftrightarrow b^x = 15$

19. $\log_3 5 = x \leftrightarrow 3^x = 5$

21. $\log_{\frac{1}{4}} x = 10 \leftrightarrow \left(\dfrac{1}{4}\right)^{10} = x$

23. $3^x = 81 \leftrightarrow \log_3 81 = x$

25. $5^2 = 25 \leftrightarrow \log_5 25 = 2$

27. $7^{-1} = \dfrac{1}{7} \leftrightarrow \log_7\left(\dfrac{1}{7}\right) = -1$

29. $b^x = y \leftrightarrow \log_b y = x$

31. $e^x = y \leftrightarrow \log_e y = x$

33. $\left(\dfrac{1}{3}\right)^{-2} = 9 \leftrightarrow \log_{\frac{1}{3}} 9 = -2$

35. $y = \log_7 49$
$7^y = 49$
$7^y = 7^2$
$y = 2$

37. $y = \log_{10} 0.1$
$10^y = 0.1$
$10^y = 10^{-1}$
$y = -1$

39. $y = \log_{16} 4$
$16^y = 4$
$\left(4^2\right)^y = 4^1$
$4^{2y} = 4^1$
$2y = 1$
$y = \dfrac{1}{2}$

41. $y = \log_{\frac{7}{2}} 1$
$\left(\dfrac{7}{2}\right)^y = 1$
$\left(\dfrac{7}{2}\right)^y = \left(\dfrac{7}{2}\right)^0$
$y = 0$

43. $y = \log_3 3^5$
$3^y = 3^5$
$y = 5$

45. $y = \log_{10} 10$

$10^y = 10$

$10^y = 10^1$

$y = 1$

47. $y = \log_a \left(a^3\right)$

$a^y = a^3$

$y = 3$

49. $y = \log_x \sqrt{x}$

$x^y = \sqrt{x}$

$x^y = x^{1/2}$

$y = \dfrac{1}{2}$

51. $y = \log 10$

$10^y = 10$

$10^y = 10^1$

$y = 1$

53. $y = \log 1000$

$10^y = 1000$

$10^y = 10^3$

$y = 3$

55. $y = \log\left(1.0 \times 10^6\right)$

$10^y = 1.0 \times 10^6$

$10^y = 10^6$

$y = 6$

57. $y = \log 0.01$

$10^y = 0.01$

$10^y = 10^{-2}$

$y = -2$

59. $\log 6 \approx 0.7782$

61. $\log \pi \approx 0.4971$

63. $\log\left(\dfrac{1}{32}\right) \approx -1.5051$

65. $\log\left(0.0054\right) \approx -2.2676$

67. $\log\left(3.4 \times 10^5\right) \approx 5.5315$

69. $\log\left(3.8 \times 10^{-8}\right) \approx -7.4202$

71. **a.** log 93 is slightly less than 2.

b. log 12 is slightly more than 1.

c. $\log 93 \approx 1.9685$

$\log 12 \approx 1.0792$

73. a. $f(x) = \log_4(x)$

$f\left(\tfrac{1}{64}\right) = \log_4\left(\tfrac{1}{64}\right) = x$

$4^x = \dfrac{1}{64}$

$4^x = 4^{-3}$

$x = -3$

$f\left(\tfrac{1}{16}\right) = \log_4\left(\tfrac{1}{16}\right) = x$

$4^x = \dfrac{1}{16}$

$4^x = 4^{-2}$

$x = -2$

$f\left(\tfrac{1}{4}\right) = \log_4\left(\tfrac{1}{4}\right) = x$

$4^x = \dfrac{1}{4}$

$4^x = 4^{-1}$

$x = -1$

$f(1) = \log_4(1) = x$

$4^x = 1$

$4^x = 4^0$

$x = 0$

$f(4) = \log_4(4) = x$

$4^x = 4$

$4^x = 4^1$

$x = 1$

$f(16) = \log_4(16) = x$

$4^x = 16$

$4^x = 4^2$

$x = 2$

$f(64) = \log_4(64) = x$

$4^x = 64$

$4^x = 4^3$

$x = 3$

b.

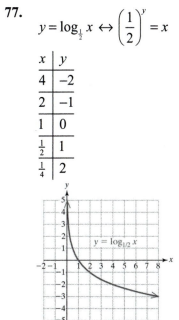

75. $y = \log_3 x \leftrightarrow 3^y = x$

x	y
$\tfrac{1}{9}$	-2
$\tfrac{1}{3}$	-1
1	0
3	1
9	2

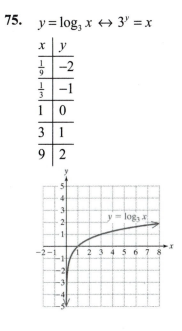

77. $y = \log_{\frac{1}{2}} x \leftrightarrow \left(\dfrac{1}{2}\right)^y = x$

x	y
4	-2
2	-1
1	0
$\tfrac{1}{2}$	1
$\tfrac{1}{4}$	2

79. $y = \log_7(x-5)$

$x - 5 > 0$

$\quad x > 5 \quad (5, \infty)$

81. $g(x) = \log(2-x)$

$2 - x > 0$

$\quad -x > -2$

$\qquad x < 2 \quad (-\infty, 2)$

83. $y = \log(3x-1)$

$3x - 1 > 0$

$\quad 3x > 1$

$\qquad x > \dfrac{1}{3} \quad \left(\dfrac{1}{3}, \infty\right)$

85. $y = \log_3(x+1.2)$

$x + 1.2 > 0$

$\qquad x > -1.2 \quad (-1.2, \infty)$

87. $k(x) = \log(4-2x)$

$4 - 2x > 0$

$\quad -2x > -4$

$\qquad x < 2 \quad (-\infty, 2)$

89. $y = \log(x^2)$

$x^2 > 0$

$\quad x > 0 \text{ or } x < 0 \quad (-\infty, 0) \cup (0, \infty)$

91. $\text{pH} = -\log(4.47 \times 10^{-8}) \approx -(-7.35) \approx 7.35$

93. a.

t	0	1	2	6	12	24
$S_1(t)$	91	82.0	76.7	65.6	57.6	49.1
$S_2(t)$	88	83.5	80.8	75.3	71.3	67.0

b. Group 1: 91

Group 2: 88

c. Method II

Problem Recognition Exercises

1. $g(x) = 3^x$ e

3. $h(x) = x^2$ j

5. $L(x) = |x|$ c

7. $B(x) = 3$ k

9. $n(x) = \sqrt[3]{x}$ i

11. $q(x) = \dfrac{1}{x}$ f

Section 8.5   Properties of Logarithms

1. a. $1; 0; x; x$

   b. $\log_b x + \log_b y$; $\log_b x - \log_b y$ $(0, \infty)$;

   c. $p \log_b x$

   d. False

   e. False

   f. False

3. $y = \log 10{,}000$

$10^y = 10{,}000$

$10^y = 10^4$

$\quad y = 4$

5. $6^{-1} = \dfrac{1}{6}$

7. $\log 8 \approx 0.9031$

9. $\pi^{\sqrt{2}} \approx 5.0475$

11. $q(x) = \left(\dfrac{1}{5}\right)^{x}$ a

13. $k(x) = \log_{\frac{1}{3}} x$ c

15. a, b, c

17. $\log_3 3 = 1$

19. $\log_5 \left(5^4\right) = 4$

21. $6^{\log_6 11} = 11$

23. $\log\left(10^3\right) = 3$

25. $\log_3 1 = 0$

27. $10^{\log 9} = 9$

29. $\log_{\frac{1}{2}} 1 = 0$

31. $\log_2 1 + \log_2\left(2^3\right) = 0 + 3 = 3$

33. $\log_4 4 + \log_2 1 = 1 + 0 = 1$

35. $\log_{\frac{1}{4}}\left(\dfrac{1}{4}\right)^{2x} = 2x$

37. $\log_a\left(a^4\right) = 4$

39. $\log 10^2 - \log_3 3^2 = 2 - 2 = 0$

41.
 a. $\log(3 \cdot 5) \approx 1.1761$
 b. $\log 3 \cdot \log 5 \approx 0.3335$
 c. $\log 3 + \log 5 \approx 1.1761$
 Expressions a and c are equivalent.

43.
 a. $\log\left(20^2\right) \approx 2.6021$
 b. $\left[\log 20\right]^2 \approx 1.6927$
 c. $2\log 20 \approx 2.6021$
 Expressions a and c are equivalent.

45. $\log_3\left(\dfrac{x}{5}\right) = \log_3 x - \log_3 5$

47. $\log(2x) = \log 2 + \log x$

49. $\log_5\left(x^4\right) = 4\log_5 x$

51. $\log_4\left(\dfrac{ab}{c}\right) = \log_4(ab) - \log_4 c$

$= \log_4 a + \log_4 b - \log_4 c$

53. $\log_b\left(\dfrac{\sqrt{x}\,y}{z^3 w}\right) = \log_b\left(\sqrt{x}\,y\right) - \log_b\left(z^3 w\right)$

$= \log_b x^{1/2} + \log_b y - \left(\log_b z^3 + \log_b w\right)$

$= \dfrac{1}{2}\log_b x + \log_b y - \left(3\log_b z + \log_b w\right)$

$= \dfrac{1}{2}\log_b x + \log_b y - 3\log_b z - \log_b w$

55. $\log_2\left(\dfrac{x+1}{y^2 \sqrt{z}}\right) = \log_2(x+1) - \log_2\left(y^2 \sqrt{z}\right)$

$= \log_2(x+1) - \left(\log_2 y^2 + \log_2 z^{1/2}\right)$

$= \log_2(x+1) - \left(2\log_2 y + \dfrac{1}{2}\log_2 z\right)$

$= \log_2(x+1) - 2\log_2 y - \dfrac{1}{2}\log_2 z$

57.

$$\log\left(\sqrt[3]{\frac{ab^2}{c}}\right) = \log\left(\frac{ab^2}{c}\right)^{1/3} = \frac{1}{3}\log\left(\frac{ab^2}{c}\right)$$

$$= \frac{1}{3}\left(\log\left(ab^2\right) - \log c\right)$$

$$= \frac{1}{3}\left(\log a + \log b^2 - \log c\right)$$

$$= \frac{1}{3}\left(\log a + 2\log b - \log c\right)$$

$$= \frac{1}{3}\log a + \frac{2}{3}\log b - \frac{1}{3}\log c$$

59.

$$\log\left(\frac{1}{w^5}\right) = \log 1 - \log w^5 = 0 - 5\log w$$

$$= -5\log w$$

61.

$$\log_b\left(\frac{\sqrt{a}}{b^3 c}\right) = \log_b \sqrt{a} - \log_b\left(b^3 c\right)$$

$$= \log_b a^{1/2} - \left(\log_b b^3 + \log_b c\right)$$

$$= \frac{1}{2}\log_b a - \left(3\log_b b + \log_b c\right)$$

$$= \frac{1}{2}\log_b a - 3\cdot 1 - \log_b c$$

$$= \frac{1}{2}\log_b a - 3 - \log_b c$$

63.

$$\log_3 270 - \log_3 2 - \log_3 5$$

$$= \log_3 270 - \log_3\left(2\cdot 5\right)$$

$$= \log_3 270 - \log_3 10$$

$$= \log_3\left(\frac{270}{10}\right)$$

$$= \log_3 27 = \log_3 3^3 = 3$$

65.

$$\log_7 98 - \log_7 2 = \log_7\left(\frac{98}{2}\right)$$

$$= \log_7 49 = \log_7 7^2 = 2$$

67.

$$2\log_3 x - 3\log_3 y + \log_3 z$$

$$= \log_3 x^2 - \log_3 y^3 + \log_3 z$$

$$= \log_3\left(x^2 z\right) - \log_3 y^3$$

$$= \log_3\left(\frac{x^2 z}{y^3}\right)$$

69.

$$2\log_3 a - \frac{1}{4}\log_3 b + \log_3 c$$

$$= \log_3 a^2 - \log_3 b^{1/4} + \log_3 c$$

$$= \log_3\left(a^2 c\right) - \log_3 b^{1/4}$$

$$= \log_3\left(\frac{a^2 c}{\sqrt[4]{b}}\right)$$

71.

$$\log_b x - 3\log_b x + 4\log_b x = 2\log_b x$$

$$= \log_b\left(x^2\right)$$

73. $5\log_8 a - \log_8 1 + \log_8 8 = 5\log_8 a - 0 + 1$
$$= \log_8(a^5) + 1$$

75. $2\log(x+6) + \dfrac{1}{3}\log y - 5\log z$
$$= \log(x+6)^2 + \log y^{1/3} - \log z^5$$
$$= \log\left[(x+6)^2 y^{1/3}\right] - \log z^5$$
$$= \log\left[\frac{(x+6)^2 \sqrt[3]{y}}{z^5}\right]$$

77. $\log_b(x+1) - \log_b(x^2-1) = \log_b\left(\dfrac{x+1}{x^2-1}\right)$
$$= \log_b\left(\frac{x+1}{(x+1)(x-1)}\right)$$
$$= \log_b\left(\frac{1}{x-1}\right)$$

79. $\log_b 6 = \log_b(3\cdot 2) = \log_b 3 + \log_b 2$
$$\approx 1.099 + 0.693 \approx 1.792$$

81. $\log_b 12 = \log_b(3\cdot 2^2) = \log_b 3 + \log_b 2^2$
$$= \log_b 3 + 2\log_b 2$$
$$\approx 1.099 + 2(0.693) \approx 2.485$$

83. $\log_b 81 = \log_b(3^4) = 4\log_b 3$
$$\approx 4(1.099) \approx 4.396$$

85. $\log_b\left(\dfrac{5}{2}\right) = \log_b 5 - \log_b 2$
$$\approx 1.609 - 0.693 \approx 0.916$$

87. $\log_b(10^6) = 6\log_b 10 = 6\log_b(2\cdot 5)$
$$= 6(\log_b 2 + \log_b 5)$$
$$\approx 6(0.693 + 1.609) \approx 13.812$$

89. $\log_b(5^{10}) = 10\log_b 5 \approx 10(1.609) \approx 16.09$

91. **a.**
$$B = 10\log\left(\frac{I}{I_0}\right) = 10(\log I - \log I_0)$$
$$= 10\log I - 10\log I_0$$

b. $B = 10\log I - 10\log 10^{-16}$
$$= 10\log I - 10\cdot(-16) = 10\log I + 160$$

Section 8.6   The Irrational Number e and Change of Base

1. **a.** e
   b. e
   c. natural; $\ln x$
   d. $0;\ 1;\ p;\ x$
   e. $\ln x + \ln y$; $\ln x - \ln y$
   f. $p\ln x$
   g. $\log_a b$

3. $\log_3 a - 5\log_3 c + \dfrac{1}{2}\log_3 d$
$$= \log_3 a - \log_3 c^5 + \log_3 d^{1/2}$$
$$= \log_3 ad^{1/2} - \log_3 c^5 = \log_3\left(\frac{a\sqrt{d}}{c^5}\right)$$

5.

$$\log_6 \sqrt[4]{\frac{xy^2}{z^3}} = \log_6\left(\frac{xy^2}{z^3}\right)^{1/4} = \frac{1}{4}\log_6\left(\frac{xy^2}{z^3}\right)$$

$$= \frac{1}{4}\left(\log_6 x + \log_6 y^2 - \log_6 z^3\right)$$

$$= \frac{1}{4}\left(\log_6 x + 2\log_6 y - 3\log_6 z\right)$$

$$= \frac{1}{4}\log_6 x + \frac{1}{2}\log_6 y - \frac{3}{4}\log_6 z$$

7. $y = e^{x+1}$

x	y
−4	0.05
−3	0.14
−2	0.37
−1	1
0	2.72
1	7.39

Domain: $(-\infty, \infty)$

9. $y = e^x + 2$

x	y
−2	2.14
−1	2.37
0	3
1	4.72
2	9.39
3	22.09

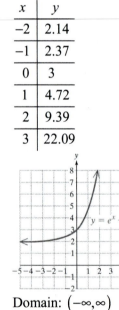

Domain: $(-\infty, \infty)$

11.

a. $A(5) = 10000\left(1 + \dfrac{0.04}{12}\right)^{12(5)} = \$12,209.97$

b. $A(5) = 10000\left(1 + \dfrac{0.06}{12}\right)^{12(5)} = \$13,488.50$

c. $A(5) = 10000\left(1 + \dfrac{0.08}{12}\right)^{12(5)} = \$14,898.46$

d. $A(5) = 10000\left(1 + \dfrac{.095}{12}\right)^{12(5)} = \$16,050.09$

An investment grows more rapidly at higher interest rates.

13.

a. $A(10) = 8000\left(1 + \dfrac{.045}{1}\right)^{1(10)} = \$12,423.76$

b. $A(10) = 8000\left(1 + \dfrac{.045}{4}\right)^{4(10)} = \$12,515.01$

c. $A(10) = 8000\left(1 + \dfrac{.045}{12}\right)^{12(10)} = \$12,535.94$

d. $A(10) = 8000\left(1 + \dfrac{.045}{365}\right)^{365(10)} = \$12,546.15$

e. $A(10) = 8000e^{0.045(10)} = \$12,546.50$

More money is earned at a greater number of compound periods per year.

15.

a. $A(5) = 5000e^{0.065(5)} = \$6920.15$

b. $A(10) = 5000e^{0.065(10)} = \$9577.70$

c. $A(15) = 5000e^{0.065(15)} = \$13,255.84$

d. $A(20) = 5000e^{0.065(20)} = \$18,346.48$

e. $A(30) = 5000e^{0.065(30)} = \$35,143.44$

More money is earned over a longer period of time.

17. $y = \ln(x - 2)$

x	y
2.25	−1.39
2.50	−0.69
2.75	−0.29
3	0
4	0.69
5	1.10
6	1.39

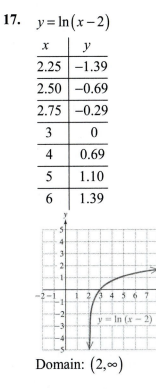

Domain: $(2, \infty)$

19. $y = \ln(x) - 1$

x	y
0.25	−2.39
0.50	−1.69
0.75	−1.29
1	−1.00
2	−0.31
3	0.10
4	0.39

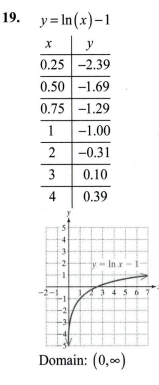

Domain: $(0, \infty)$

21. a. $f(x) = 10^x$ and $g(x) = \log x$

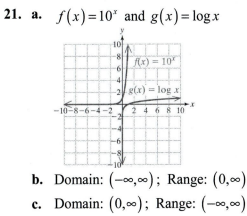

b. Domain: $(-\infty, \infty)$; Range: $(0, \infty)$

c. Domain: $(0, \infty)$; Range: $(-\infty, \infty)$

23. $\ln e = 1$

25. $\ln 1 = 0$

27. $\ln e^{-6} = -6$

29. $e^{\ln(2x+3)} = 2x + 3$

31.
$$6\ln p + \frac{1}{3}\ln q = \ln p^6 + \ln q^{1/3}$$
$$= \ln\left(p^6 q^{1/3}\right) = \ln\left(p^6 \sqrt[3]{q}\right)$$

33.
$$\frac{1}{2}\left(\ln x - 3\ln y\right) = \frac{1}{2}\left(\ln x - \ln y^3\right)$$
$$= \frac{1}{2}\ln\left(\frac{x}{y^3}\right) = \ln\left(\frac{x}{y^3}\right)^{1/2} = \ln\sqrt{\frac{x}{y^3}}$$

35.
$$2\ln a - \ln b - \frac{1}{3}\ln c = 2\ln a - \left(\ln b + \frac{1}{3}\ln c\right)$$
$$= \ln a^2 - \left(\ln b + \ln c^{1/3}\right)$$
$$= \ln a^2 - \ln\left(b \cdot c^{1/3}\right)$$
$$= \ln a^2 - \ln\left(b\sqrt[3]{c}\right)$$
$$= \ln\left(\frac{a^2}{b\sqrt[3]{c}}\right)$$

37.
$$4\ln x - 3\ln y - \ln z = 4\ln x - \left(3\ln y + \ln z\right)$$
$$= \ln x^4 - \left(\ln y^3 + \ln z\right)$$
$$= \ln x^4 - \ln\left(y^3 z\right)$$
$$= \ln\left(\frac{x^4}{y^3 z}\right)$$

39.
$$\ln\left(\frac{a}{b}\right)^2 = 2\ln\left(\frac{a}{b}\right) = 2\left(\ln a - \ln b\right)$$
$$= 2\ln a - 2\ln b$$

41. $\ln\left(b^2 \cdot e\right) = \ln b^2 + \ln e = 2\ln b + 1$

43.
$$\ln\left(\frac{a^4\sqrt{b}}{c}\right) = \ln\left(a^4\sqrt{b}\right) - \ln c$$
$$= \ln a^4 + \ln b^{1/2} - \ln c$$
$$= 4\ln a + \frac{1}{2}\ln b - \ln c$$

45.
$$\ln\left(\frac{ab}{c^2}\right)^{1/5} = \frac{1}{5}\ln\left(\frac{ab}{c^2}\right)$$
$$= \frac{1}{5}\left(\ln(ab) - \ln c^2\right)$$
$$= \frac{1}{5}\left(\ln a + \ln b - 2\ln c\right)$$
$$= \frac{1}{5}\ln a + \frac{1}{5}\ln b - \frac{2}{5}\ln c$$

47. **a.** $\log_6 200 = \dfrac{\log 200}{\log 6} \approx \dfrac{2.3010}{0.7782} \approx 2.9570$

b. $\log_6 200 = \dfrac{\ln 200}{\ln 6} \approx \dfrac{5.2983}{1.7918} \approx 2.9570$

c. They are the same.

49. $\log_2 7 = \dfrac{\log 7}{\log 2} \approx 2.8074$

51. $\log_8 24 = \dfrac{\log 24}{\log 8} \approx 1.5283$

53. $\log_8(0.012) = \dfrac{\log 0.012}{\log 8} \approx -2.1269$

55. $\log_9 1 = \dfrac{\log 1}{\log 9} = \dfrac{0}{\log 9} = 0$

57. $\log_4\left(\dfrac{1}{100}\right) = \dfrac{\log\left(\dfrac{1}{100}\right)}{\log 4} \approx -3.3219$

59. $\log_7(0.0006) = \dfrac{\log(0.0006)}{\log 7} \approx -3.8124$

61. a. $t = \dfrac{\ln 2}{0.045} \approx 15.4$ years

   b. $t = \dfrac{\ln 2}{0.10} \approx 6.9$ years

   c. Since the investment is doubled twice, the time would be 13.8 years.

63. a. $t = \dfrac{\ln 2}{0.035} \approx 19.8$ years

   b. $t = \dfrac{\ln 2}{0.05} \approx 13.9$ years

   c. Since the investment is doubled twice, the time would be 27.8 years.

Problem Recognition Exercises

1. $2^5 = 32$ $\log_2 32 = 5$

3. $z^y = x$ $\log_z x = y$

5. $10^3 = 1000$ $\log 1000 = 3$

7. $e^a = b$ $\ln b = a$

9. $\left(\tfrac{1}{2}\right)^2 = \tfrac{1}{4}$ $\log_{\frac{1}{2}}\left(\tfrac{1}{4}\right) = 2$

11. $10^{-2} = 0.01$ $\log 0.01 = -2$

13. $e^0 = 1$ $\ln 1 = 0$

15. $25^{\frac{1}{2}} = 5$ $\log_{25} 5 = \tfrac{1}{2}$

17. $e^t = s$ $\ln s = t$

19. $15^{-2} = \tfrac{1}{225}$ $\log_{15}\left(\tfrac{1}{225}\right) = -2$

Section 8.7   Logarithmic and Exponential Equations and Applications

1. a. $x = y$
   b. $x = y$

3. $\log_b x + \log_b(2x+3) = \log_b\left[x(2x+3)\right]$

5. $\log_b(x+2) - \log_b(3x-5) = \log_b\left(\dfrac{x+2}{3x-5}\right)$

7. $\log_3 x = 2$
$\quad x = 3^2 = 9 \quad \{9\}$

9. $\log p = 42$
$\quad p = 10^{42} \quad \{10^{42}\}$

11. $\ln x = 0.08$
$\quad x = e^{0.08} \quad \{e^{0.08}\}$

13. $\log(x+40) = -9.2$
$\quad 10^{-9.2} = x + 40$
$\quad\quad x = 10^{-9.2} - 40 \quad \{10^{-9.2} - 40\}$

15. $\log_x 25 = 2 \quad (x > 0)$
$\quad x^2 = 25$
$\quad\quad x = \sqrt{25} = 5 \quad \{5\}$

17. $\log_b 10,000 = 4 \quad (b > 0)$

$\qquad b^4 = 10,000$

$\qquad b^4 = 10^4$

$\qquad b = 10 \quad \{10\}$

19. $\log_y 5 = \dfrac{1}{2} \quad (y > 0)$

$\qquad y^{1/2} = 5$

$\qquad y = 5^2 = 25 \quad \{25\}$

21. $\log_4 (c + 5) = 3$

$\qquad c + 5 = 4^3$

$\qquad c + 5 = 64$

$\qquad c = 59 \quad \{59\}$

23. $\log_5 (4y + 1) = 1$

$\qquad 4y + 1 = 5^1$

$\qquad 4y + 1 = 5$

$\qquad 4y = 4$

$\qquad y = 1 \quad \{1\}$

25. $\ln(1 - x) = 0$

$\qquad 1 - x = e^0$

$\qquad 1 - x = 1$

$\qquad -x = 0$

$\qquad x = 0 \quad \{0\}$

27. $\log_3 8 - \log_3 (x + 5) = 2$

$\qquad \log_3 \left(\dfrac{8}{x + 5} \right) = 2$

$\qquad \dfrac{8}{x + 5} = 3^2$

$\qquad \dfrac{8}{x + 5} = 9$

$\qquad 8 = 9(x + 5)$

$\qquad 8 = 9x + 45$

$\qquad -37 = 9x$

$\qquad x = -\dfrac{37}{9} \quad \left\{ -\dfrac{37}{9} \right\}$

29. $\log_2 (h - 1) + \log_2 (h + 1) = 3$

$\qquad \log_2 \left[(h - 1)(h + 1) \right] = 3$

$\qquad (h - 1)(h + 1) = 2^3$

$\qquad h^2 - 1 = 8$

$\qquad h^2 - 9 = 0$

$\qquad (h - 3)(h + 3) = 0$

$\qquad h - 3 = 0 \text{ or } h + 3 = 0$

$\qquad h = 3 \text{ or } \quad h = -3$

$\{3\} \ (h = -3 \text{ does not check.})$

31. $\log(x + 2) = \log(3x - 6)$

$\qquad x + 2 = 3x - 6$

$\qquad -2x = -8$

$\qquad x = 4 \quad \{4\}$

33. $\ln x - \ln(4x - 9) = 0$

$$\ln x = \ln(4x - 9)$$
$$x = 4x - 9$$
$$-3x = -9$$
$$x = 3 \quad \{3\}$$

35. $\log_5(3t + 2) - \log_5 t = \log_5 4$

$$\log_5\left(\frac{3t + 2}{t}\right) = \log_5 4$$
$$\frac{3t + 2}{t} = 4$$
$$3t + 2 = 4t$$
$$2 = t \quad \{2\}$$

37. $\log(4m) = \log 2 + \log(m - 3)$

$$\log(4m) = \log[2(m - 3)]$$
$$4m = 2(m - 3)$$
$$4m = 2m - 6$$
$$2m = -6$$
$$m = -3$$
$$\{\,\} \ (m = -3 \text{ does not check})$$

39. $5^x = 625$

$$5^x = 5^4$$
$$x = 4 \quad \{4\}$$

41. $2^{-x} = 64$

$$2^{-x} = 2^6$$
$$-x = 6$$
$$x = -6 \quad \{-6\}$$

43. $36^x = 6$

$$(6^2)^x = 6^1$$
$$6^{2x} = 6^1$$
$$2x = 1$$
$$x = \frac{1}{2} \quad \left\{\frac{1}{2}\right\}$$

45. $4^{2x-1} = 64$

$$4^{2x-1} = 4^3$$
$$2x - 1 = 3$$
$$2x = 4$$
$$x = 2 \quad \{2\}$$

47. $81^{3x-4} = \dfrac{1}{243}$

$$(3^4)^{3x-4} = 3^{-5}$$
$$3^{12x-16} = 3^{-5}$$
$$12x - 16 = -5$$
$$12x = 11$$
$$x = \frac{11}{12} \quad \left\{\frac{11}{12}\right\}$$

49.
$$\left(\frac{2}{3}\right)^{-x+4} = \frac{8}{27}$$
$$\left(\frac{2}{3}\right)^{-x+4} = \left(\frac{2}{3}\right)^{3}$$
$$-x+4 = 3$$
$$-x = -1$$
$$x = 1 \quad \{1\}$$

51.
$$16^{-x+1} = 8^{5x}$$
$$\left(2^4\right)^{-x+1} = \left(2^3\right)^{5x}$$
$$2^{-4x+4} = 2^{15x}$$
$$-4x+4 = 15x$$
$$4 = 19x$$
$$x = \frac{4}{19} \quad \left\{\frac{4}{19}\right\}$$

53.
$$\left(4^x\right)^{x+1} = 16$$
$$4^{x^2+x} = 4^2$$
$$x^2 + x = 2$$
$$x^2 + x - 2 = 0$$
$$(x+2)(x-1) = 0$$
$$x+2 = 0 \ \text{ or } \ x-1 = 0$$
$$x = -2 \ \text{ or } \ x = 1 \quad \{-2,1\}$$

55.
$$8^a = 21$$
$$\ln 8^a = \ln 21$$
$$a \ln 8 = \ln 21$$
$$a = \frac{\ln 21}{\ln 8} \quad \left\{\frac{\ln 21}{\ln 8}\right\}$$

57.
$$e^x = 8.1254$$
$$\ln e^x = \ln 8.1254$$
$$x = \ln 8.1254 \quad \{\ln 8.1254\}$$

59.
$$10^t = 0.0138$$
$$\log 10^t = \log 0.0138$$
$$t = \log 0.0138 \quad \{\log 0.0138\}$$

61.
$$e^{0.07h} - 6 = 9$$
$$e^{0.07h} = 15$$
$$\ln e^{0.07h} = \ln 15$$
$$0.07h = \ln 15$$
$$h = \frac{\ln 15}{0.07} \quad \left\{\frac{\ln 15}{0.07}\right\}$$

63.
$$e^{1.2t} = 3$$
$$\ln e^{1.2t} = \ln 3$$
$$1.2t = \ln 3$$
$$t = \frac{\ln 3}{1.2} \quad \left\{\frac{\ln 3}{1.2}\right\}$$

65.
$$3^{x+1} = 5^x$$
$$\ln 3^{x+1} = \ln 5^x$$
$$(x+1)\ln 3 = x \ln 5$$
$$x \ln 3 + \ln 3 = x \ln 5$$
$$\ln 3 = x \ln 5 - x \ln 3$$
$$\ln 3 = x(\ln 5 - \ln 3)$$
$$x = \frac{\ln 3}{\ln 5 - \ln 3} \quad \left\{\frac{\ln 3}{\ln 5 - \ln 3}\right\}$$

67.
$$2^{x+2} = 6^x$$
$$\ln 2^{x+2} = \ln\left(6^x\right)$$
$$(x+2)\ln 2 = x \ln 6$$
$$x \ln 2 + 2 \ln 2 = x \ln 6$$
$$2 \ln 2 = x \ln 6 - x \ln 2$$
$$2 \ln 2 = x(\ln 6 - \ln 2)$$
$$x = \frac{2 \ln 2}{\ln 6 - \ln 2} \quad \left\{\frac{2 \ln 2}{\ln 6 - \ln 2}\right\}$$

69.
$$32e^{0.04m} = 128$$
$$e^{0.04m} = 4$$
$$\ln e^{0.04m} = \ln 4$$
$$0.04m = \ln 4$$
$$m = \frac{\ln 4}{0.04} \quad \left\{ \frac{\ln 4}{0.04} \right\}$$

71.
$$6e^{x/3} = 125$$
$$e^{x/3} = \frac{125}{6}$$
$$\ln e^{x/3} = \ln\left(\frac{125}{6}\right)$$
$$\frac{x}{3} = \ln\left(\frac{125}{6}\right)$$
$$x = 3\ln\left(\frac{125}{6}\right) \quad \left\{ 3\ln\left(\frac{125}{6}\right) \right\}$$

73.
$$5^{x-2} - 4 = 16$$
$$5^{x-2} = 20$$
$$\ln 5^{x-2} = \ln 20$$
$$(x-2)\ln 5 = \ln 20$$
$$x - 2 = \frac{\ln 20}{\ln 5}$$
$$x = \frac{\ln 20}{\ln 5} + 2 \quad \left\{ \frac{\ln 20}{\ln 5} + 2 \right\}$$

75. **a.**
$$P(4) = 1237(1.0095)^4 \approx 1285 \text{ million}$$
$$\approx 1{,}285{,}000{,}000 \text{ people}$$

b.
$$P(18) = 1237(1.0095)^{18} \approx 1466.5 \text{ million}$$
$$\approx 1{,}466{,}500{,}000 \text{ people}$$

c.
$$2000 = 1237(1.0095)^t$$
$$\frac{2000}{1237} = 1.0095^t$$
$$\ln(1.0095^t) = \ln\left(\frac{2000}{1237}\right)$$
$$t\ln(1.0095) = \ln\left(\frac{2000}{1237}\right)$$
$$t = \frac{\ln\left(\dfrac{2000}{1237}\right)}{\ln(1.0095)} \approx 50.8$$

The population will be 2 billion in 2049.

77. **a.**
$$A(0) = 500e^{0.0277(0)} \approx 500 \text{ bacteria}$$

b.
$$A(10) = 500e^{0.0277(10)} \approx 660 \text{ bacteria}$$
$$1000 = 500e^{0.0277t}$$
$$2 = e^{0.0277t}$$

c.
$$\ln 2 = \ln\left(e^{0.0277t}\right)$$
$$\ln 2 = 0.0277t$$
$$t = \frac{\ln 2}{0.0277} \approx 25 \text{ min}$$

79.
$$10{,}000 = 5000e^{0.07t}$$
$$2 = e^{0.07t}$$
$$\ln 2 = \ln e^{0.07t}$$
$$\ln 2 = 0.07t$$
$$t = \frac{\ln 2}{0.07} \approx 9.9 \text{ years}$$

It will take 9.9 years for the investment to double.

81.
$$10{,}000 = 8000\left(1 + \frac{0.045}{12}\right)^{12t}$$

$$1.25 = (1.00375)^{12t}$$

$$\ln 1.25 = \ln 1.00375^{12t}$$

$$\ln 1.25 = 12t \ln 1.00375$$

$$t = \frac{\ln 1.25}{12 \ln 1.00375} \approx 5 \text{ years}$$

It will take 5 years for the investment to reach $10,0000.

83. **a.** $A(5) = 10(0.5)^{5/14} \approx 7.8$ grams

b.
$$4 = 10(0.5)^{t/14}$$

$$0.4 = (0.5)^{t/14}$$

$$\log 0.4 = \log(0.5)^{t/14}$$

$$\log 0.4 = \frac{t}{14} \log(0.5)$$

$$t = \frac{14 \log 0.4}{\log 0.5} \approx 18.5 \text{ days}$$

85.
$$89.3 = 10 \log\left(\frac{I}{10^{-12}}\right)$$

$$8.93 = \log\left(\frac{I}{10^{-12}}\right)$$

$$\frac{I}{10^{-12}} = 10^{8.93}$$

$$I = 10^{8.93}\left(10^{-12}\right)$$

$$I = 10^{-3.07} \text{ W/m}^2$$

The intensity of sound of heavy traffic is $10^{-3.07}$ W/m^2.

87.
$$1{,}000{,}000 = 10{,}000 e^{0.12t}$$

$$100 = e^{0.12t}$$

$$\ln 100 = \ln e^{0.12t}$$

$$\ln 100 = 0.12t$$

$$t = \frac{\ln 100}{0.12} \approx 38.4 \text{ years}$$

It will take 38.4 years.

89. **a.** $P(43) = 2e^{-0.0079(43)} \approx 1.42$ kg

b. No, since 1.42 kg < 1.5 kg.

91.
$$(\log x)^2 - 2\log x - 15 = 0$$

Let $u = \log x$

$$u^2 - 2u - 15 = 0$$

$$(u - 5)(u + 3) = 0$$

$$u - 5 = 0 \quad \text{or} \quad u + 3 = 0$$

$$u = 5 \quad \text{or} \quad u = -3$$

$$\log x = 5 \quad \text{or} \quad \log x = -3$$

$$x = 10^5 \quad \text{or} \quad x = 10^{-3}$$

$$\left\{10^5, 10^{-3}\right\}$$

93. $\left(\log_3 w\right)^2 + 5\log_3 w + 6 = 0$

Let $u = \log_3 w$

$u^2 + 5u + 6 = 0$

$(u+2)(u+3) = 0$

$u + 2 = 0 \quad$ or $\quad u + 3 = 0$

$u = -2 \quad$ or $\qquad u = -3$

$\log_3 w = -2 \quad$ or $\quad \log_3 w = -3$

$w = 3^{-2} \quad$ or $\qquad w = 3^{-3}$

$w = \dfrac{1}{9} \quad$ or $\qquad w = \dfrac{1}{27} \quad \left\{\dfrac{1}{9}, \dfrac{1}{27}\right\}$

Chapter 8 Group Activity

1.–8. Answers will vary.

Chapter 8 Review Exercises

Section 8.1

1. $(f-g)(x) = f(x) - g(x)$

$= (x-7) - \left(-2x^3 - 8x\right)$

$= x - 7 + 2x^3 + 8x$

$= 2x^3 + 9x - 7$

3. $(f \cdot n)(x) = f(x) \cdot n(x) = (x-7)\left(\dfrac{1}{x-2}\right)$

$= \dfrac{x-7}{x-2}, x \neq 2$

5. $\left(\dfrac{f}{g}\right)(x) = \dfrac{f(x)}{g(x)} = \dfrac{x-7}{-2x^3 - 8x}, x \neq 0$

7. $(m \circ f)(x) = m(f(x)) = m(x-7) = (x-7)^2$

$= x^2 - 14x + 49$

9. $(m \circ g)(-1) = m(g(-1))$

$= m\left(-2(-1)^3 - 8(-1)\right)$

$= m(2+8) = m(10) = 10^2 = 100$

11. $(f \circ g)(4) = f(g(4)) = f\left(-2(4)^3 - 8(4)\right)$

$= f(-128 - 32) = f(-160)$

$= -160 - 7 = -167$

13. a. $(g \circ f)(x) = g(f(x)) = g(2x+1)$

$= (2x+1)^2 = 4x^2 + 4x + 1$

b. $(f \circ g)(x) = f(g(x)) = f(x^2) = 2x^2 + 1$

c. No, $f \circ g \neq g \circ f$

15. $(f \cdot g)(-2) = f(-2) \cdot g(-2) = -1 \cdot 3 = -3$

17. $(f-g)(2) = f(2) - g(2) = 2 - 3 = -1$

19. $(f \circ g)(4) = f(g(4)) = f(1) = 1$

Section 8.2

21. Yes, it is a one-to-one function.

23.
$$q(x) = \frac{3}{4}x - 2$$
$$y = \frac{3}{4}x - 2$$
$$x = \frac{3}{4}y - 2$$
$$x + 2 = \frac{3}{4}y$$
$$\frac{4}{3}(x+2) = y$$
$$q^{-1}(x) = \frac{4}{3}(x+2)$$

25.
$$f(x) = (x-1)^3$$
$$y = (x-1)^3$$
$$x = (y-1)^3$$
$$\sqrt[3]{x} = y - 1$$
$$\sqrt[3]{x} + 1 = y$$
$$f^{-1}(x) = \sqrt[3]{x} + 1$$

27.
$$f(x) = 5x - 2 \; ; \; g(x) = \frac{1}{5}x + \frac{2}{5}$$
$$(f \circ g)(x) = f(g(x)) = f\left(\frac{1}{5}x + \frac{2}{5}\right)$$
$$= 5\left(\frac{1}{5}x + \frac{2}{5}\right) - 2 = x + 2 - 2 = x$$
$$(g \circ f)(x) = g(f(x)) = g(5x - 2)$$
$$= \frac{1}{5}(5x - 2) + \frac{2}{5} = x - \frac{2}{5} + \frac{2}{5} = x$$

29. a. $h(x) = \sqrt{x+1}$

Domain: $[-1, \infty)$; Range: $[0, \infty)$

b. $k(x) = x^2 - 1, \; x \geq 0$

Domain: $[0, \infty)$; Range: $[-1, \infty)$

Section 8.3

31. $4^5 = 1024$

33. $8^{1/3} = \sqrt[3]{8} = 2$

35. $2^{\pi} \approx 8.825$

37. $\left(\sqrt{7}\right)^{1/2} \approx 1.627$

39. $f(x) = 3^x$

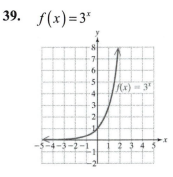

41. $h(x) = 5^{-x}$

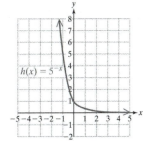

43. a. $y = b^x, b > 0, b \neq 1$ has a horizontal asymptote.

    b. $y = 0$

Section 8.4

45.
$$y = \log_3\left(\frac{1}{27}\right)$$
$$3^y = \frac{1}{27}$$
$$3^y = 3^{-3}$$
$$y = -3$$

47.
$$y = \log_7 7$$
$$7^y = 7$$
$$7^y = 7^1$$
$$y = 1$$

49.
$$y = \log_2 16$$
$$2^y = 16$$
$$2^y = 2^4$$
$$y = 4$$

51.
$$y = \log(100{,}000)$$
$$10^y = 100{,}000$$
$$10^y = 10^5$$
$$y = 5$$

53. $q(x) = \log_3 x$

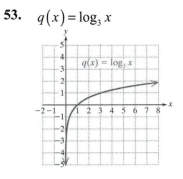

55. a. $y = \log_b x$ has a vertical asymptote.

    b. $x = 0$

Section 8.5

57. $\log_8 8 = 1$

59. $\log_{1/2} 1 = 0$

61. **a.** $\log_b(xy) = \log_b x + \log_b y$

b. $\log_b x - \log_b y = \log_b\left(\dfrac{x}{y}\right)$

c. $\log_b(x^p) = p\log_b x$

63. $\dfrac{1}{2}\log_3 a + \dfrac{1}{2}\log_3 b - 2\log_3 c - 4\log_3 d$

$= \dfrac{1}{2}\left(\log_3 a + \log_3 b\right) - \left(2\log_3 c + 4\log_3 d\right)$

$= \dfrac{1}{2}\left(\log_3(ab)\right) - \left(\log_3 c^2 + \log_3 d^4\right)$

$= \log_3(ab)^{1/2} - \log_3\left(c^2 d^4\right)$

$= \log_3\left(\dfrac{\sqrt{ab}}{c^2 d^4}\right)$

65. $-\log_4 18 + \log_4 6 + \log_4 3 - \log_4 1$

$= \left(\log_4 6 + \log_4 3\right) - \left(\log_4 18 + \log_4 1\right)$

$= \log_4(6\cdot 3) - \log_4(18\cdot 1)$

$= \log_4 18 - \log_4 18 = 0$

67. $\dfrac{\log 8^{-3}}{\log 2 + \log 4} = \dfrac{-3\log 8}{\log(2\cdot 4)} = \dfrac{-3\log 8}{\log 8} = -3$ a

Section 8.6

69. $e^{\sqrt{7}} \approx 14.0940$

71. $58e^{-0.0125} \approx 57.2795$

73. $\ln\left(\dfrac{1}{9}\right) \approx -2.1972$

75. $\log e^3 \approx 1.3029$

77. $\log_9 80 = \dfrac{\ln 80}{\ln 9} \approx 1.9943$

79. $\log_4(0.0062) = \dfrac{\ln 0.0062}{\ln 4} \approx -3.6668$

81. **a.** $S(0) = 75e^{-0.5(0)} + 20 = 95$ The student's score is 95 at the end of the course.

b. $S(6) = 75e^{-0.5(6)} + 20 \approx 23.7$ The student's score is 23.7 after 6 months.

c. $S(12) = 75e^{-0.5(12)} + 20 \approx 20.2$ The student's score is 20.2 after 1 year.

83. $g(x) = e^{x+6}$ Domain: $(-\infty, \infty)$

85. $k(x) = \ln x$ Domain: $(0, \infty)$

87. $p(x) = \ln(x-7)$ Domain: $(7, \infty)$

89. $w(x) = \ln(5-x)$ Domain: $(-\infty, 5)$

Section 8.7

91. $\log_7 x = -2$

$x = 7^{-2} = \dfrac{1}{7^2} = \dfrac{1}{49}$ $\left\{\dfrac{1}{49}\right\}$

93. $\log_3 y = \dfrac{1}{12}$

$y = 3^{1/12}$ $\left\{3^{1/12}\right\}$

95. $\log_2(3w+5)=5$

$3w+5=2^5$

$3w+5=32$

$3w=27$

$w=9 \quad \{9\}$

97. $\log_4(2+t)-3=\log_4(3-5t)$

$\log_4(2+t)-\log_4(3-5t)=3$

$\log_4\left(\dfrac{2+t}{3-5t}\right)=3$

$\dfrac{2+t}{3-5t}=4^3$

$\dfrac{2+t}{3-5t}=64$

$2+t=64(3-5t)$

$2+t=192-320t$

$321t=190$

$t=\dfrac{190}{321} \quad \left\{\dfrac{190}{321}\right\}$

99. $5^{7x}=625$

$5^{7x}=5^4$

$7x=4$

$x=\dfrac{4}{7} \quad \left\{\dfrac{4}{7}\right\}$

101. $5^a=18$

$\ln 5^a=\ln 18$

$a\ln 5=\ln 18$

$a=\dfrac{\ln 18}{\ln 5} \quad \left\{\dfrac{\ln 18}{\ln 5}\right\}$

103. $e^{-2x}=0.06$

$\ln e^{-2x}=\ln 0.06$

$-2x=\ln 0.06$

$x=-\dfrac{\ln 0.06}{2} \quad \left\{-\dfrac{\ln 0.06}{2}\right\}$

105. $10^{-3m}=\dfrac{1}{821}$

$\log 10^{-3m}=\log\left(\dfrac{1}{821}\right)$

$-3m=\log\left(\dfrac{1}{821}\right)$

$m=\dfrac{\log\left(\dfrac{1}{821}\right)}{-3}=\dfrac{\log 1-\log 821}{-3}$

$=\dfrac{\log 821}{3} \quad \left\{\dfrac{\log 821}{3}\right\}$

107. $14^{x-5}=6^x$

$\ln 14^{x-5}=\ln 6^x$

$(x-5)\ln 14=x\ln 6$

$x\ln 14-5\ln 14=x\ln 6$

$x\ln 14-x\ln 6=5\ln 14$

$x(\ln 14-\ln 6)=5\ln 14$

$x=\dfrac{5\ln 14}{\ln 14-\ln 6} \quad \left\{\dfrac{5\ln 14}{\ln 14-\ln 6}\right\}$

109. **a.** $A(0)=150e^{0.007(0)}=150$ bacteria

b. $A(30)=150e^{0.007(30)}\approx 185$ bacteria

c. $300=150e^{0.007t}$

$2=e^{0.007t}$

$\ln 2=\ln e^{0.007t}$

$\ln 2=0.007t$

$t=\dfrac{\ln 2}{0.007}\approx 99$ min

Chapter 8 Test

1.
$$\left(\frac{f}{g}\right)(x) = \frac{f(x)}{g(x)} = \frac{x-4}{x^2+2}$$

3.
$$(g \circ f)(x) = g(f(x)) = g(x-4)$$
$$= (x-4)^2 + 2$$
$$= x^2 - 8x + 16 + 2 = x^2 - 8x + 18$$

5.
$$(f-g)(7) = f(7) - g(7) = (7-4) - (7^2+2)$$
$$= 3 - (49+2) = 3 - 51 = -48$$

7.
$$(h \circ g)(4) = h(g(4)) = h(4^2+2)$$
$$= h(16+2) = h(18) = \frac{1}{18}$$

9.
$$\left(\frac{g}{f}\right)(x) = \frac{g(x)}{f(x)} = \frac{x^2+2}{x-4}, x \neq 4$$

11. Graph b is one-to-one since it passes the horizontal line test.

13.
$$g(x) = (x-1)^2, \ x \geq 1$$
$$y = (x-1)^2$$
$$x = (y-1)^2$$
$$\sqrt{x} = y - 1$$
$$\sqrt{x} + 1 = y$$
$$g^{-1}(x) = \sqrt{x} + 1, x \geq 0$$

15.
  a. $10^{2/3} \approx 4.6416$

  b. $3^{\sqrt{10}} \approx 32.2693$

  c. $8^{\pi} \approx 687.2913$

17.
  a. $16^{3/4} = 8 \leftrightarrow \log_{16} 8 = \frac{3}{4}$

  b. $\log_x 31 = 5 \leftrightarrow x^5 = 31$

19.
$$\log_b n = \frac{\log_a n}{\log_a b}$$

21.
  a.
$$-\log_3\left(\frac{3}{9x}\right) = -\left(\log_3 3 - \log_3(9x)\right)$$
$$= -\left[\log_3 3 - \left(\log_3 3^2 + \log_3 x\right)\right]$$
$$= -\left[1 - \left(2 + \log_3 x\right)\right]$$
$$= -\left[1 - 2 - \log_3 x\right]$$
$$= -1 + 2 + \log_3 x$$
$$= 1 + \log_3 x$$

  b. $\log\left(\frac{1}{10^5}\right) = \log 10^{-5} = -5$

23.
  a. $e^{1/2} \approx 1.6487$

  b. $e^{-3} \approx 0.0498$

  c. $\ln\left(\frac{1}{3}\right) \approx -1.0986$

  d. $\ln e = 1$

25.
  a. $p(4) = 92 - 20\ln(4+1) \approx 59.8$
    59.8% of the material is retained after 4 months.

  b. $p(12) = 92 - 20\ln(12+1) \approx 40.7$
    40.7% of the material is retained after 1 year.

27.
  a. $P(0) = \dfrac{1,500,000}{1 + 5000e^{-0.8(0)}} \approx 300$
    There are 300 bacteria initially.

  b. $P(6) = \dfrac{1,500,000}{1 + 5000e^{-0.8(6)}} \approx 35,588$

c. $p(0) = 92 - 20\ln(0+1) = 92$

92% of the material is retained at the end of the course.

c. $P(12) = \dfrac{1,500,000}{1 + 5000e^{-0.8(12)}} \approx 1,120,537$

d. $P(18) = \dfrac{1,500,000}{1 + 5000e^{-0.8(18)}} \approx 1,495,831$

29. $\log_{1/2} x = -5$

$$x = \left(\dfrac{1}{2}\right)^{-5} = 2^5 = 32 \quad \{32\}$$

31.
$$3^{x+4} = \dfrac{1}{27}$$
$$3^{x+4} = 3^{-3}$$
$$x + 4 = -3$$
$$x = -7 \quad \{-7\}$$

33. $e^{2.4x} = 250$

$\ln e^{2.4x} = \ln 250$

$2.4x = \ln 250$

$x = \dfrac{\ln 250}{2.4} \quad \left\{\dfrac{\ln 250}{2.4}\right\}$

35.
$$4^{x+7} = 5^x$$
$$\ln 4^{x+7} = \ln 5^x$$
$$(x+7)\ln 4 = x \ln 5$$
$$x \ln 4 + 7\ln 4 = x \ln 5$$
$$7\ln 4 = x \ln 5 - x \ln 4$$
$$7\ln 4 = x(\ln 5 - \ln 4)$$
$$\dfrac{7\ln 4}{\ln 5 - \ln 4} = x \quad \left\{\dfrac{7\ln 4}{\ln 5 - \ln 4}\right\}$$

37. a. $A(5) = 2000e^{0.075(5)} \approx \$2909.98$

b. $4000 = 2000e^{0.075t}$

$2 = e^{0.075t}$

$\ln 2 = \ln e^{0.075t}$

$\ln 2 = 0.075t$

$t = \dfrac{\ln 2}{0.075} \approx 9.24$ years

Chapters 1 – 8   Cumulative Review Exercises

1. $\dfrac{8-4\cdot 2^2+15\div 5}{|-3+7|}=\dfrac{8-4\cdot 4+15\div 5}{|4|}$

$=\dfrac{8-16+3}{4}=\dfrac{-5}{4}=-\dfrac{5}{4}$

3.

$$t-2\ \overline{\smash{\big)}\ t^4\qquad\ -13t^2\qquad +36}$$

Quotient: $t^3+2t^2-9t-18$ Remainder: 0

5. $\dfrac{4}{\sqrt[3]{40}}=\dfrac{4}{\sqrt[3]{8\cdot 5}}=\dfrac{4}{2\sqrt[3]{5}}=\dfrac{2}{\sqrt[3]{5}}\cdot\dfrac{\sqrt[3]{25}}{\sqrt[3]{25}}=\dfrac{2\sqrt[3]{25}}{5}$

7. $\dfrac{2^{2/5}c^{-1/4}d^{1/5}}{2^{-8/5}c^{3/4}d^{1/10}}=2^{(2/5)-(-8/5)}c^{(-1/4)-(3/4)}d^{(1/5)-(1/10)}$

$=2^2c^{-1}d^{1/10}=\dfrac{4d^{1/10}}{c}$

9. $\dfrac{4-3i}{2+5i}\cdot\dfrac{2-5i}{2-5i}=\dfrac{8-20i-6i+15i^2}{4-10i+10i-25i^2}$

$=\dfrac{8-26i-15}{4+25}=\dfrac{-7-26i}{29}$

$=-\dfrac{7}{29}-\dfrac{26}{29}i$

11.

	100% Solution	20% Solution	50% Solution
Amount of Solution	x	8	$x+8$
Amount of Alcohol	$1.00x$	$0.20(8)$	$0.50(x+8)$

(amt of 100%)+(amt of 20%)=(amt of 50%)

$1.00x+0.20(8)=0.50(x+8)$

$1.00x+1.6=0.50x+4$

$0.50x+1.6=4$

$0.50x=2.4$

$\dfrac{0.50x}{0.50}=\dfrac{2.4}{0.50}$

$x=4.8$

4.8 L of pure alcohol must be used.

13. $5x+10y=25$

$-2x+6y=-20$

$\begin{bmatrix}5 & 10 & | & 25\\ -2 & 6 & | & -20\end{bmatrix}\xrightarrow{\frac{1}{5}R_1\Rightarrow R_1}\begin{bmatrix}1 & 2 & | & 5\\ -2 & 6 & | & -20\end{bmatrix}\xrightarrow{2R_1+R_2\Rightarrow R_2}\begin{bmatrix}1 & 2 & | & 5\\ 0 & 10 & | & -10\end{bmatrix}$

$\xrightarrow{\frac{1}{10}R_2\Rightarrow R_2}\begin{bmatrix}1 & 2 & | & 5\\ 0 & 1 & | & -1\end{bmatrix}\xrightarrow{-2R_2+R_1\Rightarrow R_1}\begin{bmatrix}1 & 0 & | & 7\\ 0 & 1 & | & -1\end{bmatrix}$

The solution is $\{(7,-1)\}$.

15.
$$ax - c = bx + d$$
$$ax - bx - c = d$$
$$ax - bx = c + d$$
$$x(a - b) = c + d$$
$$x = \frac{c + d}{a - b}$$

17.
$$3w(w - 2) = 6 + 2w$$
$$3w^2 - 6w = 6 + 2w$$
$$3w^2 - 8w - 6 = 0$$
$$w = \frac{-(-8) \pm \sqrt{(-8)^2 - 4(3)(-6)}}{2(3)}$$
$$= \frac{8 \pm \sqrt{64 + 72}}{6} = \frac{8 \pm \sqrt{136}}{6} = \frac{8 \pm 2\sqrt{34}}{6}$$
$$= \frac{4 \pm \sqrt{34}}{3} \qquad \left\{ \frac{4 \pm \sqrt{34}}{3} \right\}$$

19. **a.** $(f \cdot g)(t) = f(t) \cdot g(t) = 6 \cdot (-5t) = -30t$

b. $(g \circ h)(t) = g(h(t)) = g(2t^2)$
$$= -5(2t^2) = -10t^2$$

c. $(h - g)(t) = h(t) - g(t)$
$$= 2t^2 - (-5t) = 2t^2 + 5t$$

21. **a.** $x = 2$ (Vertical line)

b. $y = 6$ (Horizontal line)

c. $2x + y = 4$
$$y = -2x + 4 \quad m = -2 \quad m_\perp = \frac{1}{2}$$
$$y - 6 = \frac{1}{2}(x - 2)$$
$$y - 6 = \frac{1}{2}x - 1$$
$$y = \frac{1}{2}x + 5$$

23. Multiply the first equation by 4 and add the equations:
$$\frac{1}{2}x - \frac{1}{4}y = 1 \xrightarrow{\times 4} 2x - y = 4$$
$$-2x + y = -4 \longrightarrow \underline{-2x + y = -4}$$
$$0 = 0$$
Infinitely many solutions of the form
$\{(x, y) | -2x + y = -4\}$; dependent equations.

25.
$$f(x) = 5x - \frac{2}{3}$$
$$y = 5x - \frac{2}{3}$$
$$x = 5y - \frac{2}{3}$$
$$x + \frac{2}{3} = 5y$$
$$\frac{1}{5}\left(x + \frac{2}{3}\right) = y$$
$$\frac{1}{5}x + \frac{2}{15} = y$$
$$f^{-1}(x) = \frac{1}{5}x + \frac{2}{15}$$

27. $\dfrac{5x-10}{x^2-4x+4} \div \dfrac{5x^2-125}{25-5x} \cdot \dfrac{x^3+125}{10x+5} = \dfrac{5x-10}{x^2-4x+4} \cdot \dfrac{25-5x}{5x^2-125} \cdot \dfrac{x^3+125}{10x+5}$

$$= \dfrac{5(x-2)}{(x-2)(x-2)} \cdot \dfrac{-5(x-5)}{5(x^2-25)} \cdot \dfrac{(x+5)(x^2-5x+25)}{5(2x+1)}$$

$$= \dfrac{\cancel{5}\,\cancel{(x-2)}}{\cancel{(x-2)}(x-2)} \cdot \dfrac{-1 \cdot \cancel{5}\,\cancel{(x-5)}}{\cancel{5}\,\cancel{(x-5)}\,\cancel{(x+5)}} \cdot \dfrac{\cancel{(x+5)}(x^2-5x+25)}{\cancel{5}(2x+1)}$$

$$= \dfrac{-(x^2-5x+25)}{(x-2)(2x+1)} = \dfrac{-x^2+5x-25}{(x-2)(2x+1)} = -\dfrac{x^2-5x+25}{(x-2)(2x+1)}$$

29. a. Yes; $x \neq 4, x \neq -2$

b. $\dfrac{2}{x-4} = \dfrac{5}{x+2}$

$2(x+2) = 5(x-4)$

$2x+4 = 5x-20$

$-3x+4 = -20$

$-3x = -24$

$x = 8 \quad \{8\}$

c. Boundary points are 8 and –2, 4 (undefined).
Use test points $x = -3$, $x = 0$, $x = 5$ and $x = 9$.

Test $x = -3$: $\quad \dfrac{2}{-3-4} = \dfrac{2}{-7} \geq \dfrac{5}{-3+2} = \dfrac{5}{-1} = -5 \quad$ True

Test $x = 0$: $\quad \dfrac{2}{0-4} = \dfrac{2}{-4} = -\dfrac{1}{2} \geq \dfrac{5}{0+2} = \dfrac{5}{2} \quad$ False

Test $x = 5$: $\quad \dfrac{2}{5-4} = \dfrac{2}{1} = 2 \geq \dfrac{5}{5+2} = \dfrac{5}{7} \quad$ True

Test $x = 9$: $\quad \dfrac{2}{9-4} = \dfrac{2}{5} \geq \dfrac{5}{9+2} = \dfrac{5}{11} \quad$ False

The boundary point $x = 8$ is included.

The solution is $\{x \mid x < -2 \text{ or } 4 < x \leq 8\}$ or $(-\infty, -2) \cup (4, 8]$.

31. $\sqrt{-x} = x+6$

$\left(\sqrt{-x}\right)^2 = (x+6)^2$

$-x = x^2+12x+36$

$0 = x^2+13x+36$

$(x+4)(x+9) = 0$

$x+4 = 0 \quad$ or $\quad x+9 = 0$

$x = -4 \quad$ or $\quad x = -9$

$\{-4\} \ (x = -9 \text{ does not check.})$

33. a.

$$P(6) = 4,000,000 \left(\frac{1}{2}\right)^{6/6} = 2,000,000$$

$$P(12) = 4,000,000 \left(\frac{1}{2}\right)^{12/6} = 1,000,000$$

$$P(18) = 4,000,000 \left(\frac{1}{2}\right)^{18/6} = 500,000$$

$$P(24) = 4,000,000 \left(\frac{1}{2}\right)^{24/6} = 250,000$$

$$P(30) = 4,000,000 \left(\frac{1}{2}\right)^{30/6} = 125,000$$

b.

$$15,625 = 4,000,000 \left(\frac{1}{2}\right)^{t/6}$$

$$\frac{15,625}{4,000,000} = \left(\frac{1}{2}\right)^{t/6}$$

$$\frac{4,000,000}{15,625} = 2^{t/6}$$

$$256 = 2^{t/6}$$

$$2^8 = 2^{t/6}$$

$$8 = \frac{t}{6}$$

$$t = 48 \text{ hr}$$

35. a. $\pi^{4.7} \approx 217.0723$

b. $e^{\pi} \approx 23.1407$

c. $\left(\sqrt{2}\right)^{-5} \approx 0.1768$

d. $\log 5362 \approx 3.7293$

e. $\ln(0.67) \approx -0.4005$

f. $\log_4 37 = \dfrac{\ln 37}{\ln 4} \approx 2.6047$

37.

$$e^x = 100$$

$$\ln(e^x) = \ln 100$$

$$x = \ln 100 \quad \{\ln 100\}$$

39. $\dfrac{1}{2} \log z - 2 \log x - 3 \log y$

$$= \frac{1}{2} \log z - (2 \log x + 3 \log y)$$

$$= \log z^{1/2} - \left(\log x^2 + \log y^3\right)$$

$$= \log z^{1/2} - \log\left(x^2 y^3\right)$$

$$= \log\left(\frac{z^{1/2}}{x^2 y^3}\right) = \log\left(\frac{\sqrt{z}}{x^2 y^3}\right)$$

Chapter 9   Conic Sections

Are You Prepared?

1. $x^2 + 6x + k$

$$k = \left(\frac{b}{2a}\right)^2 = \left(\frac{6}{2(1)}\right)^2 = 3^2 = 9$$

$$x^2 + 6x + 9 = (x+3)^2 \qquad \text{S}$$

3. $x^2 - 7x + k$

$$k = \left(\frac{b}{2a}\right)^2 = \left(\frac{-7}{2(1)}\right)^2 = \left(-\frac{7}{2}\right)^2 = \frac{49}{4}$$

$$x^2 - 7x + \frac{49}{4} = \left(x - \frac{7}{2}\right)^2 \qquad \text{I}$$

5. $x^2 - \frac{2}{5}x + k$

$$k = \left(\frac{b}{2a}\right)^2 = \left(\frac{-\frac{2}{5}}{2(1)}\right)^2 = \left(-\frac{1}{5}\right)^2 = \frac{1}{25}$$

$$x^2 - \frac{2}{5}x + \frac{1}{25} = \left(x - \frac{1}{5}\right)^2 \qquad \text{C}$$

7. $a = d\left[(3,\,1),\,(7,\,1)\right] = 4$

$b = d\left[(7,\,1),\,(7,\,6)\right] = 5$

$c = \sqrt{a^2 + b^2} = \sqrt{4^2 + 5^2} = \sqrt{16 + 25} = \sqrt{41} \qquad \text{E}$

The parabola, ellipse, hyperbola, are all $\underline{\text{C O N I C \quad S E C T I O N S}}$.

$\qquad\qquad\qquad\qquad\qquad\qquad\quad$ 5 4 6 3 5   1 7 5 2 3 4 6 1

Section 9.1   Distance Formula, Midpoint Formula, and Circles

1.   **(a)** $\sqrt{(x_2 - x_1)^2 + (y_2 - xy_1)^2}$

(b) circle; center

(c) radius

(d) $(x - h)^2 + (y - k)^2 = r^2$

(e) $\left(\dfrac{x_1 + x_2}{2},\, \dfrac{y_1 + y_2}{2}\right)$

3. $(x_1,\, y_1) = (1,\, 10), \quad (x_2,\, y_2) = (-2,\, 4)$

$$d = \sqrt{\left[(-2) - (1)\right]^2 + \left[(4) - (10)\right]^2}$$

$$= \sqrt{(-3)^2 + (-6)^2} = \sqrt{9 + 36}$$

$$= \sqrt{45} = 3\sqrt{5}$$

5. $(x_1,\, y_1) = (6,\, 7), \quad (x_2,\, y_2) = (3,\, 2)$

$$d = \sqrt{\left[(3) - (6)\right]^2 + \left[(2) - (7)\right]^2}$$

$$= \sqrt{(-3)^2 + (-5)^2} = \sqrt{9 + 25}$$

$$= \sqrt{34}$$

7. $(x_1,\, y_1) = \left(-\dfrac{1}{2},\, \dfrac{5}{8}\right), \quad (x_2,\, y_2) = \left(-\dfrac{3}{2},\, \dfrac{1}{4}\right)$

$$d = \sqrt{\left[\left(-\frac{3}{2}\right) - \left(-\frac{1}{2}\right)\right]^2 + \left[\left(\frac{1}{4}\right) - \left(\frac{5}{8}\right)\right]^2}$$

$$= \sqrt{(-1)^2 + \left(-\frac{3}{8}\right)^2} = \sqrt{1 + \frac{9}{64}}$$

$$= \sqrt{\frac{73}{64}} = \frac{\sqrt{73}}{8}$$

9. $(x_1, y_1) = (-2, 5), \quad (x_2, y_2) = (-2, 9)$

$d = \sqrt{[(-2)-(-2)]^2 + [(9)-(5)]^2}$

$\quad = \sqrt{(0)^2 + (4)^2} = \sqrt{0+16} = \sqrt{16} = 4$

11. $(x_1, y_1) = (7, 2), \quad (x_2, y_2) = (15, 2)$

$d = \sqrt{[(15)-(7)]^2 + [(2)-(2)]^2}$

$\quad = \sqrt{(8)^2 + (0)^2} = \sqrt{64+0} = \sqrt{64} = 8$

13. $(x_1, y_1) = (-1, -5), \quad (x_2, y_2) = (-5, -9)$

$d = \sqrt{[(-5)-(-1)]^2 + [(-9)-(-5)]^2}$

$\quad = \sqrt{(-4)^2 + (-4)^2} = \sqrt{16+16} = \sqrt{32} = 4\sqrt{2}$

15. $(x_1, y_1) = (4\sqrt{6}, -2\sqrt{2}), \quad (x_2, y_2) = (2\sqrt{6}, \sqrt{2})$

$d = \sqrt{[2\sqrt{6} - 4\sqrt{6}]^2 + [(\sqrt{2})-(-2\sqrt{2})]^2}$

$\quad = \sqrt{(-2\sqrt{6})^2 + (3\sqrt{2})^2} = \sqrt{24+18} = \sqrt{42}$

17. Subtract 5 and −7. This becomes
$5 - (-7) = 12$.

19. $(4, 7), \quad (-4, y)$

$10 = \sqrt{[-4-4]^2 + [y-7]^2}$

$10 = \sqrt{(-8)^2 + y^2 - 14y + 49}$

$10 = \sqrt{y^2 - 14y + 49 + 64}$

$100 = y^2 - 14y + 113$

$0 = y^2 - 14y + 13$

$0 = (y-13)(y-1)$

$y - 13 = 0 \quad \text{or} \quad y - 1 = 0$

$y = 13 \quad \text{or} \quad\quad y = 1$

21. $(x, 2), \quad (4, -1)$

$5 = \sqrt{[4-x]^2 + [-1-2]^2}$

$5 = \sqrt{(4-x)^2 + (-3)^2}$

$5 = \sqrt{16 - 8x + x^2 + 9}$

$25 = x^2 - 8x + 25$

$0 = x^2 - 8x$

$0 = x(x-8)$

$x = 0 \quad \text{or} \quad x - 8 = 0$

$x = 0 \quad \text{or} \quad\quad x = 8$

23. $A: (-3, 2), \quad B: (-2, -4), \quad C: (3, 3)$

$d_{AB} = \sqrt{[-2-(-3)]^2 + [-4-2]^2}$

$\quad = \sqrt{(1)^2 + (-6)^2} = \sqrt{1+36} = \sqrt{37}$

$d_{BC} = \sqrt{[3-(-2)]^2 + [3-(-4)]^2}$

$\quad = \sqrt{(5)^2 + (7)^2} = \sqrt{25+49} = \sqrt{74}$

$d_{AC} = \sqrt{[3-(-3)]^2 + [3-2]^2}$

$\quad = \sqrt{(6)^2 + (1)^2} = \sqrt{36+1} = \sqrt{37}$

$(\sqrt{37})^2 + (\sqrt{37})^2 = (\sqrt{74})^2$

$37 + 37 = 74$

$74 = 74$

The three points define a right triangle.

25. $A : (-3, -2), \quad B : (4, -3), \quad C : (1, 5)$

$$d_{AB} = \sqrt{\left[4 - (-3)\right]^2 + \left[-3 - (-2)\right]^2}$$
$$= \sqrt{(7)^2 + (-1)^2} = \sqrt{49 + 1} = \sqrt{50}$$

$$d_{BC} = \sqrt{\left[1 - 4\right]^2 + \left[5 - (-3)\right]^2}$$
$$= \sqrt{(-3)^2 + (8)^2} = \sqrt{9 + 64} = \sqrt{73}$$

$$d_{AC} = \sqrt{\left[1 - (-3)\right]^2 + \left[5 - (-2)\right]^2}$$
$$= \sqrt{(4)^2 + (7)^2} = \sqrt{16 + 49} = \sqrt{65}$$

$$\left(\sqrt{50}\right)^2 + \left(\sqrt{65}\right)^2 = \left(\sqrt{73}\right)^2$$
$$50 + 65 = 73$$
$$115 \ne 73$$

The three points do not define a right triangle.

27. $(x - 4)^2 + (y + 2)^2 = 9$

$$\left[x - 4\right]^2 + \left[y - (-2)\right]^2 = 3^2$$

$h = 4, \ k = -2, \ r = 3$

Center: $(4, -2)$; radius: 3

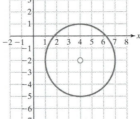

29. $(x + 1)^2 + (y + 1)^2 = 1$

$$\left[x - (-1)\right]^2 + \left[y - (-1)\right]^2 = 1^2$$

$h = -1, \ k = -1, \ r = 1$

Center: $(-1, -1)$; radius: 1

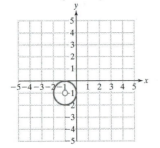

31. $x^2 + (y - 2)^2 = 4$

$$\left[x - 0\right]^2 + \left[y - 2\right]^2 = 2^2$$

$h = 0, \ k = 2, \ r = 2$

Center: $(0, 2)$; radius: 2

33. $(x - 3)^2 + y^2 = 8$

$$\left[x - 3\right]^2 + \left[y - 0\right]^2 = \left(\sqrt{8}\right)^2$$

$h = 3, \ k = 0, \ r = 2\sqrt{2}$

Center: $(3, 0)$; radius: $2\sqrt{2}$

35. $x^2 + y^2 = 6$

$$\left[x - 0\right]^2 + \left[y - 0\right]^2 = \left(\sqrt{6}\right)^2$$

$h = 0, \ k = 0, \ r = \sqrt{6}$

Center: $(0, 0)$; radius: $\sqrt{6}$

37.

$$\left(x+\frac{4}{5}\right)^2+y^2=\frac{64}{25}$$

$$\left[x-\left(-\frac{4}{5}\right)\right]^2+[y-0]^2=\left(\frac{8}{5}\right)^2$$

$$h=-\frac{4}{5},\,k=0,\,r=\frac{8}{5}$$

Center: $\left(-\frac{4}{5},\,0\right)$; radius: $\frac{8}{5}$

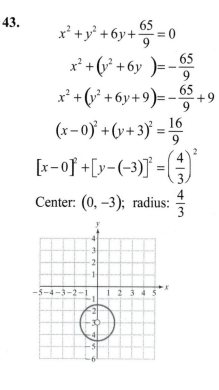

39.

$$x^2+y^2-2x-6y-26=0$$

$$\left(x^2-2x\ \right)+\left(y^2-6y\ \right)=26$$

$$\left(x^2-2x+1\right)+\left(y^2-6y+9\right)=26+1+9$$

$$(x-1)^2+(y-3)^2=36$$

$$[x-1]^2+[y-3]^2=6^2$$

Center: $(1,\,3)$;   radius: 6

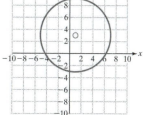

41.

$$x^2+y^2-6y+5=0$$

$$x^2+\left(y^2-6y\ \right)=-5$$

$$x^2+\left(y^2-6y+9\right)=-5+9$$

$$(x-0)^2+(y-3)^2=4$$

$$[x-0]^2+[y-3]^2=2^2$$

Center: $(0,\,3)$; radius: 2

43.

$$x^2+y^2+6y+\frac{65}{9}=0$$

$$x^2+\left(y^2+6y\ \right)=-\frac{65}{9}$$

$$x^2+\left(y^2+6y+9\right)=-\frac{65}{9}+9$$

$$(x-0)^2+(y+3)^2=\frac{16}{9}$$

$$[x-0]^2+[y-(-3)]^2=\left(\frac{4}{3}\right)^2$$

Center: $(0,\,-3)$; radius: $\frac{4}{3}$

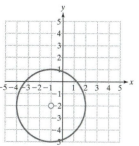

45.

$$x^2+y^2+2x+4y-4=0$$

$$\left(x^2+2x\ \right)+\left(y^2+4y\right)=4$$

$$\left(x^2+2x+1\right)+\left(y^2+4y+4\right)=4+1+4$$

$$(x+1)^2+(y+2)^2=9$$

$$[x-(-1)]^2+[y-(-2)]^2=3^2$$

Center: $(-1,\,-2)$; radius: 3

47.
$$3x^2 + 3y^2 = 3$$
$$x^2 + y^2 = 1$$
$$[x-0]^2 + [y-0]^2 = 1^2$$
Center: $(0, 0)$; radius: 1

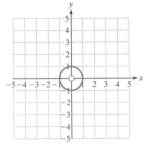

49. Center: $(0, 0)$; radius: 2
$$h = 0, \ k = 0, \ r = 2$$
$$[x-(0)]^2 + [y-(0)]^2 = 2^2$$
$$x^2 + y^2 = 4$$

51. Center: $(0, 2)$; radius: 2
$$h = 0, \ k = 2, \ r = 2$$
$$[x-(0)]^2 + [y-(2)]^2 = 2^2$$
$$x^2 + (y-2)^2 = 4$$

53. Center: $(-2, 2)$; radius: 3
$$h = -2, \ k = 2, \ r = 3$$
$$[x-(-2)]^2 + [y-(2)]^2 = 3^2$$
$$(x+2)^2 + (y-2)^2 = 9$$

55. Center: $(0, 0)$; radius: 7
$$h = 0, \ k = 0, \ r = 7$$
$$[x-(0)]^2 + [y-(0)]^2 = 7^2$$
$$x^2 + y^2 = 49$$

57. Center: $(-3, -4)$; diameter: 12
$$h = -3, \ k = -4, \ r = 6$$
$$[x-(-3)]^2 + [y-(-4)]^2 = 6^2$$
$$(x+3)^2 + (y+4)^2 = 36$$

59. Center: $(5, 3)$; radius: 1.5
$$h = 5, \ k = 3, \ r = 1.5$$
$$(x-5)^2 + (y-3)^2 = 1.5^2$$
$$(x-5)^2 + (y-3)^2 = 2.25$$

61. $\left(\dfrac{4+(-2)}{2}, \dfrac{3+1}{2} \right) = \left(\dfrac{2}{2}, \dfrac{4}{2} \right) = (1, \ 2)$

63. $\left(\dfrac{-4+2}{2}, \dfrac{-2+2}{2} \right) = \left(\dfrac{-2}{2}, \dfrac{0}{2} \right) = (-1, \ 0)$

65. $\left(\dfrac{4+(-6)}{2}, \dfrac{0+12}{2} \right) = \left(\dfrac{-2}{2}, \dfrac{12}{2} \right) = (-1, 6)$

67. $\left(\dfrac{-3+3}{2}, \dfrac{8+(-2)}{2} \right) = \left(\dfrac{0}{2}, \dfrac{6}{2} \right) = (0, 3)$

69. $\left(\dfrac{5+(-6)}{2}, \dfrac{2+1}{2} \right) = \left(\dfrac{-1}{2}, \dfrac{3}{2} \right) = \left(-\dfrac{1}{2}, \dfrac{3}{2} \right)$

71. $\left(\dfrac{-2.4+1.6}{2}, \dfrac{-3.1+1.1}{2} \right) = \left(\dfrac{-0.8}{2}, \dfrac{-2}{2} \right)$
$$= (-0.4, \ -1)$$

73. $(x_1, y_1) = (30, 20), \ (x_2, y_2) = (50, -5)$
$$\left(\dfrac{30+50}{2}, \dfrac{20+(-5)}{2} \right) = \left(\dfrac{80}{2}, \dfrac{15}{2} \right) = (40, 7.5)$$
They should meet 40 miles east and 7.5 miles north of the warehouse.

75. a. Midpoint: $\left(\dfrac{-1+3}{2}, \dfrac{2+4}{2}\right) = (1,3)$

Center: $(1,3)$

b. $d = \sqrt{\left[3-(-1)\right]^2 + (4-2)^2} = \sqrt{4^2 + 2^2} = 2\sqrt{5}$

$r = \dfrac{d}{2} = \sqrt{5}$

$h = 1, k = 3, r = \sqrt{5}$

$(x-1)^2 + (y-3)^2 = \left(\sqrt{5}\right)^2$

$(x-1)^2 + (y-3)^2 = 5$

77. a. Midpoint: $\left(\dfrac{-2+2}{2}, \dfrac{3+3}{2}\right) = (0,3)$

Center: $(0,3)$

b. $d = \sqrt{\left[2-(-2)\right]^2 + (3-3)^2} = \sqrt{4^2 + 0^2} = 4$

$r = \dfrac{d}{2} = 2$

$h = 0, k = 3, r = 2$

$(x-0)^2 + (y-3)^2 = 2^2$

$x^2 + (y-3)^2 = 4$

79. Center: $(4, 4)$; tangent to x- and y-axes.

$h = 4, k = 4, r = 4$

$\left[x-(4)\right]^2 + \left[y-(4)\right]^2 = 4^2$

$(x-4)^2 + (y-4)^2 = 16$

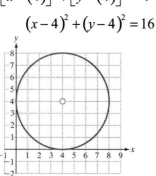

81. $d = \sqrt{(-4-1)^2 + (3-1)^2}$

$= \sqrt{(-5)^2 + 2^2} = \sqrt{25+4} = \sqrt{29}$

Center: $(1, 1)$; radius: $\sqrt{29}$

$h = 1, k = 1, r = \sqrt{29}$

$\left[x-(1)\right]^2 + \left[y-(1)\right]^2 = \left(\sqrt{29}\right)^2$

$(x-1)^2 + (y-1)^2 = 29$

Section 9.2   More on the Parabola

1. **(a)** conic; plane

(b) parabola; directrix; focus

(c) vertex; symmetry

(d) $y = k$

3. $(x_1, y_1) = (0, 0), \quad (x_2, y_2) = (4, -3)$

$d = \sqrt{\left[4-0\right]^2 + \left[-3-0\right]^2}$

$= \sqrt{(4)^2 + (-3)^2}$

$= \sqrt{16+9}$

$= \sqrt{25} = 5$

5. $x^2 + (y+1)^2 = 16$

$\left[x-0\right]^2 + \left[y-(-1)\right]^2 = 4^2$

$h = 0, k = -1, r = 4$

Center: $(0, -1)$; radius: 4

7. $\left(\dfrac{7+4}{2}, \dfrac{-3+5}{2}\right) = \left(\dfrac{11}{2}, \dfrac{2}{2}\right) = \left(\dfrac{11}{2}, 1\right)$

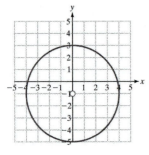

9. $y = (x + 2)^2 + 1$

Vertex: $(-2, 1)$ Axis of symmetry: $x = -2$

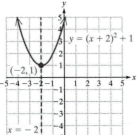

11. $y = x^2 - 4x + 3$

$y = (x^2 - 4x + 4 - 4) + 3$

$y = (x^2 - 4x + 4) - 4 + 3$

$y = (x - 2)^2 - 1$

Vertex: $(2, -1)$ Axis of symmetry: $x = 2$

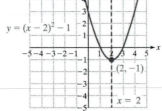

13. $y = -2x^2 + 8x$

$y = -2(x^2 - 4x)$

$y = -2(x^2 - 4x + 4 - 4)$

$y = -2(x^2 - 4x + 4) + 8$

$y = -2(x - 2)^2 + 8$

Vertex: $(2, 8)$ Axis of symmetry: $x = 2$

15. $y = -x^2 - 3x + 2$

$y = -(x^2 + 3x) + 2$

$y = -\left(x^2 + 3x + \dfrac{9}{4} - \dfrac{9}{4}\right) + 2$

$y = -\left(x^2 + 3x + \dfrac{9}{4}\right) + \dfrac{9}{4} + 2$

$y = -\left(x + \dfrac{3}{2}\right)^2 + \dfrac{17}{4}$

Vertex: $\left(-\dfrac{3}{2}, \dfrac{17}{4}\right)$

Axis of symmetry: $x = -\dfrac{3}{2}$

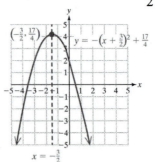

410

17. $x = y^2 - 3$

Vertex: $(-3, 0)$ Axis of symmetry: $y = 0$

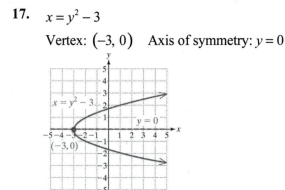

19. $x = -(y - 3)^2 - 3$

Vertex: $(-3, 3)$ Axis of symmetry: $y = 3$

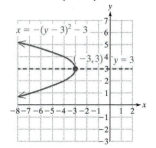

21. $x = -y^2 + 4y - 4$

$x = -(y^2 - 4y) - 4$

$x = -(y^2 - 4y + 4 - 4) - 4$

$x = -(y^2 - 4y + 4) + 4 - 4$

$x = -(y - 2)^2$

Vertex: $(0, 2)$

Axis of symmetry: $y = 2$

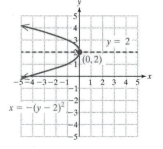

23. $x = y^2 - 2y + 2$

$x = (y^2 - 2y + 1 - 1) + 2$

$x = (y^2 - 2y + 1) - 1 + 2$

$x = (y - 1)^2 + 1$

Vertex: $(1, 1)$ Axis of symmetry: $y = 1$

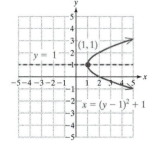

25. $y = x^2 - 4x + 3$

$a = 1, \ b = -4, \ c = 3$

$x = -\dfrac{b}{2a} = -\dfrac{-4}{2(1)} = -\dfrac{-4}{2} = 2$

$y = (2)^2 - 4(2) + 3 = 4 - 8 + 3 = -1$

Vertex: $(2, -1)$

27. $x = y^2 + 2y + 6$

$a = 1, \ b = 2, \ c = 6$

$y = -\dfrac{b}{2a} = -\dfrac{2}{2(1)} = -\dfrac{2}{2} = -1$

$x = (-1)^2 + 2(-1) + 6 = 1 - 2 + 6 = 5$

Vertex: $(5, -1)$

29. $y = -\frac{1}{4}x^2 + x + \frac{3}{4}$

$a = -\frac{1}{4}, \ b = 1, \ c = \frac{3}{4}$

$x = -\frac{b}{2a} = -\frac{1}{2\left(-\frac{1}{4}\right)} = -\frac{1}{-\frac{1}{2}} = 2$

$y = -\frac{1}{4}(2)^2 + 2 + \frac{3}{4} = -1 + 2 + \frac{3}{4} = \frac{7}{4}$

Vertex: $\left(2, \frac{7}{4}\right)$

31. $y = x^2 - 3x + 2$

$a = 1, \ b = -3, \ c = 2$

$x = -\frac{b}{2a} = -\frac{-3}{2(1)} = -\frac{-3}{2} = \frac{3}{2}$

$y = \left(\frac{3}{2}\right)^2 - 3\left(\frac{3}{2}\right) + 2 = \frac{9}{4} - \frac{9}{2} + 2 = -\frac{1}{4}$

Vertex: $\left(\frac{3}{2}, -\frac{1}{4}\right)$

33. $x = -3y^2 - 6y + 7$

$a = -3, \ b = -6, \ c = 7$

$y = -\frac{b}{2a} = -\frac{-6}{2(-3)} = -\frac{-6}{-6} = -1$

$x = -3(-1)^2 - 6(-1) + 7 = -3 + 6 + 7 = 10$

Vertex: $(10, -1)$

35. $h(x) = -x^2 + 10x - 3$

$a = -1, \ b = 10, \ c = -3$

$x = -\frac{b}{2a} = -\frac{10}{2(-1)} = -\frac{10}{-2} = 5$

$h(x) = -(5)^2 + 10(5) - 3 = -25 + 50 - 3 = 22$

Vertex: $(5, 22)$

The maximum height of the water is 22 ft.

37. A parabola whose equation is written in the form $y = a(x-h)^2 + k$ has a vertical axis of symmetry.

A parabola whose equation is written in the form $x = a(y-k)^2 + h$ has a horizontal axis of symmetry.

39. $y = (x-4)^2 + 2$

Vertical axis of symmetry; opens upward.

41. $y = -3(x+2)^2 - 1$

Vertical axis of symmetry; opens downward.

43. $x = y^2 - 2$

Horizontal axis of symmetry; opens right.

45. $x = -2(y-1)^2 - 3$

Horizontal axis of symmetry; opens left.

47. $y = -x^2 + 3$

Vertical axis of symmetry; opens downward.

49. $x = y^2 - 5y + 1$

Horizontal axis of symmetry; opens right.

Section 9.3 The Ellipse and Hyperbola

1. **(a)** ellipse; focus

(b) ellipse

(c) hyperbola; foci

(d) transverse

(e) $\dfrac{x^2}{a^2} - \dfrac{y^2}{b^2} = 1$

(f) $\dfrac{y^2}{b^2} - \dfrac{x^2}{a^2} = 1$

3.
$$x^2 + y^2 - 16x + 12y = 0$$
$$\left(x^2 - 16x \right) + \left(y^2 + 12y \right) = 0$$
$$\left(x^2 - 16x + 64\right) + \left(y^2 + 12y + 36\right)$$
$$= 0 + 64 + 36$$
$$(x - 8)^2 + (y + 6)^2 = 100$$
$$[x - 8]^2 + [y - (-6)]^2 = 10^2$$
Center: $(8, -6)$; radius: 10

5. $y = 3(x + 3)^2 - 1$

Vertex: $(-3, -1)$ Axis of symmetry: $x = -3$

7. Center: $\left(\dfrac{1}{2}, \dfrac{5}{2}\right)$; radius: $\dfrac{1}{2}$

$$h = \frac{1}{2}, \; k = \frac{5}{2}, \; r = \frac{1}{2}$$

$$\left[x - \left(\frac{1}{2}\right)\right]^2 + \left[y - \left(\frac{5}{2}\right)\right]^2 = \left(\frac{1}{2}\right)^2$$

$$\left(x - \frac{1}{2}\right)^2 + \left(y - \frac{5}{2}\right)^2 = \frac{1}{4}$$

9.
$$\frac{x^2}{4} + \frac{y^2}{9} = 1$$
$$\frac{x^2}{2^2} + \frac{y^2}{3^2} = 1$$
$$a = 2; \; b = 3$$
Intercepts: $(2, 0), (-2, 0), (0, 3), (0, -3)$

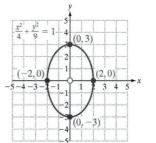

11.
$$\frac{x^2}{16} + \frac{y^2}{9} = 1$$
$$\frac{x^2}{4^2} + \frac{y^2}{3^2} = 1$$
$$a = 4; \; b = 3$$
Intercepts: $(4, 0), (-4, 0), (0, 3), (0, -3)$

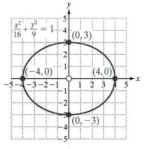

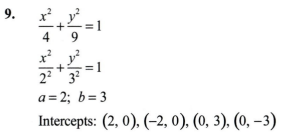

13. $4x^2 + y^2 = 4$

$$\frac{4x^2}{4} + \frac{y^2}{4} = \frac{4}{4}$$

$$\frac{x^2}{1^2} + \frac{y^2}{2^2} = 1$$

$a = 1;\ \ b = 2$

Intercepts: $(1, 0), (-1, 0), (0, 2), (0, -2)$

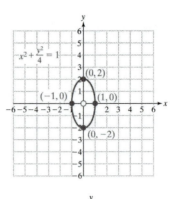

15. $x^2 + 25y^2 - 25 = 0$

$$x^2 + 25y^2 = 25$$

$$\frac{x^2}{25} + \frac{25y^2}{25} = \frac{25}{25}$$

$$\frac{x^2}{25} + \frac{y^2}{1} = 1$$

$$\frac{x^2}{5^2} + \frac{y^2}{1^2} = 1$$

$a = 5;\ \ b = 1$

Intercepts: $(5, 0), (-5, 0), (0, 1), (0, -1)$

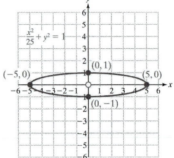

17. $\dfrac{(x-4)^2}{4} + \dfrac{(y-5)^2}{9} = 1$

Center: $(4, 5)$ $a = 3,\ b = 2$

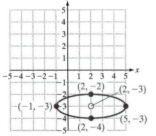

19. $\dfrac{(x+1)^2}{25} + \dfrac{(y-2)^2}{9} = 1$

Center: $(-1, 2)$ $a = 5,\ b = 3$

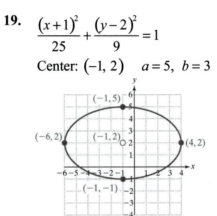

21. $\dfrac{(x-2)^2}{9} + (y+3)^2 = 1$

Center: $(2, -3)$ $a = 3,\ b = 1$

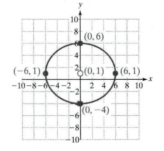

23. $\dfrac{x^2}{36} + \dfrac{(y-1)^2}{25} = 1$

Center: $(0, 1)$ $a = 6,\ b = 5$

414

25. $\dfrac{y^2}{6} - \dfrac{x^2}{18} = 1$ Vertical

27. $\dfrac{x^2}{20} - \dfrac{y^2}{15} = 1$ Horizontal

29. $x^2 - y^2 = 12$

$\dfrac{x^2}{12} - \dfrac{y^2}{12} = 1$ Horizontal

31. $x^2 - 3y^2 = -9$

$\dfrac{x^2}{-9} - \dfrac{3y^2}{-9} = 1$

$\dfrac{y^2}{3} - \dfrac{x^2}{9} = 1$ Vertical

33. $\dfrac{x^2}{25} - \dfrac{y^2}{16} = 1$

$\dfrac{x^2}{5^2} - \dfrac{y^2}{4^2} = 1$

Transverse axis is horizontal.

$a = 5; \ b = 4$

Reference rectangle has corners at:

$$(5, 4), (5, -4), (-5, 4), (-5, -4)$$

Vertices: $(5, 0), (-5, 0)$

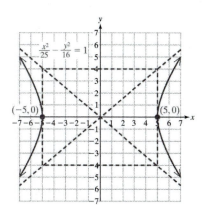

35. $\dfrac{y^2}{4} - \dfrac{x^2}{4} = 1$

$\dfrac{y^2}{2^2} - \dfrac{x^2}{2^2} = 1$

Transverse axis is vertical.

$a = 2; \ b = 2$

Reference rectangle has corners at:

$$(2, 2), (2, -2), (-2, 2), (-2, -2)$$

Vertices: $(0, 2), (0, -2)$

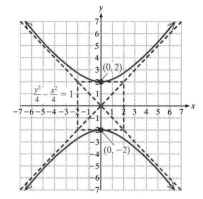

37. $36x^2 - y^2 = 36$

$\dfrac{36x^2}{36} - \dfrac{y^2}{36} = \dfrac{36}{36}$

$\dfrac{x^2}{1^2} - \dfrac{y^2}{6^2} = 1$

Transverse axis is horizontal.

$a = 1; \ b = 6$

Reference rectangle has corners at:

$$(1, 6), (1, -6), (-1, 6), (-1, -6)$$

Vertices: $(1, 0), (-1, 0)$

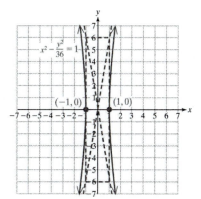

39. $y^2 - 4x^2 - 16 = 0$

$$y^2 - 4x^2 = 16$$

$$\frac{y^2}{16} - \frac{4x^2}{16} = \frac{16}{16}$$

$$\frac{y^2}{4^2} - \frac{x^2}{2^2} = 1$$

Transverse axis is vertical.

$a = 2; \ b = 4$

Reference rectangle has corners at:

$$(2, 4), (2, -4), (-2, 4), (-2, -4)$$

Vertices: $(0, 4), (0, -4)$

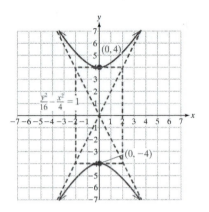

41. $\frac{x^2}{6} - \frac{y^2}{10} = 1$ Hyperbola

43. $\frac{y^2}{4} + \frac{x^2}{16} = 1$ Ellipse

45. $4x^2 + y^2 = 16$

$\frac{x^2}{4} + \frac{y^2}{16} = 1$ Ellipse

47. $-y^2 + 2x^2 = -10$

$\frac{y^2}{10} - \frac{x^2}{5} = 1$ Hyperbola

49. $5x^2 + y^2 - 10 = 0$

$5x^2 + y^2 = 10$ Ellipse

$\frac{x^2}{2} + \frac{y^2}{10} = 1$

51. $y^2 - 6x^2 = 6$

$\frac{y^2}{6} - \frac{x^2}{1} = 1$ Hyperbola

53. Find the equation of the ellipse with $a = 60$ and $b = 50$.

$$\frac{x^2}{60^2} + \frac{y^2}{50^2} = 1$$

Find y when $x = 10$.

$$\frac{10^2}{60^2} + \frac{y^2}{50^2} = 1$$

$$\frac{100}{3600} + \frac{y^2}{2500} = 1$$

$$\frac{y^2}{2500} = 1 - \frac{100}{3600}$$

$$y^2 = 2500\left(1 - \frac{100}{3600}\right) \approx 2430.5556$$

$$y \approx 49 \text{ ft}$$

The height 10 ft from the center is approximately 49 ft.

55. $\frac{(x-1)^2}{9} - \frac{(y+2)^2}{4} = 1$

Center: $(1, -2)$ $a = 3, \ b = 2$

Vertices: $(4, -2), (-2, -2)$

57. $\dfrac{(y-1)^2}{4} - (x+3)^2 = 1$

Center: $(-3, 1)$ $a = 2,\ b = 1$

Vertices: $(-3, 3), (-3, -1)$

Problem Recognition Exercises

1. $(x-h)^2 + (y-k)^2 = r^2$

Standard equation of a circle

3. $\sqrt{(x_2 - x_1)^2 + (y_2 - y_1)^2}$

Distance between two points

5. $y = a(x-h)^2 + k$

Parabola with vertical axis of symmetry

7. $\dfrac{y^2}{b^2} - \dfrac{x^2}{a^2} = 1$

Hyperbola with vertical transverse

9. $y = -2(x-3)^2 + 4$ Parabola

11. $(x+3)^2 + (y+2)^2 = 4$ Circle

13. $\dfrac{x^2}{9} - \dfrac{y^2}{9} = 1$ Hyperbola

15. $\dfrac{x^2}{16} + \dfrac{y^2}{4} = 0$ None of the above

17. $y = \dfrac{1}{2}(x+2)^2 - 3$ Parabola

19. $x = (y+2)^2 - 4$ Parabola

21. $(x-1)^2 + (y+1)^2 = 0$ None of the above

23. $\dfrac{x^2}{25} + \dfrac{y^2}{4} = 1$ Ellipse

25. $y = (x-6)^2 + 4$ Parabola

27. $\dfrac{y^2}{3} - \dfrac{x^2}{3} = 1$ Hyperbola

29. $\dfrac{x^2}{9} + \dfrac{y^2}{12} = 1$ Ellipse

Section 9.4   Nonlinear Systems of Equations in Two Variables

1.   **(a)** nonlinear

   (b) ordered

3. $(x_1, y_1) = (8, -1), \quad (x_2, y_2) = (1, -8)$

$$d = \sqrt{[1-8]^2 + [-8-(-1)]^2}$$
$$= \sqrt{(-7)^2 + (-7)^2}$$
$$= \sqrt{49+49}$$
$$= \sqrt{98} = 7\sqrt{2}$$

5.   Center: $(-5, 3)$; radius: 8

   $h = -5, k = 3, r = 8$

$$[x-(-5)]^2 + [y-(3)]^2 = 8^2$$
$$(x+5)^2 + (y-3)^2 = 64$$

7. $\left(\dfrac{3+(-4)}{2}, \dfrac{-9+2}{2}\right) = \left(\dfrac{-1}{2}, \dfrac{-7}{2}\right) = \left(-\dfrac{1}{2}, -\dfrac{7}{2}\right)$

9.   Zero, one, or two

11.   Zero, one, or two

13.   Zero, one, two, three, or four

15.   Zero, one, two, three, or four

17. $y = x+3$

   $x^2 + y = 9$

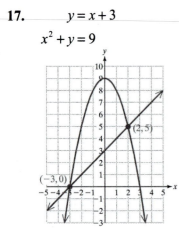

Substitute the first equation into the second and solve for x:

$$x^2 + x + 3 = 9$$
$$x^2 + x - 6 = 0$$
$$(x-2)(x+3) = 0$$
$$x - 2 = 0 \quad \text{or} \quad x + 3 = 0$$
$$x = 2 \quad \text{or} \quad x = -3$$
$$y = 2+3 = 5 \quad y = -3+3 = 0$$

Solutions: $\{(2, 5), (-3, 0)\}$

19. $x^2 + y^2 = 1$

   $y = x+1$

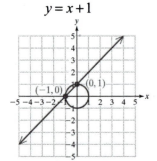

Substitute the second equation into the first and solve for x:

21. $x^2 + y^2 = 6$

   $y = x^2$

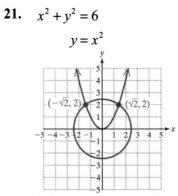

Substitute the second equation into the first and solve for y:

$$x^2 + (x+1)^2 = 1$$
$$x^2 + x^2 + 2x + 1 = 1$$
$$2x^2 + 2x = 0$$
$$2x(x+1) = 0$$

$2x = 0$ or $\quad x + 1 = 0$

$x = 0$ or $\qquad x = -1$

$y = 0 + 1 = 1 \quad y = -1 + 1 = 0$

Solutions: $\{(0,\ 1),\ (-1,\ 0)\}$

$$y + y^2 = 6$$
$$y^2 + y - 6 = 0$$
$$(y-2)(y+3) = 0$$

$y - 2 = 0$ or $\quad y + 3 = 0$

$y = 2$ or $\qquad y = -3$

$x = \pm\sqrt{2} \qquad x = \pm\sqrt{-3}$

no solution

Solutions: $\left\{\left(\sqrt{2},\ 2\right), \left(-\sqrt{2},\ 2\right)\right\}$

23.
$$y = \sqrt{x}$$
$$2x^2 - y^2 = 1$$

Substitute the first equation into the second and solve for x:

$$2x^2 - \left(\sqrt{x}\right)^2 = 1$$
$$2x^2 - x = 1$$
$$2x^2 - x - 1 = 0$$
$$(2x+1)(x-1) = 0$$

$2x + 1 = 0 \quad$ or $\ x - 1 = 0$

$x = -\dfrac{1}{2} \quad$ or $\quad x = 1$

$y = \sqrt{-\dfrac{1}{2}} \qquad y = \sqrt{1} = 1$

no solution

Solution: $\{(1,\ 1)\}$

25.
$$y = x^2$$
$$y = -\sqrt{x}$$

Substitute the second equation into the first and solve for x:

$$-\sqrt{x} = x^2$$
$$\left(-\sqrt{x}\right)^2 = \left(x^2\right)^2$$
$$x = x^4$$
$$x^4 - x = 0$$
$$x\left(x^3 - 1\right) = 0$$

$x = 0$ or $\quad x^3 - 1 = 0$

$x = 0$ or $\qquad x^3 = 1$

$x = 0$ or $\qquad x = 1$

$y = 0^2 = 0 \qquad y = 1^2 = 1$

Solutions: $\{(0,\ 0)\} \left(1 = -\sqrt{1} \text{ does not check.}\right)$

27.
$$y = x^2$$
$$y = (x-3)^2$$

Substitute the second equation into the first and solve for x:

29.
$$y = x^2 + 6x$$
$$y = 4x$$

Substitute the second equation into the first and solve for x:

$$x^2 + 6x = 4x$$
$$x^2 + 2x = 0$$
$$x(x+2) = 0$$

$x = 0$ or $\quad x + 2 = 0$

$x = 0$ or $\qquad x = -2$

$y = 4(0) = 0 \quad y = 4(-2) = -8$

Solutions: $\{(0,\ 0),\ (-2,\ -8)\}$

$$(x-3)^2 = x^2$$
$$x^2 - 6x + 9 = x^2$$
$$-6x + 9 = 0$$
$$-6x = -9$$
$$x = \frac{3}{2}$$
$$y = \left(\frac{3}{2}\right)^2 = \frac{9}{4}$$

Solutions: $\left\{\left(\frac{3}{2}, \frac{9}{4}\right)\right\}$

31. $x^2 - 5x + y = 0$

$$y = 3x + 1$$

Substitute the second equation into the first and solve for x:

$$x^2 - 5x + 3x + 1 = 0$$
$$x^2 - 2x + 1 = 0$$
$$(x-1)^2 = 0$$
$$x - 1 = 0$$
$$x = 1$$
$$y = 3(1) + 1 = 4$$

Solution: $\{(1, 4)\}$

33. Add the equations to eliminate y^2 and solve:

$$4x^2 - y^2 = 4$$
$$\underline{4x^2 + y^2 = 4}$$
$$8x^2 \quad\quad = 8$$
$$x^2 = 1$$
$$x = \pm 1$$

$x = 1:$ $x = -1:$

$$4(1)^2 + y^2 = 4 \quad\quad 4(-1)^2 + y^2 = 4$$
$$4 + y^2 = 4 \quad\quad\quad 4 + y^2 = 4$$
$$y^2 = 0 \quad\quad\quad\quad y^2 = 0$$
$$y = 0 \quad\quad\quad\quad y = 0$$

Solutions: $\{(1, 0), (-1, 0)\}$

35. Multiply the first equation by -1 and add to the second equation to eliminate y^2 and solve:

$$x^2 + y^2 = 4 \xrightarrow{\times -1} -x^2 - y^2 = -4$$
$$2x^2 + y^2 = 8 \longrightarrow \underline{2x^2 + y^2 = 8}$$
$$x^2 \quad\quad = 4$$
$$x = \pm 2$$

$x = 2:$ $x = -2$

$$2^2 + y^2 = 4 \quad\quad (-2)^2 + y^2 = 4$$
$$4 + y^2 = 4 \quad\quad\quad 4 + y^2 = 4$$
$$y^2 = 0 \quad\quad\quad\quad y^2 = 0$$
$$y = 0 \quad\quad\quad\quad y = 0$$

Solutions: $\{(2, 0), (-2, 0)\}$

37. Multiply the first equation by 2 and the second equation by 5 and add the results to eliminate y^2 and solve:

$$2x^2 - 5y^2 = -2 \xrightarrow{\times 2} 4x^2 - 10y^2 = -4$$
$$3x^2 + 2y^2 = 35 \xrightarrow{\times 5} \underline{15x^2 + 10y^2 = 175}$$
$$19x^2 \qquad = 171$$
$$x^2 = 9$$
$$x = \pm 3$$

$x = 3:$

$$2(3)^2 - 5y^2 = -2$$
$$18 - 5y^2 = -2$$
$$-5y^2 = -20$$
$$y^2 = 4$$
$$y = \pm 2$$

$x = -3:$

$$2(-3)^2 - 5y^2 = -2$$
$$18 - 5y^2 = -2$$
$$-5y^2 = -20$$
$$y^2 = 4$$
$$y = \pm 2$$

Solutions: $\{(3, 2), (3, -2), (-3, 2), (-3, -2)\}$

39. Solve the second equation for y, substitute into the first equation and solve:

$$x^2 + y^2 = 100$$

$$4y - 3x = 0 \rightarrow y = \frac{3}{4}x$$

$$x^2 + \left(\frac{3}{4}x\right)^2 = 100$$
$$x^2 + \frac{9}{16}x^2 = 100$$
$$16x^2 + 9x^2 = 1600$$
$$25x^2 = 1600$$
$$x^2 = 64$$
$$x = \pm 8$$

$$x = 8: \quad y = \frac{3}{4}(8) = 6$$

$$x = -8: \quad y = \frac{3}{4}(-8) = -6$$

Solutions: $\{(8, 6), (-8, -6)\}$

41. Multiply the first equation by 16 and the second equation by -4 and add the results to eliminate y^2 and solve:

$$\frac{x^2}{16} + \frac{y^2}{4} = 1 \xrightarrow{\times 16} x^2 + 4y^2 = 16$$
$$x^2 + y^2 = 4 \xrightarrow{\times -4} \underline{-4x^2 - 4y^2 = -16}$$
$$-3x^2 \qquad = 0$$
$$x^2 = 0$$
$$x = 0$$

$x = 0:$

$$0^2 + y^2 = 4$$
$$0 + y^2 = 4$$
$$y^2 = 4$$
$$y = \pm 2$$

Solutions: $\{(0, 2), (0, -2)\}$

43. Solve the first equation for y, substitute into the second equation and solve:

45. Multiply the second equation by -1 and add to the first equation to eliminate xy and solve:

$2y = -x + 2 \rightarrow y = -\dfrac{1}{2}x + 1$

$x^2 + 4y^2 = 4$

$x^2 + 4\left(-\dfrac{1}{2}x + 1\right)^2 = 4$

$x^2 + 4\left(\dfrac{1}{4}x^2 - x + 1\right) = 4$

$x^2 + x^2 - 4x + 4 = 4$

$2x^2 - 4x = 0$

$2x(x - 2) = 0$

$2x = 0 \text{ or } x - 2 = 0$

$x = 0 \text{ or }\quad x = 2$

$x = 0: \quad y = -\dfrac{1}{2}(0) + 1 = 1$

$x = 2: \quad y = -\dfrac{1}{2}(2) + 1 = -1 + 1 = 0$

Solutions: $\{(0, 1), (2, 0)\}$

$x^2 - xy = -4 \longrightarrow \qquad x^2 - xy = -4$

$2x^2 - xy = 12 \xrightarrow{\times -1} -2x^2 + xy = -12$

$\qquad\qquad\qquad\qquad \dfrac{}{-x^2 \qquad = -16}$

$x^2 = 16$

$x = \pm 4$

$x = 4: \qquad\qquad x = -4$

$4^2 - 4y = -4 \qquad (-4)^2 - (-4)y = -4$

$16 - 4y = -4 \qquad 16 + 4y = -4$

$-4y = -20 \qquad\qquad 4y = -20$

$y = 5 \qquad\qquad\quad y = -5$

Solutions: $\{(4, 5), (-4, -5)\}$

47. Multiply the second equation by 2 and add the equations to eliminate xy and solve:

$3x^2 + 2xy = 4 \longrightarrow 3x^2 + 2xy = 4$

$\dfrac{x^2 - xy = 3 \xrightarrow{\times 2} 2x^2 - 2xy = 6}{ 5x^2 \qquad\quad = 10}$

$x^2 = 2$

$x = \pm\sqrt{2}$

$x = \sqrt{2}: \qquad\qquad x = -\sqrt{2}:$

$\left(\sqrt{2}\right)^2 - \left(\sqrt{2}\right)y = 3 \quad \left(-\sqrt{2}\right)^2 - \left(-\sqrt{2}\right)y = 3$

$2 - \sqrt{2}y = 3 \qquad\qquad 2 + \sqrt{2}y = 3$

$-\sqrt{2}y = 1 \qquad\qquad\quad \sqrt{2}y = 1$

$y = -\dfrac{1}{\sqrt{2}} = -\dfrac{\sqrt{2}}{2} \qquad y = \dfrac{1}{\sqrt{2}} = \dfrac{\sqrt{2}}{2}$

Solutions: $\left\{\left(\sqrt{2}, -\dfrac{\sqrt{2}}{2}\right), \left(-\sqrt{2}, \dfrac{\sqrt{2}}{2}\right)\right\}$

49. Let $x =$ one number

$y =$ the second number

$x + y = 7 \quad \rightarrow \quad y = 7 - x$

$x^2 + y^2 = 25$

Substitute the first equation into the second and solve for x:

$x^2 + (7 - x)^2 = 25$

$x^2 + 49 - 14x + x^2 = 25$

$2x^2 - 14x + 24 = 0$

$2(x^2 - 7x + 12) = 0$

$2(x - 4)(x - 3) = 0$

$x - 4 = 0 \text{ or } \quad x - 3 = 0$

$x = 4 \text{ or } \qquad x = 3$

$y = 7 - 4 = 3 \quad y = 7 - 3 = 4$

The numbers are 3 and 4.

51. Let $x =$ one number

$y =$ the second number

Add the equations to eliminate y^2 and solve:

$$x^2 + y^2 = 32$$
$$\underline{x^2 - y^2 = 18}$$
$$2x^2 \quad\;\; = 50$$
$$x^2 = 25$$
$$x = \pm 5$$

$x = 5:$ $x = -5:$

$5^2 + y^2 = 32$ $(-5)^2 + y^2 = 32$

$25 + y^2 = 32$ $25 + y^2 = 32$

$y^2 = 7$ $y^2 = 7$

$y = \pm\sqrt{7}$ $y = \pm\sqrt{7}$

The numbers are 5 and $\sqrt{7}$, 5 and $-\sqrt{7}$, -5 and $\sqrt{7}$, or -5 and $-\sqrt{7}$.

Section 9.5   Nonlinear Inequalities and Systems of Inequalities

1. $y = \left(\dfrac{1}{3}\right)^x$ Graph k

3. $y = -4x^2$ Graph e

5. $y = x^3$ Graph i

7. $\dfrac{x^2}{4} - \dfrac{y^2}{9} = 1$ Graph j

9. $y = \log_2(x)$ Graph a

11. $(x+2)^2 + (y-1)^2 = 4$ Graph d

13. $-x^2 + y^3 > 1$

$-(2)^2 + (3)^3 = -4 + 27 = 23 > 1$

The statement is true.

15. $\dfrac{x^2}{36} + \dfrac{y^2}{25} < 1$

$\dfrac{5^2}{36} + \dfrac{4^2}{25} = \dfrac{25}{36} + \dfrac{16}{25} < 1$ False

$x^2 + y^2 \geq 4$

$5^2 + 4^2 = 25 + 16 = 41 \geq 4$ True

The statement is false.

17. a. $x^2 + y^2 \leq 9$

Graph the related equation $x^2 + y^2 = 9$ (a circle) using a solid curve.
The circle divides the xy-plane into two regions.
Select test points from each region:

Test: $(0, 0)$ Test: $(0, 4)$

$0^2 + 0^2 \leq 9$ $0^2 + 4^2 \leq 9$

$0 + 0 \leq 9$ $0 + 16 \leq 9$

$0 \leq 9$ True $16 \leq 9$ False

Shade the region inside the circle.

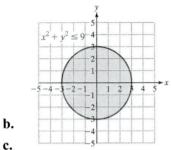

b.

c.

The set of points on and outside the circle $x^2 + y^2 = 9$.

The set of points on the circle $x^2 + y^2 = 9$.

19. a. $y \geq x^2 + 1$

Graph the related equation $y = x^2 + 1$ (a parabola) using a solid curve.

The parabola divides the xy-plane into two regions.

Select test points from each region:

Test: $(0, 0)$ Test: $(0, 2)$

$\quad 0 \geq 0^2 + 1$ $\quad 2 \geq 0^2 + 1$

$\quad 0 \geq 0 + 1$ $\quad 2 \geq 0 + 1$

$\quad 0 \geq 1$ False $\quad 2 \geq 1$ True

Shade the region above the parabola.

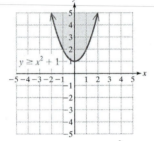

b.   The parabola $y = x^2 + 1$ would be drawn as a dashed curve.

21.   The area described is the interior of a circle
centered at (3, –4) with a radius of 25 miles.
The inequality that describes this area is:

$(x - 3)^2 + (y + 4)^2 \leq 25^2$

$(x - 3)^2 + (y + 4)^2 \leq 625$

23. $2x + y \geq 1$

Graph the related equation $2x + y = 1$ (a line) using a solid line.

The line divides the xy-plane into two regions.

Select test points from each region:

Test: $(0, 0)$ Test: $(0, 4)$

$\quad 2(0) + 0 \geq 1$ $\quad 2(0) + 4 \geq 1$

$\quad 0 + 0 \geq 1$ $\quad 0 + 4 \geq 1$

$\quad 0 \geq 1$ False $\quad 4 \geq 1$ True

Shade the region above the line.

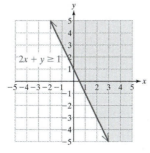

25. $x \le y^2$

Graph the related equation $x = y^2$ (a parabola) using a solid curve.

The curve divides the xy-plane into two regions.

Select test points from each region:

Test: $(-1, 0)$ Test: $(1, 0)$

$\quad -1 \le (0)^2$ $\quad 1 \le (0)^2$

$\quad -1 \le 0$ True $\quad 1 \le 0$ False

Shade the region to the left of the curve.

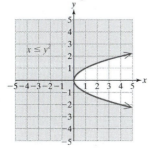

27. $(x-1)^2 + (y+2)^2 > 9$

Graph the related equation $(x-1)^2 + (y+2)^2 = 9$ (a circle) using a dashed curve. The curve divides the xy-plane into two regions.

Select test points from each region:

Test: $(0, 0)$ Test: $(0, 2)$

$\quad (0-1)^2 + (0+2)^2 > 9$ $\quad (0-1)^2 + (2+2)^2 > 9$

$\quad\quad\quad 1 + 4 > 9$ $\quad\quad\quad\quad 1 + 16 > 9$

$\quad\quad\quad\quad 5 > 9$ False $\quad\quad\quad\quad 17 > 9$ True

Shade the region outside the circle.

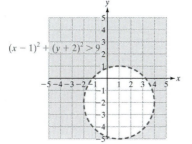

29. $x + y^2 \geq 4$

Graph the related equation $x + y^2 = 4$ (a parabola) using a solid curve. The curve divides the xy-plane into two regions.

Select test points from each region:

Test: $(0, 0)$

$0 + (0)^2 \geq 4$

$0 + 0 \geq 1$

$0 \geq 1$ False

Test: $(0, 4)$

$0 + (4)^2 \geq 4$

$0 + 16 \geq 4$

$16 \geq 4$ True

Shade the region to the right of the parabola.

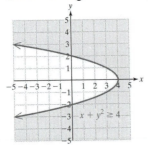

31. $9x^2 - y^2 > 9$

Graph the related equation $9x^2 - y^2 = 9$ (a hyperbola) using a dashed curve. The curve divides the xy-plane into three regions.

Select test points from each region:

Test: $(-2, 0)$

$9(-2)^2 - (0)^2 > 9$

$36 - 0 > 9$

$36 > 9$ True

Test: $(0, 0)$

$9(0)^2 - (0)^2 > 9$

$0 - 0 > 9$

$0 > 9$ False

Test: $(2, 0)$

$9(2)^2 - (0)^2 > 9$

$36 - 0 > 9$

$36 > 9$ True

Shade the region to the left of the left branch and to the right of the right branch of the hyperbola.

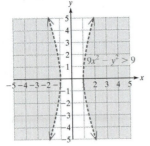

33. $x^2 + 16y^2 \leq 16$

Graph the related equation $x^2 + 16y^2 = 16$ (an ellipse) using a solid curve. The curve divides the xy-plane into two regions.

Select test points from each region:

Test: $(0, 0)$ Test: $(0, 2)$

$(0)^2 + 16(0)^2 \leq 16$ $(0)^2 + 16(2)^2 \leq 16$

$\quad 0 + 0 \leq 16$ $0 + 64 \leq 16$

$\quad\quad 0 \leq 16$ True $64 \leq 16$ False

Shade the region inside the ellipse.

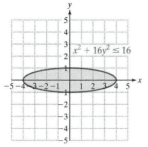

35. $y \leq \ln x$

Graph the related equation $y = \ln x$ using a solid curve. The curve divides the xy-plane into two regions.

Select test points from each region:

Test: $(0, 0)$ Test: $(2, 0)$

$\quad 0 \leq \ln(0)$ $0 \leq \ln(2)$

$\quad 0 \leq DNE$ False $0 \leq 0.6931$ True

Shade the region below the curve.

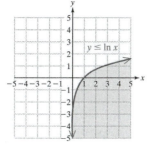

37. $y > 5^x$

Graph the related equation $y = 5^x$ using a dashed curve. The curve divides the xy-plane into two regions.

Select test points from each region:

Test: $(0, 0)$ Test: $(0, 2)$

$\quad 0 > 5^0$ $2 > 5^0$

$\quad 0 > 1$ False $2 > 1$ True

Shade the region above the curve.

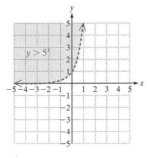

39. $y \geq \sqrt{x}, \quad x \geq 0$

Graph the related equation $y = \sqrt{x}$ using a solid curve.

The curve divides the xy-plane into two regions.

Select test points from each region:

Test: $(2, 0)$ Test: $(0, 2)$

$\quad\quad 0 \geq \sqrt{2}$ $2 \geq \sqrt{0}$

$\quad\quad 0 \geq 1.414$ False $2 \geq 0$ True

Shade the region above the curve.

Graph the related equation $x = 0$, a vertical line using a solid line.

Shade the region to the right of the line.

The solution is the intersection of the two regions.

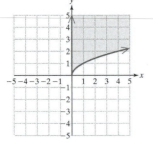

41. $x^2 - y^2 \geq 1, \quad x \leq 0$

Graph the related equation $x^2 - y^2 = 1$ (a hyperbola) using a solid curve. The curve divides the xy-plane into three regions.

Select test points from each region:

Test: $(-2, 0)$ Test: $(0, 0)$ Test: $(2, 0)$

$\quad (-2)^2 - 0^2 \geq 1$ $0^2 - 0^2 \geq 1$ $(2)^2 - 0^2 \geq 1$

$\quad\quad 4 - 0 \geq 1$ $0 - 0 \geq 1$ $4 - 0 \geq 1$

$\quad\quad\quad 4 \geq 1$ True $0 \geq 1$ False $4 \geq 1$ True

Shade the region to the left of the left branch and to the right of the right branch of the hyperbola.

Graph the related equation $x = 0$, a vertical line using a solid line.

Shade the region to the left of the line.

The solution is the intersection of the two regions.

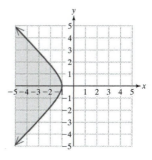

43. $y^2 - x^2 \geq 1, \quad y \geq 0$

Graph the related equation $y^2 - x^2 = 1$ (a hyperbola) using a solid curve. The hyperbola divides the xy-plane into three regions.

Select test points from each region:

Test: $(0, -2)$ Test: $(0, 0)$ Test: $(0, 2)$

$(-2)^2 - 0^2 \geq 1$ $0^2 - 0^2 \geq 1$ $(2)^2 - 0^2 \geq 1$

$4 - 0 \geq 1$ $0 - 0 \geq 1$ $4 - 0 \geq 1$

$4 \geq 1$ True $0 \geq 1$ False $4 \geq 1$ True

Shade the region below the lower branch and above the upper branch of the hyperbola.

Graph the related equation $y = 0$, a horizontal line using a solid line. Shade the region above the line.

The solution is the intersection of the two regions.

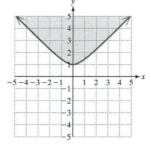

45. $y > x^3, \quad y < 8, \quad x > 0$

Graph the related equation $y = x^3$ using a dashed curve.

The curve divides the xy-plane into two regions.

Select test points from each region:

Test: $(0, -2)$ Test: $(0, 2)$

$\quad\quad -2 > 0^3$ $\quad\quad 2 > 0^3$

$\quad\quad -2 > 0$ False $\quad\quad 2 > 0$ True

Shade the region above the curve.

Graph the related equation $y = 8,$ a horizontal line using a dashed line. Shade the region below the line.

Graph the related equation $x = 0,$ a vertical line using a dashed line. Shade the region to the right of the line.

The solution is the intersection of the three regions.

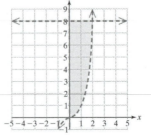

47. $\dfrac{x^2}{4} + \dfrac{y^2}{25} \geq 1, \quad x^2 + \dfrac{y^2}{4} \leq 1$

Graph the related equation $\dfrac{x^2}{4} + \dfrac{y^2}{25} = 1$ (an ellipse) using a solid curve. The ellipse divides the xy-plane into two regions.

Select test points from each region:

Test: $(0, 0)$ Test: $(0, 6)$

$\quad \dfrac{0^2}{4} + \dfrac{0^2}{25} \geq 1$ $\quad \dfrac{0^2}{4} + \dfrac{36^2}{25} \geq 1$

$\quad\quad 0 + 0 \geq 1$ $\quad\quad 0 + \dfrac{36}{25} \geq 1$

$\quad\quad 0 \geq 1$ False $\quad\quad \dfrac{36}{25} \geq 1$ True

Shade the region outside the ellipse.

Graph the related equation $x^2 + \dfrac{y^2}{4} = 1$ (an ellipse) using a solid curve.

The ellipse divides the xy-plane into two regions.

Select test points from each region:

Test: $(0, 0)$　　　　　　Test: $(0, 4)$

$$0^2 + \frac{0^2}{4} \leq 1 \qquad\qquad 0^2 + \frac{4^2}{4} \leq 1$$

$$0 + 0 \leq 1 \qquad\qquad\qquad 0 + 4 \leq 1$$

$$0 \leq 1 \quad \text{True} \qquad\qquad 4 \leq 1 \quad \text{False}$$

Shade the region inside the ellipse

The solution is the intersection of the two regions, which is empty.

There is no solution. $\{\,\}$

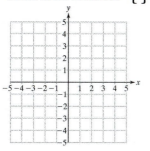

49.　$x > (y - 2)^2 + 1, \quad x - y < 1$

Graph the related equation $x = (y - 2)^2 + 1$ (a parabola) using a dashed curve. The parabola divides the xy-plane into two regions.

Select test points from each region:

Test: $(0, 0)$　　　　Test: $(2, 2)$

$$0 > (0 - 2)^2 + 1 \qquad\qquad 2 > (2 - 2)^2 + 1$$

$$0 > 4 + 1 \qquad\qquad\qquad 2 > 0 + 1$$

$$0 > 5 \quad \text{False} \qquad\qquad 2 > 1 \quad \text{True}$$

Shade the region to the right of the parabola.

Graph the related equation $x - y = 1$ using a dashed line. The line divides the xy-plane into two regions.

Select test points from each region:

Test: $(0, -2)$　　　　　　Test: $(0, 0)$

$$0 - (-2) < 1 \qquad\qquad 0 - 0 < 1$$

$$2 < 1 \quad \text{False} \qquad\qquad 0 < 1 \quad \text{True}$$

Shade the region above the line.

The solution is the intersection of the two regions.

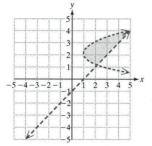

51. $y < e^x, \quad y > 1, \quad x < 2$

Graph the related equation $y = e^x$ using a dashed curve.
The curve divides the xy-plane into two regions.
Select test points from each region:

Test: $(0, 0)$ Test: $(0, 2)$

$\qquad 0 < e^0 \qquad\qquad\qquad 2 < e^0$

$\qquad 0 < 1 \quad$ True $\qquad\quad 2 < 1 \quad$ False

Shade the region below the curve.
Graph the related equation $y = 1$, a horizontal line using a
dashed line. Shade the region above the line.
Graph the related equation $x = 2$, a vertical line using a
dashed line. Shade the region to the left of the line.
The solution is the intersection of the three regions.

53. $y \le -x^2 + 4 \quad$ or $\quad y \ge x^2 - 4$

Graph the related equation $y = -x^2 + 4$ (a parabola) using a solid
curve. The parabola divides the xy-plane into two regions.
Select test points from each region:

Test: $(0, 0)$ Test: $(0, 5)$

$\qquad 0 \le -(0)^2 + 4 \qquad\qquad 5 \le -(0)^2 + 4$

$\qquad 0 \le 4 \qquad$ True $\qquad\quad 5 \le 4 \qquad$ False

Shade the region below the parabola.

Graph the related equation $y = x^2 - 4$ (a parabola) using a solid
curve. The parabola divides the xy-plane into two regions.
Select test points from each region:

Test: $(0, -5)$ Test: $(0, 0)$

$\qquad -5 \ge 0^2 - 4 \qquad\qquad 0 \ge 0^2 - 4$

$\qquad -5 \ge -4 \quad$ False $\qquad\quad 0 \ge -4 \quad$ True

Shade the region above the parabola.

The solution is the union of the two regions.

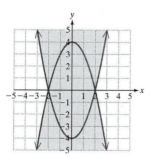

55. $(x+2)^2 + (y+3)^2 \leq 4$ or $x \geq y^2$

Graph the related equation $(x+2)^2 + (y+3)^2 = 4$ (a circle) using a solid curve. The circle divides the xy-plane into two regions. Select test points from each region:

Test: $(-2, -3)$ Test: $(0, 0)$

$(-2+2)^2 + (-3+3)^2 \leq 4$ $(0+2)^2 + (0+3)^2 \leq 4$

$\qquad\qquad 0+0 \leq 4$ $\qquad\qquad 4+9 \leq 4$

$\qquad\qquad\quad 0 \leq 4$ True $\qquad\qquad\quad 13 \leq 4$ False

Shade the region inside the circle.

Graph the related equation $x = y^2$ (a parabola) using a solid curve. The parabola divides the xy-plane into two regions. Select test points from each region:

Test: $(0, -2)$ Test: $(2, 0)$

$\qquad 0 \geq (-2)^2$ $\qquad 2 \geq (0)^2$

$\qquad 0 \geq 4$ False $\qquad 2 \geq 0$ True

Shade the region to the right of the parabola.

The solution is the union of the two regions.

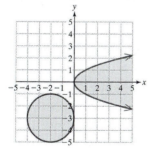

Chapter 9 Group Activity

1. a. $(-5, 0), (5, 0); (0, -5), (0, 5)$

   b. $x^2 + y^2 = 25$

$$y^2 = 25 - x^2$$
$$y = \pm\sqrt{25 - x^2}$$
$$Y_1 = \sqrt{25 - x^2}$$
$$Y_2 = -\sqrt{25 - x^2}$$

3. a. $\dfrac{x^2}{4} + \dfrac{y^2}{16} = 1$

$$\dfrac{y^2}{16} = 1 - \dfrac{x^2}{4}$$
$$y^2 = 16 - 4x^2$$
$$y = \pm\sqrt{16 - 4x^2}$$
$$Y_1 = \sqrt{16 - 4x^2}$$
$$Y_2 = -\sqrt{16 - 4x^2}$$

   b. $(-2, 0), (2, 0)$

Chapter 9 Review Exercises

Section 9.1

1. $(x_1, y_1) = (-6, 3), \quad (x_2, y_2) = (0, 1)$

$$d = \sqrt{[0 - (-6)]^2 + [1 - 3]^2}$$
$$= \sqrt{(6)^2 + (-2)^2}$$
$$= \sqrt{36 + 4}$$
$$= \sqrt{40} = 2\sqrt{10}$$

3. $(x, 5), \quad (2, 9)$

$$5 = \sqrt{[2 - x]^2 + [9 - 5]^2}$$
$$5 = \sqrt{(2 - x)^2 + (4)^2}$$
$$5 = \sqrt{4 - 4x + x^2 + 16}$$
$$25 = x^2 - 4x + 20$$
$$0 = x^2 - 4x - 5$$
$$0 = (x - 5)(x + 1)$$
$$x - 5 = 0 \text{ or } x + 1 = 0$$
$$x = 5 \text{ or } \quad x = -1$$

5. $(x - 12)^2 + (y - 3)^2 = 16$

   $h = 12, \, k = 3, \, r = 4$

   Center: $(12, 3)$; radius: 4

7. $(x + 3)^2 + (y + 8)^2 = 20$

   $[x - (-3)]^2 + [y - (-8)]^2 = (2\sqrt{5})^2$

   $h = -3, \, k = -8, \, r = 2\sqrt{5}$

   Center: $(-3, -8)$; radius: $2\sqrt{5}$

9. a.   Center: $(0, 0)$; radius: 8

   $h = 0, \, k = 0, \, r = 8$

   $[x - 0]^2 + [y - 0]^2 = (8)^2$

   $x^2 + y^2 = 64$

   b.   Center: $(8, 8)$; radius: 8

   $h = 8, \, k = 8, \, r = 8$

   $(x - 8)^2 + (y - 8)^2 = (8)^2$

   $(x - 8)^2 + (y - 8)^2 = 64$

11.
$$x^2 + y^2 + 4x + 16y + 60 = 0$$
$$(x^2 + 4x \quad) + (y^2 + 16y \quad) = -60$$
$$(x^2 + 4x + 4) + (y^2 + 16y + 64) = -60 + 4 + 64$$
$$(x+2)^2 + (y+8)^2 = 8$$

13.
$$x^2 + y^2 - 6x - \frac{2}{3}y + \frac{1}{9} = 0$$
$$(x^2 - 6x \quad) + \left(y^2 - \frac{2}{3}y \quad\right) = -\frac{1}{9}$$
$$(x^2 - 6x + 9) + \left(y^2 - \frac{2}{3}y + \frac{1}{9}\right) = -\frac{1}{9} + 9 + \frac{1}{9}$$
$$(x-3)^2 + \left(y - \frac{1}{3}\right)^2 = 9$$

15.   Center: $(0, 2)$; radius: 3
$$h = 0, \ k = 2, \ r = 3$$
$$(x-0)^2 + (y-2)^2 = (3)^2$$
$$x^2 + (y-2)^2 = 9$$

17.
$$\left(\frac{0+(-2)}{2}, \frac{9+7}{2}\right) = \left(\frac{-2}{2}, \frac{16}{2}\right) = (-1, 8)$$

Section 9.2

19. $x = 3(y-9)^2 + 1$

Horizontal axis of symmetry.
Parabola opens right.

21. $y = (x+3)^2 - 10$

Vertical axis of symmetry.
Parabola opens upward.

23. $y = (x+2)^2$

Vertex: $(-2, 0)$ Axis of symmetry: $x = -2$

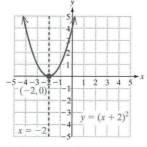

25. $x = 2y^2 - 1$

Vertex: $(-1, 0)$ Axis of symmetry: $y = 0$

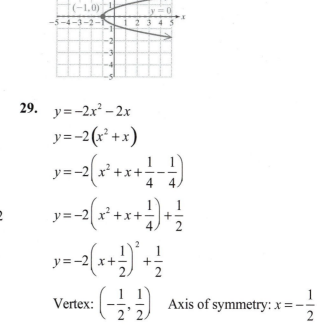

27. $x = y^2 + 4y + 2$
$$x = (y^2 + 4y + 4 - 4) + 2$$
$$x = (y^2 + 4y + 4) - 4 + 2$$
$$x = (y+2)^2 - 2$$

Vertex: $(-2, -2)$ Axis of symmetry: $y = -2$

29. $y = -2x^2 - 2x$
$$y = -2(x^2 + x)$$
$$y = -2\left(x^2 + x + \frac{1}{4} - \frac{1}{4}\right)$$
$$y = -2\left(x^2 + x + \frac{1}{4}\right) + \frac{1}{2}$$
$$y = -2\left(x + \frac{1}{2}\right)^2 + \frac{1}{2}$$

Vertex: $\left(-\frac{1}{2}, \frac{1}{2}\right)$ Axis of symmetry: $x = -\frac{1}{2}$

Section 9.3

31. $x^2 + 4y^2 = 36$

$\dfrac{x^2}{36} + \dfrac{4y^2}{36} = \dfrac{36}{36}$

$\dfrac{x^2}{6^2} + \dfrac{y^2}{3^2} = 1$

$a = 6; \ b = 3$

Intercepts: $(6, 0), (-6, 0), (0, 3), (0, -3)$

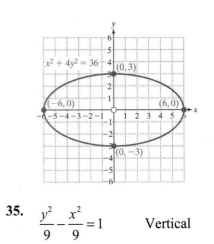

33. $\dfrac{x^2}{25} + \dfrac{(y-2)^2}{9} = 1$

Center: $(0, 2)$

35. $\dfrac{y^2}{9} - \dfrac{x^2}{9} = 1$ Vertical

37. $3x^2 - y^2 = 18$

$\dfrac{x^2}{6} - \dfrac{y^2}{18} = 1$ Horizontal

39. $y^2 - x^2 = 16$

$\dfrac{y^2}{16} - \dfrac{x^2}{16} = \dfrac{16}{16}$

$\dfrac{y^2}{4^2} - \dfrac{x^2}{4^2} = 1$

Transverse axis is vertical

$a = 4; \ b = 4$

Reference rectangle has corners at:

$(4, 4), (4, -4), (-4, 4), (-4, -4)$

Vertices: $(0, 4), (0, -4)$

41. $\dfrac{x^2}{16} + \dfrac{y^2}{9} = 1$ Ellipse

43. $\dfrac{y^2}{1} - \dfrac{x^2}{16} = 1$ Hyperbola

Section 9.4

45. a. $4x + 2y = 10$ Line

$y = x^2 - 10$ Parabola

b.

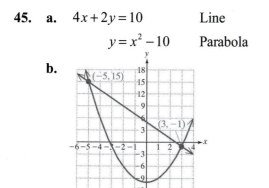

c.   Substitute the second equation into the first and solve for x:

$$4x + 2(x^2 - 10) = 10$$
$$4x + 2x^2 - 20 = 10$$
$$2x^2 + 4x - 30 = 0$$
$$2(x^2 + 2x - 15) = 0$$
$$2(x - 3)(x + 5) = 0$$
$$x - 3 = 0 \ \text{ or } \ x + 5 = 0$$
$$x = 3 \ \text{ or } \qquad x = -5$$
$$y = (3)^2 - 10 \quad y = (-5)^2 - 10$$
$$y = -1 \qquad\qquad y = 15$$

Solutions: $\{(3, -1), (-5, 15)\}$

47. a. $x^2 + y^2 = 16$ Circle

$x - 2y = 8 \ \rightarrow \ x = 2y + 8$ Line

b.

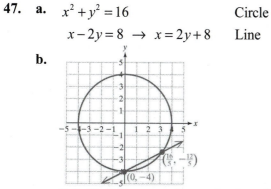

c.   Substitute the second equation into the first and solve for y:

$$(2y + 8)^2 + y^2 = 16$$
$$4y^2 + 32y + 64 + y^2 = 16$$
$$5y^2 + 32y + 48 = 0$$
$$(5y + 12)(y + 4) = 0$$
$$5y + 12 = 0 \ \text{ or } \ y + 4 = 0$$
$$5y = -12 \ \text{ or } \ y = -4$$
$$y = -\frac{12}{5} \ \text{ or } \ y = -4$$
$$x = 2\left(-\frac{12}{5}\right) + 8 \quad x = 2(-4) + 8$$
$$x = \frac{16}{5} \qquad\qquad x = 0$$

Solutions: $\left\{\left(\dfrac{16}{5}, -\dfrac{12}{5}\right), (0, -4)\right\}$

49. $x^2 + 4y^2 = 29$

$x - y = -4 \ \rightarrow \ x = y - 4$

Substitute the second equation into the first and solve for y:

$$(y - 4)^2 + 4y^2 = 29$$
$$y^2 - 8y + 16 + 4y^2 = 29$$
$$5y^2 - 8y - 13 = 0$$
$$(5y - 13)(y + 1) = 0$$
$$5y - 13 = 0 \ \text{ or } \ y + 1 = 0$$
$$5y = 13 \ \text{ or } \qquad y = -1$$
$$y = \frac{13}{5} \ \text{ or } \qquad y = -1$$
$$x = \frac{13}{5} - 4 \qquad x = -1 - 4$$
$$x = -\frac{7}{5} \qquad\qquad x = -5$$

Solutions: $\left\{\left(-\dfrac{7}{5}, \dfrac{13}{5}\right), (-5, -1)\right\}$

51. $y = x^2$

$6x^2 - y^2 = 8$

Substitute the first equation into the second and solve for y:

$$6y - y^2 = 8$$

$$y^2 - 6y + 8 = 0$$

$$(y - 4)(y - 2) = 0$$

$y - 4 = 0$ or $y - 2 = 0$

$y = 4$ or $y = 2$

$x = \pm\sqrt{4} = \pm 2$ $x = \pm\sqrt{2}$

Solutions: $\left\{(2, 4), (-2, 4), \left(\sqrt{2}, 2\right), \left(-\sqrt{2}, 2\right)\right\}$

53.   Add the equations to eliminate y^2 and solve:

$$x^2 + y^2 = 61$$

$$\underline{x^2 - y^2 = 11}$$

$$2x^2 \qquad = 72$$

$$x^2 = 36$$

$$x = \pm 6$$

$x = 6:$ $x = -6$

$6^2 + y^2 = 61$ $(-6)^2 + y^2 = 61$

$36 + y^2 = 61$ $36 + y^2 = 61$

$y^2 = 25$ $y^2 = 25$

$y = \pm 5$ $y = \pm 5$

Solutions: $\left\{(6, 5), (6, -5), (-6, 5), (-6, -5)\right\}$

Section 9.5

55. $\dfrac{x^2}{25} + \dfrac{y^2}{4} > 1$

Graph the related equation $\dfrac{x^2}{25} + \dfrac{y^2}{4} = 1$ (an ellipse) using

a dashed curve. The curve divides the xy-plane into two regions.

Select test points from each region:

Test: $(0, 0)$ Test: $(0, 3)$

$\dfrac{0^2}{25} + \dfrac{0^2}{4} > 1$ $\dfrac{0^2}{25} + \dfrac{3^2}{4} > 1$

$0 + 0 > 1$ $0 + \dfrac{9}{4} > 1$

$0 > 1$ False $\dfrac{9}{4} > 1$ True

Shade the region outside the ellipse.

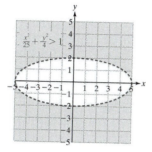

57. $(x+2)^2 + (y+1)^2 \leq 4$

Graph the related equation $(x+2)^2 + (y+1)^2 = 4$ (a circle) using a solid curve. The curve divides the xy-plane into two regions. Select test points from each region:

Test: $(-2, -1)$

$(-2+2)^2 + (-1+1)^2 \leq 4$

$0 + 0 \leq 4$

$0 \leq 4$ True

Test: $(0, 3)$

$(0+2)^2 + (3+1)^2 \leq 4$

$4 + 16 \leq 4$

$20 \leq 4$ False

Shade the region inside the circle.

59. $x^2 - \dfrac{y^2}{4} \leq 1$

Graph the related equation $x^2 - \dfrac{y^2}{4} = 1$ (a hyperbola) using a solid curve. The curve divides the xy-plane into three regions. Select test points from each region:

Test: $(0, -2)$

$(-2)^2 - \dfrac{0^2}{4} \leq 1$

$4 - 0 \leq 1$

$4 \leq 1$ False

Test: $(0, 0)$

$(0)^2 - \dfrac{0^2}{4} \leq 1$

$0 - 0 \leq 1$

$0 \leq 1$ True

Test: $(0, 2)$

$(2)^2 - \dfrac{0^2}{4} \leq 1$

$4 - 0 \leq 1$

$4 \leq 1$ False

Shade the region in the middle of the hyperbola.

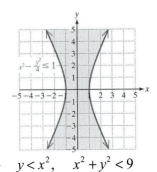

61. $y < x^2, \quad x^2 + y^2 < 9$

Graph the related equation $y = x^2$ using a dashed curve.

The curve divides the xy-plane into two regions.

Select test points from each region:

Test: $(0, -1)$ Test: $(0, 3)$

$\qquad -1 < 0^2 \qquad\qquad\qquad 3 < 0^2$

$\qquad -1 < 0 \quad$ True $\qquad\quad 3 < 0 \quad$ False

Shade the region below the curve.

Graph the related equation $x^2 + y^2 = 9$ (a circle) using a dashed curve. The circle divides the xy-plane into two regions.

Select test points from each region:

Test: $(0, 0)$ Test: $(0, 4)$

$\qquad 0^2 + 0^2 < 9 \qquad\qquad 0^2 + 4^2 < 9$

$\qquad 0 + 0 < 9 \qquad\qquad 0 + 16 < 9$

$\qquad\quad 0 < 9 \quad$ True $\qquad\quad 16 < 9 \quad$ False

Shade the region inside the circle.

The solution is the intersection of the two regions.

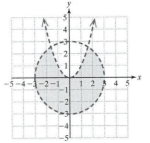

Chapter 9 Test

1. $(x_1, y_1) = (-5, 19), \quad (x_2, y_2) = (-3, 13)$

$d = \sqrt{[-3 - (-5)]^2 + [13 - 19]^2}$

$\quad = \sqrt{(2)^2 + (-6)^2}$

$\quad = \sqrt{4 + 36}$

$\quad = \sqrt{40}$

$\quad = 2\sqrt{10}$

3. $x^2 + y^2 - 4y - 5 = 0$

$x^2 + (y^2 - 4y + 4) = 5 + 4$

$(x - 0)^2 + (y - 2)^2 = 9$

$[x - 0]^2 + [y - 2]^2 = (3)^2$

$h = 0, \ k = 2, \ r = 3$

Center: $(0, 2)$; radius: 3

5. $\left(\dfrac{7.3 + 0.3}{2}, \dfrac{-1.2 + 5.1}{2} \right) = \left(\dfrac{7.6}{2}, \dfrac{3.9}{2} \right)$

$\qquad\qquad\qquad\qquad = (3.8, 1.95)$

The center is the midpoint $(3.8, 1.95)$.

7. $y = x^2 + 4x + 5$

$y = (x^2 + 4x + 4 - 4) + 5$

$y = (x^2 + 4x + 4) - 4 + 5$

$y = (x + 2)^2 + 1$

Vertex: $(-2, 1)$ Axis of symmetry: $x = -2$

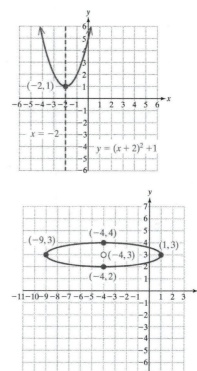

9. $\dfrac{(x + 4)^2}{25} + (y - 3)^2 = 1$

$\dfrac{(x + 4)^2}{5^2} + \dfrac{(y - 3)^2}{1^2} = 1$

$a = 5; \ b = 1$

Center: $(-4, 3)$

Intercepts: $(-9, 3), (1, 3), (-4, 4), (-4, 2)$

11. $16x^2 + 9y^2 = 144$

$4x - 3y = -12 \ \rightarrow \ y = \dfrac{4}{3}x + 4$

Substitute the second equation into the first and solve for x:

13. The addition method can be used if the equations have corresponding like terms.

$$16x^2 + 9\left(\frac{4}{3}x + 4\right)^2 = 144$$

$$16x^2 + 9\left(\frac{16}{9}x^2 + \frac{32}{3}x + 16\right) = 144$$

$$16x^2 + 16x^2 + 96x + 144 = 144$$

$$32x^2 + 96x = 0$$

$$32x(x + 3) = 0$$

$$32x = 0 \quad \text{or} \quad x + 3 = 0$$

$$x = 0 \quad \text{or} \qquad x = -3$$

$$y = \frac{4}{3}(0) + 4 \quad y = \frac{4}{3}(-3) + 4$$

$$y = 4 \qquad\qquad y = 0$$

Solutions: $\{(0, 4), (-3, 0)\}$

Graph b

15. $x \le y^2 + 1$

Graph the related equation $x = y^2 + 1$ (a parabola) using a solid curve. The curve divides the xy-plane into two regions.

Select test points from each region:

Test: $(0, 0)$ Test: $(2, 0)$

$0 \le 0^2 + 1$ $2 \le 0^2 + 1$

$0 \le 1$ True $2 \le 1$ False

Shade the region to the left of the parabola.

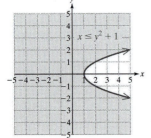

17.

$$x < y^2 + 1, \quad y > -\frac{1}{3}x + 1$$

Graph the related equation $x = y^2 + 1$ (a parabola) using a

dashed curve. The parabola divides the xy-plane into two regions.

Select test points from each region:

Test: $(0, 0)$ Test: $(2, 0)$

     $0 < 0^2 + 1$ $2 < 0^2 + 1$

     $0 < 1$ True $2 < 1$ False

Shade the region to the left of the parabola.

Graph the related equation $y = -\frac{1}{3}x + 1$ using a dashed

line. The line divides the xy-plane into two regions.

Select test points from each region:

Test: $(0, 0)$ Test: $(0, 2)$

     $0 > -\frac{1}{3}(0) + 1$ $2 > -\frac{1}{3}(0) + 1$

     $0 > 1$ False $2 > 1$ True

Shade the region above the line.

The solution is the intersection of the two regions.

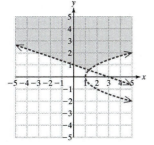

Chapters 1 – 9   Cumulative Review Exercises

1.

$$5(2y - 1) = 2y - 4 + 8y - 1$$
$$10y - 5 = 10y - 5$$
$$-5 = -5$$
$$\{x \mid x \text{ is a real number}\}$$

3. Let x = one integer

$2x - 5$ = the other integer

$$x(2x - 5) = 150$$
$$2x^2 - 5x = 150$$
$$2x^2 - 5x - 150 = 0$$
$$(2x + 15)(x - 10) = 0$$
$$x - 10 = 0 \quad \text{or} \quad 2x + 15 = 0$$
$$x = 10 \quad \text{or} \quad 2x = -15$$
$$x = 10 \quad \text{or} \quad x = -\frac{15}{2}$$
$$2x - 5 = 2(10) - 5 = 15$$

The integers are 10 and 15.

5. $3x - 4y = 6$

$-4y = -3x + 6$

$y = \dfrac{3}{4}x - \dfrac{3}{2}$

Slope: $\dfrac{3}{4}$; y-intercept: $\left(0, -\dfrac{3}{2}\right)$

7. $x + y = -1$

$2x - z = 3$

$y + 2z = -1$

Multiply the third equation by -1 and add to the first equation to eliminate y:

$x + y = -1 \longrightarrow x + y = -1$

$y + 2z = -1 \xrightarrow{\;\times -1\;} \underline{ -y - 2z = 1}$

$ x - 2z = 0$

Multiply the second equation by -2 and add to this result to eliminate z:

$2x - z = 3 \xrightarrow{\;\times -2\;} -4x + 2z = -6$

$x - 2z = 0 \longrightarrow \underline{ x - 2z = 0}$

$ -3x = -6$

$ x = 2$

Substitute and solve for y and z:

$x + y = -1 \qquad 2x - z = 3$

$2 + y = -1 \qquad 2(2) - z = 3$

$y = -3 \qquad\;\; 4 - z = 3$

$ 1 = z$

The solution is $\{(2, -3, 1)\}$.

9. $\begin{bmatrix} 3 & -4 & | & 6 \\ 1 & 2 & | & 12 \end{bmatrix} \xrightarrow{R_1 \leftrightarrow R_2} \begin{bmatrix} 1 & 2 & | & 12 \\ 3 & -4 & | & 6 \end{bmatrix} \xrightarrow{-3R_1 + R_2} \begin{bmatrix} 1 & 2 & | & 12 \\ 0 & -10 & | & -30 \end{bmatrix} \xrightarrow{-\frac{1}{10}R_2} \begin{bmatrix} 1 & 2 & | & 12 \\ 0 & 1 & | & 3 \end{bmatrix}$

$\xrightarrow{-2R_2 + R_1} \begin{bmatrix} 1 & 0 & | & 6 \\ 0 & 1 & | & 3 \end{bmatrix}$

The solution is $\{(6, 3)\}$.

11. $x^2 + 6x < -8$

$x^2 + 6x + 8 = 0$

$(x + 2)(x + 4) = 0$

$x + 2 = 0$ or $x + 4 = 0$

$x = -2$ or $\;\; x = -4$

The boundary points are -4 and -2.

Use test points $x = -5$, $x = -3$, and $x = 0$.

Test $x = -5$: $(-5)^2 + 6(-5) = 25 - 30 = -5 < -8$ False

Test $x = -3$: $(-3)^2 + 6(-3) = 9 - 18 = -9 < -8$ True

Test $x = 0$: $\;\; 0^2 + 6(0) = 0 + 0 = 0 < -8$ False

The boundary points are not included.

The solution is $\{x \mid -4 < x < -2\}$ or $(-4, -2)$

13. $(g \circ f)(x) = g(f(x))$

$= g\left(\sqrt{x+1}\right)$

$= \left(\sqrt{x+1}\right)^2 + 6$

$= x + 1 + 6$

$= x + 7, \;\; x \geq -1$

15. $x^2 - y^2 - 6x - 6y = (x+y)(x-y) - 6(x+y)$
$$= (x+y)(x-y-6)$$

17.
$$2x(x-7) = x - 18$$
$$2x^2 - 14x = x - 18$$
$$2x^2 - 15x + 18 = 0$$
$$(2x-3)(x-6) = 0$$
$$2x - 3 = 0 \text{ or } x - 6 = 0$$
$$2x = 3 \text{ or } \quad x = 6$$
$$x = \frac{3}{2} \text{ or } \quad x = 6 \quad \left\{\frac{3}{2}, 6\right\}$$

19. $\dfrac{2}{x+3} - \dfrac{x}{x-2} = \dfrac{x-2}{x-2} \cdot \dfrac{2}{x+3} - \dfrac{x}{x-2} \cdot \dfrac{x+3}{x+3}$
$$= \frac{2x - 4 - x^2 - 3x}{(x-2)(x+3)}$$
$$= \frac{-x^2 - x - 4}{(x-2)(x+3)}$$

21. a.
$$\sqrt{2x-5} = -3$$
$$\left(\sqrt{2x-5}\right)^2 = (-3)^2$$
$$2x - 5 = 9$$
$$2x = 14$$
$$x = 7$$

Check:
$$\sqrt{2(7)-5} = -3$$
$$\sqrt{9} = -3$$
$$3 = -3$$
7 does not check.
No solution. $\{\ \}$

b.
$$\sqrt[3]{2x-5} = -3$$
$$\left(\sqrt[3]{2x-5}\right)^3 = (-3)^3$$
$$2x - 5 = -27$$
$$2x = -22$$
$$x = -11 \quad \{-11\}$$

23. $\dfrac{3}{4-5i} = \dfrac{3}{4-5i} \cdot \dfrac{4+5i}{4+5i}$
$$= \frac{12 + 15i}{16 + 20i - 20i - 25i^2}$$
$$= \frac{12 + 15i}{16 + 25} = \frac{12 + 15i}{41} = \frac{12}{41} + \frac{15}{41}i$$

25. a. $d(t) = 4.4t^2$
$$d(2) = 4.4(2)^2 = 4.4(4) = 17.6 \text{ ft}$$
$$d(3) = 4.4(3)^2 = 4.4(9) = 39.6 \text{ ft}$$
$$d(4) = 4.4(4)^2 = 4.4(16) = 70.4 \text{ ft}$$
The car has traveled 17.6 ft, 39.6 ft, and 70.4 ft.

b. $281.6 = 4.4t^2$
$$64 = t^2$$
$$t = 8 \text{ sec}$$
It will take 8 sec.

27. $\dfrac{x}{x+2} - \dfrac{3}{x-1} = \dfrac{1}{x^2+x-2}$

$(x+2)(x-1)\left(\dfrac{x}{x+2} - \dfrac{3}{x-1}\right)$

$\qquad = \left(\dfrac{1}{(x+2)(x-1)}\right)(x+2)(x-1)$

$x(x-1) - 3(x+2) = 1$

$x^2 - x - 3x - 6 = 1$

$\quad x^2 - 4x - 7 = 0$

$x = \dfrac{-(-4) \pm \sqrt{(-4)^2 - 4(1)(-7)}}{2(1)}$

$\quad = \dfrac{4 \pm \sqrt{16+28}}{2} = \dfrac{4 \pm \sqrt{44}}{2} = \dfrac{4 \pm 2\sqrt{11}}{2}$

$\quad = 2 \pm \sqrt{11} \qquad \left\{2 \pm \sqrt{11}\right\}$

29. $g(x) = -x^2 - 2x + 3$

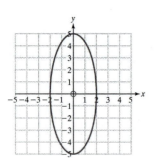

a. x-intercepts: $(-3, 0)$, $(1, 0)$
b. y-intercept: $(0, 3)$
c. Vertex: $(-1, 4)$

31. $|2x - 5| \ge 4$

$2x - 5 \le -4 \ \text{ or } \ 2x - 5 \ge 4$

$\quad 2x \le 1 \quad \text{ or } \quad 2x \ge 9$

$\quad x \le \dfrac{1}{2} \quad \text{ or } \quad x \ge \dfrac{9}{2}$

$\left(-\infty, \dfrac{1}{2}\right] \cup \left[\dfrac{9}{2}, \infty\right)$

33. $5^2 = 125^x$

$5^2 = \left(5^3\right)^x$

$5^2 = 5^{3x}$

$2 = 3x$

$x = \dfrac{2}{3} \qquad \left\{\dfrac{2}{3}\right\}$

35. $(x-0)^2 + (y-5)^2 = 4^2$

$\qquad x^2 + (y-5)^2 = 16$

37. $\left(\dfrac{-3+2}{2}, \dfrac{-2+2}{2}\right) = \left(\dfrac{-1}{2}, \dfrac{0}{2}\right) = \left(-\dfrac{1}{2}, 0\right)$

39. $y^2 - x^2 < 1$

Graph the related equation $y^2 - x^2 = 1$ (a hyperbola) using
a dashed curve. The curve divides the xy-plane into three regions.
Select test points from each region:

Test: $(0, -3)$ Test: $(0, 0)$ Test: $(0, 3)$

$(-3)^2 - 0^2 < 1$ $0^2 - 0^2 < 1$ $(3)^2 - 0^2 < 1$

$9 - 0 < 1$ $0 - 0 < 1$ $9 - 0 < 1$

$9 < 1$ False $0 < 1$ True $9 < 1$ False

Shade the region between the branches of the hyperbola.

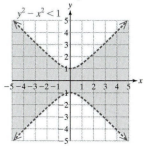

Chapter 10   Binomial Expansions, Sequences, and Series

Are You Prepared?

1. $1, 4, 7, 10, \underline{13}, \underline{16}, \underline{19}$

3. $3, 6, 12, 24, \underline{48}, \underline{96}, \underline{192}$

5. $1, 4, 9, 16, \underline{25}, \underline{36}, \underline{49}$

7. $x^7, x^6 y, x^5 y^2, x^4 y^3, \underline{x^3 y^4}, \underline{x^2 y^5}, \underline{xy^6}$

9. $(a+b)^2 = a^2 + 2ab + b^2$

11. $n+1$

Section 10.1   Binomial Expansions

1.
  a. $a^2 + 2ab + b^2$; $a^3 + 3a^2 b + 3ab^2 + b^3$; binomial
  b. $n(n-1)(n-2)\cdots(2)(1)$; factorial
  c. $6; 2; 1; 1$
  d. binomial
  e. Pascal's

3. $(a+b)^3 = a^3 + 3a^2 b + 3ab^2 + b^3$

5. $(1+g)^4 = 1^4 + 4 \cdot 1^3 g + 6 \cdot 1^2 g^2 + 4 \cdot 1 g^3 + g^4$
$$= 1 + 4g + 6g^2 + 4g^3 + g^4$$

7. $(p+q^2)^7 = p^7 + 7p^6 (q^2) + 21p^5 (q^2)^2 + 35p^4 (q^2)^3 + 35p^3 (q^2)^4 + 21p^2 (q^2)^5 + 7p(q^2)^6 + (q^2)^7$
$$= p^7 + 7p^6 q^2 + 21p^5 q^4 + 35p^4 q^6 + 35p^3 q^8 + 21p^2 q^{10} + 7pq^{12} + q^{14}$$

9. $(s-t)^5 = [s+(-t)]^5 = s^5 + 5s^4(-t) + 10s^3(-t)^2 + 10s^2(-t)^3 + 5s(-t)^4 + (-t)^5$
$$= s^5 - 5s^4 t + 10s^3 t^2 - 10s^2 t^3 + 5st^4 - t^5$$

11. $(5-u^3)^4 = (5+(-u^3))^4 = 5^4 + 4 \cdot 5^3 (-u^3) + 6 \cdot 5^2 (-u^3)^2 + 4 \cdot 5(-u^3)^3 + (-u^3)^4$
$$= 625 - 500u^3 + 150u^6 - 20u^9 + u^{12}$$

13. $(x^2 - 4)^3 = [x^2 + (-4)]^3 = (x^2)^3 + 3(x^2)^2(-4) + 3(x^2)(-4)^2 + (-4)^3$
$$= x^6 - 12x^4 + 48x^2 - 64$$

15. $5! = 5 \cdot 4 \cdot 3 \cdot 2 \cdot 1 = 120$

17. $0! = 1$ by definition

19. False

21. True

23. $6! = 6 \cdot (5 \cdot 4 \cdot 3 \cdot 2 \cdot 1) = 6 \cdot 5!$

25. $\dfrac{8!}{4!} = \dfrac{8 \cdot 7 \cdot 6 \cdot 5 \cdot 4!}{4!} = 8 \cdot 7 \cdot 6 \cdot 5 = 1680$

27. $\dfrac{3!}{0!} = \dfrac{3 \cdot 2 \cdot 1}{1} = 6$

29. $\dfrac{8!}{3! 5!} = \dfrac{8 \cdot 7 \cdot 6 \cdot 5!}{3 \cdot 2 \cdot 1 \cdot 5!} = \dfrac{8 \cdot 7 \cdot 6}{3 \cdot 2 \cdot 1} = \dfrac{336}{6} = 56$

31. $\dfrac{4!}{0! \, 4!} = \dfrac{4!}{1 \cdot 4!} = 1$

33. $(m+n)^{11} = \dfrac{11!}{11! \cdot 0!} m^{11} + \dfrac{11!}{10! \cdot 1!} m^{10} n + \dfrac{11!}{9! \cdot 2!} m^9 n^2 = m^{11} + 11 m^{10} n + 55 m^9 n^2$

35. $(u^2 - v)^{12} = (u^2 + (-v))^{12}$

$\quad\quad = \dfrac{12!}{12! \cdot 0!} (u^2)^{12} + \dfrac{12!}{11! \cdot 1!} (u^2)^{11}(-v) + \dfrac{12!}{10! \cdot 2!} (u^2)^{10}(-v)^2 = u^{24} - 12 u^{22} v + 66 u^{20} v^2$

37. $(a+b)^8$ has 9 terms.

39. $(s+t)^6 = \dfrac{6!}{6! \cdot 0!} s^6 + \dfrac{6!}{5! \cdot 1!} s^5 t + \dfrac{6!}{4! \cdot 2!} s^4 t^2 + \dfrac{6!}{3! \cdot 3!} s^3 t^3 + \dfrac{6!}{2! \cdot 4!} s^2 t^4 + \dfrac{6!}{1! \cdot 5!} s^1 t^5 + \dfrac{6!}{0! \cdot 6!} t^6$

$\quad\quad = s^6 + 6 s^5 t + 15 s^4 t^2 + 20 s^3 t^3 + 15 s^2 t^4 + 6 s t^5 + t^6$

41. $(b-3)^3 = (b + (-3))^3 = \dfrac{3!}{3! \cdot 0!} b^3 + \dfrac{3!}{2! \cdot 1!} b^2 (-3) + \dfrac{3!}{1! \cdot 2!} b(-3)^2 + \dfrac{3!}{0! \cdot 3!} (-3)^3$

$\quad\quad = b^3 + 3 b^2 (-3) + 3 b(9) + (-27)$

$\quad\quad = b^3 - 9 b^2 + 27 b - 27$

43. $(2x+y)^4 = \dfrac{4!}{4! \cdot 0!} (2x)^4 + \dfrac{4!}{3! \cdot 1!} (2x)^3 y + \dfrac{4!}{2! \cdot 2!} (2x)^2 y^2 + \dfrac{4!}{1! \cdot 3!} (2x) y^3 + \dfrac{4!}{0! \cdot 4!} y^4$

$\quad\quad = 1 \cdot 16 x^4 + 4 \cdot 8 x^3 y + 6 \cdot 4 x^2 y^2 + 4 \cdot 2 x y^3 + 1 \cdot y^4$

$\quad\quad = 16 x^4 + 32 x^3 y + 24 x^2 y^2 + 8 x y^3 + y^4$

45. $(c^2 - d)^7 = (c^2 + (-d))^7$

$\quad\quad = \dfrac{7!}{7! \cdot 0!} (c^2)^7 + \dfrac{7!}{6! \cdot 1!} (c^2)^6 (-d) + \dfrac{7!}{5! \cdot 2!} (c^2)^5 (-d)^2 + \dfrac{7!}{4! \cdot 3!} (c^2)^4 (-d)^3 + \dfrac{7!}{3! \cdot 4!} (c^2)^3 (-d)^4$

$\quad\quad\quad + \dfrac{7!}{2! \cdot 5!} (c^2)^2 (-d)^5 + \dfrac{7!}{1! \cdot 6!} (c^2)(-d)^6 + \dfrac{7!}{0! \cdot 7!} (-d)^7$

$\quad\quad = c^{14} - 7 c^{12} d + 21 c^{10} d^2 - 35 c^8 d^3 + 35 c^6 d^4 - 21 c^4 d^5 + 7 c^2 d^6 - d^7$

47.

$$\left(\frac{a}{2}-b\right)^5 = \left(\frac{a}{2}+(-b)\right)^5$$

$$= \frac{5!}{5!\cdot 0!}\left(\frac{a}{2}\right)^5 + \frac{5!}{4!\cdot 1!}\left(\frac{a}{2}\right)^4(-b) + \frac{5!}{3!\cdot 2!}\left(\frac{a}{2}\right)^3(-b)^2 + \frac{5!}{2!\cdot 3!}\left(\frac{a}{2}\right)^2(-b)^3$$

$$+ \frac{5!}{1!\cdot 4!}\left(\frac{a}{2}\right)(-b)^4 + \frac{5!}{0!\cdot 5!}(-b)^5$$

$$= \frac{a^5}{32} + 5\left(\frac{a^4}{16}\right)(-b) + 10\left(\frac{a^3}{8}\right)b^2 + 10\left(\frac{a^2}{4}\right)(-b^3) + 5\left(\frac{a}{2}\right)b^4 + \left(-b^5\right)$$

$$= \frac{1}{32}a^5 - \frac{5}{16}a^4b + \frac{5}{4}a^3b^2 - \frac{5}{2}a^2b^3 + \frac{5}{2}ab^4 - b^5$$

49.

$$(x+4y)^4 = \frac{4!}{4!\cdot 0!}x^4 + \frac{4!}{3!\cdot 1!}x^3(4y) + \frac{4!}{2!\cdot 2!}x^2(4y)^2 + \frac{4!}{1!\cdot 3!}x(4y)^3 + \frac{4!}{0!\cdot 4!}(4y)^4$$

$$= x^4 + 4\cdot 4x^3y + 6\cdot 16x^2y^2 + 4\cdot 64xy^3 + 1\cdot 256y^4$$

$$= x^4 + 16x^3y + 96x^2y^2 + 256xy^3 + 256y^4$$

51. $\dfrac{11!}{6!\cdot 5!}m^6(-n)^5 = -462m^6n^5$

53. $\dfrac{12!}{8!\cdot 4!}(u^2)^8(-v)^4 = 495u^{16}v^4$

55. $\dfrac{9!}{0!\cdot 9!}g^9 = g^9$

Section 10.2   Sequences and Series

1.
   a.   infinite; finite
   b.   terms; nth
   c.   alternating
   d.   series
   e.   summation
   f.   index; $(3)^2 + (4)^2 + (5)^2$

3. $\dfrac{8!}{2!6!} = \dfrac{8\cdot 7\cdot 6!}{2\cdot 1\cdot 6!} = \dfrac{8\cdot 7}{2\cdot 1} = \dfrac{56}{2} = 28$

5.

$$(2x+z)^4 = \frac{4!}{4!\cdot 0!}(2x)^4 + \frac{4!}{3!\cdot 1!}(2x)^3z + \frac{4!}{2!\cdot 2!}(2x)^2z^2 + \frac{4!}{1!\cdot 3!}(2x)z^3 + \frac{4!}{0!\cdot 4!}z^4$$

$$= 1\cdot 16x^4 + 4\cdot 8x^3z + 6\cdot 4x^2z^2 + 4\cdot 2xz^3 + 1\cdot z^4$$

$$= 16x^4 + 32x^3z + 24x^2z^2 + 8xz^3 + z^4$$

7. $a_n = 3n+1$
$a_1 = 3(1)+1 = 3+1 = 4$
$a_2 = 3(2)+1 = 6+1 = 7$
$a_3 = 3(3)+1 = 9+1 = 10$
$a_4 = 3(4)+1 = 12+1 = 13$
$a_5 = 3(5)+1 = 15+1 = 16$

9. $a_n = \sqrt{n+2}$
$a_1 = \sqrt{1+2} = \sqrt{3}$
$a_2 = \sqrt{2+2} = \sqrt{4} = 2$
$a_3 = \sqrt{3+2} = \sqrt{5}$
$a_4 = \sqrt{4+2} = \sqrt{6}$

11.
$$a_n = (-1)^n \frac{n+1}{n+2}$$
$$a_1 = (-1)^1 \frac{1+1}{1+2} = -1\left(\frac{2}{3}\right) = -\frac{2}{3}$$
$$a_2 = (-1)^2 \frac{2+1}{2+2} = 1\left(\frac{3}{4}\right) = \frac{3}{4}$$
$$a_3 = (-1)^3 \frac{3+1}{3+2} = -1\left(\frac{4}{5}\right) = -\frac{4}{5}$$
$$a_4 = (-1)^4 \frac{4+1}{4+2} = 1\left(\frac{5}{6}\right) = \frac{5}{6}$$

13.
$$a_n = (-1)^{n+1}(n^2 - 1)$$
$$a_1 = (-1)^{1+1}(1^2 - 1) = (-1)^2(1-1) = 1(0) = 0$$
$$a_2 = (-1)^{2+1}(2^2 - 1) = (-1)^3(4-1) = -1(3) = -3$$
$$a_3 = (-1)^{3+1}(3^2 - 1) = (-1)^4(9-1) = 1(8) = 8$$

15.
$$a_n = n^2 - n$$
$$a_1 = 1^2 - 1 = 1 - 1 = 0$$
$$a_2 = 2^2 - 2 = 4 - 2 = 2$$
$$a_3 = 3^2 - 3 = 9 - 3 = 6$$
$$a_4 = 4^2 - 4 = 16 - 4 = 12$$
$$a_5 = 5^2 - 5 = 25 - 5 = 20$$
$$a_6 = 6^2 - 6 = 36 - 6 = 30$$
$$a_n = n^2 - n$$
$$a_1 = 1^2 - 1 = 1 - 1 = 0$$
$$a_2 = 2^2 - 2 = 4 - 2 = 2$$
$$a_3 = 3^2 - 3 = 9 - 3 = 6$$
$$a_4 = 4^2 - 4 = 16 - 4 = 12$$

17.
$$a_n = (-1)^n 3^n$$
$$a_1 = (-1)^1 3^1 = -1(3) = -3$$
$$a_2 = (-1)^2 3^2 = 1(9) = 9$$
$$a_3 = (-1)^3 3^3 = -1(27) = -27$$
$$a_4 = (-1)^4 3^4 = 1(81) = 81$$

19. When n is odd, the term is negative. When n is even, the term is positive.

21. For example: $a_n = 2n$

23. For example: $a_n = 2n - 1$

25. For example: $a_n = \dfrac{1}{n^2}$

27. For example: $a_n = (-1)^{n+1}$

29. For example: $a_n = (-1)^n 2^n$

31. For example: $a_n = \dfrac{3}{5^n}$

33.
$$p_1 = 50 + 0.02(500) = 50 + 10 = 60$$
$$p_2 = 50 + 0.02(500 - 60) = 50 + 0.02(440) = 50 + 8.80 = 58.80$$
$$p_3 = 50 + 0.02(440 - 58.80) = 50 + 0.02(381.20) = 50 + 7.62 = 57.62$$
$$p_4 = 50 + 0.02(381.20 - 57.62) = 50 + 0.02(323.58) = 50 + 6.47 = 56.47$$
$$\$60.00, \ \$58.80, \ \$57.62, \ \$56.47$$

35.
$a_1 = 25,000$
$a_2 = 2(25,000) = 50,000$
$a_3 = 2(50,000) = 100,000$
$a_4 = 2(100,000) = 200,000$
$a_5 = 2(200,000) = 400,000$
$a_6 = 2(400,000) = 800,000$
$a_7 = 2(800,000) = 1,600,000$
$25,000; 50,000; 100,000; 200,000;$
$\qquad 400,000; 800,000; 1,600,000$

37. A sequence is an ordered list of terms. A series is the sum of the terms of a sequence.

39.
$$\sum_{i=1}^{4}(3i^2) = 3(1)^2 + 3(2)^2 + 3(3)^2 + 3(4)^2$$
$$= 3 + 12 + 27 + 48 = 90$$

41.
$$\sum_{j=0}^{4}\left(\frac{1}{2}\right)^j = \left(\frac{1}{2}\right)^0 + \left(\frac{1}{2}\right)^1 + \left(\frac{1}{2}\right)^2 + \left(\frac{1}{2}\right)^3 + \left(\frac{1}{2}\right)^4 = 1 + \frac{1}{2} + \frac{1}{4} + \frac{1}{8} + \frac{1}{16} = \frac{31}{16}$$

43.
$$\sum_{i=1}^{6} 5 = 5 + 5 + 5 + 5 + 5 + 5 = 30$$

45.
$$\sum_{j=1}^{4}(-1)^j(5j) = (-1)^1(5\cdot1) + (-1)^2(5\cdot2) + (-1)^3(5\cdot3) + (-1)^4(5\cdot4) = -1(5) + 1(10) - 1(15) + 1(20)$$
$$= -5 + 10 - 15 + 20 = 10$$

47.
$$\sum_{i=1}^{4}\frac{i+1}{i} = \frac{1+1}{1} + \frac{2+1}{2} + \frac{3+1}{3} + \frac{4+1}{4} = \frac{2}{1} + \frac{3}{2} + \frac{4}{3} + \frac{5}{4} = \frac{73}{12}$$

49.
$$\sum_{j=1}^{3}(j+1)(j+2) = (1+1)(1+2) + (2+1)(2+2) + (3+1)(3+2) = 2(3) + 3(4) + 4(5) = 6 + 12 + 20 = 38$$

51.
$$\sum_{k=1}^{7}(-1)^k = (-1)^1 + (-1)^2 + (-1)^3 + (-1)^4 + (-1)^5 + (-1)^6 + (-1)^7 = -1 + 1 - 1 + 1 - 1 + 1 - 1 = -1$$

53.
$$\sum_{k=1}^{5} k^2 = 1^2 + 2^2 + 3^2 + 4^2 + 5^2 = 1 + 4 + 9 + 16 + 25 = 55$$

55.
$$1 + 2 + 3 + 4 + 5 + 6 = \sum_{n=1}^{6} n$$

57.
$$4 + 4 + 4 + 4 + 4 = \sum_{i=1}^{5} 4$$

59.
$$4+8+12+16+20 = \sum_{j=1}^{5} 4j$$

61.
$$\frac{1}{3} - \frac{1}{9} + \frac{1}{27} - \frac{1}{81} = \sum_{k=1}^{4} (-1)^{k+1} \frac{1}{3^k}$$

63.
$$\frac{5}{11} + \frac{6}{22} + \frac{7}{33} + \frac{8}{44} + \frac{9}{55} + \frac{10}{66} = \sum_{i=1}^{6} \frac{i+4}{11i}$$

65.
$$x + x^2 + x^3 + x^4 + x^5 = \sum_{n=1}^{5} x^n$$

67. $a_1 = -3,\ a_n = a_{n-1} + 5$
$a_2 = a_1 + 5 = -3 + 5 = 2$
$a_3 = a_2 + 5 = 2 + 5 = 7$
$a_4 = a_3 + 5 = 7 + 5 = 12$
$a_5 = a_4 + 5 = 12 + 5 = 17$
$-3, 2, 7, 12, 17$

69. $a_1 = 5,\ a_n = 4a_{n-1} + 1$
$a_2 = 4a_1 + 1 = 4(5) + 1 = 21$
$a_3 = 4a_2 + 1 = 4(21) + 1 = 84 + 1 = 85$
$a_4 = 4a_3 + 1 = 4(85) + 1 = 340 + 1 = 341$
$a_5 = 4a_4 + 1 = 4(341) + 1 = 1364 + 1 = 1365$
$5, 21, 85, 341, 1365$

71. $a_1 = 1$
$a_2 = 1$
$a_3 = a_2 + a_1 = 1 + 1 = 2$
$a_4 = a_3 + a_2 = 2 + 1 = 3$
$a_5 = a_4 + a_3 = 3 + 2 = 5$
$a_6 = a_5 + a_4 = 5 + 3 = 8$

$a_7 = a_6 + a_5 = 8 + 5 = 13$
$a_8 = a_7 + a_6 = 13 + 8 = 21$
$a_9 = a_8 + a_7 = 21 + 13 = 34$
$a_{10} = a_9 + a_8 = 34 + 21 = 55$
1, 1, 2, 3, 5, 8, 13, 21, 34, 55

Section 10.3   Arithmetic Sequences and Series

1.
  a. arithmetic
  b. difference
  c. $a_1 + (n-1)d\ ;\ d$
  e. series
  f. $\dfrac{n}{2}(a_1 + a_n)$

3.
$d_n = (-1)^n$
$d_1 = (-1)^1 = -1$
$d_2 = (-1)^2 = 1$
$d_3 = (-1)^3 = -1$
$d_4 = (-1)^4 = 1$

5.
$$\sum_{n=1}^{5}(1-4n) = (1-4\cdot 1) + (1-4\cdot 2) + (1-4\cdot 3) + (1-4\cdot 4) + (1-4\cdot 5)$$
$$= -3 + (-7) + (-11) + (-15) + (-19) = -55$$

7. $a_n = a_1 + (n-1)d$
$a_1 = 3,\ d = 8$

$a_1 = 3$
$a_2 = a_1 + (2-1)d = 3 + (1)8 = 11$
$a_3 = a_1 + (3-1)d = 3 + (2)8 = 19$
$a_4 = a_1 + (4-1)d = 3 + (3)8 = 27$
$a_5 = a_1 + (5-1)d = 3 + (4)8 = 35$
The first five terms of the sequence are 3, 11, 19, 27, and 35.

9. $a_n = a_1 + (n-1)d$
$a_1 = 80,\ d = -20$

$a_1 = 80$
$a_2 = a_1 + (2-1)d = 80 + (1)(-20) = 60$
$a_3 = a_1 + (3-1)d = 80 + (2)(-20) = 40$
$a_4 = a_1 + (4-1)d = 80 + (3)(-20) = 20$
$a_5 = a_1 + (5-1)d = 80 + (4)(-20) = 0$
The first five terms of the sequence are 80, 60, 40, 20, and 0.

11. $a_n = a_1 + (n-1)d$

$a_1 = 3, d = \dfrac{3}{4}$

$a_1 = 3$

$a_2 = a_1 + (2-1)d = 3 + (1)\left(\dfrac{3}{4}\right) = \dfrac{15}{4}$

$a_3 = a_1 + (3-1)d = 3 + (2)\left(\dfrac{3}{4}\right) = \dfrac{9}{2}$

$a_4 = a_1 + (4-1)d = 3 + (3)\left(\dfrac{3}{4}\right) = \dfrac{21}{4}$

$a_5 = a_1 + (5-1)d = 3 + (4)\left(\dfrac{3}{4}\right) = 6$

The first five terms of the sequence are

$3, \dfrac{15}{4}, \dfrac{9}{2}, \dfrac{21}{4},$ and 6.

13. $d = 3 - 1 = 2$

15. $d = 3 - 6 = -3$

17. $d = -9 - (-7) = -2$

19. $3, 8, 13, 18, 23$

21. $2, \dfrac{5}{2}, 3, \dfrac{7}{2}, 4$

23. $2, -2, -6, -10, -14$

25. $a_1 = 0, \quad d = 5 - 0 = 5$

$a_n = a_1 + (n-1)d$

$a_n = 0 + (n-1)5 = 5n - 5 = -5 + 5n$

27. $a_1 = -2, \quad d = -4 - (-2) = -2$

$a_n = a_1 + (n-1)d$

$a_n = -2 + (n-1)(-2) = -2 - 2n + 2 = -2n$

29. $a_1 = 2, \quad d = \dfrac{5}{2} - 2 = \dfrac{1}{2}$

$a_n = a_1 + (n-1)d$

$a_n = 2 + (n-1)\left(\dfrac{1}{2}\right) = 2 + \dfrac{1}{2}n - \dfrac{1}{2} = \dfrac{3}{2} + \dfrac{1}{2}n$

31. $a_1 = 21, \quad d = 17 - 21 = -4$

$a_n = a_1 + (n-1)d$

$a_n = 21 + (n-1)(-4) = 21 - 4n + 4 = 25 - 4n$

33. $a_1 = -8, \quad d = -2 - (-8) = 6$

$a_n = a_1 + (n-1)d$

$a_n = -8 + (n-1)(6) = -8 + 6n - 6 = -14 + 6n$

35. $a_1 = -3, \quad d = 4, \quad n = 6$

$a_n = a_1 + (n-1)d$

$a_6 = -3 + (6-1)(4) = -3 + 5(4)$

$= -3 + 20 = 17$

37. $a_1 = -1, \quad d = 6, \quad n = 9$

$a_n = a_1 + (n-1)d$

$a_9 = -1 + (9-1)(6) = -1 + 8(6)$

$= -1 + 48 = 47$

39. $a_1 = 0, \quad a_{10} = -45, \quad n = 10$

$a_n = a_1 + (n-1)d$

$-45 = 0 + (10-1)d$

$-45 = 9d$

$-5 = d$

For $n = 7$

$a_7 = 0 + (7-1)(-5) = 6(-5) = -30$

41. $a_1 = 12, \quad a_6 = -18, \quad n = 6$

$a_n = a_1 + (n-1)d$

$-18 = 12 + (6-1)d$

$-30 = 5d$

$-6 = d$

For $n = 11$

$a_{11} = 12 + (11-1)(-6) = 12 + 10(-6)$

$= 12 - 60 = -48$

43. $a_1 = 8, \quad d = 13 - 8 = 5, \quad a_n = 98$

$a_n = a_1 + (n-1)d$

$98 = 8 + (n-1)(5)$

$90 = 5n - 5$

$95 = 5n$

$19 = n$

45. $a_1 = 1, \quad d = 5 - 1 = 4, \quad a_n = 85$

$a_n = a_1 + (n-1)d$

$85 = 1 + (n-1)(4)$

$84 = 4n - 4$

$88 = 4n$

$22 = n$

47. $a_1 = 2, \quad d = \dfrac{5}{2} - 2 = \dfrac{1}{2}, \quad a_n = 13$

$a_n = a_1 + (n-1)d$

$13 = 2 + (n-1)\left(\dfrac{1}{2}\right)$

$11 = \dfrac{1}{2}n - \dfrac{1}{2}$

$\dfrac{23}{2} = \dfrac{1}{2}n$

$23 = n$

49. $a_1 = \dfrac{13}{3}, \quad d = \dfrac{19}{3} - \dfrac{13}{3} = \dfrac{6}{3} = 2, \quad a_n = \dfrac{73}{3}$

$a_n = a_1 + (n-1)d$

$\dfrac{73}{3} = \dfrac{13}{3} + (n-1)(2)$

$20 = 2n - 2$

$22 = 2n$

$11 = n$

51. $d = -11 - (-8) = -3$

$a_2 = -8 - (-3) = -5$

$a_1 = -5 - (-3) = -2$

53. $\displaystyle\sum_{i=1}^{20}(3i + 2)$

$a_1 = 3(1) + 2 = 3 + 2 = 5$

$a_{20} = 3(20) + 2 = 60 + 2 = 62$

$S_n = \dfrac{n}{2}(a_1 + a_n)$

$S_{20} = \dfrac{20}{2}(5 + 62) = 10(67) = 670$

55. $\displaystyle\sum_{i=1}^{20}(i+4)$

$a_1 = 1 + 4 = 5$

$a_{20} = 20 + 4 = 24$

$S_n = \dfrac{n}{2}(a_1 + a_n)$

$S_{20} = \dfrac{20}{2}(5 + 24) = 10(29) = 290$

57. $\displaystyle\sum_{j=1}^{10}(4-j)$

$a_1 = 4 - 1 = 3$

$a_{10} = 4 - 10 = -6$

$S_n = \dfrac{n}{2}(a_1 + a_n)$

$S_{10} = \dfrac{10}{2}(3 + (-6)) = 5(-3) = -15$

59. $\displaystyle\sum_{j=1}^{15}\left(\dfrac{2}{3}j + 1\right)$

$a_1 = \dfrac{2}{3}(1) + 1 = \dfrac{2}{3} + 1 = \dfrac{5}{3}$

$a_{15} = \dfrac{2}{3}(15) + 1 = 10 + 1 = 11$

$S_n = \dfrac{n}{2}(a_1 + a_n)$

$S_{15} = \dfrac{15}{2}\left(\dfrac{5}{3} + 11\right) = \dfrac{15}{2}\left(\dfrac{38}{3}\right) = 95$

61. $4 + 8 + 12 + \ldots + 84$

$a_1 = 4, \; a_n = 84, \; d = 8 - 4 = 4$

$a_n = a_1 + (n-1)d$

$84 = 4 + (n-1)4$

$80 = 4n - 4$

$84 = 4n$

$21 = n$

$S_n = \dfrac{n}{2}(a_1 + a_n)$

$S_{21} = \dfrac{21}{2}(4 + 84) = \dfrac{21}{2}(88) = 924$

63. $6 + 8 + 10 + \ldots + 34$

$a_1 = 6, \; a_n = 34, \; d = 8 - 6 = 2$

$a_n = a_1 + (n-1)d$

$34 = 6 + (n-1)2$

$28 = 2n - 2$

$30 = 2n$

$15 = n$

$S_n = \dfrac{n}{2}(a_1 + a_n)$

$S_{15} = \dfrac{15}{2}(6 + 34) = \dfrac{15}{2}(40) = 300$

65. $4 + 8 + 12 + \ldots + 84$

$a_1 = 4, \; a_n = 84, \; d = 8 - 4 = 4$

$a_n = a_1 + (n-1)d$

$84 = 4 + (n-1)4$

$80 = 4n - 4$

$84 = 4n$

$21 = n$

$S_n = \dfrac{n}{2}(a_1 + a_n)$

$S_{21} = \dfrac{21}{2}(4 + 84) = \dfrac{21}{2}(88) = 924$

67. $a_1 = 1, \ a_n = 100, \ n = 100$

$$S_n = \frac{n}{2}(a_1 + a_n)$$

$$S_{100} = \frac{100}{2}(1 + 100) = 50(101) = 5050$$

69. $30 + 32 + 34 + \ldots$

$a_1 = 30, \ d = 32 - 30 = 2, \ n = 20$

$a_n = a_1 + (n-1)d$

$a_{20} = 30 + (20-1)2 = 30 + 19(2)$

$\qquad = 30 + 38 = 68$

$$S_n = \frac{n}{2}(a_1 + a_n)$$

$$S_{20} = \frac{20}{2}(30 + 68) = 10(98) = 980 \text{ seats}$$

$$R = 15(980) = \$14,700$$

Section 10.4   Geometric Sequences and Series

1.
    a. geometric
    b. ratio
    c. $a_1 r^{n-1}$
    d. $\dfrac{a_1(1 - r^n)}{1 - r}$
    e. $\dfrac{a_1}{1 - r}; 1$

3. $a_1 = -4, \ d = 0 - (-4) = 4$

$a_n = a_1 + (n-1)d$

$a_n = -4 + (n-1)4 = -4 + 4n - 4 = 4n - 8$

5. $a_1 = -15, \ d = 10, \ n = 5$

$a_n = a_1 + (n-1)d$

$a_5 = -15 + (5-1)(10)$

$\qquad = -15 + 4(10) = -15 + 40 = 25$

7. $a_n = a_1 r^{n-1}$

$a_1 = 1, \ r = 10$

$a_1 = 1$

$a_2 = a_1 r^{2-1} = (1)(10) = 10$

$a_3 = a_1 r^{3-1} = (1)(10)^2 = 100$

$a_4 = a_1 r^{4-1} = (1)(10)^3 = 1000$

The first four terms of the sequence are 1, 10, 100, and 1000.

9. $a_n = a_1 r^{n-1}$

$a_1 = 64, \ r = \dfrac{1}{2}$

$a_1 = 64$

$a_2 = a_1 r^{2-1} = (64)\left(\dfrac{1}{2}\right) = 32$

$a_3 = a_1 r^{3-1} = (64)\left(\dfrac{1}{2}\right)^2 = 16$

$a_4 = a_1 r^{4-1} = (64)\left(\dfrac{1}{2}\right)^3 = 8$

The first four terms of the sequence are 64,

11. $a_n = a_1 r^{n-1}$

$a_1 = 8, \ r = -\dfrac{1}{4}$

$a_1 = 8$

$a_2 = a_1 r^{2-1} = (8)\left(-\dfrac{1}{4}\right) = -2$

$a_3 = a_1 r^{3-1} = (8)\left(-\dfrac{1}{4}\right)^2 = \dfrac{1}{2}$

$a_4 = a_1 r^{4-1} = (8)\left(-\dfrac{1}{4}\right)^3 = -\dfrac{1}{8}$

The first four terms of the sequence are

32, 16, and 8.

$8, -2, \dfrac{1}{2}, \text{ and } -\dfrac{1}{8}$

13. $r = \dfrac{a_{n+1}}{a_n} = \dfrac{10}{5} = 2$

15. $r = \dfrac{a_{n+1}}{a_n} = \dfrac{-2}{8} = -\dfrac{1}{4}$

17. $r = \dfrac{a_{n+1}}{a_n} = \dfrac{-6}{3} = -2$

19. $-3, 6, -12, 24, -48$

21. $6, 3, \dfrac{3}{2}, \dfrac{3}{4}, \dfrac{3}{8}$

23. $-1, -6, -36, -216, -1296$

25. $a_1 = 3, \quad r = \dfrac{12}{3} = 4$

$a_n = a_1 \cdot r^{n-1}$

$a_n = 3(4)^{n-1}$

27. $a_1 = -5, \quad r = \dfrac{15}{-5} = -3$

$a_n = a_1 \cdot r^{n-1}$

$a_n = -5(-3)^{n-1}$

29. $a_1 = \dfrac{1}{2}, \quad r = \dfrac{2}{\dfrac{1}{2}} = 4$

$a_n = a_1 \cdot r^{n-1}$

$a_n = \dfrac{1}{2}(4)^{n-1}$

31. $a_n = 2\left(\dfrac{1}{2}\right)^{n-1}$

$a_8 = 2\left(\dfrac{1}{2}\right)^{8-1} = 2\left(\dfrac{1}{2}\right)^{7} = 2\left(\dfrac{1}{128}\right) = \dfrac{1}{64}$

33. $a_n = 4\left(-\dfrac{3}{2}\right)^{n-1}$

$a_6 = 4\left(-\dfrac{3}{2}\right)^{6-1} = 4\left(-\dfrac{3}{2}\right)^{5} = 4\left(-\dfrac{243}{32}\right)$

$= -\dfrac{243}{8}$

35. $a_n = -3(2)^{n-1}$

$a_5 = -3(2)^{5-1} = -3(2)^{4} = -3(16) = -48$

37.
$$a_5 = -\frac{16}{9}, \quad r = -\frac{2}{3}$$
$$a_n = a_1 \cdot r^{n-1}$$
$$-\frac{16}{9} = a_1 \left(-\frac{2}{3}\right)^{5-1}$$
$$-\frac{16}{9} = a_1 \left(-\frac{2}{3}\right)^{4}$$
$$-\frac{16}{9} = a_1 \left(\frac{16}{81}\right)$$
$$-\frac{16}{9} \left(\frac{81}{16}\right) = a_1$$
$$-9 = a_1$$

39.
$$a_7 = 8, \quad r = 2$$
$$a_n = a_1 \cdot r^{n-1}$$
$$8 = a_1 (2)^{7-1}$$
$$8 = a_1 (2)^6$$
$$8 = a_1 (64)$$
$$\frac{8}{64} = a_1$$
$$\frac{1}{8} = a_1$$

41.
$$a_2 = 16, \quad a_3 = 64, \quad r = \frac{64}{16} = 4$$
$$a_n = a_1 \cdot r^{n-1}$$
$$64 = a_1 (4)^{3-1}$$
$$64 = a_1 (4)^2$$
$$64 = a_1 (16)$$
$$4 = a_1$$

43. A geometric sequence is an ordered list of numbers in which the ratio between a term and its predecessor is constant. A geometric series is the sum of the terms of such a sequence.

45. a.
$$\sum_{i=1}^{4} 3(4)^{i-1} = 3 \cdot 4^0 + 3 \cdot 4^1 + 3 \cdot 4^2 + 3 \cdot 4^3$$
$$= 3 \cdot 1 + 3 \cdot 4 + 3 \cdot 16 + 3 \cdot 64$$
$$= 3 + 12 + 48 + 192$$

b. 255

47.
$$a_1 = 10, \quad r = \frac{2}{10} = \frac{1}{5}, \quad n = 5$$

$$S_n = \frac{a_1\left(1 - r^n\right)}{1 - r}$$

$$S_5 = \frac{10\left(1 - \left(\frac{1}{5}\right)^5\right)}{1 - \frac{1}{5}} = \frac{10\left(1 - \frac{1}{3125}\right)}{\frac{4}{5}}$$

$$= \frac{10\left(\frac{3124}{3125}\right)}{\frac{4}{5}} = 10\left(\frac{3124}{3125}\right)\left(\frac{5}{4}\right) = \frac{1562}{125}$$

49.
$$a_1 = -2, \quad r = \frac{1}{-2} = -\frac{1}{2}, \quad n = 5$$

$$S_n = \frac{a_1\left(1 - r^n\right)}{1 - r}$$

$$S_5 = \frac{-2\left(1 - \left(-\frac{1}{2}\right)^5\right)}{1 - \left(-\frac{1}{2}\right)} = \frac{-2\left(1 + \frac{1}{32}\right)}{\frac{3}{2}}$$

$$= \frac{-2\left(\frac{33}{32}\right)}{\frac{3}{2}} = -2\left(\frac{33}{32}\right)\left(\frac{2}{3}\right) = -\frac{11}{8}$$

51.
$$a_1 = 12, \quad r = \frac{16}{12} = \frac{4}{3}, \quad n = 5$$

$$S_n = \frac{a_1\left(1 - r^n\right)}{1 - r}$$

$$S_5 = \frac{12\left(1 - \left(\frac{4}{3}\right)^5\right)}{1 - \left(\frac{4}{3}\right)} = \frac{12\left(1 - \frac{1024}{243}\right)}{-\frac{1}{3}}$$

$$= \frac{12\left(-\frac{781}{243}\right)}{-\frac{1}{3}} = 12\left(-\frac{781}{243}\right)\left(-\frac{3}{1}\right) = \frac{3124}{27}$$

53.
$$a_1 = 1, \quad r = \frac{\frac{2}{3}}{1} = \frac{2}{3}, \quad a_n = \frac{32}{243}$$

$$a_n = a_1 \cdot r^{n-1}$$

$$\frac{32}{243} = 1\left(\frac{2}{3}\right)^{n-1}$$

$$\left(\frac{2}{3}\right)^5 = \left(\frac{2}{3}\right)^{n-1}$$

$$5 = n - 1$$

$$6 = n$$

$$S_n = \frac{a_1\left(1 - r^n\right)}{1 - r}$$

$$S_6 = \frac{1\left(1 - \left(\frac{2}{3}\right)^6\right)}{1 - \left(\frac{2}{3}\right)} = \frac{1\left(1 - \frac{64}{729}\right)}{\frac{1}{3}}$$

$$= \frac{\left(\frac{665}{729}\right)}{\frac{1}{3}} = \left(\frac{665}{729}\right)\left(\frac{3}{1}\right) = \frac{665}{243}$$

55.

$$a_1 = -4, \ r = \frac{8}{-4} = -2, \ a_n = -256$$

$$a_n = a_1 \cdot r^{n-1}$$

$$-256 = -4(-2)^{n-1}$$

$$64 = (-2)^{n-1}$$

$$(-2)^6 = (-2)^{n-1}$$

$$6 = n - 1$$

$$7 = n$$

$$S_n = \frac{a_1(1 - r^n)}{1 - r}$$

$$S_7 = \frac{-4\left(1 - (-2)^7\right)}{1 - (-2)} = \frac{-4(1 + 128)}{3}$$

$$= \frac{-4(129)}{3} = -172$$

57. a.

$$a_n = 1000(1.05)^n$$

$$a_1 = 1000(1.05)^1 = \$1050$$

$$a_2 = 1000(1.05)^2 = \$1102.50$$

$$a_3 = 1000(1.05)^3 = \$1157.63$$

$$a_4 = 1000(1.05)^4 = \$1215.51$$

b.

$$a_{10} = 1000(1.05)^{10} = \$1628.89$$

$$a_{20} = 1000(1.05)^{20} = \$2653.30$$

$$a_{40} = 1000(1.05)^{40} = \$7039.99$$

59.

$$a_1 = 1, \ r = \frac{\frac{1}{6}}{1} = \frac{1}{6}$$

$$S = \frac{a_1}{1 - r} = \frac{1}{1 - \frac{1}{6}} = \frac{1}{\frac{5}{6}} = \frac{6}{5}$$

$$r = \frac{1}{6}; \text{ sum is } \frac{6}{5}$$

61.

$$\sum_{i=1}^{\infty}\left(-\frac{1}{4}\right)^{i-1}$$

$$a_1 = \left(-\frac{1}{4}\right)^{1-1} = 1, \ r = -\frac{1}{4}$$

$$S = \frac{a_1}{1 - r} = \frac{1}{1 - \left(-\frac{1}{4}\right)} = \frac{1}{\frac{5}{4}} = 1\left(\frac{4}{5}\right) = \frac{4}{5}$$

$$r = -\frac{1}{4}; \text{ sum is } \frac{4}{5}$$

63.

$$a_1 = \frac{2}{3}, \ r = \frac{-1}{\frac{2}{3}} = -\frac{3}{2}$$

Sum does not exist because $|r| \ge 1$.

65.

$$a_1 = 200, \ r = 0.75$$

$$S = \frac{a_1}{1 - r} = \frac{200}{1 - (0.75)} = \frac{200}{0.25} = 800$$

The sum is \$800 million.

67.

$$a_1 = 2\left(\tfrac{3}{4}\right)(4), \ r = \tfrac{3}{4}$$

$$S = \frac{a_1}{1 - r} = \frac{2\left(\tfrac{3}{4}\right)(4)}{1 - \tfrac{3}{4}} = \frac{6}{\tfrac{1}{4}} = 24$$

The total vertical distance is $24 + 4 = 28$ ft.

69. a.

$$a_1 = \frac{7}{10}$$

b.

$$r = \frac{1}{10}$$

c.

$$S = \frac{a_1}{1 - r} = \frac{\frac{7}{10}}{1 - \frac{1}{10}} = \frac{\frac{7}{10}}{\frac{9}{10}} = \frac{7}{10} \cdot \frac{10}{9} = \frac{7}{9}$$

71. **a.** $a_1 = 48,000; \quad r = 1.04, \quad n = 20$

$$S_n = \frac{a_1\left(1 - r^n\right)}{1 - r}$$

$$S_{20} = \frac{48,000\left(1 - (1.04)^{20}\right)}{1 - 1.04} = \frac{48,000\left(1 - 2.191123\right)}{-0.04} = \frac{48,000\left(-1.191123\right)}{-0.04} = \$1,429,348$$

b. $a_1 = 48,000; \quad r = 1.045, \quad n = 20$

$$S_n = \frac{a_1\left(1 - r^n\right)}{1 - r}$$

$$S_{20} = \frac{48,000\left(1 - (1.045)^{20}\right)}{1 - 1.045} = \frac{48,000\left(1 - 2.411714\right)}{-0.045} = \frac{48,000\left(-1.411714\right)}{-0.045} = \$1,505,828$$

c.   Difference: $1,505,828 - 1,429,348 = \$76,480$

Problem Recognition Exercises

1.   Geometric; $r = \dfrac{-\frac{5}{2}}{5} = -\dfrac{5}{2} \cdot \dfrac{1}{5} = -\dfrac{1}{2}$

3.   Geometric; $r = \dfrac{-4}{-2} = 2$

5.   Arithmetic: $d = \dfrac{1}{3} - \left(-\dfrac{1}{3}\right) = \dfrac{2}{3}$

7.   Neither

9.   Arithmetic; $d = -4 - (-2) = -2$

11.   Neither

13.   Arithmetic; $d = \dfrac{1}{2}\pi - \dfrac{1}{4}\pi = \dfrac{1}{4}\pi$

15.   Neither

17.   Neither

Chapter 10   Group Activity

1. $\bar{x} = \dfrac{160 + 120 + 140 + 240 + 180 + 210 + 380 + 320}{8} = \dfrac{1750}{8} = 218.75$

The mean, rounded to one decimal place, is $\bar{x} = 218.8$.

3.   Answers will vary.

5.   Answers will vary.

Chapter 10   Review Exercises

Section 10.1

1. $8! = 8 \cdot 7 \cdot 6 \cdot 5 \cdot 4 \cdot 3 \cdot 2 \cdot 1 = 40,320$

3. $\dfrac{12!}{10!2!} = \dfrac{12 \cdot 11 \cdot 10!}{10! \cdot 2 \cdot 1} = 66$

5.
$$\left(x^2 + 4\right)^5 = \left(x^2\right)^5 + 5\left(x^2\right)^4(4) + 10\left(x^2\right)^3(4)^2 + 10\left(x^2\right)^2(4)^3 + 5\left(x^2\right)(4)^4 + 4^5$$
$$= x^{10} + 20x^8 + 160x^6 + 640x^4 + 1280x^2 + 1024$$

7.
$$(a-2b)^{11} = \left(a+(-2b)\right)^{11} = a^{11} + \frac{11!}{1!10!}a^{10}(-2b) + \frac{11!}{2!9!}a^9(-2b)^2$$
$$= a^{11} + 11a^{10}(-2b) + 55a^9\left(4b^2\right) = a^{11} - 22a^{10}b + 220a^9b^2$$

9.
$$\frac{10!}{7!3!}(5x)^3\left(-y^3\right)^7 = \frac{10 \cdot 9 \cdot 8 \cdot 7!}{3 \cdot 2 \cdot 1 \cdot 7!}\left(5^3x^3\right)\left(-y^{21}\right)$$
$$= 120\left(125x^3\right)\left(-y^{21}\right)$$
$$= -15,000x^3y^{21}$$

Section 10.2

11. $a_n = -3n + 4$

$a_1 = -3(1) + 4 = -3 + 4 = 1$

$a_2 = -3(2) + 4 = -6 + 4 = -2$

$a_3 = -3(3) + 4 = -9 + 4 = -5$

$a_4 = -3(4) + 4 = -12 + 4 = -8$

$a_5 = -3(5) + 4 = -15 + 4 = -11$

$1, -2, -5, -8, -11$

13. $a_n = (-1)^{n+1}\dfrac{n}{n+2}$

$a_1 = (-1)^{1+1}\dfrac{1}{1+2} = (-1)^2\dfrac{1}{3} = 1 \cdot \dfrac{1}{3} = \dfrac{1}{3}$

$a_2 = (-1)^{2+1}\dfrac{2}{2+2} = (-1)^3\dfrac{2}{4} = -1 \cdot \dfrac{1}{2} = -\dfrac{1}{2}$

$a_3 = (-1)^{3+1}\dfrac{3}{3+2} = (-1)^4\dfrac{3}{5} = 1 \cdot \dfrac{3}{5} = \dfrac{3}{5}$

$a_4 = (-1)^{4+1}\dfrac{4}{4+2} = (-1)^5\dfrac{4}{6} = -1 \cdot \dfrac{2}{3} = -\dfrac{2}{3}$

$\dfrac{1}{3}, -\dfrac{1}{2}, \dfrac{3}{5}, -\dfrac{2}{3}$

15. For example: $a_n = \dfrac{n}{n+1}$

17. The index is k.

19.
$$\sum_{i=1}^{7} 5(-1)^{i-1} = 5(-1)^{1-1} + 5(-1)^{2-1} + 5(-1)^{3-1} + 5(-1)^{4-1} + 5(-1)^{5-1} + 5(-1)^{6-1} + 5(-1)^{7-1}$$
$$= 5(-1)^0 + 5(-1)^1 + 5(-1)^2 + 5(-1)^3 + 5(-1)^4 + 5(-1)^5 + 5(-1)^6$$
$$= 5 \cdot 1 + 5(-1) + 5 \cdot 1 + 5(-1) + 5 \cdot 1 + 5(-1) + 5 \cdot 1 = 5 - 5 + 5 - 5 + 5 - 5 + 5 = 5$$

21.
$$\frac{4}{1}+\frac{5}{2}+\frac{6}{3}+\frac{7}{4}+\frac{8}{5}+\frac{9}{6}+\frac{10}{7}=\sum_{i=1}^{7}\frac{3+i}{i}$$

Section 10.3

23. $-12,-10.5,-9,-7.5,-6$

25. $a_1=1,\ d=11-1=10$

$a_n=a_1+(n-1)d$

$a_n=1+(n-1)(10)=1+10n-10=10n-9$

27.
$a_n=\frac{1}{2}+(n-1)\cdot\frac{1}{4},\ n=17$

$a_{17}=\frac{1}{2}+(17-1)\cdot\frac{1}{4}=\frac{1}{2}+16\cdot\frac{1}{4}=\frac{1}{2}+4=\frac{9}{2}$

29. $a_1=3,\ d=8-3=5,\ a_n=118$

$a_n=a_1+(n-1)d$

$118=3+(n-1)5$

$118=3+5n-5$

$118=5n-2$

$120=5n$

$24=n$

31. $a_8=30,\ a_{19}=140,\ n=8$

$a_n=a_1+(n-1)d$

$30=a_1+(8-1)d$

$30=a_1+7d\rightarrow a_1=30-7d$

Let $n=19$

$140=a_1+(19-1)d$

$140=a_1+18d$

$140=30-7d+18d$

$110=11d$

$10=d$

33.
$$\sum_{k=1}^{40}(5-k)$$

$a_1=5-1=4$

$a_{40}=5-40=-35$

$S_n=\frac{n}{2}(a_1+a_n)$

$S_{40}=\frac{40}{2}(4+(-35))=20(-31)=-620$

35. $-6+(-4)+(-2)+\ldots+34$

$a_1=-6,\ a_n=34,\ d=-4-(-6)=2$

$a_n=a_1+(n-1)d$

$34=-6+(n-1)2$

$40=2n-2$

$42=2n$

$21=n$

$S_n=\frac{n}{2}(a_1+a_n)$

$S_{21}=\frac{21}{2}(-6+34)=\frac{21}{2}(28)=294$

Section 10.4

37.
$$r = \frac{15}{5} = 3$$

39.
$$-1, \frac{1}{4}, -\frac{1}{16}, \frac{1}{64}$$

41.
$$a_1 = -4, \ r = \frac{-8}{-4} = 2$$
$$a_n = a_1 \cdot r^{n-1}$$
$$a_n = -4(2)^{n-1}$$

43.
$$a_n = 4\left(\frac{2}{3}\right)^{n-1}$$
$$a_6 = 4\left(\frac{2}{3}\right)^{6-1} = 4\left(\frac{2}{3}\right)^5 = 4\left(\frac{32}{243}\right) = \frac{128}{243}$$

45.
$$a_7 = \frac{1}{16}, \ r = \frac{1}{2}, \ n = 7$$
$$a_n = a_1 \cdot r^{n-1}$$
$$\frac{1}{16} = a_1\left(\frac{1}{2}\right)^{7-1}$$
$$\frac{1}{16} = a_1\left(\frac{1}{64}\right)$$
$$4 = a_1$$

47.
$$a_1 = 5, \ r = 2, \ n = 8$$
$$S_n = \frac{a_1\left(1-r^n\right)}{1-r}$$
$$S_8 = \frac{5\left(1-(2)^8\right)}{1-2} = \frac{5(1-256)}{-1}$$
$$= \frac{5(-255)}{-1} = 1275$$

49.
$$a_1 = 6, \ r = \frac{4}{6} = \frac{2}{3}$$
$$S = \frac{a_1}{1-r} = \frac{6}{1-\frac{2}{3}} = \frac{6}{\frac{1}{3}} = 18$$

51. a.
$$b_n = 10,000(1.07)^n$$
$$b_0 = 10,000(1.07)^0 = 10,000 \cdot 1 = 10,000$$
$$b_1 = 10,000(1.07)^1 = 10,000(1.07) = 10,700$$
$$b_2 = 10,000(1.07)^2 = 10,000(1.1449) = 11,449$$
$$10,000; 10,700; 11,449$$

b.
$$b_3 = 10,000(1.07)^3 = 10,000(1.225043)$$
$$= 12,250.43$$
The account is worth \$12,250.43 after 3 years.

c.
$$a_{10} = 10,000(1.07)^{10} = \$19,671.51$$

Chapter 10   Test

1. $0! = 1$

3. $(a+b)^4 = a^4 + 4a^3b + 6a^2b^2 + 4ab^3 + b^4$

5.
$$\frac{8!}{5!3!}a^3\left(-c^3\right)^5 = \frac{8\cdot7\cdot6\cdot5!}{5!\cdot3\cdot2\cdot1}a^3\left(-c^{15}\right)$$
$$= -56a^3c^{15}$$

7.
$$\sum_{i=1}^{5} i^2 + 2 = \left(1^2 + 2\right) + \left(2^2 + 2\right) + \left(3^2 + 2\right) + \left(4^2 + 2\right) + \left(5^2 + 2\right) = 3 + 6 + 11 + 18 + 27 = 65$$

9.
$$\sum_{n=1}^{4} x^{3n}$$

11.
$$r = \frac{3}{9} = \frac{1}{3}$$

13. **a.** $4, -8, 16, -32$

   b. $a_n = 4(-2)^{n-1}$

   c. $a_{10} = 4(-2)^{10-1} = 4(-2)^9 = 4(-512)$
$$= -2048$$

15. $a_1 = 20, \quad d = 18 - 20 = -2$
$$a_n = a_1 + (n-1)d$$
$$a_n = 20 + (n-1)(-2) = 20 + -2n + 2$$
$$= -2n + 22$$

17.
$$a_1 = 1, \quad r = \frac{\frac{2}{3}}{1} = \frac{2}{3}, \quad a_n = \frac{64}{729}$$
$$a_n = a_1 r^{n-1}$$
$$\frac{64}{729} = 1\left(\frac{2}{3}\right)^{n-1}$$
$$\left(\frac{2}{3}\right)^6 = \left(\frac{2}{3}\right)^{n-1}$$
$$6 = n - 1$$
$$7 = n$$

19.
$$a_1 = 8, \quad r = \frac{1}{2}, \quad n = 10$$
$$S_n = \frac{a_1\left(1 - r^n\right)}{1 - r}$$
$$S_{10} = \frac{8\left(1 - \left(\frac{1}{2}\right)^{10}\right)}{1 - \frac{1}{2}} = \frac{8\left(1 - \frac{1}{1024}\right)}{\frac{1}{2}}$$
$$= \frac{8\left(\frac{1023}{1024}\right)}{\frac{1}{2}} = 8\left(\frac{1023}{1024}\right)\left(\frac{2}{1}\right) = \frac{1023}{64}$$

21. $a_6 = 9, \quad r = 3, \quad n = 6$
$$a_n = a_1 r^{n-1}$$
$$9 = a_1 (3)^{6-1}$$
$$9 = a_1 (3)^5$$
$$9 = a_1 (243)$$
$$\frac{1}{27} = a_1$$

23. **a.** $3(365) = \$1095$

   b. $1095 + 1095(0.06) = 1095 + 65.70$
$$= \$1160.70$$

   c. $a_{30} = 1095(1.06)^{30-1} = 1095(1.06)^{29}$
$$= \$5933.13$$

   d. $S_{30} = \dfrac{1095\left(1 - 1.06^{30}\right)}{1 - 1.06} = \dfrac{1095(-4.74349)}{-0.06}$
$$= \$86,568.71$$

Chapters 1–10   Cumulative Review Exercises

1.
$$(2x - 3)(x - 4) - (x - 5)^2$$
$$= 2x^2 - 8x - 3x + 12 - \left(x^2 - 10x + 25\right)$$
$$= 2x^2 - 11x + 12 - x^2 + 10x - 25$$
$$= x^2 - x - 13$$

3.
$$64^{-2/3} = \frac{1}{64^{2/3}} = \frac{1}{\left(\sqrt[3]{64}\right)^2} = \frac{1}{4^2} = \frac{1}{16}$$

5.

$$\frac{2c^2 - 3c - 20}{2c^2 - 32} = \frac{(2c+5)(c-4)}{2(c^2 - 16)}$$

$$= \frac{(2c+5)\cancel{(c-4)}}{2\cancel{(c-4)}(c+4)} = \frac{2c+5}{2(c+4)}$$

7.

$$y = \log_3 81$$

$$3^y = 81$$

$$3^y = 3^4$$

$$y = 4$$

9.

$$\frac{a^3 + 64}{16 - a^2} \div \frac{a^3 - 4a^2 + 16a}{a^2 - 3a - 4}$$

$$= \frac{a^3 + 64}{-(a^2 - 16)} \cdot \frac{a^2 - 3a - 4}{a^3 - 4a^2 + 16a}$$

$$= \frac{\cancel{(a+4)}\cancel{(a^2 - 4a + 16)}}{-\cancel{(a-4)}\cancel{(a+4)}} \cdot \frac{\cancel{(a-4)}(a+1)}{a\cancel{(a^2 - 4a + 16)}}$$

$$= -\frac{a+1}{a}$$

11.

$$\sqrt{50x^5} + 4x\sqrt{2x^3} - x^2\sqrt{8x}$$

$$= \sqrt{25 \cdot 2 \cdot x^4 \cdot x} + 4x\sqrt{2x^2 \cdot x} - x^2\sqrt{4 \cdot 2x}$$

$$= 5x^2\sqrt{2x} + 4x \cdot x\sqrt{2x} - x^2 \cdot 2\sqrt{2x}$$

$$= 5x^2\sqrt{2x} + 4x^2\sqrt{2x} - 2x^2\sqrt{2x}$$

$$= 7x^2\sqrt{2x}$$

13.

$$\begin{array}{r|rrrrr} 2 & 2 & -1 & 0 & 5 & -7 \\ & & 4 & 6 & 12 & 34 \\ \hline & 2 & 3 & 6 & 17 & \underline{27} \end{array}$$

Quotient: $2x^3 + 3x^2 + 6x + 17 + \dfrac{27}{x-2}$

15.

$$75c^3 - 12c = 3c(25c^2 - 4)$$

$$= 3c(5c - 2)(5c + 2)$$

17.

$$w^3 + 9w^2 - 4w - 36 = w^2(w+9) - 4(w+9)$$

$$= (w+9)(w^2 - 4)$$

$$= (w+9)(w-2)(w+2)$$

19.

$$\frac{9}{x-4} + \frac{3}{x+2} = \frac{6}{x^2 - 2x - 8}$$

$$\frac{9}{x-4} + \frac{3}{x+2} = \frac{6}{(x-4)(x+2)}$$

$$(x-4)(x+2)\left(\frac{9}{x-4} + \frac{3}{x+2}\right) = (x-4)(x+2)\left(\frac{6}{(x-4)(x+2)}\right)$$

$$9(x+2) + 3(x-4) = 6$$

$$9x + 18 + 3x - 12 = 6$$

$$12x + 6 = 6$$

$$12x = 0$$

$$x = 0 \quad \{0\}$$

21. $(5y-2)^2 + 7 = 4$

$\qquad (5y-2)^2 = -3$

$\qquad 5y - 2 = \pm\sqrt{-3}$

$\qquad 5y = 2 \pm i\sqrt{3}$

$\qquad y = \dfrac{2 \pm i\sqrt{3}}{5} \qquad \left\{\dfrac{2}{5} \pm \dfrac{\sqrt{3}}{5}i\right\}$

23. $|x+5| = |3x-2|$

$\qquad x+5 = 3x-2 \text{ or } x+5 = -(3x-2)$

$\qquad -2x+5 = -2 \quad \text{ or } x+5 = -3x+2$

$\qquad -2x = -7 \quad \text{ or } \quad 4x = -3$

$\qquad x = \dfrac{7}{2} \quad \text{ or } \quad x = -\dfrac{3}{4} \quad \left\{\dfrac{7}{2}, -\dfrac{3}{4}\right\}$

25. $\log x + \log(x+3) = 1$

$\qquad \log(x)(x+3) = 1$

$\qquad x^2 + 3x = 10^1$

$\qquad x^2 + 3x - 10 = 0$

$\qquad (x+5)(x-2) = 0$

$\qquad\qquad x+5 = 0 \text{ or } x-2 = 0$

$\qquad\qquad\qquad x = -5 \text{ or } \quad x = 2$

Check:

$\log(-5) + \log(-5+3) = 1$

No solution

$\log 2 + \log(2+3) = 1$

$\qquad \log 2 + \log 5 = 1$

$\qquad\quad \log(2 \cdot 5) = 1$

$\qquad\qquad \log 10 = 1$

$\qquad\qquad\quad 1 = 1$

$\{2\}$ ($x = -5$ does not check.)

27. $\log_7 895 = \dfrac{\log 895}{\log 7} \approx 3.4929$

29. $3x(x+5)(x-1) \geq 0$

$3x(x+5)(x-1) = 0$

$\qquad\qquad 3x = 0 \quad \text{or } x+5 = 0 \quad \text{or } x-1 = 0$

$\qquad\qquad x = 0 \quad \text{or } \quad x = -5 \text{ or } \quad x = 1$

The boundary points are -5, 0, and 1.

Use test points $x = -6$, $x = -1$, $x = 0.5$, and $x = 2$.

Test $x = -6$: $\quad 3(-6)(-6+5)(-6-1) = -18(-1)(-7) = -126 \geq 0$ $\qquad$ False

Test $x = -1$: $\quad 3(-1)(-1+5)(-1-1) = -3(4)(-2) = 24 \geq 0$ $\qquad$ True

Test $x = 0.5$: $\quad 3(0.5)(0.5+5)(0.5-1) = 1.5(5.5)(-0.5) = -4.125 \geq 0$ $\quad$ False

Test $x = 2$: $\quad 3(2)(2+5)(2-1) = 6(7)(1) = 42 \geq 0$ $\qquad$ True

The boundary points are included.

The solution is $\{x \mid -5 \leq x \leq 0 \text{ or } x \geq 1\}$ or $[-5,0] \cup [1,\infty)$.

31. $\dfrac{3}{x+4} \le 1$

$\dfrac{3}{x+4} = 1$

$3 = 1(x+4)$

$3 = x+4$

$x = -1$

Boundary points are –1 and –4 (undefined).
Use test points $x = -5$, $x = -2$, and $x = 0$.

Test $x = -5$: $\dfrac{3}{-5+4} = \dfrac{3}{-1} = -3 \le 1$ True

Test $x = -2$: $\dfrac{3}{-2+4} = \dfrac{3}{2} \le 1$ False

Test $x = 0$: $\dfrac{3}{0+4} = \dfrac{3}{4} \le 1$ True

The boundary point $x = -1$ is included.
The solution is

$\{x \mid x < -4 \text{ or } x \ge -1\}$ or $(-\infty, -4) \cup [-1, \infty)$

33. $3x - 2y = 6$

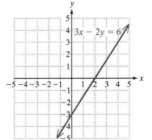

35. $y = 2^x$

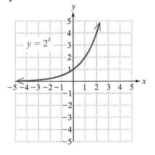

37. **a.** $f(x) = x^2 + 6x + 5 = 0$

$(x+5)(x+1) = 0$

$x+5 = 0$ or $x+1 = 0$

$x = -5$ or $x = -1$

x-intercepts: $(-5, 0), (-1, 0)$

b. $f(0) = 0^2 + 6(0) + 5 = 0 + 0 + 5 = 5$

y-intercept: $(0, 5)$

c. $f(x) = x^2 + 6x + 5$

$f(x) = (x^2 + 6x + 9) + 5 - 9$

$f(x) = (x+3)^2 - 4$

Vertex: $(-3, -4)$

39. $x - 5y = 2$

$-5y = -x + 2$

$y = \dfrac{1}{5}x - \dfrac{2}{5}$

$m = \dfrac{1}{5}, \quad m_\perp = -5$

$y - (-2) = -5(x - 4)$

$y + 2 = -5x + 20$

$y = -5x + 18$

41. $5 - x \ge 0$

$5 \ge x$ $(-\infty, 5]$

43.

$$x \quad +2z = 7$$
$$2x + 4y \quad = 10$$
$$-3y + 5z = 9$$

Multiply the first equation by –2 and add to the second equation:

$$x \quad + 2z = 7 \xrightarrow{\times -2} -2x \quad -4z = -14$$
$$2x + 4y \quad = 10 \longrightarrow \underline{2x + 4y \quad = 10}$$
$$4y - 4z = -4 \rightarrow y - z = -1$$

Multiply this result by 3, add to the third equation and solve for z:

$$y - z = -1 \xrightarrow{\times 3} 3y - 3z = -3$$
$$-3y + 5z = 9 \longrightarrow \underline{-3y + 5z = 9}$$
$$2z = 6$$
$$z = 3$$

Substitute and solve:

$$x + 2z = 7 \qquad 2x + 4y = 10$$
$$x + 2(3) = 7 \qquad 2(1) + 4y = 10$$
$$x + 6 = 7 \qquad 2 + 4y = 10$$
$$x = 1 \qquad 4y = 8$$
$$y = 2$$

The solution is: $\{(1, 2, 3)\}$.

45.

$$A = \frac{1}{2}(b_1 + b_2)h$$
$$2A = b_1 h + b_2 h$$
$$2A - b_2 h = b_1 h$$
$$\frac{2A - b_2 h}{h} = b_1$$

47.

$$s = \frac{k}{t}$$
$$60 = \frac{k}{10}$$
$$600 = k$$
$$s = \frac{600}{8} = 75 \text{ mph}$$

The car is moving 75 mph.

49.

	5% Account	6.5% Account	
Total Amount Invested	$x - 3000$	x	
Interest Earned	$0.05(x - 3000)$	$0.065x$	770

(int at 5%) + (int at 6.5%) = (total int)

$$0.05(x - 3000) + 0.065(x) = 770$$
$$0.05x - 150 + 0.065x = 770$$
$$0.115x = 920$$
$$\frac{0.115x}{0.115} = \frac{920}{0.115}$$
$$x = 8000$$
$$x - 3000 = 8000 - 3000 = 5000$$

$5000 was invested at 5% and $8000 was invested at 6.5%.

Additional Topics Appendix

Section A.1   Determinants and Cramer's Rule

1. **a.** determinant; $ad - bc$
    b. minor
    c. $a_2; \begin{vmatrix} b_1 & c_1 \\ b_2 & c_2 \end{vmatrix}$

3. $\begin{vmatrix} 5 & 6 \\ 4 & 8 \end{vmatrix} = 5(8) - 6(4) = 40 - 24 = 16$

5. $\begin{vmatrix} 5 & -1 \\ 1 & 0 \end{vmatrix} = 5(0) - (-1)(1) = 0 + 1 = 1$

7. $\begin{vmatrix} -3 & \frac{1}{4} \\ 8 & -2 \end{vmatrix} = -3(-2) - \left(\frac{1}{4}\right)(8) = 6 - 2 = 4$

9. $\begin{vmatrix} 2 & 0 \\ -7 & 3 \end{vmatrix} = 2(3) - (0)(-7) = 6 - 0 = 6$

11. $\begin{vmatrix} 4 & -1 \\ 2 & 6 \end{vmatrix} = 4(6) - (-1)(2) = 24 + 2 = 26$

13. $\begin{vmatrix} 6 & 0 \\ -2 & 1 \end{vmatrix} = 6(1) - (0)(-2) = 6 - 0 = 6$

15. $\begin{vmatrix} 4 & -2 \\ 5 & 9 \end{vmatrix} = 4(9) - (-2)(5) = 36 + 10 = 46$

17. **a.**
$$\begin{vmatrix} 0 & 1 & 2 \\ 3 & -1 & 2 \\ 3 & 2 & -2 \end{vmatrix} = 0 \cdot \begin{vmatrix} -1 & 2 \\ 2 & -2 \end{vmatrix} - 3 \cdot \begin{vmatrix} 1 & 2 \\ 2 & -2 \end{vmatrix} + 3 \cdot \begin{vmatrix} 1 & 2 \\ -1 & 2 \end{vmatrix}$$
$$= 0\left[-1(-2) - 2(2)\right] - 3\left[1(-2) - 2(2)\right] + 3\left[1(2) - 2(-1)\right] = 0 - 3(-6) + 3(4)$$
$$= 18 + 12 = 30$$

    b.
$$\begin{vmatrix} 0 & 1 & 2 \\ 3 & -1 & 2 \\ 3 & 2 & -2 \end{vmatrix} = -3 \cdot \begin{vmatrix} 1 & 2 \\ 2 & -2 \end{vmatrix} + (-1) \cdot \begin{vmatrix} 0 & 2 \\ 3 & -2 \end{vmatrix} - 2 \cdot \begin{vmatrix} 0 & 1 \\ 3 & 2 \end{vmatrix}$$
$$= -3\left[1(-2) - 2(2)\right] - 1\left[0(-2) - 2(3)\right] - 2\left[0(2) - 1(3)\right] = -3(-6) - 1(-6) - 2(-3)$$
$$= 18 + 6 + 6 = 30$$

19. Choosing the row or column with the most zero elements simplifies the arithmetic when evaluating a determinant.

21. About the third column:
$$\begin{vmatrix} 5 & 2 & 1 \\ 3 & -6 & 0 \\ -2 & 8 & 0 \end{vmatrix} = 1 \cdot \begin{vmatrix} 3 & -6 \\ -2 & 8 \end{vmatrix} - 0 \cdot \begin{vmatrix} 5 & 2 \\ -2 & 8 \end{vmatrix} + 0 \cdot \begin{vmatrix} 5 & 2 \\ 3 & -6 \end{vmatrix} = 1\left[3(8) - (-6)(-2)\right] - 0 + 0 = 1(12) = 12$$

23. About the third row:
$$\begin{vmatrix} 3 & 2 & 1 \\ 1 & -1 & 2 \\ 1 & 0 & 4 \end{vmatrix} = 1 \cdot \begin{vmatrix} 2 & 1 \\ -1 & 2 \end{vmatrix} - 0 \cdot \begin{vmatrix} 3 & 1 \\ 1 & 2 \end{vmatrix} + 4 \cdot \begin{vmatrix} 3 & 2 \\ 1 & -1 \end{vmatrix} = 1\left[2(2) - 1(-1)\right] - 0 + 4\left[3(-1) - 2(1)\right]$$
$$= 1(5) + 4(-5) = 5 - 20 = -15$$

25. About the first column:
$$\begin{vmatrix} 0 & 5 & -8 \\ 0 & -4 & 1 \\ 0 & 3 & 6 \end{vmatrix} = 0$$
Since all the elements in the first column are zero, the determinant will be zero.

27.
$$\begin{vmatrix} a & 2 \\ b & 8 \end{vmatrix} = a(8) - 2(b) = 8a - 2b$$

29.
$$\begin{vmatrix} x & 0 & 3 \\ y & -2 & 6 \\ z & -1 & 1 \end{vmatrix} = x \cdot \begin{vmatrix} -2 & 6 \\ -1 & 1 \end{vmatrix} - y\begin{vmatrix} 0 & 3 \\ -1 & 1 \end{vmatrix} + z\begin{vmatrix} 0 & 3 \\ -2 & 6 \end{vmatrix}$$
$$= x\left[-2(1) - 6(-1)\right] - y\left[0(1) - 3(-1)\right] + z\left[0(6) - 3(-2)\right] = x(4) - y(3) + z(6)$$
$$= 4x - 3y + 6z$$

31.
$$\begin{vmatrix} f & e & 0 \\ d & c & 0 \\ b & a & 0 \end{vmatrix} = 0$$
Since all the elements in the third column are zero, the determinant will be zero.

33.
$$D = \begin{vmatrix} 4 & 6 \\ -2 & 1 \end{vmatrix} = 4(1) - 6(-2) = 4 + 12 = 16$$
$$D_x = \begin{vmatrix} 9 & 6 \\ 12 & 1 \end{vmatrix} = 9(1) - 6(12) = 9 - 72 = -63$$
$$D_y = \begin{vmatrix} 4 & 9 \\ -2 & 12 \end{vmatrix} = 4(12) - 9(-2) = 48 + 18 = 66$$

35.
$$D = \begin{vmatrix} 2 & 1 \\ 1 & -4 \end{vmatrix} = 2(-4) - 1(1) = -8 - 1 = -9$$
$$D_x = \begin{vmatrix} 3 & 1 \\ 6 & -4 \end{vmatrix} = 3(-4) - 1(6) = -12 - 6 = -18$$
$$D_y = \begin{vmatrix} 2 & 3 \\ 1 & 6 \end{vmatrix} = 2(6) - 3(1) = 12 - 3 = 9$$
$$x = \frac{D_x}{D} = \frac{-18}{-9} = 2 \quad y = \frac{D_y}{D} = \frac{9}{-9} = -1$$
The solution is $\{(2, -1)\}$.

37. $4y = x - 8 \;\rightarrow\; x - 4y = 8$
$3x = -7y + 5 \rightarrow 3x + 7y = 5$
$$D = \begin{vmatrix} 1 & -4 \\ 3 & 7 \end{vmatrix} = 1(7) - (-4)(3) = 7 + 12 = 19$$
$$D_x = \begin{vmatrix} 8 & -4 \\ 5 & 7 \end{vmatrix} = 8(7) - (-4)(5) = 56 + 20 = 76$$
$$D_y = \begin{vmatrix} 1 & 8 \\ 3 & 5 \end{vmatrix} = 1(5) - 8(3) = 5 - 24 = -19$$
$$x = \frac{D_x}{D} = \frac{76}{19} = 4 \quad y = \frac{D_y}{D} = \frac{-19}{19} = -1$$
The solution is $\{(4, -1)\}$.

39.
$$D = \begin{vmatrix} 4 & -3 \\ 2 & 5 \end{vmatrix} = 4(5) - (-3)(2) = 20 + 6 = 26$$
$$D_x = \begin{vmatrix} 5 & -3 \\ 7 & 5 \end{vmatrix} = 5(5) - (-3)(7) = 25 + 21 = 46$$
$$D_y = \begin{vmatrix} 4 & 5 \\ 2 & 7 \end{vmatrix} = 4(7) - 5(2) = 28 - 10 = 18$$
$$x = \frac{D_x}{D} = \frac{46}{26} = \frac{23}{13} \quad y = \frac{D_y}{D} = \frac{18}{26} = \frac{9}{13}$$
The solution is $\left\{\left(\dfrac{23}{13}, \dfrac{9}{13}\right)\right\}$.

41.

$$D = \begin{vmatrix} 2 & -1 & 3 \\ 1 & 4 & 4 \\ 3 & 2 & 2 \end{vmatrix} = 2 \cdot \begin{vmatrix} 4 & 4 \\ 2 & 2 \end{vmatrix} - 1 \cdot \begin{vmatrix} -1 & 3 \\ 2 & 2 \end{vmatrix} + 3 \cdot \begin{vmatrix} -1 & 3 \\ 4 & 4 \end{vmatrix} = 2(8-8) - 1(-2-6) + 3(-4-12)$$

$$= 2(0) - 1(-8) + 3(-16) = 8 - 48 = -40$$

$$D_x = \begin{vmatrix} 9 & -1 & 3 \\ 5 & 4 & 4 \\ 5 & 2 & 2 \end{vmatrix} = 9 \cdot \begin{vmatrix} 4 & 4 \\ 2 & 2 \end{vmatrix} - 5 \cdot \begin{vmatrix} -1 & 3 \\ 2 & 2 \end{vmatrix} + 5 \cdot \begin{vmatrix} -1 & 3 \\ 4 & 4 \end{vmatrix} = 9(8-8) - 5(-2-6) + 5(-4-12)$$

$$= 9(0) - 5(-8) + 5(-16) = 40 - 80 = -40$$

$$x = \frac{D_x}{D} = \frac{-40}{-40} = 1$$

43.

$$D = \begin{vmatrix} 3 & -2 & 2 \\ 6 & 3 & -4 \\ 3 & -1 & 2 \end{vmatrix} = 2 \cdot \begin{vmatrix} 6 & 3 \\ 3 & -1 \end{vmatrix} - (-4) \cdot \begin{vmatrix} 3 & -2 \\ 3 & -1 \end{vmatrix} + 2 \cdot \begin{vmatrix} 3 & -2 \\ 6 & 3 \end{vmatrix} = 2(-6-9) + 4(-3+6) + 2(9+12)$$

$$= 2(-15) + 4(3) + 2(21) = -30 + 12 + 42 = 24$$

$$D_z = \begin{vmatrix} 3 & -2 & 5 \\ 6 & 3 & -1 \\ 3 & -1 & 4 \end{vmatrix} = 5 \cdot \begin{vmatrix} 6 & 3 \\ 3 & -1 \end{vmatrix} - (-1) \cdot \begin{vmatrix} 3 & -2 \\ 3 & -1 \end{vmatrix} + 4 \cdot \begin{vmatrix} 3 & -2 \\ 6 & 3 \end{vmatrix} = 5(-6-9) + 1(-3+6) + 4(9+12)$$

$$= 5(-15) + 1(3) + 4(21) = -75 + 3 + 84 = 12$$

$$z = \frac{D_z}{D} = \frac{12}{24} = \frac{1}{2}$$

45.

$$D = \begin{vmatrix} 5 & 0 & 6 \\ -2 & 1 & 0 \\ 0 & 3 & -1 \end{vmatrix} = -0 \cdot \begin{vmatrix} -2 & 0 \\ 0 & -1 \end{vmatrix} + 1 \cdot \begin{vmatrix} 5 & 6 \\ 0 & -1 \end{vmatrix} - 3 \cdot \begin{vmatrix} 5 & 6 \\ -2 & 0 \end{vmatrix} = 0(2-0) + 1(-5-0) - 3(0+12)$$

$$= 0 + 1(-5) - 3(12) = -5 - 36 = -41$$

$$D_y = \begin{vmatrix} 5 & 5 & 6 \\ -2 & -6 & 0 \\ 0 & 3 & -1 \end{vmatrix} = -5 \cdot \begin{vmatrix} -2 & 0 \\ 0 & -1 \end{vmatrix} + (-6) \cdot \begin{vmatrix} 5 & 6 \\ 0 & -1 \end{vmatrix} - 3 \cdot \begin{vmatrix} 5 & 6 \\ -2 & 0 \end{vmatrix} = -5(2-0) - 6(-5-0) - 3(0+12)$$

$$= -5(2) - 6(-5) - 3(12) = -10 + 30 - 36 = -16$$

$$y = \frac{D_y}{D} = \frac{-16}{-41} = \frac{16}{41}$$

47. Cramer's rule does not apply when the determinant $D = 0$.

49.

$$D = \begin{vmatrix} 4 & -2 \\ -2 & 1 \end{vmatrix} = 4(1) - (-2)(-2) = 4 - 4 = 0$$

Cramer's rule is not possible since $D = 0$. Use the elimination method. Eliminate x by multiplying the second equation by 2.

$$\begin{array}{r} 4x - 2y = 3 \\ -4x + 2y = 2 \\ \hline 0 \neq 5 \end{array}$$

The system is inconsistent. There is no solution. $\{\ \}$

51.
$$D = \begin{vmatrix} 4 & 1 \\ 1 & -7 \end{vmatrix} = 4(-7) - 1(1) = -28 - 1 = -29$$

$$D_x = \begin{vmatrix} 0 & 1 \\ 0 & -7 \end{vmatrix} = 0$$

$$D_y = \begin{vmatrix} 4 & 0 \\ 1 & 0 \end{vmatrix} = 0$$

$$x = \frac{D_x}{D} = \frac{0}{-29} = 0 \qquad y = \frac{D_y}{D} = \frac{0}{-29} = 0$$

The solution is $\{(0, 0)\}$.

53.
$$D = \begin{vmatrix} 1 & 5 \\ 2 & 10 \end{vmatrix} = 1(10) - 5(2) = 10 - 10 = 0$$

Cramer's rule is not possible since $D = 0$. Use the elimination method. Eliminate x by multiplying the first equation by -2.

$$-2x - 10y = -6$$
$$\underline{2x + 10y = 6}$$
$$0 = 0$$

The equations are dependent. There are infinitely many solutions.

The solution is $\{(x, y) \mid x + 5y = 3\}$.

55.
$$D = \begin{vmatrix} 1 & 0 & 0 \\ -1 & 3 & 0 \\ 0 & 1 & 2 \end{vmatrix} = 0 \cdot \begin{vmatrix} -1 & 3 \\ 0 & 1 \end{vmatrix} - 0 \cdot \begin{vmatrix} 1 & 0 \\ 0 & 1 \end{vmatrix} + 2 \cdot \begin{vmatrix} 1 & 0 \\ -1 & 3 \end{vmatrix} = 0 - 0 + 2(3 - 0) = 6$$

$$D_x = \begin{vmatrix} 3 & 0 & 0 \\ 3 & 3 & 0 \\ 4 & 1 & 2 \end{vmatrix} = 3 \cdot \begin{vmatrix} 3 & 0 \\ 1 & 2 \end{vmatrix} - 3 \cdot \begin{vmatrix} 0 & 0 \\ 1 & 2 \end{vmatrix} + 4 \cdot \begin{vmatrix} 0 & 0 \\ 3 & 0 \end{vmatrix} = 3(6 - 0) - 3(0 - 0) + 4(0 - 0) = 18$$

$$D_y = \begin{vmatrix} 1 & 3 & 0 \\ -1 & 3 & 0 \\ 0 & 4 & 2 \end{vmatrix} = -3 \cdot \begin{vmatrix} -1 & 0 \\ 0 & 2 \end{vmatrix} + 3 \cdot \begin{vmatrix} 1 & 0 \\ 0 & 2 \end{vmatrix} - 4 \cdot \begin{vmatrix} 1 & 0 \\ -1 & 0 \end{vmatrix} = -3(-2 - 0) + 3(2 - 0) - 4(0 - 0)$$
$$= -3(-2) + 3(2) - 4(0) = 6 + 6 = 12$$

$$D_z = \begin{vmatrix} 1 & 0 & 3 \\ -1 & 3 & 3 \\ 0 & 1 & 4 \end{vmatrix} = 3 \cdot \begin{vmatrix} -1 & 3 \\ 0 & 1 \end{vmatrix} - 3 \cdot \begin{vmatrix} 1 & 0 \\ 0 & 1 \end{vmatrix} + 4 \cdot \begin{vmatrix} 1 & 0 \\ -1 & 3 \end{vmatrix} = 3(-1 - 0) - 3(1 - 0) + 4(3 - 0)$$
$$= 3(-1) - 3(1) + 4(3) = -3 - 3 + 12 = 6$$

$$x = \frac{D_x}{D} = \frac{18}{6} = 3 \qquad y = \frac{D_y}{D} = \frac{12}{6} = 2 \qquad z = \frac{D_z}{D} = \frac{6}{6} = 1$$

The solution is $\{(3, 2, 1)\}$.

57.
$$D = \begin{vmatrix} 1 & 1 & 8 \\ 2 & 1 & 11 \\ 1 & 0 & 3 \end{vmatrix} = -1 \cdot \begin{vmatrix} 2 & 11 \\ 1 & 3 \end{vmatrix} + 1 \cdot \begin{vmatrix} 1 & 8 \\ 1 & 3 \end{vmatrix} - 0 \cdot \begin{vmatrix} 1 & 8 \\ 2 & 11 \end{vmatrix} = -1(6 - 11) + 1(3 - 8) - 0 = -1(-5) + 1(-5)$$
$$= 5 - 5 = 0$$

Cramer's rule does not apply. $\{\ \}$

59.
$$\begin{vmatrix} 6 & x \\ 2 & -4 \end{vmatrix} = 14$$
$$-24 - 2x = 14$$
$$-2x = 38$$
$$x = -19$$

61.
$$\begin{vmatrix} 3 & 1 & 0 \\ 0 & 4 & -2 \\ 1 & 0 & w \end{vmatrix} = 10$$
$$1 \cdot \begin{vmatrix} 1 & 0 \\ 4 & -2 \end{vmatrix} - 0 \cdot \begin{vmatrix} 3 & 0 \\ 0 & -2 \end{vmatrix} + w \cdot \begin{vmatrix} 3 & 1 \\ 0 & 4 \end{vmatrix} = 10$$
$$1(-2) - 0 + w(12) = 10$$
$$-2 + 12w = 10$$
$$12w = 12$$
$$w = 1$$

63.

$$\begin{vmatrix} 1 & 0 & 3 & 0 \\ 0 & 1 & 2 & 4 \\ -2 & 0 & 0 & 1 \\ 4 & -1 & -2 & 0 \end{vmatrix} = 1 \cdot \begin{vmatrix} 1 & 2 & 4 \\ 0 & 0 & 1 \\ -1 & -2 & 0 \end{vmatrix} - 0 + (-2) \cdot \begin{vmatrix} 0 & 3 & 0 \\ 1 & 2 & 4 \\ -1 & -2 & 0 \end{vmatrix} - 4 \cdot \begin{vmatrix} 0 & 3 & 0 \\ 1 & 2 & 4 \\ 0 & 0 & 1 \end{vmatrix}$$

$$= 1\left[0 - 0 + 1\begin{vmatrix} 1 & 2 \\ -1 & -2 \end{vmatrix} \right] - 2\left[0 - 3\begin{vmatrix} 1 & 4 \\ -1 & 0 \end{vmatrix} + 0 \right] - 4\left[0 - 3\begin{vmatrix} 1 & 4 \\ 0 & 1 \end{vmatrix} + 0 \right]$$

$$= 1[1(-2+2)] - 2[-3(0+4)] - 4[-3(1-0)] = 1(0) - 2(-12) - 4(-3) = 24 + 12 = 36$$

65. a.

$$\begin{vmatrix} 1 & 1 & 1 & 1 \\ 2 & 0 & -1 & 1 \\ 2 & 1 & 0 & -1 \\ 0 & 1 & 1 & 0 \end{vmatrix} = 1 \cdot \begin{vmatrix} 0 & -1 & 1 \\ 1 & 0 & -1 \\ 1 & 1 & 0 \end{vmatrix} - 2 \cdot \begin{vmatrix} 1 & 1 & 1 \\ 1 & 0 & -1 \\ 1 & 1 & 0 \end{vmatrix} + 2 \cdot \begin{vmatrix} 1 & 1 & 1 \\ 0 & -1 & 1 \\ 1 & 1 & 0 \end{vmatrix} - 0$$

$$= 1\left[0 - 1\begin{vmatrix} -1 & 1 \\ 1 & 0 \end{vmatrix} + 1\begin{vmatrix} -1 & 1 \\ 0 & -1 \end{vmatrix} \right] - 2\left[-1\begin{vmatrix} 1 & -1 \\ 1 & 0 \end{vmatrix} + 0 - 1\begin{vmatrix} 1 & 1 \\ 1 & -1 \end{vmatrix} \right] + 2\left[1\begin{vmatrix} -1 & 1 \\ 1 & 0 \end{vmatrix} - 0 + 1\begin{vmatrix} 1 & 1 \\ -1 & 1 \end{vmatrix} \right]$$

$$= 1[(0 - 1(0 - 1) + 1(1 - 0)] - 2[-1(0 + 1) + 0 - 1(-1 - 1)] + 2[1(0 - 1) - 0 + 1(1 + 1)]$$

$$= 1(0 + 1 + 1) - 2(-1 + 0 + 2) + 2(-1 - 0 + 2) = 1(2) - 2(1) + 2(1) = 2$$

b.

$$\begin{vmatrix} 0 & 1 & 1 & 1 \\ 5 & 0 & -1 & 1 \\ 0 & 1 & 0 & -1 \\ -1 & 1 & 1 & 0 \end{vmatrix} = 0 - 5 \cdot \begin{vmatrix} 1 & 1 & 1 \\ 1 & 0 & -1 \\ 1 & 1 & 0 \end{vmatrix} + 0 - (-1) \cdot \begin{vmatrix} 1 & 1 & 1 \\ 0 & -1 & 1 \\ 1 & 0 & -1 \end{vmatrix}$$

$$= -5\left[-1\begin{vmatrix} 1 & -1 \\ 1 & 0 \end{vmatrix} + 0 - 1\begin{vmatrix} 1 & 1 \\ 1 & -1 \end{vmatrix} \right] + 1\left[1\begin{vmatrix} -1 & 1 \\ 0 & -1 \end{vmatrix} - 0 + 1\begin{vmatrix} 1 & 1 \\ -1 & 1 \end{vmatrix} \right]$$

$$= -5[-1(0 + 1) + 0 - 1(-1 - 1)] + 1[1(1 - 0) - 0 + 1(1 + 1)]$$

$$= -5(-1 + 0 + 2) + 1(1 - 0 + 2) = -5(1) + 1(3) = -5 + 3 = -2$$

c.

$$x = \frac{D_x}{D} = \frac{-2}{2} = -1$$

67. Let x and y be the measures of the two angles.

$$x + y = 90$$

$$x = \frac{5}{7}y \ \rightarrow 7x - 5y = 0$$

$$D = \begin{vmatrix} 1 & 1 \\ 7 & -5 \end{vmatrix} = 1(-5) - 1(7) = -5 - 7 = -12$$

$$D_x = \begin{vmatrix} 90 & 1 \\ 0 & -5 \end{vmatrix} = 90(-5) - 1(0) = -450$$

$$D_y = \begin{vmatrix} 1 & 90 \\ 7 & 0 \end{vmatrix} = 1(0) - 90(7) = -630$$

$$x = \frac{D_x}{D} = \frac{-450}{-12} = 37.5$$

$$y = \frac{D_y}{D} = \frac{-630}{-12} = 52.5$$

The measures are 37.5° and 52.5°.

69. Let x be the number of iPods, y be the number of iPads, and z be the number of iPhones.

$x = 2z \rightarrow x - 2z = 0$

$x = 5 + y + z \rightarrow x - y - z = 5$

$x + y + z = 75$

$$D = \begin{vmatrix} 1 & 0 & -2 \\ 1 & -1 & -1 \\ 1 & 1 & 1 \end{vmatrix} = 1 \cdot \begin{vmatrix} -1 & -1 \\ 1 & 1 \end{vmatrix} - 0 \cdot \begin{vmatrix} 1 & -1 \\ 1 & 1 \end{vmatrix} + (-2) \cdot \begin{vmatrix} 1 & -1 \\ 1 & 1 \end{vmatrix}$$

$$= 1(-1+1) - 0 + (-2)(1+1)$$

$$= 0 - 0 - 4 = -4$$

$$D_x = \begin{vmatrix} 0 & 0 & -2 \\ 5 & -1 & -1 \\ 75 & 1 & 1 \end{vmatrix} = 0 \cdot \begin{vmatrix} -1 & -1 \\ 1 & 1 \end{vmatrix} - 0 \cdot \begin{vmatrix} 5 & -1 \\ 75 & 1 \end{vmatrix} + (-2) \cdot \begin{vmatrix} 5 & -1 \\ 75 & 1 \end{vmatrix}$$

$$= 0 - 0 + (-2)(5 + 75)$$

$$= 0 - 0 - 160 = -160$$

$$D_y = \begin{vmatrix} 1 & 0 & -2 \\ 1 & 5 & -1 \\ 1 & 75 & 1 \end{vmatrix} = 1 \cdot \begin{vmatrix} 5 & -1 \\ 75 & 1 \end{vmatrix} - 0 \cdot \begin{vmatrix} 1 & -1 \\ 2 & 1 \end{vmatrix} + (-2) \cdot \begin{vmatrix} 1 & 5 \\ 1 & 75 \end{vmatrix}$$

$$= 1(5 + 75) - 0 + (-2)(75 - 5)$$

$$= 80 - 0 + (-140) = -60$$

$$D_z = \begin{vmatrix} 1 & 0 & 0 \\ 1 & -1 & 5 \\ 1 & 1 & 75 \end{vmatrix} = 1 \cdot \begin{vmatrix} -1 & 5 \\ 1 & 75 \end{vmatrix} - 0 \cdot \begin{vmatrix} 1 & 5 \\ 1 & 75 \end{vmatrix} + 0 \cdot \begin{vmatrix} 1 & -1 \\ 1 & 1 \end{vmatrix}$$

$$= 1(-75 - 5) - 0 + 0$$

$$= -80 - 0 + 0 = -80$$

$$x = \frac{D_x}{D} = \frac{-160}{-4} = 40$$

$$y = \frac{D_y}{D} = \frac{-60}{-4} = 15$$

$$z = \frac{D_z}{D} = \frac{-80}{-4} = 20$$

40 iPods, 15 iPads, and 20 iPhones were sold.

71. Let x be the number of women and y be the number of men.

$x + y = 1000$

$\dfrac{1}{2}x + \dfrac{3}{8}y = 445 \rightarrow 4x + 3y = 3560$

$$D = \begin{vmatrix} 1 & 1 \\ 4 & 3 \end{vmatrix} = 1(3) - 1(4) = 3 - 4 = -1$$

$$D_x = \begin{vmatrix} 1000 & 1 \\ 3560 & 3 \end{vmatrix} = 1000(3) - 1(3560)$$

$$= 3000 - 3560 = -560$$

$$D_y = \begin{vmatrix} 1 & 1000 \\ 4 & 3560 \end{vmatrix} = 1(3560) - 1000(4)$$

$$= 3560 - 1000 = -440$$

$$x = \frac{D_x}{D} = \frac{-560}{-1} = 560$$

$$y = \frac{D_y}{D} = \frac{-440}{-1} = 440$$

There were 560 women and 440 men in the survey.